Selected Problems
in
Differential Geometry
and Topology

D1363964

Selected Problems in Differential Geometry and Topology

A.T. Fomenko, A.S. Mishchenko and Yu.P. Solovyev

Faculty of Mechanics and Mathematics, Moscow State University

C S P

Cambridge Scientific Publishers

British Library Cataloguing in Publication Data
A catalogue record for this book has been requested

Library of Congress Cataloguing in Publication Data
A catalogue record has been requested

ISBN 978-1-904868-33-0

Cambridge Scientific Publishers Ltd
45 Margett Street
Cottenham, Cambridge CB24 8QY
UK
www.cambridgescientificpublishers.com
Printed and bound by CPI Group (UK) Ltd, Croydon, CR0 4YY

Contents

Preface to the English Edition

Selected Problems in Differential Geometry and Topology by Professor A.T. Fomenko, Professor A.S. Mishchenko and Professor Yu.P. Solovyev is designed as an associated companion volume to *A Short Course in Differential Geometry and Topology* by Professor A.T. Fomenko and Professor A.S. Mishchenko (ISBN 978-1-904868-32-3: Cambridge Scientific Publishers) which was published in 2009 and is based on seminars conducted at the Faculty of Mechanics and Mathematics of Moscow State University for second and third year mathematics and mechanics students. Although there is no detailed cross referencing, the two volumes complement each other and can be used as a two volume set; the volumes are intended for students of mathematics, mechanics and physics and also provide useful reference for postgraduates and researchers specialising in modern geometry and its applications.

Selected Problems in Differential Geometry and Topology is structured in two parts. Part 1 consists of problems on standard sections of differential geometry and topology. Part 2 consists of problems which require more advanced level of modern geometry and its applications. The book includes the following topics: theory of curves, theory of surfaces, coordinate systems, Riemannian geometry, classical metrics (on a sphere, a Lobachevskii plane), topological spaces, manifolds, topology of two-dimensional surfaces, Lie groups and algebras, vector fields and tensors, differential forms (including de Rham integration and de Rham theory), curvature tensor, geodesics, elements of algebraic topology, Lie connectivities and groups. There are Answers and Solutions given to the problems presented in Parts 1 and 2 and there is also a Bibliography.

Preface to the First Russian Edition

Selected Problems in Differential Geometry and Topology is intended to provide for academic activity in courses of differential geometry and topology at the specialties of mechanics and mathematics of universities and pedagogical institutes. The aim of the collection is to maximally reflect the existing requirements of the courses of differential geometry and topology both as part of new programmes and as part of other courses of mathematics, physics and mechanics. Besides, publication of the collection will make available for the wider mathematical community new scientific and methodological developments of leading scientists in the field of differential geometry, topology, algebra and mechanics.

Selected Problems in Differential Geometry and Topology provides the basis for practical exercises in courses of differential geometry and topology at the specialties of mathematics and mechanics of universities and pedagogical institutes both in Russia and CIS countries, such as Belorussia, Ukraine, Georgia, Kazakhstan, Turkmenistan, Lithuania – from where applications for collections of problems of such kind have repeatedly come in the recent years.

The book can also be used to support diverse special courses on various sections of modern geometry and its applications to mechanics and mathematical physics.

Selected Problems in Differential Geometry and Topology consists of two parts. Part 1 contains problems on the standard sections of differential geometry and topology. This material surpasses the necessary minimum of problems on the standard courses of geometry and topology. Part 2 contains problems intended for a more profound and advanced level of modern geometry and its applications.

The book presents the following topics: theory of curves (including evolutes and evolvents), theory of surfaces, coordinate systems, Riemannian geometry, classical metrics (on a sphere, a Lobachevskii plane, etc.), topological spaces, manifolds (including elements of fibrations, phase spaces and configuration spaces), topology of two-dimensional surfaces, two-dimensional surfaces in a three-dimensional Euclidean space, Lie groups and algebras (including

ix

small Lie groups, their parameterizations frequently used in mechanics), vector fields and tensors, differential forms (including de Rham integration and de Rham theory, etc.), connectivities and parallel translation, geodesics, curvature tensor, elements of algebraic topology (Euler characteristic, vector field index, intersection index etc.), Lie connectivities and groups.

The book also contains additional problems on the subjects reflected in the first part and problems on some new subjects, touching upon more profound issues of differential geometry and topology. Among the new subjects presented in the second part, mention should be made of computer geometry and topology, kinematics and geometry, geometrical constructions (such as jets, Stiefel manifolds and Grassmann manifolds, etc.), the Lie derivative, packing problems, combinatorial geometry in the plane and in space, elements of Hamiltonian mechanics.

The present collection is a natural supplement to the textbook by A.S. Mishchenko and A.T. Fomenko *A Course of Differential Geometry and Topology* re-published in 2000. To a significant extent, the present book of problems is based on the book by A.S. Mishchenko, Yu.P. Solovyev and A.T. Fomenko *A Collection of Problems in Differential Geometry and Topology* published in 1981 by Moscow University Press. It should be noted here that after some time, in 1998, an electronic version of that previous book of problems was created by A.A. Oshemkov. Then, in 1998–1999, on the initiative of the Chair of Differential Geometry and Applications a special scientific-methods seminar was organized at the Faculty of Mechanics and Mathematics of the Moscow State University to prepare a new "Collection of Problems in Differential Geometry and Topology". The Seminar was led by Professors A.S. Mishchenko and Yu.P. Solovyev and Academician A.T. Fomenko. Due to these circumstances, these three names are given as the authors on the cover of the book. However, in actual fact, the Collection was created with the participation and contribution of a large group of notable scientists, distinguished specialists in the field of modern geometry, topology, algebra, mechanics and applications: Academician V.V. Kozlov, Prof. V.V. Fedorchuk, Prof. A.V. Bolsinov, Prof. E.R. Rozendorn, Prof. V.V. Trofimov, Prof. A.A. Borisenko (Kharkov, Ukraine), Prof. I.Kh. Sabitov, Prof. E.V. Troitsky, Prof. A.O. Ivanov, Prof. A.A. Tuzhilin, Senior Research Assistant G.V. Nosovsky, Research Assistant A.I. Shafarevich, Associate Professor A.A. Oshemkov, Junior Research Assistant F.Yu. Popelensky, Assistant E.A. Kudryavtseva.

The most active part in the work of the Seminar, the preparation of problems and their solution, was taken by practically all students and postgraduates of the Chair of Differential Geometry and Applications of the Faculty of Mechanics and Mathematics of the Moscow State University. We express our deep gratitude to all of them.

In the field of differential geometry and topology there are several collections of problems, textbooks and manuals. A list of the most well known editions is given at the end of the book, in the Bibliography.

It should be noted that the collections of problems in differential geometry and topology published recently had small printruns and for this reason are practically unobtainable. Books published earlier and in larger printruns have already been sold out and have become rarities. Besides, the content and material of some books became obsolete to a significant degree and needed to be updated. This is due to both the improvement of the curriculums of mathematical courses, and to the fact that the requirements of geometrical courses as part of other courses making use or based upon the methods of differential geometry and topology have changed or have been enhanced significantly. All of this progress made the publication of *Selected Problems in Differential Geometry and Topology* especially topical.

We collected extensive scientific-methods material, in the most part unknown to the wide mathematical community. Unknown due to several reasons, the main reasons are the following. First, many problems emerged in the research of the participants of the above mentioned Seminar. Second, some problems appeared as the result of discussions at Seminar's meetings. Finally, a sufficiently large number of problems was retrieved from old mathematical literature, which has become a bibliographic rarity and is therefore inaccessible for modern students and teachers.

The scientific-methods material thus collected formed the present *Selected Problems in Differential Geometry and Topology.*

We especially note the absolutely invaluable role of F.Yu. Popelensky in the preparation of the book for publication. The volume of work he performed on ordering problems, organizing the verification of their conditions and solutions, as well as typesetting etc. was so great that it is only due to his untiring efforts that the book finally saw the light.

A.T. Fomenko
A.S. Mishchenko
Yu.P. Solovyev

Moscow University, May 2000

Preface to the Second Russian Edition

The second edition of *Selected Problems in Differential Geometry and Topology* is in front of the reader. Its first edition was published in 2001 by Fizmatlit Publishers. Over the last two years, active work was underway to improve the book. An especially active part in this work was taken by members of the Chair of Differential Geometry and Applications, Faculty of Mechanics and Mathematics, Moscow State University. We took into account numerous comments by both academics and students accumulated in the use of the first edition. Noticed typos and inaccuracies were corrected, and new problems were added. This was done especially with respect to the first part of the book, which was used the most actively in the seminar classes on the obligatory courses "Classical Differential Geometry" and "Differential Geometry and Topology" at the Faculty of Mechanics and Mathematics of the Moscow State University. We also changed the structure of some paragraphs and added new ones, striving for a more clearcut distribution of problems with respect to topics. Also, many new figures were added to make the presentation more illustrative.

Despite the fact that differential geometry is a mature science, the book improvement process proved "divergent". The deeper we immersed ourselves in the work, the more comparatively elementary and at the same time rather useful problems were found, which simply "could not but be included" in the book. We interrupted the process consciously, having tried to achieve a significantly large content in a reasonable volume. Besides, we are aware that it is virtually impossible to incorporate all the diversity and richness of modern geometry into one book. Still, it appears to us that the present edition is not only the most modern but also the most complete book of problems for university courses of differential geometry and topology.

A.T. Fomenko and F.Yu. Popelensky

Commentary by A.T. Fomenko: I would like to emphasize the important role of F.Yu. Popelensky in the preparation of this edition. He not only deeply analyzed most problems and their solutions, found and corrected many typos and inaccuracies, but also proposed an exceptionally valuable and elegant new series of problems.

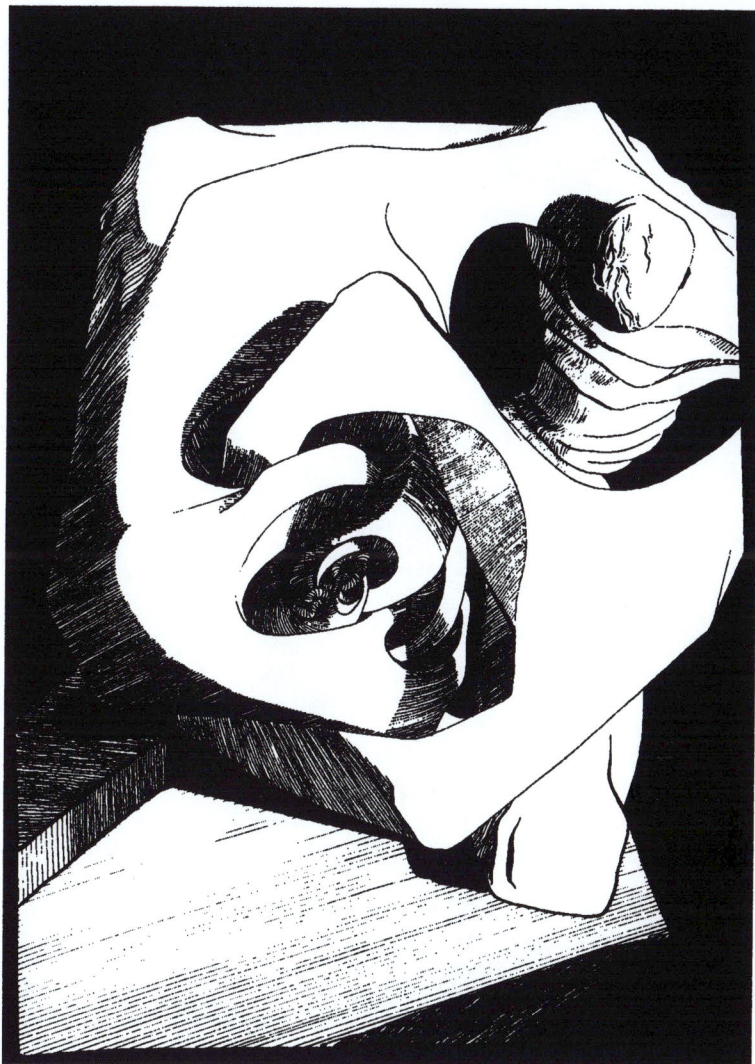

Alexander's sphere: a topological embedding of a two-dimensional sphere in $\mathbb{R}^3$, in which the image of the sphere divides $\mathbb{R}^3$ into two open domains. One of them is a ball and the other is nonsimply connected (a drawing by A.T. Fomenko)

Part 1

§ 1. Coordinate Systems

Consider a domain U of the space $\mathbb{R}^n$ in which the Cartesian coordinates $(x^1, \ldots, x^n)$ are given. Assume that in another copy of the space $\mathbb{R}^n$ with the coordinates $(q^1, \ldots, q^n)$, a domain V is given, and there is a one-to-one correspondence between points of the domains U and V. In this case, to define a point of the domain U, we can use the tuple $(q^1, \ldots, q^n)$, the Cartesian coordinates of the corresponding point of the domain V. We say that $(q^1, \ldots, q^n)$ are *curvilinear coordinates in the domain* U if:

(1) the functions

$$x^i = x^i(q^1, \ldots, q^n)$$

defining the one-to-one correspondence between points of the domains U and V have continuous derivatives of all orders in the domain V;

(2) the Jacobian

$$J = \left| \frac{\partial x^i}{\partial q^j} \right|$$

is different from zero in all points of the domain V.

Note that in the vast majority of cases it suffices to assume that the functions $x^i(q^1, \ldots, q^n)$ have continuous derivatives up to the third order inclusive.

It follows from the definition of curvilinear coordinates that the inverse functions $q^i(x^1, \ldots, x^n)$ also have continuous derivatives of all orders in the domain U, the Jacobian

$$J' = \left| \frac{\partial q^i}{\partial x^j} \right|$$

being different from zero (equal to J^{-1}).

It is convenient to consider the functions $x^i = x^i(q^1, \ldots, q^n)$ simultaneously for all $i = 1, \ldots, n$, using the vector function

$$\mathbf{r} = \mathbf{r}(q^1, \ldots, q^n), \quad \text{where} \quad \mathbf{r} = (x^1, \ldots, x^n).$$

In the domain U, the conditions $q^i = \text{const}$ define n families of *coordinate hypersurfaces*. Coordinate hypersurfaces of the same family are disjoint.

1

Any $n-1$ coordinate hypersurfaces belonging to different families intersect along a curve. Such curves are said to be *coordinate*. We will also call them *coordinate lines*.

The vectors $\mathbf{r}_k = \dfrac{\partial \mathbf{r}}{\partial q^k}$ have directions of the tangents to the coordinate lines. At each point of the domain U, these vectors are linearly independent. In a neighborhood of a point $M(q^1, q^2, \ldots, q^n)$, they define the infinitely small vector

$$d\mathbf{r} = \sum_{i=1}^{n} \mathbf{r}_i dq^i.$$

The square of its length expressed in the curvilinear coordinates is found from the equation

$$ds^2 = \langle d\mathbf{r}, d\mathbf{r} \rangle = \left\langle \sum_{i=1}^{n} \mathbf{r}_i dq^i, \sum_{j=1}^{n} \mathbf{r}_j dq^j \right\rangle = \sum_{i,j=1}^{n} g_{ij} dq^i dq^j,$$

where $\langle \, , \, \rangle$ is the inner product in $\mathbb{R}^n$.

The quantities $g_{ij} = g_{ji} = \langle \mathbf{r}_i, \mathbf{r}_j \rangle$ define the metric in the curvilinear coordinate system $(q^1, \ldots, q^n)$.

The *orthogonal curvilinear coordinate system* is a coordinate system for which

$$g_{ij} = \langle \mathbf{r}_i, \mathbf{r}_j \rangle = \begin{cases} 0, & i \neq j, \\ H_i^2, & i = j. \end{cases}$$

The quantities $H_i > 0$ are called the Lamé coefficients. They are equal to the modules of length of the vectors $\mathbf{r}_i$:

$$H_i = |\mathbf{r}_i| = \sqrt{\left(\frac{\partial x^1}{\partial q^i}\right)^2 + \left(\frac{\partial x^2}{\partial q^i}\right)^2 + \cdots + \left(\frac{\partial x^n}{\partial q^i}\right)^2}.$$

The square of the linear element in the orthogonal curvilinear coordinates is given by the expression

$$ds^2 = H_1^2 \left(dq^1\right)^2 + H_2^2 \left(dq^2\right)^2 + \cdots + H_n^2 \left(dq^n\right)^2.$$

In Problems **1.1–1.5**,

(a) find the formulas expressing the curvilinear coordinates of a point of the plane $\mathbb{R}^2$ through the Cartesian coordinates and vice versa;

(b) find the coordinate lines;

(c) calculate the determinants

$$\begin{vmatrix} \dfrac{\partial x_1}{\partial u_1} & \dfrac{\partial x_1}{\partial u_2} \\[2mm] \dfrac{\partial x_2}{\partial u_1} & \dfrac{\partial x_2}{\partial u_2} \end{vmatrix}, \qquad \begin{vmatrix} \dfrac{\partial u_1}{\partial x_1} & \dfrac{\partial u_1}{\partial x_2} \\[2mm] \dfrac{\partial u_2}{\partial x_1} & \dfrac{\partial u_2}{\partial x_2} \end{vmatrix}$$

and reveal at which points of the plane $\mathbb{R}^2$ the one-to-one property between the curvilinear and Cartesian coordinates of a point on the plane is violated for the following curvilinear coordinates (u_1, u_2).

1.1. For the generalized polar coordinate system defined by the equation $\dfrac{x_1}{a_1} + i\dfrac{x_2}{a_2} = u_1 e^{iu_2}$, where $0 \le u_1 < \infty$, $-\pi < u_2 \le \pi$, $a_1 > 0$, $a_2 > 0$. Under which conditions these coordinates coincide with the usual polar coordinates?

1.2. For the elliptic coordinate system defined by the equation $x_1 + ix_2 = \cosh(u_1 + iu_2)$, where $0 \le u_1 < \infty$, $-\pi < u_2 \le \pi$.

1.3. For the parabolic coordinate system defined by the equation $x_1 + ix_2 = (u_1 + iu_2)^2$, where $-\infty < u_1 < \infty$, $0 \le u_2 < \infty$.

1.4. For the bipolar coordinate system defined by the equation $x_1 + ix_2 = \tanh\left(\dfrac{u_1 + iu_2}{2}\right)$, where $-\infty < u_1 < \infty$, $-\pi < u_2 \le \pi$, without points $(u_1 = 0, u_2 = \pi) = (x_1 = x_2 = \infty)$, $(u_1 = +\infty) = (x_1 = 1, x_2 = 0)$, $(u_1 = -\infty) = (x_1 = -1, x_2 = 0)$.

1.5. For the coordinate system defined by the equation $x_1 + ix_2 = (u_1 + iu_2)^3$, where $u_2 \ge 0$ and $u_1 + u_2\sqrt{3} \ge 0$.

In Problems **1.6–1.12**,

(a) find the coordinate surfaces and coordinate lines;

(b) calculate the determinants $\left|\dfrac{\partial x_i}{\partial u_j}\right|$ and $\left|\dfrac{\partial u_i}{\partial x_j}\right|$ and find at which points of the space $\mathbb{R}^3$ the one-to-one correspondence between the curvilinear and rectangular Cartesian coordinates is violated for the following curvilinear coordinates u_1, u_2, u_3 of the space $\mathbb{R}^3$;

(c) are these coordinate systems orthogonal?

1.6. For the generalized cylindrical coordinate system defined by the equations

$$x_1 = a_1 u_1 \cos u_2, \quad x_2 = a_2 u_1 \sin u_2, \quad x_3 = u_3,$$

where $u_1 \ge 0$, $0 < u_2 \le 2\pi$, $-\infty < u_3 < \infty$, $a_1 > 0$, $a_2 > 0$ (see Fig. 1).

1.7. For the generalized spherical coordinate system defined by the equations

$$x_1 = a_1 u_1 \sin u_2 \cos u_3, \quad x_2 = a_2 u_1 \sin u_2 \sin u_3, \quad x_3 = a_3 u_1 \cos u_2,$$

where $u_1 \ge 0$, $0 \le u_2 \le \pi$, $0 \le u_3 < 2\pi$, $a_1 > 0$, $a_2 > 0$ and $a_3 > 0$ (see Fig. 2).

1.8. For the ellipsoidal coordinate system defined by the equations

$$x_1^2 = \frac{(a_1 - u_1)(a_1 - u_2)(a_1 - u_3)}{(a_2 - a_1)(a_3 - a_1)},$$

$$x_2^2 = \frac{(a_2 - u_1)(a_2 - u_2)(a_2 - u_3)}{(a_3 - a_2)(a_1 - a_2)},$$

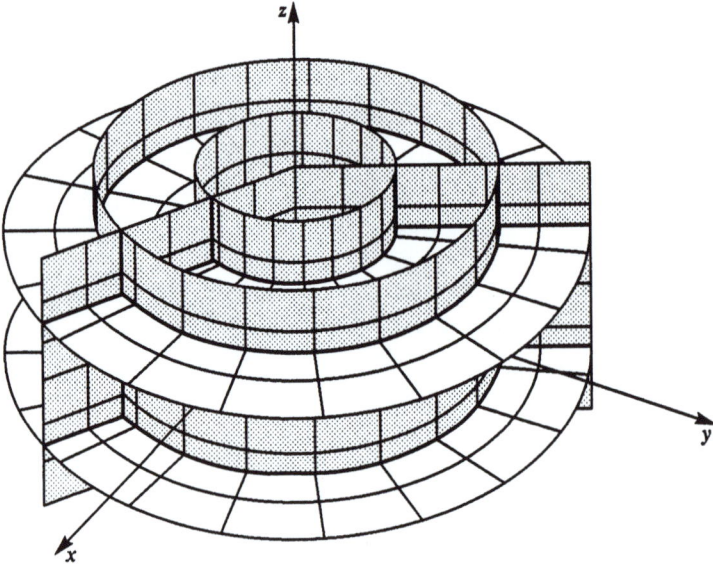

Figure 1 Cylindrical coordinate system, $a_1 = a_2 = 1$

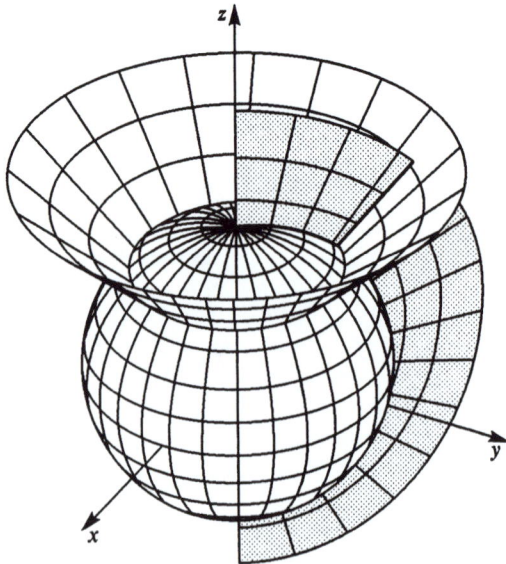

Figure 2 Spherical coordinate system, $a_1 = a_2 = a_3 = 1$

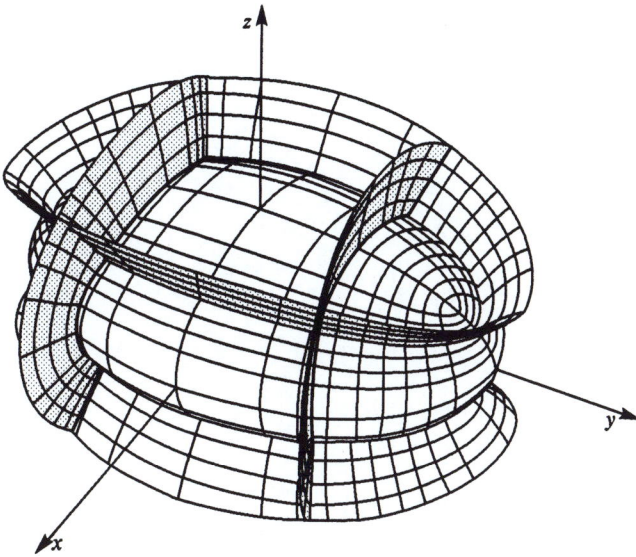

Figure 3 Ellipsoidal coordinate system

$$x_3^2 = \frac{(a_3 - u_1)(a_3 - u_2)(a_3 - u_3)}{(a_1 - a_3)(a_2 - a_3)},$$

where $a_1 > a_2 > a_3 > 0$ and $u_1 < a_3 < u_2 < a_2 < u_3 < a_1$ (see Figs. 3, 4).

1.9. For the parabolic coordinate system defined by the equations $x_1 = u_1 u_2 \cos u_3$, $x_2 = u_1 u_2 \sin u_3$, $x_3 = \frac{1}{2}(u_1^2 - u_2^2)$, where $0 \le u_1 < \infty$, $0 \le u_2 < \infty$ and $-\pi < u_3 \le \pi$.

1.10. For the degenerate ellipsoidal coordinate system defined by the equations $x_1 = \sinh u_1 \sin u_2 \cos u_3$, $x_2 = \sinh u_1 \sin u_2 \sin u_3$, $x_3 = \cosh u_1 \cos u_2$,

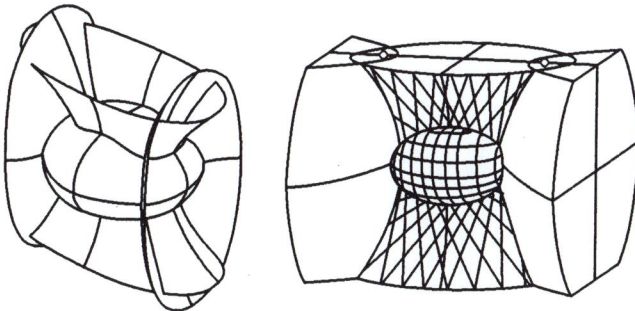

Figure 4 Ellipsoidal coordinate system

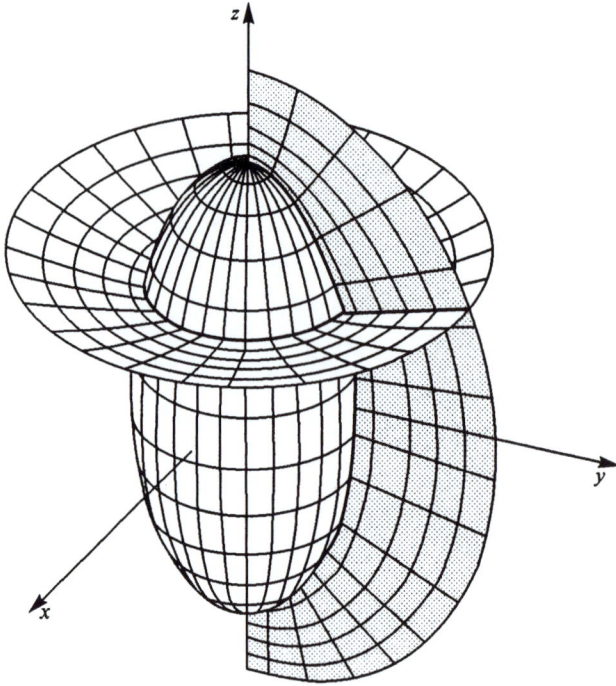

Figure 5 Degenerate ellipsoidal coordinate system

where $0 \le u_1 < \infty, 0 \le u_2 \le \pi, -\pi < u_3 \le \pi$ (see Fig. 5).

1.11. For the degenerate ellipsoidal coordinate system defined by the equations $x_1 = \cosh u_1 \sin u_2 \cos u_3$, $x_2 = \cosh u_1 \sin u_2 \sin u_3$, $x_3 = \sinh u_1 \cos u_2$, where $0 \le u_1 < \infty, 0 \le u_2 \le \pi$, and $-\pi < u_3 \le \pi$ (see Fig. 6).

1.12. For the toroidal coordinate system defined by the equations

$$x_1 = \frac{\sinh u_1 \cos u_3}{\cosh u_1 - \cos u_2}, \quad x_2 = \frac{\sinh u_1 \sin u_3}{\cosh u_1 - \cos u_2}, \quad x_3 = \frac{\sin u_2}{\cosh u_1 - \cos u_2},$$

where $0 \le u_1 < \infty, -\pi < u_2 \le \pi, -\pi < u_3 \le \pi$ (see Fig. 7).

1.13. Transform the expression $y\dfrac{\partial z}{\partial x} - x\dfrac{\partial z}{\partial y}$ to the new coordinates u, v connected with x, y by the relations $u = x, v = x^2 + y^2$. Verify whether or not the proposed system u, v is a curvilinear coordinate system. Find its domain and range.

1.14. Transform the following expressions to the polar coordinates $x = r \cos \varphi, y = r \sin \varphi$:

(a) $x\dfrac{\partial u}{\partial y} - y\dfrac{\partial u}{\partial x}$; (b) $x\dfrac{\partial u}{\partial x} - y\dfrac{\partial u}{\partial y}$; (c) $\left(\dfrac{\partial u}{\partial x}\right)^2 + \left(\dfrac{\partial u}{\partial y}\right)^2$;

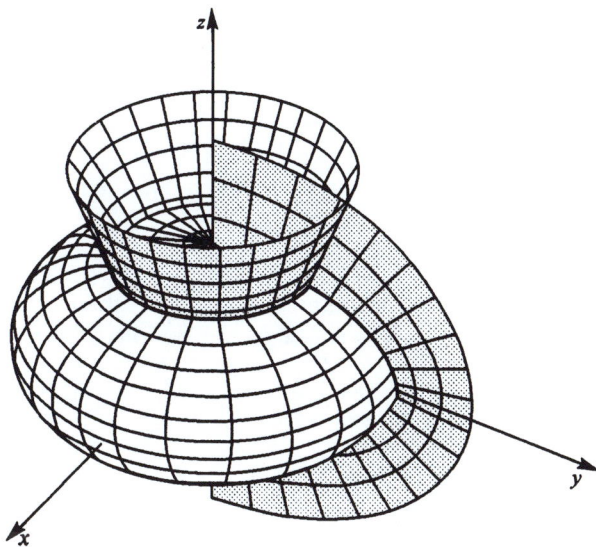

Figure 6 Degenerate ellipsoidal coordinate system

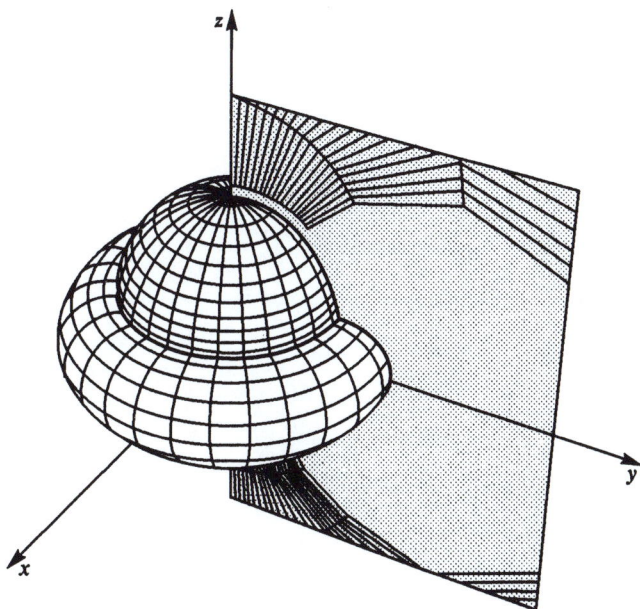

Figure 7 Toroidal coordinate system

(d) $\dfrac{\partial^2 u}{\partial x^2} + \dfrac{\partial^2 u}{\partial y^2}$ (Laplace operator).

1.15. Transform the expression $\dfrac{\partial^2 z}{\partial x^2} - a^2 \dfrac{\partial^2 z}{\partial y^2}$ to the new coordinates u, v, where $u = y + ax$ and $v = y - ax$. For which a does the proposed change yield a regular coordinate system?

1.16. Transform the expression $\dfrac{\partial^2 z}{\partial x^2} + \dfrac{\partial^2 z}{\partial y^2} + k^2 z = 0$ to the new coordinates u, v, where $x = \dfrac{1}{2}(u^2 - v^2)$ and $y = uv$.

1.17. Transform the following expressions to the spherical coordinates r, θ, φ connected with x, y, z by the equations $x = r \sin\theta \cos\varphi$, $y = r \sin\theta \sin\varphi$ and $z = r \cos\theta$:

(a) $\left(\dfrac{\partial V}{\partial x}\right)^2 + \left(\dfrac{\partial V}{\partial y}\right)^2 + \left(\dfrac{\partial V}{\partial z}\right)^2$; (b) $\dfrac{\partial^2 V}{\partial x^2} + \dfrac{\partial^2 V}{\partial y^2} + \dfrac{\partial^2 V}{\partial z^2}$.

1.18. Let $x = f(u, v)$, $y = \varphi(u, v)$ be a coordinate system such that $\dfrac{\partial f}{\partial u} = \dfrac{\partial \varphi}{\partial v}$ and $\dfrac{\partial f}{\partial v} = -\dfrac{\partial \varphi}{\partial u}$. Prove that the following equation holds:

$$\frac{\partial^2 V}{\partial u^2} + \frac{\partial^2 V}{\partial v^2} = \left(\frac{\partial^2 V}{\partial x^2} + \frac{\partial^2 V}{\partial y^2}\right)\left(\left(\frac{\partial f}{\partial u}\right)^2 + \left(\frac{\partial f}{\partial v}\right)^2\right).$$

1.19. Calculate the Laplace operator $\dfrac{\partial^2 V}{\partial x^2} + \dfrac{\partial^2 V}{\partial y^2} + \dfrac{\partial^2 V}{\partial z^2}$ in the cylindrical coordinate system r, φ, z such that $x = r \cos\varphi$, $y = r \sin\varphi$ and $z = z$.

1.20. Show that in passing from the Cartesian coordinates x, y to the polar coordinates ρ, φ the Cauchy–Riemann conditions $\dfrac{\partial u}{\partial x} = \dfrac{\partial v}{\partial y}, \dfrac{\partial u}{\partial y} = -\dfrac{\partial v}{\partial x}$ become $\dfrac{\partial u}{\partial \rho} = \dfrac{1}{\rho}\dfrac{\partial v}{\partial \varphi}, \dfrac{\partial v}{\partial \rho} = -\dfrac{1}{\rho}\dfrac{\partial u}{\partial \varphi}$.

1.21. Calculate the Laplace operator $\dfrac{\partial^2 V}{\partial x^2} + \dfrac{\partial^2 V}{\partial y^2}$ in the coordinate system u, v such that $w = z^2$, where $w = x + iy$ and $z = u + iv$.

1.22. Calculate the Laplace operator $\dfrac{\partial^2 V}{\partial x^2} + \dfrac{\partial^2 V}{\partial y^2}$ in the coordinate system u, v such that $w = a \cosh z$, where $w = x + iy$, $z = u + iv$.

1.23. Calculate the Laplace operator $\dfrac{\partial^2 V}{\partial x^2} + \dfrac{\partial^2 V}{\partial y^2}$ in the coordinate system u, v such that $w = e^z$, where $w = x + iy$, $z = u + iv$.

§ 2. Equations of Curves and Surfaces

2.1. A point M uniformly moves along a line ON uniformly rotating around a point O. Compose the equation of the trajectory of the point M (Archimedean spiral).

2.2. A line OL rotates around a point O with a constant angular velocity ω. A point M moves along the line OL with a speed proportional to the distance $|OM|$. Compose the equation of the trajectory circumscribed by the point M (logarithmic spiral).

2.3. A disk of radius a rolls along a line without sliding. Compose the equation of the trajectory of a point M rigidly linked with the disk and being at a distance d from its centre. The curve obtained for $d = a$ is called a *cycloid*; for $d < a$, a *shortened cycloid*; for $d > a$, an *extended cycloid*.

2.4. A circle of radius r rolls without sliding along a circle of radius R remaining outside it. Compose the equation for the trajectory of a point M of the rolling circle (*epicycloid*).

2.5. A circle of radius r rolls without sliding along a circle of radius R remaining inside it. Compose the equation for the trajectory of a point M of the rolling circle (*hypocycloid*).

2.6. Find the curve given by the equation $\mathbf{r} = \mathbf{r}(t)$, $c < t < d$, knowing that $\mathbf{r}'(t) = \lambda(t)\,\mathbf{a}$, where $\lambda(t) > 0$ is a continuous function and $\mathbf{a}$ is a constant nonzero vector.

2.7. Find the curve given by the equation $\mathbf{r} = \mathbf{r}(t)$, $-\infty < t < \infty$, if $\mathbf{r}''(t) = \mathbf{a}$ is a constant nonzero vector.

2.8. Let a vector function $\mathbf{r}(t)$ satisfy the differential equation $\mathbf{r}'' = \mathbf{r}' \times \mathbf{a}$, where $\mathbf{a}$ is a constant vector. Express the following quantities through $\mathbf{a}$ and $\mathbf{r}'$: (a) $|\mathbf{r}' \times \mathbf{r}''|^2$; (b) $(\mathbf{r}', \mathbf{r}'', \mathbf{r}''')$.

2.9. A plane curve is given by the equation $\mathbf{r}(t) = (\varphi(t),\, t\varphi(t))$. Under which condition does this equation define a straight line or its part?

2.10. Find the function $r = r(\varphi)$ knowing that this equation defines a straight line in polar coordinates on a plane.

2.11. (a) Prove that, under the action of a central force $\mathbf{F} = F\mathbf{r}$, a material point M circumscribes a trajectory lying in a fixed plane that passes through the origin. Note that the function F can depend both on the length of the vector $\mathbf{r}$ and on its direction.

(b) Compose the equation of motion of the point M in this plane in the polar coordinates.

(c) Show that for the central force $\mathbf{F}$ defined by the formula

$$\mathbf{F} = -k\frac{m\mathbf{r}}{r^3},$$

the point M moves along a curve of the second order. Here, m is the mass of the material point M, r is the length of the vector $\mathbf{r}$ and $k > 0$.

2.12. The motion of an electron in a constant magnetic field is defined by the differential equation

$$\mathbf{r}'' = \mathbf{r}' \times \mathbf{H}, \quad \mathbf{H} = \text{const}.$$

Prove that the trajectory of the electron is a helical line.

2.13. Find the curves defined by the differential equation

$$\mathbf{r}' = \omega \times \mathbf{r},$$

where ω is a constant vector.

2.14. Find the curves defined by the differential equation

$$\mathbf{r}' = \mathbf{e} \times (\mathbf{r} \times \mathbf{e}),$$

where $\mathbf{e}$ is a fixed vector of unit length.

2.15. Find the curves defined by the differential equation

$$\mathbf{r}' = a\mathbf{e} + \mathbf{e} \times \mathbf{r},$$

where the number a and the vector $\mathbf{e}$ are constant.

2.16. Find the curves defined by the differential equation

$$\mathbf{r}' = \frac{1}{2}|\mathbf{r}|^2\mathbf{e} - \mathbf{r}\,\langle \mathbf{r}, \mathbf{e}\rangle,$$

where $\mathbf{e}$ is a fixed vector of unit length.

2.17. At which angle do the curves

$$x^2 + y^2 = 8 \quad \text{and} \quad y^2 = 2x$$

intersect?

2.18. At which angle do the curves

$$x^2 + y^2 = 8x \quad \text{and} \quad y^2 = \frac{x^3}{2 - x}$$

intersect?

2.19. At which angle do the curves

$$x^2 = 4y \quad \text{and} \quad y = \frac{8}{x^2 + 4}$$

intersect?

2.20. Find the natural parameter (or the length) of the following curves:
(a) $y = a\cosh(x/a)$, the catenary;
(b) $y = x^{3/2}$; (c) $y = x^2$;
(d) $y = \ln x$; (e) $r = a(1 + \cos\varphi)$;
(f) $\mathbf{r}(t) = (a\,(t - \sin t),\, a\,(1 - \cos t))$;
(g) $\mathbf{r}\,(t) = (a\,(\cos t + t\sin t),\, a\,(\sin t - t\cos t))$;
(h) $\mathbf{r}(t) = \left(\dfrac{a}{3}\,(2\cos t + \cos 2t),\, \dfrac{a}{3}\,(2\sin t + \sin 2t)\right)$;
(i) $\mathbf{r}\,(t) = (a\cos^3 t,\, a\sin^3 t)$; (j) $y = e^x$;
(k) $\mathbf{r}\,(t) = \left(a\left(\ln\cot\dfrac{t}{2} - \cos t\right),\, a\sin t\right)$.

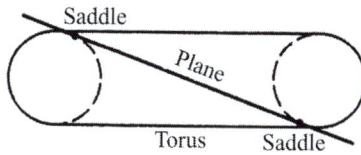

Figure 8 Cross-section of the torus by the plane

2.21. Find the arclength of the curve

$$x = -f'(\alpha)\sin\alpha - f''(\alpha)\cos\alpha,$$

$$y = f'(\alpha)\cos\alpha - f''(\alpha)\sin\alpha.$$

2.22. The circle $x = a + b\cos v$, $z = b\sin v$ $(0 < b < a)$ rotates around the axis Oz. Compose the equation of the surface of revolution.

2.23. A line moves translationally with constant velocity to intersect another line at a right angle and simultaneously uniformly rotates around this line. Compose the equation of the surface which the moving line circumscribes. This surface is called a *right helicoid*.

2.24. Compose the equation of the surface formed by rotating the catenary $y = a\cosh\dfrac{x}{a}$ around the axis Ox. This surface is called a *catenoid*.

2.25. Compose the equation of the surface formed by rotating the tractrix

$$\rho = \left(a\ln\tan\left(\frac{\pi}{4} + \frac{t}{2}\right) - a\sin t,\, a\cos t\right), \qquad -\frac{\pi}{2} < t < \frac{\pi}{2},$$

around its asymptote. This surface is called a *pseudosphere*.

2.26. On the circular torus of revolution, along with parallels and meridians being plane circles, there exist two more families of plane circles called the *Vilarseaux circles*. They are formed by the intersection of the torus by its tangent plane tangent to the torus at two points. Obtain the equations of these circles, verify that they have the same radius and intersect all parallels of the torus at a constant angle (see Fig. 8).

§ 3. Classical Metrics on the Sphere and on the Lobachevskii Plane; Their Properties

3.1. Calculate the metrics on the standard unit sphere in $\mathbb{R}^3$ in the following coordinates:

(a) the Cartesian coordinates x, y;

(b) the spherical coordinates θ, φ;

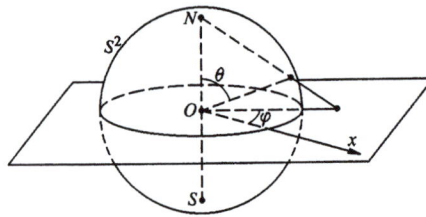

Figure 9 Stereographic projection

(c) the Cartesian coordinates u, v on the plane $z = 0$, which is the image of the sphere under the stereographic projection from the north pole of the sphere (see Fig. 9);

(d) the polar coordinates ρ, φ on the plane $z = 0$, which is the image of the sphere under the stereographic projection (see the previous item).

(e) the complex coordinates $z = x + iy$, $\bar{z} = x - iy$ on the plane $z = 0$, which is the image of the sphere under the stereographic projection.

3.2. Consider a pseudo-Euclidean space with the coordinates t, x, y and the inner product $ds^2 = -dt^2 + dx^2 + dy^2$. Calculate the metric of the pseudosphere $-t^2 + x^2 + y^2 = -1$ of imaginary radius in the following coordinates:

(a) the Cartesian coordinates x, y;

(b) the coordinates φ, χ, where ρ, φ, χ are the pseudo-spherical coordinates in our pseudo-Euclidean space;

(c) the Cartesian coordinates u, v on the plane $t = 0$, which is the image of the stereographic projection of the pseudosphere from its south pole $(0, 0, -1)$ (see Fig. 10);

(d) the polar coordinates r, φ on the plane $t = 0$, which is the image of the stereographic projection of the pseudosphere from its south pole $(0, 0, -1)$;

(e) the complex coordinates $z = x + iy$, $\bar{z} = x - iy$ on the plane $t = 0$, which is the image of the stereographic projection of the pseudosphere;

(f) pass from the complex coordinates of the previous item to new complex coordinates in the upper half-plane using the linear-fractional transformation that transforms a disk of unit radius into the upper half-plane;

(g) for two previous items, draw the images of both connected components of the pseudosphere at its projections on the plane indicated above;

(h) prove that the length ρ of the line segment connecting the points with radius-vectors $\mathbf{a}$ and $\mathbf{b}$ on the upper half of the pseudosphere is given by the formula $\cosh \rho = \langle \mathbf{a}, \mathbf{b} \rangle$, where the inner product is calculated in the pseudo-Euclidean space;

(i) prove that the distance ρ from the point with radius-vector $\mathbf{a}$ on the upper half of the pseudosphere to the line defined by the condition $\langle \mathbf{x}, \mathbf{e} \rangle = 0$, $|\mathbf{e}| = 1$, is given by the formula $\sinh \rho = |\langle \mathbf{a}, \mathbf{e} \rangle|$. Here, the inner product is calculated in the pseudo-Euclidean space.

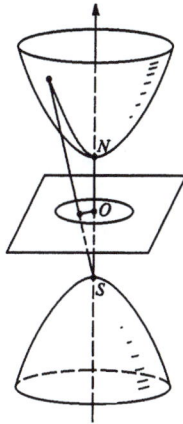

Figure 10 Stereographic projection

3.3. In the models of the Lobachevskii metric in the unit disk (Poincaré model) and on the upper half-plane, show that the angle between intersecting curves is equal to the angle between the same curves in the Euclidean metric.

3.4. Prove that under the stereographic projection of the sphere S^2 on the plane an arbitrary circle passes either into a line or a circle.

3.5. Show that the group of linear-fractional transformations being the motions of the metric

(a) of Problem 3.1e is isomorphic to $SU(2)/\{\pm E\}$;

(b) of Problem 3.2e is isomorphic to $SU(1,1)/\{\pm E\}$;

(c) of Problem 3.2f is isomorphic to $SL(2,\mathbb{R})/\{\pm E\}$.

3.6. Show that the group of linear-fractional transformations being the motions of the metric of the Lobachevskii plane in the model on the upper half-plane (see Problem 3.2f) is generated by the transformations $z \mapsto z + a$ and $z \mapsto -\dfrac{1}{z}$, where $a \in \mathbb{R}$.

3.7. On the Lobachevskii plane, consider arbitrary lines l_1 and l_2 with points A_1, B_1 and A_2, B_2 on them, respectively; the distance from A_1 to B_1 is equal to the distance from A_2 to B_2. Show that there exists a motion which transforms

(a) l_1 into l_2;

(b) l_1 into l_2 and A_1 into A_2;

(c) l_1 into l_2, A_1 into A_2, and B_1 into B_2.

3.8. Find a linear-fractional transformation of the upper half-plane being a motion of the Lobachevskii plane, which transforms

(a) the line $x^2 + y^2 = 1$ into the line $x = 0$;

(b) the line $(x - 1)^2 + y^2 = 4$ and the point $(1, 2)$ into the line $x = 3$ and the point $(3, 2)$.

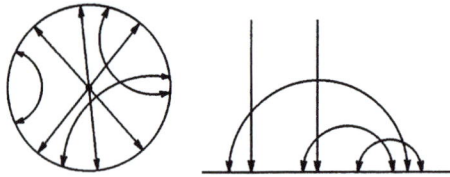

Figure 11 Lines on the Lobachevskii plane (models in the unit disk and on the upper half-plane)

3.9. Let points A_1, A_2, A_3, B_1, B_2 and B_3 of the Lobachevskii plane be such that $\rho(A_i, A_j) = \rho(B_i, B_j)$ for all i and j. Show that there exists a unique motion of the Lobachevskii plane transforming the point A_i into the point B_i, $i = 1, 2, 3$. Deduce from this that the group of all motions of the Lobachevskii plane is generated by the transformations $z \mapsto z + a$, $z \mapsto -\dfrac{1}{z}$ and $z \mapsto -\bar{z}$, where $a \in \mathbb{R}$.

3.10. (a) Show that, on the Lobachevskii plane, there exists a unique line perpendicular to a given line and passing through a given point.

(b) Using the previous item, find the symmetry with respect to a line on the Lobachevskii plane and show that this transformation is a motion.

(c) Find the formula for the symmetry with respect to the line
$$(x - 1)^2 + y^2 = 4.$$

(d) Find the formula for the symmetry with respect to the line $x = 0$.

3.11. (a) Show that a motion given as a linear-fractional transformation decomposes into a composition of an even number of symmetries with respect to a line. Show that two symmetries are sufficient.

(b) Show that other motions, i.e., those which are not represented as linear-fractional transformations, decompose into a composition of an odd number of symmetries with respect to a line.

3.12. Show that the group of all isometries of the Lobachevskii plane in the model on the upper half-plane consists of transformations of the form
$$z \mapsto \frac{az + b}{cz + d}, \quad a, b, c, d \in \mathbb{R}, \quad ad - bc = 1;$$
$$z \mapsto \frac{a\bar{z} + b}{c\bar{z} + d}, \quad a, b, c, d \in \mathbb{R}, \quad ad - bc = -1.$$

3.13. Show that under an arbitrary linear-fractional transformation $z \mapsto \dfrac{az + b}{cz + d}$ of the complex plane, where $a, b, c, d \in \mathbb{C}$, $ad - bc \neq 0$, circles and lines transform into circles or lines.

3.14. Prove that for any pairwise distinct points z_1, z_2, z_3 and pairwise distinct points w_1, w_2, w_3, belonging to an extended complex plane, there exists a unique linear-fractional transformation mapping z_i into w_j, $i = 1, 2, 3$.

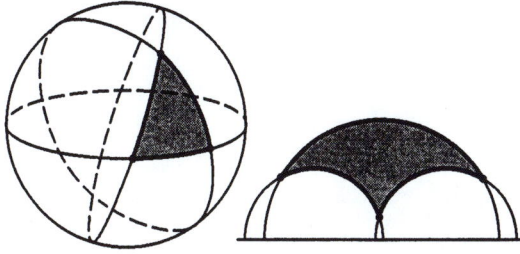

Figure 12 Triangles on the Lobachevskii plane and on the sphere

3.15. Show that the length of the segment connecting any two fixed points on the Lobachevskii plane is less than the length of any other curve connecting these points (see Fig. 11).

3.16. Prove that a unique line passes through any two points of the Lobachevskii plane (see Fig. 11).

3.17. Prove that in the models of the Lobachevskii plane in the unit disk and on the upper half-plane, circles are depicted by usual circles.

3.18. Find the length of the circle and the area of the disk of radius R (a) on the Lobachevskii plane; (b) on the sphere of radius 1. Compare with the formulas on the plane.

3.19. (a) Find the centre of the circle in the Lobachevskii metric on the upper half-plane, which is given by the equation

$$x^2 + (y - 2)^2 = 1.$$

(b) Prove that under the stereographic projection of the sphere on the plane, the circles on the sphere pass to circles or lines.

(c) Consider the circle $x^2 + (y - 2)^2 = 1$, which is the image of a circle on the sphere under the stereographic projection (see above). Take the centre of this spherical circle, which (the centre) lies on the sphere. Find its image on the plane.

3.20. Find the midpoint of the segment AB on the plane, where $A = \left(\frac{1}{2}, \frac{1}{2}\right)$ and $B = (0.9, 0.3)$, in the following metrics:

(a) the metric of the Lobachevskii plane in the upper half-plane;

(b) the metric of the sphere (under the stereographic projection).

3.21. Express the area of a triangle (see Fig. 12) through its angles (a) on the sphere of unit radius; (b) on the Lobachevskii plane.

3.22. Prove the following formulas for a rectangular triangle on the Lobachevskii plane:

(a) $\cosh c = \cosh a \cosh b$; (b) $\sinh b = \sinh c \sin \beta$;

(c) $\tanh a = \tanh c \cos \beta$; (d) $\cosh c = \cot \alpha \cot \beta$;

(e) $\cos\alpha = \cosh a \sin\beta$; (f) $\tanh a = \sinh b \tan\alpha$.

Here, c is the length of the side subtending the right angle.

3.23. On the sphere of radius 1, consider a rectangular triangle ABC for which $\angle C = \dfrac{\pi}{2}$, $\angle B = \beta$, $\angle A = \alpha$, $AC = b$, $AB = c$, and $BC = a$. Prove the following relations:

(a) $\cos c = \cos a \cos b$; (b) $\cos\alpha = \cos a \sin\beta$;

(c) $\sin a = \sin c \sin\alpha$; (d) $\tan a = \tan c \cos\beta$;

(e) $\cos c = \cot\alpha \cot\beta$; (f) $\tan a = \sin b \tan\alpha$.

3.24. Prove the cosine theorems for triangles on the Lobachevskii plane:

(a) $\sin\beta \sin\gamma \cosh a = \cos\alpha + \cos\beta \cos\gamma$;

(b) $\cos\alpha \sinh b \sinh c = \cosh b \cosh c - \cosh a$.

3.25. Prove the cosine theorems for triangles on the sphere of unit radius:

(a) $\cos a = \cos b \cos c + \sin b \sin c \cos\alpha$;

(b) $\cos\alpha = -\cos\beta \cos\gamma + \sin\beta \sin\gamma \cos a$.

3.26. Prove the sine theorems for

(a) the Lobachevskii plane:

$$\frac{\sinh a}{\sin\alpha} = \frac{\sinh b}{\sin\beta} = \frac{\sinh c}{\sin\gamma} = \frac{\sqrt{Q}}{\sin\alpha \sin\beta \sin\gamma},$$

where $Q = \cos^2\alpha + \cos^2\beta + \cos^2\gamma + 2\cos\alpha \cos\beta \cos\gamma - 1$;

(b) the sphere of radius 1:

$$\frac{\sin a}{\sin\alpha} = \frac{\sin b}{\sin\beta} = \frac{\sin c}{\sin\gamma}.$$

3.27. For which integer n there exist regular polygons with the angle $\dfrac{2\pi}{n}$ (a) on the sphere; (b) on the Lobachevskii plane?

3.28. Prove that on a complex plane, points z_1, z_2, z_3 and z_4 lie on the same circle (or line) if and only if the cross ratio $w = \dfrac{z_3 - z_1}{z_2 - z_3} : \dfrac{z_4 - z_1}{z_2 - z_4}$ is a real number.

3.29. Prove that on the Lobachevskii plane, the sine of the angle α in a rectangular triangle is equal to the ratio of the length of the circle of radius equal to the opposite leg and the length of the circle of radius equal to the hypotenuse. Prove the same for the sphere.

3.30. Prove that in a triangle with equal sides, all angles (a) are less than $\pi/3$ on the Lobachevskii plane; (b) are greater than $\pi/3$ on the sphere.

3.31. Prove the following formulas for the distance between points A and B (see Fig. 13) in the metric of the Lobachevskii plane on the upper half-plane:

(a) if they have the same abscissa, then $\rho(A, B) = \left|\ln \dfrac{OB}{OA}\right|$, where O is the intersection point of the line connecting them and of the absolute;

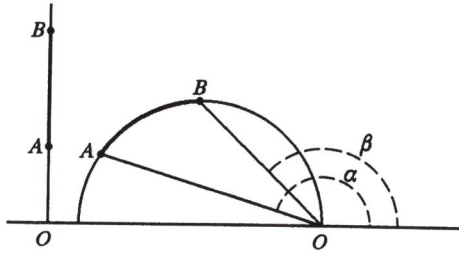

Figure 13

(b) if they have different abscissas, then $\rho\left(A, B\right) = \left|\ln\dfrac{\tan\alpha}{\tan\beta}\right|$, where O is the intersection point of the absolute and the line of the Lobachevskii plane passing through A and B, and α and β are the angles between the positive direction of the real line and the rays OA and OB, respectively;

(c) in the general case: if z_1 and z_2 are arbitrary points of the upper half-plane, then

$$\rho\left(z_1, z_2\right) = \ln\left(\frac{1 + \left|\left(z_1 - \bar{z}_2\right)/\left(z_1 - z_2\right)\right|}{-1 + \left|\left(z_1 - \bar{z}_2\right)/\left(z_1 - z_2\right)\right|}\right),$$

or, which is the same,

$$\rho\left(z_1, z_2\right) = \ln\left(\frac{1 + \left|\left(z_1 - z_2\right)/\left(z_1 - \bar{z}_2\right)\right|}{1 - \left|\left(z_1 - z_2\right)/\left(z_1 - \bar{z}_2\right)\right|}\right).$$

3.32. Is it true that it is possible to circumscribe a circle around any triangle (a) on the sphere? (b) on the Lobachevskii plane?

3.33. Show that it is not always possible to draw a line through a point lying inside an angle on the Lobachevskii plane, which (the line) intersects both sides of the angle.

3.34. Compare a side of a regular hexagon with the radius of the circle circumscribed around it (a) on the sphere; (b) on the Lobachevskii plane.

3.35. Find the area of a regular triangle with side a (a) on the Lobachevskii plane (see Fig. 14); (b) on the sphere.

3.36. Find the area of the disk and the length of the circle given by the equation $(x + 1)^2 + (y - 5)^2 = 1$ on (a) the Lobachevskii plane in its model on the upper half-plane; (b) the sphere in the coordinates of the stereographic projection.

3.37. Express the radius of the circle circumscribed around a triangle through its sides and angles (a) on the Lobachevskii plane; (b) on the sphere.

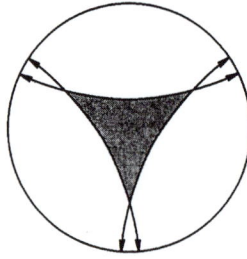

Figure 14 Regular triangle on the Lobachevskii plane

§4. Theory of Curves

Generally speaking, we will assume that the curvature of a plane curve $\mathbf{r} = \mathbf{r}\,(s)$, where s is the natural parameter, is defined by the formula $k = |\ddot{\mathbf{r}}| = |\dot{\mathbf{v}}|$. In this case, the normal to the curve is given by the formula $\mathbf{n} = \dot{\mathbf{v}}/\,|\dot{\mathbf{v}}|$. It is not defined at points where $k = 0$. Moreover, at inflection points, the normal field is discontinuous, i.e., the left and right limits are different (see Fig. 15).

In some problems on the global behaviour of a curve, it is more convenient to use a continuous field of normals. In such cases, the normal fields can be defined by turning the velocity vector at each point by the angle $\pi/2$ in the positive direction. In particular, at each point of the curve, the pair of vectors $\mathbf{v}\,(s)\,,\mathbf{n}\,(s)$ forms a positively oriented frame on the plane. The Frénet formulas remain valid; herewith, the curvature k can change the sign.

4.1. Calculate the curvature of the following curves:

(a) $y = \sin x$ in the top (sinusoid);

(b) $y = a\cosh\,(x/a)$ (catenary);

(c) $r^2 = a^2\cos 2\varphi$ (lemniscate);

(d) $r = a\,(1 + \cos\varphi)$ (cardioid, see Fig. 16);

(e) $r = a\varphi$ (Archimedean spiral, see Fig. 17);

(f) $\mathbf{r}\,(t) = \left(a\cos^3 t, a\sin^3 t\right)$ (astroid, see Fig. 18);

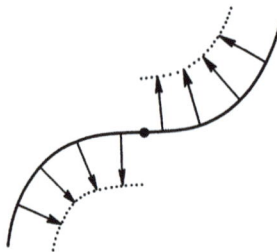

Figure 15

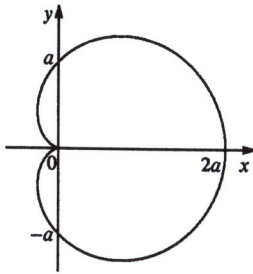

Figure 16 Cardioid

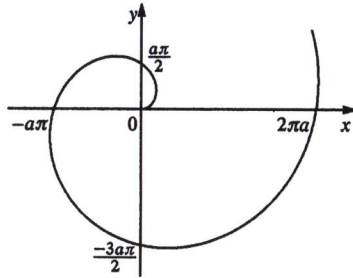

Figure 17 Archimedean spiral

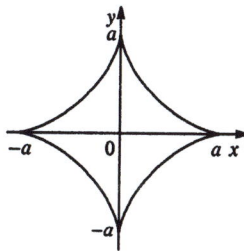

Figure 18 Astroid

(g) $\mathbf{r}(t) = (a\,(t - \sin t), a\,(1 - \cos t))$ (cycloid, see Fig. 19).

4.2. For the Bernoulli lemniscate given by the equation $r^2 = 2a^2 \cos 2\varphi$ in the polar coordinates (r, φ), calculate the integrals

$$\int_\gamma k\,(s)\,ds \quad \text{and} \quad \int_\gamma |k\,(s)|\,ds,$$

where s is the arclength and $k\,(s)$ is the curvature of the curve. See the explanation in the beginning of the section.

4.3. Find the curvature of the ellipse with semiaxes a and b at its vertices.

4.4. Find the curvature of the curve given by the equation $F\,(x, y) = 0$.

4.5. The curves are given by their differential equation $P\,(x, y)\,dx + Q\,(x, y)\,dy = 0$. Find their curvature.

4.6. Deduce the formula for the curvature of a plane curve given by the equation $r = r\,(\varphi)$ in the polar coordinates.

4.7. Replace parameter t on the helical line

$$\mathbf{r}\,(t) = (a \cos t, a \sin t, bt)\,, \quad b > 0,$$

(see Fig. 20) by the natural parameter s.

4.8. Replace parameter t on the curve

$$\mathbf{r}\,(t) = \left(e^t \cos t, e^t \sin t, e^t\right)$$

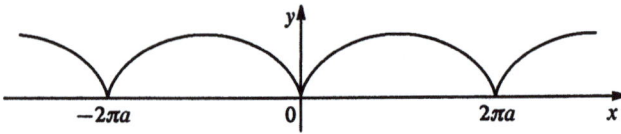

Figure 19 Cycloid

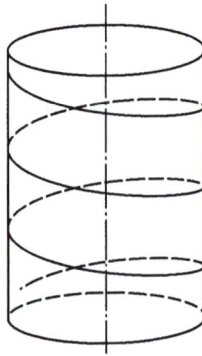

Figure 20 Helical line

by the natural parameter.

4.9. Replace parameter t on the curve $\mathbf{r}\,(t) = (\cosh t, \sinh t, t)$ by the natural parameter.

4.10. Find the curvature and torsion of the following curves at an arbitrary point:

(a) $\mathbf{r}\,(t) = \left(e^t, e^{-t}, t\sqrt{2}\right)$; (b) $\mathbf{r}\,(t) = \left(2t, \ln t, t^2\right)$;

(c) $\mathbf{r}\,(t) = (e^t \sin t, e^t \cos t, e^t)$; (d) $\mathbf{r}\,(t) = \left(3t - t^3, 3t^2, 3t + t^3\right)$;

(e) $\mathbf{r}\,(t) = \left(\cos^3 t, \sin^3 t, \cos 2t\right)$.

4.11. Find the curvature and torsion of the curve given by the equations

$$x^2 + z^2 - y^2 = 1,$$

$$y^2 - 2x + z = 0$$

at point $M\,(1, 1, 1)$.

4.12. Find the curvature and torsion of the curve given by the equations

$$x + \sinh x = \sin y + y,$$

$$z + e^z = x + \ln\,(1 + x) + 1$$

at point $M\,(0, 0, 0)$.

4.13. Deduce the formulas for calculating the curvature and torsion of the curve given by the equations $y = y\,(x)$, $z = z\,(x)$ and find the Frénet frame of this curve.

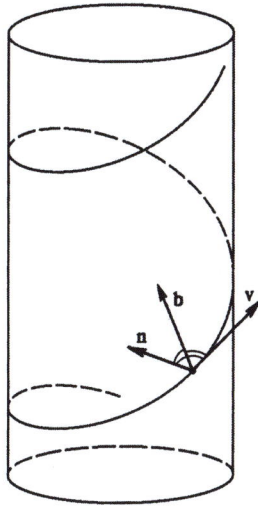

Figure 21 Frénet frame of the helix

4.14. Given the curve

$$\mathbf{r}\left(t\right) = \left(t^2, 1 - t, t^3\right),$$

find its Frénet frame. Calculate the curvature and torsion of this curve.

4.15. (a) Prove that the curvature and torsion of the helical line are constant.

(b) Find for which values of h the helix $x = a \cos t$, $y = a \sin t$, $z = ht$ has the maximum torsion.

(c) Find the Frénet frame of the helix (see Fig. 21)

$$\mathbf{r}\left(t\right) = \left(a \cos t, a \sin t, ht\right).$$

(d) Find all curves of constant curvature k and torsion $\varkappa$.

4.16. Under which condition does the curvature centre of the helix lie on the same cylinder as the curve itself?

4.17. Let γ be a plane curve. Denote by S the area of the domain bounded by curve γ and the secant drawn at a distance h from the tangent at a fixed point $P \in \gamma$. Express $\lim\limits_{h \to 0} \left(S^2/h^3\right)$ through the curvature of the curve.

4.18. Prove that the operator $Y : \mathbf{x} \mapsto \mathbf{y} \times \mathbf{x}$ acting on vectors $\mathbb{R}^3$ is written by a skew-symmetric matrix. Recall that $\mathbf{y} \times \mathbf{x}$ is the vector product of vectors $\mathbf{x}$ and $\mathbf{y}$. Find the relation of the entries of this matrix with the coordinates of vector $\mathbf{y}$. Show that for any skew-symmetric matrix there exists a vector $\mathbf{y}$ such that its action as a linear operator in $\mathbb{R}^3$ is represented in the form $\mathbf{x} \mapsto \mathbf{y} \times \mathbf{x}$. The vector $\mathbf{y}$ is called the Darboux vector of the skew-symmetric matrix.

4.19. Let Y and Z be the matrices of the operators of vector multiplication by vectors $\mathbf{y}$ and $\mathbf{z}$. Prove that the matrix of the operator of vector multiplication by $\mathbf{y} \times \mathbf{z}$ is equal to $[Y, Z] = YZ - ZY$.

4.20. Let the curvature k of a regular spatial curve $\mathbf{r}(s)$ turn to zero at finitely many points. Assume that on the curve there exists a smooth vector field $\mathbf{n}^*(s)$ such that at those points where $k \neq 0$ it coincides either with $\mathbf{n}(s)$ or with $-\mathbf{n}(s)$, where $\mathbf{n}(s)$ is the principal normal to the curve. Define $\mathbf{b}^*(s)$ by the formula $\mathbf{b}^* = \mathbf{v} \times \mathbf{n}^*$ and find the curvature k and torsion $\varkappa$ from the formulas $\dot{\mathbf{v}} = k\mathbf{n}^*$, $\dot{\mathbf{b}}^* = -\varkappa\mathbf{n}^*$ (where $\mathbf{v} = \dot{\mathbf{r}}(s)$). Prove that then the Frénet formulas also hold at points at which $k = 0$ but, in this case, the curvature of the spatial curve can prove alternating.

The natural equation of a plane curve is the equation of one of the following forms:

$$(1)\ k = k(s),$$

$$(2)\ F(k, s) = 0,$$

$$(3)\ k = k(t),\ s = s(t).$$

If we know the natural equation of the curve, then the parameterization of the curve can be given in the form

$$x = \int \cos\alpha(s)\,ds, \quad y = \int \sin\alpha(s)\,ds,$$

where $\alpha(s) = \int k(s)\,ds$.

4.21. Compose the natural equations of the curves
(a) $x = a\cos^3 t$, $y = a\sin^3 t$;
(b) $y = x^{3/2}$;
(c) $y = x^2$;
(d) $y = \ln x$;
(e) $y = a\cosh\left(\dfrac{x}{a}\right)$;
(f) $y = e^x$;
(g) $x = a\left(\ln\tan\dfrac{t}{2} + \cos t\right)$, $y = a\sin t$;
(h) $r = a(1 + \cos\varphi)$;
(i) $x = a(\cos t + t\sin t)$, $y = a(\sin t - t\cos t)$;
(j) $\mathbf{r}(t) = (a(t - \sin t), a(1 - \cos t))$.

4.22. Find the parametric equations of the curves knowing their natural equations (here, $R = 1/k$):
(a) $R = as$; (b) $\dfrac{s^2}{a^2} + \dfrac{R^2}{b^2} = 1$; (c) $Rs = a^2$;

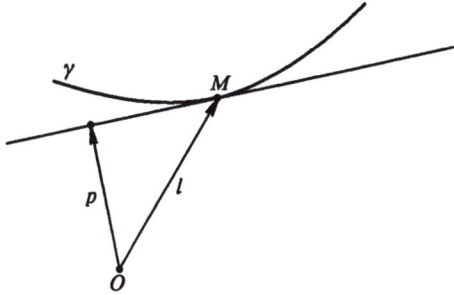

Figure 22

(d) $R = a + \dfrac{s^2}{a}$; (e) $R^2 = 2as$.

4.23. In which cases does a curve have the following parametric equations: $x = s, y = y(s), z = z(s)$, where s is the natural parameter?

4.24. Consider the cycloid with its convexity "inverted" downward:

$$x(t) = R(t + \pi + \sin t), \quad y(t) = R - R\cos t, \quad t \in [-\pi, \pi].$$

Here, the point $t = 0$ corresponds to the lowest point of the cycloid. Denote by g the acceleration of free fall. Place a material point of mass m in the cycloid. Show that the period of oscillations of the material point moving along the cycloid without friction in the gravity force field is independent of its initial position. Write the differential equation for the distance from the material point to the lowest point of the cycloid.

4.25. Find a plane curve whose tangent makes a constant angle α with the radius-vector of the curve.

4.26. Let p be the distance from the origin of radius-vectors to the tangent of a curve γ at point M, and let l be the distance from point O to point M (see Fig. 22). Prove that

$$k = \left| \frac{dp}{l\,dl} \right|.$$

4.27. Let γ be a smooth regular closed curve. Prove that

$$\int_\gamma r\,dk + \int_\gamma \varkappa b\,ds = 0.$$

4.28. Let a plane convex arc L_1 be tangent to a plane strictly convex arc L_2 remaining to one side from it. Prove that at the tangent point the curvature of curve L_1 is no less than the curvature of curve L_2.

4.29. At a point $\mathbf{r}_0 = \mathbf{r}(s_0)$ of the curve $\mathbf{r} = \mathbf{r}(s)$, let $k_0 = k(s_0) \neq 0$, $\dot{k}(s_0) \neq 0$. Consider the equation of the osculating circle $|\boldsymbol{\rho} - \mathbf{r}_0 - R_0\mathbf{n}_0| =$

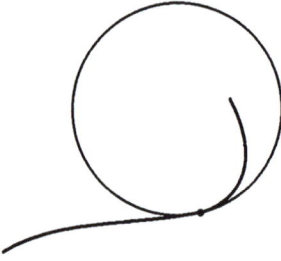

Figure 23 Curve and the osculating circle

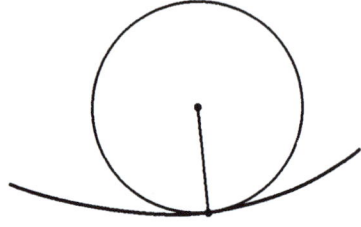

Figure 24 Curve and the osculating circle

R_0 at this point of the initial curve, $R_0 = \dfrac{1}{k_0}$. Here, $\boldsymbol{\rho}$ is the radius-vector
of a point on the osculating circle. Prove that the osculating circle intersects
the given curve in a neighborhood of the mentioned point, i.e., a small arc
of the curve corresponding to the values of the parameter from the interval
$(s_0 - \varepsilon, s_0)$ and a small arc of the curve corresponding to the values of the
parameter from the interval $(s_0, s_0 + \varepsilon)$ lie to different sides of the osculating
circle (see Fig. 23).

Explanation. The order of tangency of the curve and of its osculating circle is
equal to 3. The conditions imposed on the curve guarantee that in this case the
order of tangency at the given point is exactly equal to 3 and is not higher.

4.30. At a point of a curve, the conditions $k_0 \neq 0$, $\dot{k}_0 = 0$ and $\ddot{k}_0 \neq 0$ hold.
Prove that the osculating circle of the curve at this point does not intersect
the curve in a sufficiently small neighborhood of this point (see Fig. 24).

4.31. Let α be the angle between a constant vector $\mathbf{a}$ and the tangent
vector $\mathbf{v}$ to the curve (see Fig. 25). Compose the parametric equation of the
curve knowing the following dependence:

(a) $R = f(\alpha)$, where R is the radius of curvature of the curve;

(b) $\alpha = f(R)$;

(c) $s = f(\alpha)$, where s is an arc of the curve; (d) $\alpha = f(s)$.

4.32. The curve along which the sphere intersects the circular cylinder of
radius two times less, the cylinder passing through the centre of the sphere, is
called the Viviani curve (see Fig. 26). Compose the equation of the Viviani
curve in an implicit and parametric form. Find the equations of the tangent,
normal plane, binormal, principal normal and osculating circle. Also, find the
Frénet frame, the curvature and the torsion.

4.33. Prove that the curve

$$x^2 = 2az, \quad y^2 = 2bz$$

is a plane curve.

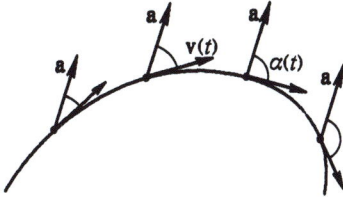

Figure 25

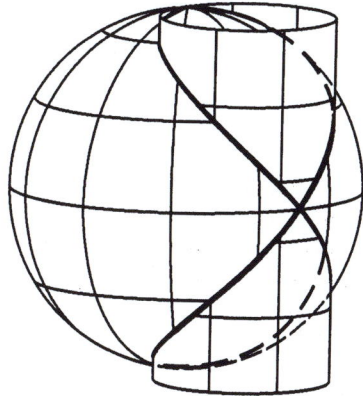

Figure 26 Viviani curve

4.34. Prove that if at a point M of a curve C the curvature and torsion are different from zero, parts of the curve close to the point M lie to different sides of the osculating plane.

4.35. Prove that if all osculating planes of a curve pass through the same point, the curve is plane on each part of biregularity (i.e., on the part where $k \neq 0$ at all points).

4.36. Express $\dfrac{d}{ds}\mathbf{r}$, $\dfrac{d^2}{ds^2}\mathbf{r}$ and $\dfrac{d^3}{ds^3}\mathbf{r}$ through $\mathbf{v}$, $\mathbf{n}$, $\mathbf{b}$, k, and $\varkappa$.

4.37. Prove that $\left(\mathbf{v}, \mathbf{b}, \dfrac{d}{ds}\mathbf{b} \right) = \varkappa$, where $(\cdot, \cdot, \cdot)$ denotes the mixed product of three vectors.

4.38. Calculate $\left(\dfrac{d}{ds}\mathbf{b}, \dfrac{d^2}{ds^2}\mathbf{b}, \dfrac{d^3}{ds^3}\mathbf{b} \right)$.

4.39. Prove that $\left(\dfrac{d}{ds}\mathbf{v}, \dfrac{d^2}{ds^2}\mathbf{v}, \dfrac{d^3}{ds^3}\mathbf{v} \right) = k^5 \dfrac{d}{ds} \left(\dfrac{\varkappa}{k} \right)$.

4.40. Prove that if the principal normals of a curve make a constant angle with the direction of a vector $\mathbf{e}$, then

$$\frac{d}{ds} \left(\frac{k^2 + \varkappa^2}{k\,(d/ds)\,(\varkappa/k)} \right) + \varkappa = 0,$$

and vice versa, if the latter relation holds, the principal normals of the curve make a constant angle with the direction of a vector. Find this vector.

4.41. Prove that if a curve is biregular (i.e., $k(s) \neq 0$ for all s) and all normal planes of the curve contain a vector $\mathbf{e}$, this curve is a plane curve.

4.42. Prove that if all osculating planes of a curve γ being not a straight line contain the same vector, the curve is plane (on any connected part of biregularity).

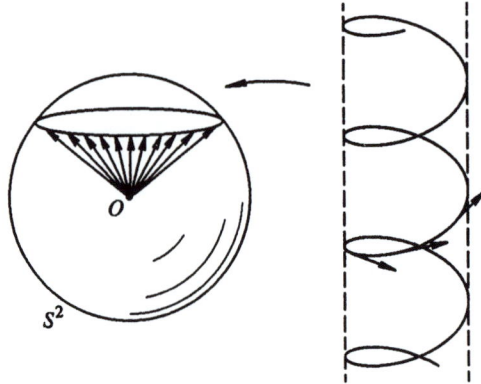

Figure 27 Tangent spherical image of a curve

4.43. (a) Prove that if a curve is regular and $k = 0$, it is a straight line.

(b) Prove that if $\mathbf{b} = $ const, the curve is plane. Write the equation of the plane.

(c) Prove that if $k \neq 0$ at all points of a curve, the curve is plane if and only if $\varkappa$ turns to zero everywhere.

(d) Give an example of a non-plane curve for which $\varkappa \equiv 0$.

4.44. Prove that if osculating planes of a curve have the same inclination (i.e., all of them are orthogonal to a fixed nonzero vector $\mathbf{e}$), the curve is a plane curve.

4.45. For which functions $f(t)$, the curves

$$(a) \quad \mathbf{r}(t) = \left(e^t, 2e^{-t}, f(t)\right) \quad \text{and} \quad (b) \quad \mathbf{r}(t) = (a\cos t, a\sin t, f(t))$$

are plane curves?

4.46. Prove that the following curves are plane and find their planes:

(a) $\mathbf{r}(t) = \left(\dfrac{1+t}{1-t}, \dfrac{1}{1-t^2}, \dfrac{1}{1+t}\right);$

(b) $\mathbf{r}(t) = \left(a_1t^2 + b_1t + c_1, a_2t^2 + b_2t + c_2, a_3t^2 + b_3t + c_3\right).$

4.47. Let $\bar{s}$ be the length of the tangent spherical image of the curve $\mathbf{r} = \mathbf{r}(s)$ (see Fig. 27):

$$\bar{s} = \int_0^s |\mathbf{v}'(\sigma)| \, d\sigma.$$

(a) Prove that $\dfrac{d\bar{s}}{ds} = k$;

(b) Find necessary and sufficient conditions under which the tangent spherical image is a regular curve.

(c) Prove that the inequality $\int k\,ds \geq 2\pi$ holds for a closed curve.

4.48. Let s^* be the length along the normal (resp. the binormal) to the spherical image of the curve $\mathbf{r} = \mathbf{r}(s)$. Prove that

$$\frac{ds^*}{ds} = \sqrt{k^2 + \varkappa^2} \ \text{(resp. } |\varkappa|\text{)}.$$

A *spherical curve* is a curve $\mathbf{r} = \mathbf{r}(t)$ for which there exists a constant vector $\mathbf{m}$ and a real number R such that

$$\langle \mathbf{r}(t) - \mathbf{m}, \mathbf{r}(t) - \mathbf{m} \rangle = R^2.$$

4.49. Prove that if $\mathbf{r} = \mathbf{r}(s)$ is a curve parameterized by the natural parameter $k \neq 0$ such that $\varkappa \neq 0$, then $\mathbf{r}(s)$ is a spherical curve if and only if

$$\frac{\varkappa}{k} = \frac{d}{ds}\left(\frac{dk/ds}{\varkappa k^2}\right).$$

4.50. Let a non-plane curve γ have a constant nonzero curvature, and let $\varkappa \neq 0$. Consider the set γ^* of its curvature centres. Prove that the curvature of γ^* is also constant. Find the torsion of γ^*.

Let $\mathbf{m}$ be a constant vector, $\mathbf{r} = \mathbf{r}(s)$ be a curve, $c(s) = |\mathbf{r}(s) - \mathbf{m}|^2$, and let a be a positive number. We say that the curve $\mathbf{r}(s)$ has, at a point $s = s_0$, a spherical contact of a j-th order with the sphere of radius a centred at the point $\mathbf{m}$ if

$$c(s_0) = a^2, \quad c'(s_0) = c''(s_0) = \cdots = c^{(j)}(s_0) = 0, \quad c^{(j+1)}(s_0) \neq 0.$$

4.51. If $k \neq 0$, express the first three derivatives of the function $c(s)$ (for its definition, see above) through $\mathbf{v}$, $\mathbf{n}$, $\mathbf{b}$, k and $\varkappa$ and their derivatives.

4.52. Let $\mathbf{r} = \mathbf{r}(s)$ be a curve parameterized by the natural parameter, let $k \neq 0$, $\varkappa \neq 0$, $\rho = 1/k$, and let $\sigma = 1/\varkappa$. Assume that $\rho^2 + (\rho'\sigma)^2 = a^2 = $ const, $a > 0$. Prove that the image of the curve $\mathbf{r} = \mathbf{r}(s)$ lies on the sphere of radius a.

4.53. Prove that a curve $\mathbf{r} = \mathbf{r}(s)$ has a spherical contact of the second or higher order at a point $s = s_0$ if and only if $k(s_0) > 0$ and the radius-vector of the centre of the sphere is given by the formula

$$\mathbf{m} = \mathbf{r}(s_0) + \frac{1}{k(s_0)}\mathbf{n}(s_0) + \lambda \mathbf{b}(s_0),$$

where λ is an arbitrary number.

4.54. Let $k(s_0) \neq 0$, and let $\varkappa(s_0) \neq 0$. Prove that the curve $\mathbf{r} = \mathbf{r}(s)$ has a spherical contact of the third or higher order at the point $s = s_0$ if and only if the centre of the sphere is given by

$$\mathbf{m} = \mathbf{r}(s_0) + \frac{1}{k(s_0)}\mathbf{n}(s_0) - \frac{k'(s_0)}{k^2(s_0)\,\varkappa(s_0)}\mathbf{b}(s_0),$$

and its radius is given by

$$R^* = \sqrt{\frac{1}{k^2} + \frac{(dk/ds)^2}{k^4 \varkappa^2}}.$$

4.55. Prove that for any closed curve on the sphere there exists a point at which the torsion of the curve is equal to 0.

4.56. At a point M, let a smooth regular curve be tangent to a circle Γ whose centre lies on the same ray normal to the curve as the centre of the curvature disk. Show that

(a) if in a neighborhood of the point M the curve is located outside (inside) the circle Γ, the curvature radius of the curve is not less (not greater) than the radius R of the circle Γ;

(b) if the curvature radius of the curve is greater (less) than R, the curve lies outside (inside) Γ in a neighborhood of M;

(c) if the circle Γ coincides with the curvature circle and the curvature radius of the curve has a local maximum (minimum) at the point M, the curve lies inside (outside) the circle Γ in a neighborhood of M.

4.57. Let a regular smooth simple closed curve L on the plane have a positive curvature everywhere. Prove that L is convex as a whole. Obtain the same assertion under the assumption that the curvature is nonnegative.

4.58. Can a point move along a regular curve in such a way that the velocity of motion is proportional to the length of the path passed from the start of the motion?

§ 5. Riemannian Metric

5.1. Calculate the first quadratic form of the following surfaces:

(a) $\mathbf{r}(u, v) = (a \cos u \cos v, a \sin u \cos v, a \sin v)$ (sphere, Fig. 28);

(b) $\mathbf{r}(u, v) = (a \cos u \cos v, b \sin u \cos v, c \sin v)$ (ellipsoid, Fig. 29);

(c) $\mathbf{r}(u, v) = (av \cos u, bv \sin u, cv)$ (cone, Fig. 30);

(d) $\mathbf{r}(u, v) = (a \cos u, b \sin u, cv)$ (cylinder, Fig. 31).

5.2. Calculate the first quadratic form of the following surfaces (here, s is the natural parameter on the curve $\boldsymbol{\rho}(s)$):

(a) $\mathbf{r}(s, \lambda) = \boldsymbol{\rho}(s) + \lambda \mathbf{e}$, $\mathbf{e} = \mathrm{const}$ (cylindrical surface, Fig. 32);

(b) $\mathbf{r}(s, v) = v \boldsymbol{\rho}(s)$ (conical surface);

(c) $\mathbf{r}(s, \lambda) = \boldsymbol{\rho}(s) + \lambda \mathbf{e}(s)$ ($|\mathbf{e}(s)| = 1$) (ruled surface, Fig. 33);

(d) $\mathbf{r}(s, \varphi) = \boldsymbol{\rho}(s) + \mathbf{n}(s) \cos \varphi + \mathbf{b}(s) \sin \varphi$ (canal surface, Fig. 34);

(e) $\mathbf{r}(u, v) = (\varphi(v) \cos u, \varphi(v) \sin u, \psi(v))$ (surface of revolution);

(f) $\mathbf{r}(u, v) = ((a + b \cos v) \cos u, (a + b \cos v) \sin u, b \sin v)$ (torus);

(g) $\mathbf{r}(u, v) = (u \cos v, u \sin v, av)$ (helicoid);

(h) $\mathbf{r}(s, \lambda) = \boldsymbol{\rho}(s) + \lambda \mathbf{n}(s)$ (principal normal surface);

Figure 28 Sphere   **Figure 29** Ellipsoid

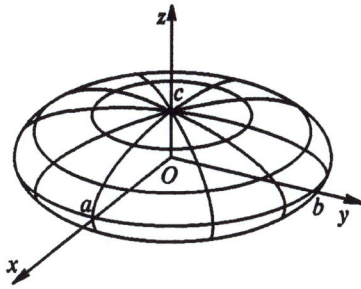

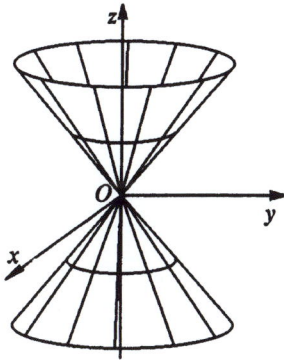

Figure 30 Cone   **Figure 31** Cylinder

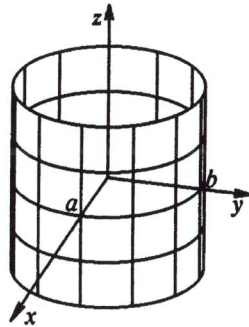

(i) $\mathbf{r}\,(s, \lambda) = \boldsymbol{\rho}\,(s) + \lambda \mathbf{b}\,(s)$ (binormal surface);

(j) $\mathbf{r}\,(z, \varphi) = \left(a \cosh \dfrac{z}{a} \cos \varphi, a \cosh \dfrac{z}{a} \sin \varphi, z \right)$ (catenoid, Fig. 35).

5.3. Find the first quadratic form of the surface (Beltrami pseudosphere)

$$x = a \sin u \cos v, \quad y = a \sin u \sin v, \quad z = a \left(\ln \tan \dfrac{u}{2} + \cos u \right),$$

where $u \neq \pi/2$ and $a = $ const (see Fig. 36).

5.4. Find the angle between the curves $v = u + 1$ and $v = 3 - u$ on the surface $x = u \cos v, y = u \sin v, z = u^2$.

5.5. The metric $ds^2 = du^2 + 2dv^2$ is given on the plane with coordinates (u, v). Find the angle between the curves $v = 2u$ and $v = -2u$.

5.6. On the surface $(u \cos v, u \sin v, av)$, find the angle between the intersecting curves (see Fig. 37)

$$u + v = 0, \quad u - v = 0.$$

5.7. Find the angle between the curves $v = 2u + 1$ and $v = -2u + 1$ on

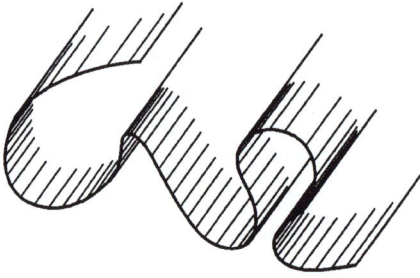

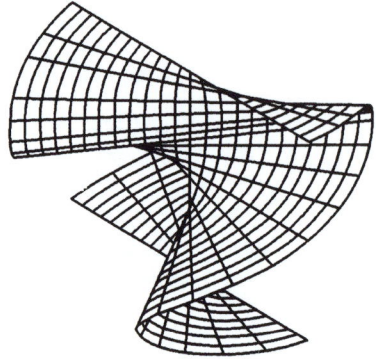

Figure 32 Cylindrical surface   **Figure 33** Ruled surface

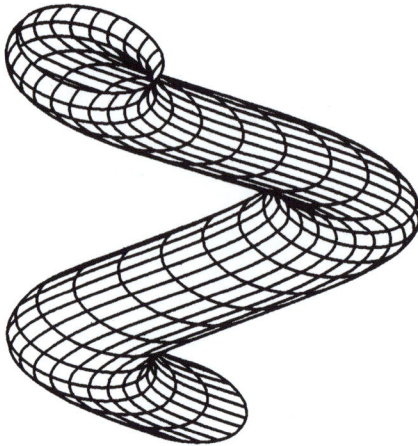

Figure 34 Canal surface

the plane with coordinates (u, v) if the metric is given by the matrix-valued function

$$G = \begin{pmatrix} 2 & 1 \\ 1 & 4 \end{pmatrix}.$$

5.8. Verify that on the plane with coordinates (u, v) the matrix-valued function

$$G = \frac{R^2}{(1 + u^2 + v^2)^2} \begin{pmatrix} 1 + v^2 & -uv \\ -uv & 1 + u^2 \end{pmatrix}$$

defines a metric. Find the length of the curve $u = v$.

5.9. Verify that the matrix-valued function

$$G = \frac{R^2}{1 - u^2 - v^2} \begin{pmatrix} 1 - v^2 & uv \\ uv & 1 - u^2 \end{pmatrix}$$

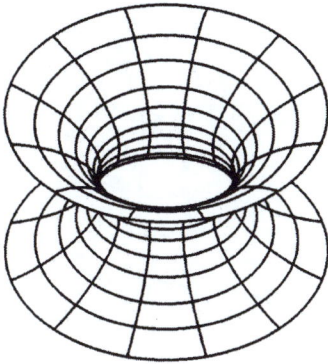

Figure 35 Catenoid   **Figure 36** Beltrami pseudosphere

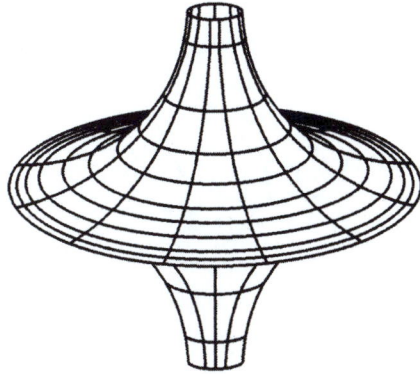

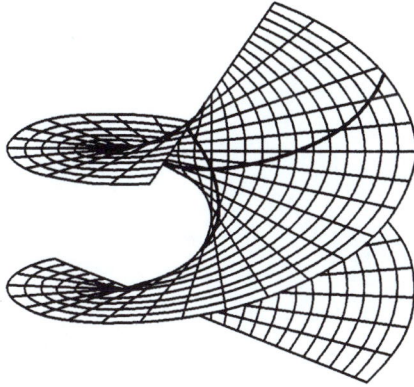

Figure 37 Two lines on the helicoid

defines a metric in the unit disk on the plane with coordinates (u, v);

(a) in this metric, find the length of the curve $-1 < u < 1$, $v = 0$;

(b) in the same metric, find the length of the curve S_α: $u^2 + v^2 = \alpha =$ const. Find the angle at which the curve $v = ku$ intersects the curves S_α.

5.10. Verify that the matrix-valued function

$$G = \frac{R^2}{1 - u^2 - v^2} \begin{pmatrix} 1 - v^2 & -uv \\ -uv & 1 - u^2 \end{pmatrix}$$

on the plane with coordinates (u, v) defines a metric in the unit disk. In this metric, find the length of the curve $-1 < u < 1$, $v = 0$.

5.11. Verify that on the plane with coordinates (x, y) the matrix-valued function

$$G = \frac{1}{(x^2 + y^2)^4} \begin{pmatrix} 4x^2 + (x^2 + y^2)^4 & 4xy \\ 4xy & 4y^2 + (x^2 + y^2)^4 \end{pmatrix}$$

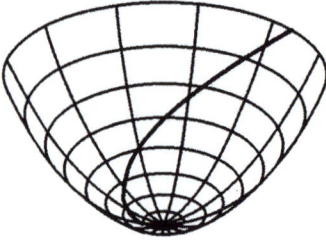

Figure 38 Curves on the paraboloid of revolution

Figure 39 Curves on the paraboloid of revolution

defines a metric. Find the length of the curve $x^2 + y^2 = \alpha$, where α is a fixed number. Calculate the angle at which these curves intersect the curves $y = kx$.

5.12. Find the intersection angle of the curves $u + 2v = 0$ and $4u - v = 0$ on the right helicoid

$$x = u \cos v, \quad y = u \sin v, \quad z = av.$$

5.13. On the surface

$$x = u \left(3v^2 - u^2 - \frac{1}{3} \right), \quad y = v \left(3u^2 - v^2 - \frac{1}{3} \right), \quad z = 2uv,$$

find the angle between the coordinate lines.

5.14. Find the equations of the curves dividing into two the angles between the coordinate lines of the paraboloid of revolution

$$x = u \cos v, \quad y = u \sin v, \quad z = \frac{1}{2} u^2.$$

(see Figs. 38 and 39).

5.15. On the surface

$$x = u \cos v, \quad y = u \sin v, \quad z = a \ln \left(u + \sqrt{u^2 - a^2} \right)$$

find the curves intersecting the curves $v = $ const at a constant angle θ (see Fig. 40).

5.16. Find the curves intersecting the rectilinear rulings of the hyperbolic paraboloid $xy = az$ at right angles.

5.17. A curve on the sphere intersecting all its meridians at a given angle is called a loxodrome (see Figs. 41 and 42). Compose the equation of the loxodrome. Find the vectors $\mathbf{v}$, $\mathbf{n}$ and $\mathbf{b}$ of the Frénet frame of this curve at an arbitrary point. Calculate its curvature and torsion.

5.18. Write the equation of the loxodrome in the polar coordinates of the stereographic projection of the sphere on the plane.

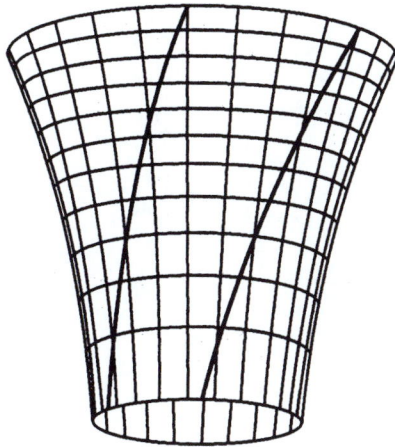

Figure 40 Loxodromes

5.19. Let the first quadratic form of a surface have the form

$$ds^2 = du^2 + \left(u^2 + a^2\right) dv^2.$$

(a) Find the perimeter of the curvilinear triangle formed by the intersection of the curves

$$u = \pm\frac{1}{2}av^2, \quad v = 1;$$

(b) Find the angles of this curvilinear triangle;

(c) Calculate the area of the triangle formed by the intersection of the curves

$$u = \pm av, \quad v = 1.$$

5.20. Given the surface $\mathbf{r}\left(u, v\right) = \left(u\sin v, u\cos v, v\right)$, find

(a) the area of the curvilinear triangle $0 \le u \le \sinh v$, $0 \le v \le v_0$ (see Fig. 43);

(b) the lengths of the sides of this triangle;

(c) the angles of this triangle.

5.21. The spherical digon is a figure formed by two large semicircles having common endpoints (see Fig. 44). Calculate the area S of the spherical digon with angle α at the vertex.

5.22. A torus is constructed by rotating a circle around a line lying in the plane of the circle. The radius of the circle is equal to r, and the distance from the line to the centre of the circle is equal to R, $R > r$. Find the area of the torus in the induced metric.

5.23. Prove that a mapping being conformal and area-preserving is a local isometry.

Figure 41 Loxodrome on the sphere   **Figure 42** Loxodrome on the sphere

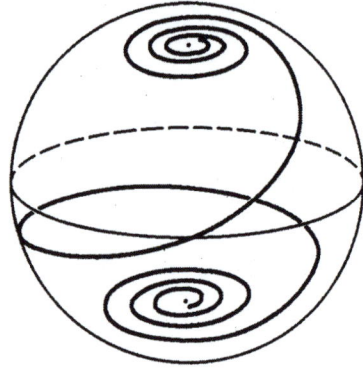

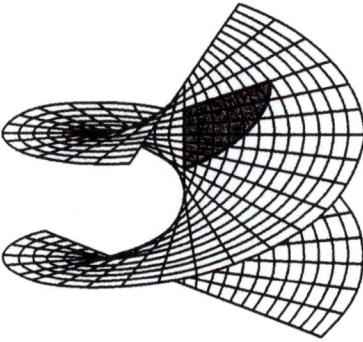

Figure 43 Triangle on the helicoid   **Figure 44** Spherical digon

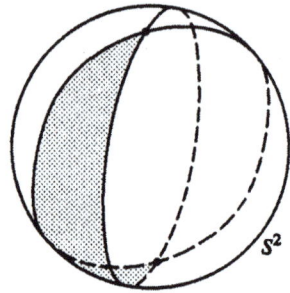

5.24. Fix a point on the plane. Find all conformal transformations of the surface that leave this point fixed.

5.25. Write the general form of a conformal self-mapping of the sphere S: $x^2 + y^2 + z^2 = 1$ with the orientation preservation condition such that it leaves the south and north poles fixed.

5.26. Prove that the deformation of the hyperbolic paraboloid defined by the formulas

$$\begin{cases} x = u, \\ y = v, \\ z = \dfrac{1}{2}\left(u^2 - v^2\right), \end{cases} \longmapsto \begin{cases} x = u, \\ y = v, \\ z = \dfrac{\sin t}{2}\left(u^2 - v^2\right) + uv\cos t \end{cases}$$

preserves the area.

5.27. Prove that the catenoid is a surface of revolution locally isometric to the helicoid

$$\mathbf{r}\left(u, v\right) = \left(u\sin v, u\cos v, av\right).$$

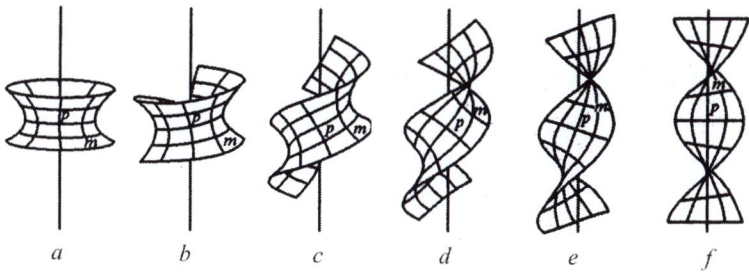

Figure 45 Bending the catenoid to obtain a part of the helicoid

Explanation. In fact, a more general fact is true. Namely, after cutting the catenoid along a meridian, the resulting surface can be bent as shown in Fig. 45, so that a part of the helicoid is obtained.

5.28. Show that the helical surface (conoid)

$$x = \rho \cos v, \quad y = \rho \sin v, \quad z = \rho + v$$

is locally isometrically mapped on the hyperboloid of revolution

$$x = r \cos \varphi, \quad y = r \sin \varphi, \quad z = \sqrt{r^2 - 1}$$

if the correspondence between their points is given by the equations

$$\varphi = v + \arctan \rho, \quad r^2 = \rho^2 + 1.$$

5.29. Prove that the helical surface

$$x = \rho \cos v, \quad y = \rho \sin v, \quad z = a \left(\ln \frac{\rho}{a} + v \right)$$

is locally isometrically mapped onto the surface of revolution

$$x = r \cos \varphi, \quad y = r \sin \varphi, \quad z = a\sqrt{2} \ln \left(r + \sqrt{r^2 - a^2} \right).$$

5.30. Show that every helical surface

$$x = u \cos v, \quad y = u \sin v, \quad z = F(u) + av$$

is locally isometrically mapped onto a surface of revolution in such a way that helical lines pass to parallels.

5.31. (a) Prove that any (even local) isometry of $\mathbb{R}^n$ is given by a linear mapping $x \mapsto Ax + x_0$.

(b) Under the conditions of (a), show that A is an orthogonal matrix.

5.32. Prove that any cylindrical surface is locally isometric to a plane.

5.33. Show that any conical surface is locally isometric to a plane.

5.34. Does there exist an isometric mapping of the domain on the right circular cylinder, given by $x^2 + y^2 = R^2$, $0 \leq z \leq H$, onto a domain on a convex conical surface?

§6. Second Quadratic Form, Gaussian Curvature and Mean Curvature

In this paragraph, the mean curvature is meant to be the sum of principal curvatures.

6.1. Calculate the second quadratic form of the following surfaces:

(a) $\mathbf{r}(u, v) = (R \cos u \cos v, R \cos u \sin v, R \sin u)$ (sphere);

(b) $\mathbf{r}(u, v) = (a \cos u \cos v, a \cos u \sin v, c \sin u)$ (ellipsoid of revolution);

(c) $\mathbf{r}(u, v) = ((a + b \cos u) \cos v, (a + b \cos u) \sin v, b \sin u)$ (torus);

(d) $\mathbf{r}(u, v) = \left(a \cosh \dfrac{u}{a} \cos v, a \cosh \dfrac{u}{a} \sin v, u \right)$ (catenoid);

(e) $\mathbf{r}(u, v) = \left(a \sin u \cos v, a \sin u \sin v, a \left(\ln \tan \dfrac{u}{2} + \cos u \right) \right)$;

(f) $\mathbf{r}(u, v) = (u \cos v, u \sin v, av)$ (right helicoid);

(h) $xyz = a^3$.

6.2. Prove that for each parameterization of the plane, its second quadratic form vanishes.

6.3. Show that for any parameterization of the sphere, its first quadratic form is proportional to the second.

6.4. Let a surface of revolution

$$\mathbf{r}(u, \varphi) = (x(u), \rho(u) \cos \varphi, \rho(u) \sin \varphi)$$

be given.

(a) Find its second quadratic form.

(b) Find the Gaussian curvature K at an arbitrary point of the surface. Reveal the dependence of the sign of K on the direction of convexity of a meridian.

(c) Find the Gaussian curvature K in a particular case where $\rho(u) = u$,

$$x(u) = \pm \left(a \ln \dfrac{a + \sqrt{a^2 - u^2}}{u} - \sqrt{a^2 - u^2} \right), \quad a > 0$$

(pseudosphere). This surface is also called the Beltrami surface.

Prove that the Beltrami surface is locally isometric to the Lobachevskii plane.

(d) Find the mean curvature H of the surface of revolution at an arbitrary point.

(e) In a particular case where $x(u) = u$, choose the function $\rho = \rho(u)$ so that $H = 0$ on the whole surface.

(f) Find principal curvatures of the surface of revolution without calculating its second form.

6.5. Prove that the ratio of the principal curvatures is constant for the surface obtained by rotating a parabola around its directrix.

6.6. Prove that a surface is a plane or its part if its Gaussian and mean curvatures are identically equal to zero.

6.7. Let a curve $\rho = \rho(s)$ with the natural parameter s, curvature $k = k(s) \neq 0$, and torsion $\varkappa = \varkappa(s) \neq 0$ be given. Let $\mathbf{v} = \mathbf{v}(s)$ be a unit vector of the tangent to this curve. For the surface composed of the tangents to this curve, i.e., for

$$\mathbf{r}(s, u) = \rho(s) + u\mathbf{v}(s), \quad u > 0,$$

find the curvatures K and H.

6.8. Find the Gaussian and mean curvatures of the surface given by the equation

(a) $z = f(x, y)$; (b) $x = f\left(\sqrt{x^2 + y^2}\right)$.

6.9. Find the Gaussian and mean curvatures of the surface given by the equation $F(x, y, z) = 0$.

6.10. Show that the surface

$$\mathbf{r}(u, v) = \left(\sqrt{u^2 + a^2} \cos v, \sqrt{u^2 + a^2} \sin v, a \ln\left(u + \sqrt{u^2 + a^2}\right)\right)$$

is the catenoid. Find its second form.

6.11. Find the principal curvature radii of the surface $y = x \tan \dfrac{z}{a}$.

6.12. Find the principal curvature radii of the surface

$$x = \cos v - u \sin v, \quad y = \sin v + u \cos v, \quad z = u + v.$$

6.13. Find the Gaussian and mean curvatures of the helical surface

$$x = u \cos v, \quad y = u \sin v, \quad z = u + v.$$

6.14. Calculate the Gaussian and mean curvatures of the surface

$$x = 3u + 3uv^2 - u^3, \quad y = v^3 - 3v - 3u^2v, \quad z = 3\left(u^2 - v^2\right).$$

6.15. Find the principal curvatures and principal directions
(a) of the helicoid $(u \cos v, u \sin v, kv)$;
(b) of the catenoid $\left(a \cosh \dfrac{u}{a} \cos v, a \cosh \dfrac{u}{a} \sin v, u\right)$;
(c) of the surface $z = xy$ at the point $(1, 1, 1)$.

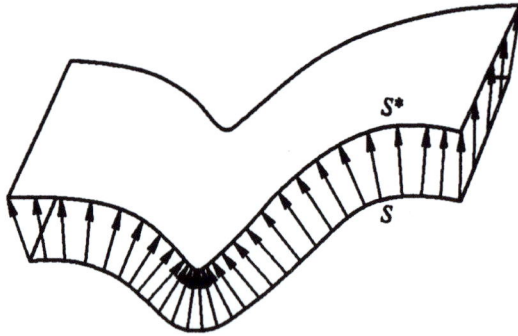

Figure 46 Parallel surfaces

6.16. For the surface $\left(u^2 + v^2, u^2 - v^2, uv\right)$, find the principal curvatures at the point P: $(u, v) = (1, 1)$. Find the curvature of the normal section at the point P tangent to the curve $u = v^2$.

6.17. Let a surface in $\mathbb{R}^3$ be given by the function $\mathbf{r}\left(u, v\right)$. Consider the expression $d\mathbf{n}^2 = \langle d\mathbf{n}, d\mathbf{n} \rangle$, where $\mathbf{n}\left(u, v\right)$ is the vector of the unit normal to the surface. Verify that this expression is a quadratic form relative to the differentials du and dv. This form is called the *third quadratic form of the surface*. Express it through the first and second forms of the surface and also through the Gaussian and mean curvatures.

6.18. Show that the spherical mapping of a minimal surface is conformal, and, vice versa, if the spherical mapping of a smooth regular surface is conformal, this surface is either minimal or is a sphere (part of a sphere).

6.19. Prove that for a ruled surface, either $K = 0$ everywhere or $K < 0$ everywhere. Besides, prove that $K = 0$ on a ruled surface if and only if the surface is developable.

6.20. Prove that on a nondevelopable ruled surface the curvature of the surface tends to zero when a point goes to infinity along any ruling.

6.21. Show that the mean curvature of the helicoid (Problem 6.1) is equal to zero.

6.22. Let S be a given surface. Draw segments of constant length on normals to the surface S in the same direction. The endpoints of these segments circumscribe the surface S^* "parallel" to the surface S (see Fig. 46). If the surface S is given in the form $\mathbf{r} = \mathbf{r}\left(u, v\right)$, the surface S^* is given in the form

$$\boldsymbol{\rho} = \mathbf{r}\left(u, v\right) + a\mathbf{n}\left(u, v\right),$$

where $\mathbf{n}\left(u, v\right)$ is the unit normal vector of S.

Express the coefficients of the first and second quadratic forms of the surface S^* through the coefficients of the first and second quadratic forms of the surface S. Prove that the parallelism property of two surfaces is reciprocal.

6.23. For which lengths of segments on the normals is a surface parallel to the given surface regular?

6.24. Express the Gaussian curvature K^* of the surface S^* "parallel" to a surface S through the Gaussian and mean curvatures of S.

6.25. Express the mean curvature H^* of the surface S^* "parallel" to a surface S through the Gaussian and mean curvatures of S.

6.26. (a) Prove that for parallel surfaces the Gaussian and mean curvatures are connected by the relation

$$\frac{H^2 - 4K}{K^2} = \frac{H^{*2} - 4K^*}{K^{*2}}.$$

(b) Compose the equation of the minimal surface S^* "parallel" to a surface S if the ratio $H/K = \text{const}$ for the surface S.

(c) Let a surface of constant mean curvature H be given. On each of its normals, segments of length $1/H$ are drawn. Prove that for the surface constructed in this way and "parallel" to a given surface, the Gaussian curvature is constant.

(d) On all normals of a surface of constant positive Gaussian curvature K, take segments of length $1/\sqrt{K}$. Prove that the mean curvature of the surface constructed in this way is constant. Calculate this mean curvature.

6.27. Prove that on parallel surfaces the points corresponding to one another along normals are simultaneously umbilical or not.

6.28. Prove that if a closed surface has a constant nonzero mean curvature and a positive Gaussian curvature, then it is a sphere.

6.29. Prove that $H^2 \geq 4K$. When is the equality attained?

6.30. Let $\mathbf{e}_1$ and $\mathbf{e}_2$ be orthogonal tangent vectors of unit length at a point of a surface. Prove that

$$H = \mathbf{II}\,(\mathbf{e}_1, \mathbf{e}_1) + \mathbf{II}\,(\mathbf{e}_2, \mathbf{e}_2),$$

where $\mathbf{II}\,(\,,)$ is the second quadratic form of the surface.

6.31. On the surface $z = x^2$, find the curvature of the normal section at the point $(2, 2, 4)$ in the direction of the curve $y = x^2/2$, $z = x^2$.

6.32. On the surface $z = 2z^2 + 9y^2$, find the curvature of the normal section at the origin in the direction of the vector that made the angle $\pi/4$ with the axis Ox.

6.33. Assume that two surfaces M_1 and M_2 intersect along a curve γ. Let k be the curvature of γ, λ_i be normal curvatures of γ in M_i, and let θ be the angle between the normals of M_1 and M_2. Prove that

$$k^2 \sin^2 \theta = \lambda_1^2 + \lambda_2^2 - 2\lambda_1\lambda_2 \cos \theta.$$

6.34. Proceeding from the fact that an ellipse can be projected on a circle and using the Meusnier theorem and the Euler formulas, find the curvatures of the ellipse at its vertices.

6.35. Prove that the curvature of a curve on a surface of positive curvature does not vanish.

6.36. Prove that for a point on a surface to be spherical, it is necessary and sufficient that the conditions $K \neq 0$ and $4K = H^2$ hold.

6.37. Determine the type of points of the torus, ellipsoid, hyperboloid and paraboloid of revolution.

6.38. Determine the type of points on the following surfaces:

(a) $z = a^2 x^4 + b^2 y^4$; (b) $z = x^4 + y^4 + x^2 y^2$; (c) $y = x^4$.

6.39. Prove that if the metric of a surface $\mathbf{r}(u, v)$ in $\mathbb{R}^3$ is written in the isothermal coordinates, i.e., $ds^2 = \Lambda^2 (du^2 + dv^2)$, then $\Delta \mathbf{r} = -H\Lambda^2 \mathbf{n}$, where $\mathbf{n}$ is the normal to the surface and H is the mean curvature.

6.40. Prove that if a surface in $\mathbb{R}^3$ is minimal, its radius-vector is a harmonic function with respect to the conformal coordinates.

6.41. Prove that if the Gaussian curvatures of two surfaces are constant and equal, these surfaces are locally isometric.

6.42. Show that there exist analytically diffeomorphic surfaces that are not isometric but, nevertheless, their Gaussian curvatures at the corresponding points are equal. In other words, the coincidence of the Gaussian curvatures of two surfaces at the corresponding points is not sufficient even for their local isometry.

6.43. Find a smooth regular connected surface of revolution that passes to itself under the action of the central symmetry with respect to the origin and its Gaussian curvature is constant and satisfies $K = -1$.

In Problems 6.44, 6.45 and 6.47, $g_{ij} dx^i dx^j$ denotes the first quadratic form of a surface, and $b_{ij} dx^i dx^j$ denotes the second quadratic form. Besides, (g^{ij}) is the matrix inverse to the matrix (g_{ij}). The summation with respect to the subscript and superscript is meant.

6.44. What is the surface for which $g^{kl} b_{ik} = \delta_i^l$?

6.45. (a) Show that the coefficients of the third quadratic form of a surface are equal to $b_{ik} b_{jl} g^{kl}$.

(b) Show that $g^{ij} \gamma_{ij} = H^2 - 2K$, where H is the mean curvature and K is the Gaussian curvature of the surface, and γ_{ij} are the coefficients of the third quadratic form of the surface.

6.46. What are surfaces for which the first and third quadratic forms are proportional?

6.47. Prove that the mean curvature of a surface is found by the formula $H = g^{ij} b_{ij}$.

6.48. Let $k_1, \ldots, k_m$ be normal curvatures of a surface in directions dividing the plane into angles π/m. Prove that $k_1 + \cdots + k_m = mH/2$.

6.49. Prove that if on a smooth regular surface there exist three distinct families of rectilinear rulings, this surface is a plane or a domain on it.

6.50. Prove that if on a smooth regular surface there exist two distinct families of rectilinear rulings, this surface is a surface of the second order.

6.51. (a) Show that an umbilical point at which the Gaussian curvature vanishes is necessarily a point of flattening.

(b) Show that on surfaces of negative curvature there are no umbilical points.

6.52. Show that the mean curvature H of a surface can be considered as the integral mean of all normal curvatures, i.e.,

$$H = \frac{1}{\pi} \int\limits_{0}^{2\pi} k\left(\varphi\right) d\varphi,$$

where $k(\varphi)$ is the normal curvature in the direction φ calculated from one of the principal directions.

§ 7. Manifolds

7.1. Prove that the n-dimensional sphere S^n given in $\mathbb{R}^{n+1}$ by the equation $x_0^2 + x_1^2 + \cdots + x_n^2 = 1$ is a smooth manifold.

(a) Construct an atlas of charts.

(b) Construct an atlas of a minimum number of charts.

(c) Construct an atlas of a minimum number of charts under the condition that each chart is homeomorphic to the disk.

(d) Construct a minimal atlas of charts homeomorphic to the disk such that all possible nonempty intersections of any number of charts are homeomorphic to the disk.

7.2. Prove that the two-dimensional torus T^2 given as a surface of revolution around the axis Oz of a circle lying in the plane Oxz and disjoint with the axis Oz is a smooth manifold.

(a) Construct an atlas of charts.

(b) Construct an atlas of a minimal number of charts.

(c) Construct an atlas of a minimal number of charts under the condition that each of the charts is homeomorphic to the disk.

(d) Construct a minimal atlas of charts homeomorphic to the disk such that all possible nonempty intersections of any number of charts are homeomorphic to the disk.

7.3. Prove that the n-dimensional projective space $\mathbb{R}P^n$ is a smooth (and real-analytic) manifold.

7.4. Prove that the n-dimensional projective space $\mathbb{C}P^n$ is a smooth (and real-analytic) manifold.

7.5. Prove that

(a) the graph of a continuous function $x_{n+1} = f\left(x_1, \ldots, x_n\right)$ is a smooth manifold;

(b) the graph of a smooth function $x_{n+1} = f(x_1, \ldots, x_n)$ is a smooth submanifold in $\mathbb{R}^{n+1}$.

(c) Give an example of a continuous function $x_{n+1} = f(x_1, \ldots, x_n)$ whose graph is not a smooth submanifold in $\mathbb{R}^{n+1}$.

7.6. Introduce a smooth manifold structure on the set of all lines in $\mathbb{R}^2$. Prove that the obtained manifold is homeomorphic to the Möbius band.

7.7. Find a diffeomorphism between S^2 and $\mathbb{C}P^1$.

7.8. Prove that the formulas

$$y^k = \frac{x^k}{\sqrt{\varepsilon^2 - (x^1)^2 - \cdots - (x^n)^2}}, k = 1, \ldots, n,$$

$$x^k = \frac{y^k}{\sqrt{\varepsilon^2 - (y^1)^2 - \cdots - (y^n)^2}}, k = 1, \ldots, n,$$

define mutually inverse diffeomorphisms of $\mathbb{R}^n$ and the ball of radius ε centred at the origin of the space $\mathbb{R}^n$.

7.9. Show that the stereographic projection of the sphere on the tangent space from the pole opposite to the tangent point is a diffeomorphism everywhere except for the pole of the projection.

7.10. Prove that the Jacobi matrix of a composition of smooth mappings is the product of the Jacobi matrices of the factors.

7.11. Prove that the rank of the Jacobi matrix is independent of the choice of a local coordinate system.

7.12. Calculate the rank of the Jacobi matrix of the mapping

$$f\colon \mathbb{R}^2 \longrightarrow \mathbb{R}^2, \quad f(x, y) = (x, 0).$$

7.13. Prove that the smoothness of a function in a chart implies its smoothness in an arbitrary coordinate system.

7.14. Let the torus $T^2 \subset \mathbb{R}^3$ be formed by rotating a circle around its axis (standard embedding). Prove that the coordinates x, y, z are smooth functions on the torus T^2.

7.15. Let the torus $T^2 \subset \mathbb{R}^3$ be standardly embedded in $\mathbb{R}^3$, and let the function $f\colon T^2 \to S^2$ assign a vector of unit length normal to the torus T^2 at a point $p \in T^2$ to each point p. Write the mapping f in coordinates. Prove that f is a smooth mapping.

7.16. Prove that the mapping $f\colon S^2 \to \mathbb{R}P^2$, which assigns a line passing through the origin and a point p on the sphere S^2 to the point p, is a smooth mapping. Write the mapping f in coordinates.

7.17. Consider $\mathbb{R}^4 = \mathbb{C}^2$ with the coordinates (z, w). Let the surface M^2 be the intersection of the three-dimensional sphere $|z|^2 + |w|^2 = 1$ and the cone $|z| = |w|$.

(a) Prove that M^2 is diffeomorphic to the two-dimensional torus T^2.

(b) Prove that the metric induced on M^2 is locally Euclidean.

(c) Prove that each of the parts of the three-dimensional sphere $|z|^2 + |w|^2 = 1$ defined by the inequalities $|z| \le |w|$ or $|z| \ge |w|$ is diffeomorphic to the solid torus, i.e., to $S^1 \times D^2$.

7.18. In $\mathbb{R}^5$, consider a two-dimensional surface defined by the formulas

$$x_1 = \frac{xy}{\sqrt{3}}, \quad x_2 = \frac{yz}{\sqrt{3}}, \quad x_3 = \frac{xz}{\sqrt{3}},$$

$$x_4 = \frac{x^2 - y^2}{2\sqrt{3}}, \quad x_5 = \frac{x^2 + y^2 - 2z^2}{6},$$

where $x^2 + y^2 + z^2 = 1$. This surface is called the *Veronese surface*. Prove that the Veronese surface is a smooth embedding of $\mathbb{R}P^2$ in $\mathbb{R}^5$.

7.19. Prove that the group $SO\,(2)$ is diffeomorphic to the circle. To which manifold is the group $O\,(2)$ diffeomorphic?

7.20. Prove that the group $SO\,(3)$ is homeomorphic to the projective space $\mathbb{R}P^3$. Construct the diffeomorphism.

7.21. Prove that $SO\,(3)$ is a smooth submanifold in the space $\mathbb{R}^9$ of all square matrices of the third order.

7.22. Prove that $\left\{ (x_1, \dots, x_n) \in \mathbb{C}^n : \sum x_i^2 = 0, \sum |x_i|^2 = 2 \right\}$ is the set of unit tangent vectors to the sphere of unit radius.

7.23. Prove that the set of unit tangent vectors to S^2 is homeomorphic to $SO\,(3)$. Construct the diffeomorphism.

7.24. In $SO\,(3)$, consider the subset of matrices $A = (a_{ij})$ such that

$$\begin{vmatrix} a_{11}+1 & a_{12} & a_{13} \\ a_{21} & a_{22}+1 & a_{23} \\ a_{31} & a_{32} & a_{33}+1 \end{vmatrix} = 0.$$

Prove that this subset is diffeomorphic to $\mathbb{R}P^2$.

7.25. Prove that the equation $x_1^2 + x_2^2 + x_3^2 = 0$ defines a submanifold in $\mathbb{C}P^2$ diffeomorphic to S^2.

7.26. Prove that the group $SU\,(2)$ is diffeomorphic to the sphere S^3.

7.27. Prove that the groups $GL\,(n, \mathbb{R})$ and $GL\,(n, \mathbb{C})$ are smooth manifolds.

7.28. The group $O\,(n)$ can be considered as a subset of $\mathbb{R}^{n^2}$. Prove that $O\,(n)$ lies in the sphere S^{n^2-1} of radius $\sqrt{n}$.

7.29. Consider the "complex circle" $X = \left\{ (z_1, z_2) \in \mathbb{C}^2 : z_1^2 + z_2^2 = 1 \right\}$. Prove that the space X is homeomorphic to the cylinder without its boundary.

7.30. Find the tangent space to the group $SO(3)$ at the point

$$\begin{pmatrix} \dfrac{\sqrt{3}}{2} & \dfrac{1}{2} & 0 \\[2ex] -\dfrac{1}{2} & \dfrac{\sqrt{3}}{2} & 0 \\[2ex] 0 & 0 & 1 \end{pmatrix}.$$

7.31. Find the coordinates of the vector $\left(\sqrt{2}, \sqrt{2}, 2\right)$ tangent to the sphere $x^2 + y^2 + z^2 = 1$ at the point $\left(-\dfrac{1}{\sqrt{2}}, \dfrac{1}{\sqrt{2}}, 0\right)$:

(a) in the spherical coordinates;

(b) in the coordinates of the stereographic projection.

7.32. Find the coordinates of an arbitrary vector tangent to the large disk of the unit sphere and lying in the plane

$$x + 2y - 3z = 0 :$$

(a) in the spherical coordinates;

(b) in the coordinates of the stereographic projection.

7.33. For the projective plane $\mathbb{R}P^2$, find the coordinates of the velocity vector of the curve $\mathbf{r}(t) = (\cos t : \sin t : t)$ for $t = \pi/4$ in affine charts.

7.34. Prove that the union of two coordinate axes in $\mathbb{R}^2$ is not a manifold.

7.35. Are the following curves on the plane smooth manifolds?

(a) the triangle;

(b) two triangles having only one common point, the vertex.

7.36. Are the boundary of the square and the figure eight curve in $\mathbb{R}^2$ smooth submanifolds?

7.37. (a) Prove that for $n \neq m$ the spaces $\mathbb{R}^n$ and $\mathbb{R}^m$ are not diffeomorphic.

(b) Prove that smooth manifolds of different dimensions are not diffeomorphic.

7.38. Give an example of a smooth homeomorphism which is not a diffeomorphism.

7.39. Give an example of a non-Hausdorff manifold.

Two smooth structures (smooth atlases) on a manifold are said to be compatible if each chart of one atlas is compatible with all charts of the other, and vice versa.

7.40. Give an example of a manifold with two incompatible smooth structures.

7.41. Prove that two smooth structures on a manifold are compatible if and only if the spaces of smooth functions of these structures coincide.

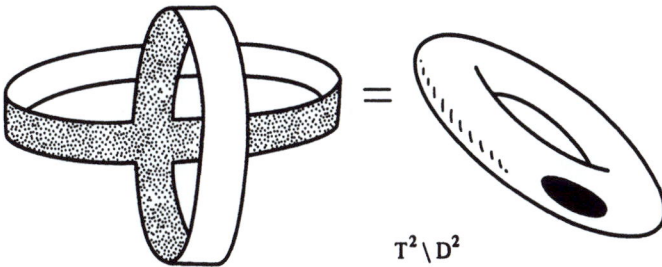

$$T^2 \setminus D^2$$

Figure 47

7.42. On $\mathbb{R}^1$, consider the following mappings: $\varphi_k \colon \mathbb{R}^1 \to \mathbb{R}^1$, $\varphi_k(x) = x^{2k-1}$, where k is a positive integer. Prove that each φ_k defines a chart on $\mathbb{R}^1$ whose domain is the whole $\mathbb{R}^1$. Prove that these charts are not pairwise compatible. In particular, this means that the atlas A_k defined by a chart φ_k is not compatible with the atlas A_l for $l \neq k$. It would seem to imply that on $\mathbb{R}^1$ there exist infinitely many pairwise non-equivalent smooth structures. But this is not true: prove that all the manifolds $(\mathbb{R}^1, A_k)$ are pairwise diffeomorphic.

7.43. Prove that any smooth manifold admits an atlas such that each of its charts is homeomorphic to the Euclidean space.

7.44. Show that it is possible to introduce the structure of a two-dimensional manifold on the plane $\mathbb{R}^2$ such that the set given by $y = x^2$ is not a smooth one-dimensional submanifold.

7.45. Prove that the product of smooth manifolds is a smooth manifold, the projections being smooth regular mappings.

7.46. Prove that the realification of an n-dimensional complex manifold is a smooth orientable real manifold of dimension $2n$.

7.47. Prove that the boundary of a smooth manifold M^n is a smooth $(n-1)$-dimensional manifold. Herewith, as a smooth atlas of the boundary of the manifold, we could take the restrictions of the charts of the manifold to its boundary. Show that the boundary is an orientable manifold irrespective of whether or not M^n is orientable.

7.48. Show that the circle, the two-dimensional sphere, and the torus are orientable manifolds.

7.49. Verify whether or not the following manifolds are orientable: (a) the sphere S^n; (b) the torus T^n.

7.50. Prove that the Möbius band, the projective plane and the Klein bottle are non-orientable manifolds.

7.51. Construct an immersion of the Möbius band into the three-dimensional Euclidean space such that its boundary circle is standardly embedded in the two-dimensional Euclidean plane.

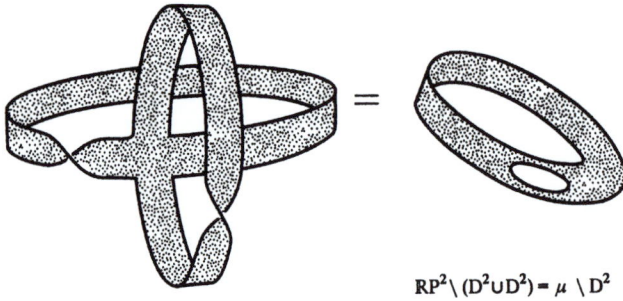

$$RP^2 \setminus (D^2 \cup D^2) = \mu \setminus D^2$$

Figure 48

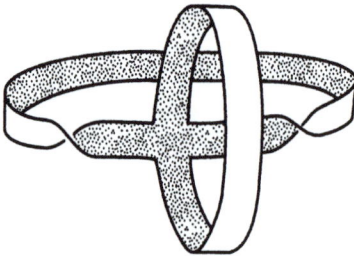   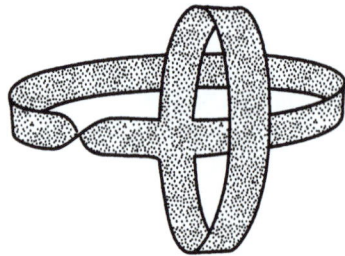

Figure 49   Figure 50

7.52. Prove that a two-dimensional manifold is orientable if and only if it contains no Möbius band.

7.53. Prove that the Möbius band is not homeomorphic to the direct product of the segment by the circle.

7.54. Prove that the Cartesian product of two manifolds is orientable if and only if both factors are orientable.

7.55. Prove the existence theorem of a Riemannian metric on any smooth manifold (a) by using the partition of unity; (b) by using the Whitney theorem.

7.56. Let $i\colon N \to M$ be an immersion and g be a Riemannian metric on M. Prove that i^*g is a Riemannian metric on N. Why is this not true for an arbitrary smooth mapping i?

7.57. (a) Prove that two annuli glued together as shown in Fig. 47 form a surface homeomorphic to the torus with a small disk removed.

(b) Prove that the surface obtained by gluing two Möbius bands as shown in Fig. 48 is homeomorphic to the Möbius band with a hole or, which is the same, to the projective plane with two holes.

(c) Prove that the surface depicted in Fig. 49 is homeomorphic to the torus with a hole.

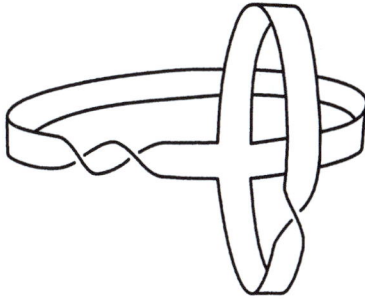

Figure 51

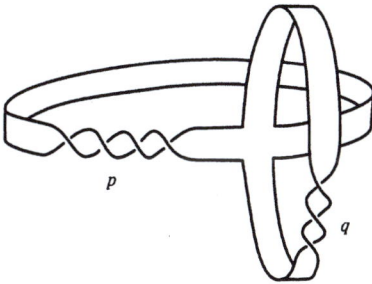

Figure 52

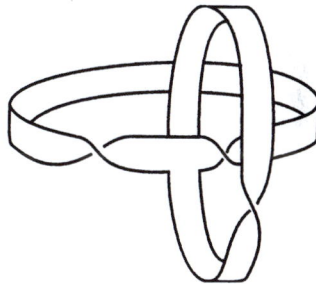

Figure 53

(d) Prove that the surface depicted in Fig. 50 is homeomorphic to the Klein bottle with a hole.

(e) Prove that the surface depicted in Fig. 51 is homeomorphic to the Klein bottle with a hole.

(f) To what is the surface depicted in Fig. 52 homeomorphic?

(g) To what is the surface depicted in Fig. 53 homeomorphic?

7.58. Prove that a bicycle bladder with nipple removed (torus with a hole) can be turned inside out.

7.59. Cut the standard torus of revolution in $\mathbb{R}^3$ by two parallel planes such that the part of the torus lying between them is homeomorphic to the surface with the boundary shown in Fig. 54. Prove that this surface is homeomorphic to the torus with two holes.

7.60. Prove that the surface depicted in Fig. 55 is homeomorphic to the Klein bottle with two holes.

7.61. Prove that the plane with a handle glued to it is homeomorphic to two annuli glued as shown in Fig. 56.

7.62. Prove that two of the three surfaces depicted in Fig. 57 are homeomorphic to the projective plane with three holes, and the remaining surface

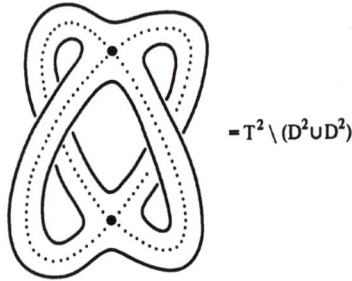

$$= T^2 \setminus (D^2 \cup D^2)$$

Figure 54

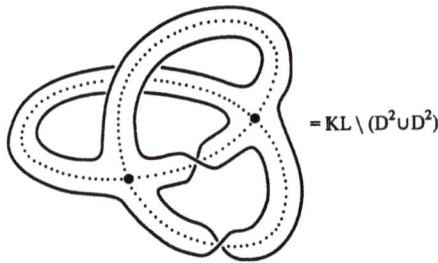

$$= KL \setminus (D^2 \cup D^2)$$

Figure 55

is homeomorphic to the Klein bottle with two holes.

7.63. Draw a cut in the torus such that the surface depicted in Fig. 58 is obtained.

7.64. Consider two surfaces in $\mathbb{R}^3$ with curves drawn on it as depicted in Fig. 59. Is it possible to transform one surface into the other by using a smooth deformation of a surface in $\mathbb{R}^3$ without self-intersections and discontinuities so that the curve drawn on one surface transforms into the other drawn on the other surface?

7.65. Consider two surfaces in $\mathbb{R}^3$ with pairs of curves drawn on them as depicted in Fig. 60. Is it possible to transform one surface to the other by a smooth deformation of a smooth surface in $\mathbb{R}^3$ without self-intersections and discontinuities (by isotopy) so that the curves drawn on one surface transform into the curves drawn on the other surface?

7.66. Consider a "topological human", i.e., a sphere with two handles.

(a) Can the topological human "unhook" his hands by a smooth isotopy (see Fig. 61)?

(b) The same question under the condition that one of the hands is linked with the circle (a watch is on one hand); see Fig. 62.

7.67. Reveal whether or not the mapping of a real line into the plane with the Cartesian coordinates (x, y) is an embedding or immersion, if

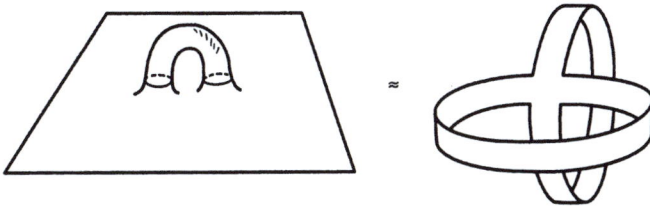

Figure 56

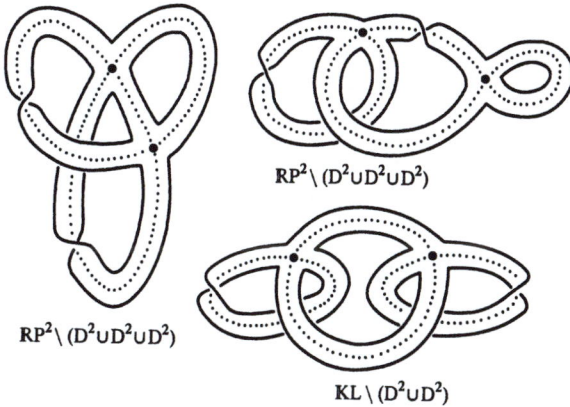

RP²\ (D²∪D²∪D²)

RP²\ (D²∪D²∪D²)

KL\ (D²∪D²)

Figure 57

(a) $x(t) = \dfrac{2 + t^2}{1 + t^2}$, $y(t) = \dfrac{2t + t^2}{1 + t^2}$;

(b) $x(t) = \dfrac{t^2}{t^2 + 1}$, $y(t) = \dfrac{t^2 + 1}{t^2 + 2}$.

7.68. Reveal whether or not the set given by the following equations is a smooth submanifold in $\mathbb{R}^3$ with the Cartesian coordinates (x, y, z):

(a) $x^2(z - 1) + y^2 z = 0$;

(b) $z(x^2 + y^2)^n = y^{2n}$.

7.69. Reveal whether or not the set given by the following equation is a smooth submanifold in $\mathbb{R}^2$ with the Cartesian coordinates (x, y):

(a) $x^4 + y^4 = 8xy^2$;

(b) $x^6 + 18x^3 y - y^3 = 0$.

7.70. On the two-dimensional sphere S^2, consider the coordinates (u, v) of the stereographic projection. Is the closure $\overline{M}$ a smooth submanifold in S^2 if the set M is given by

(a) $y^2 - x^3 + 2x^2 - x = 0$;

(b) $y^3 - x^3 + y - 2x = 0$;

Figure 58

Figure 59

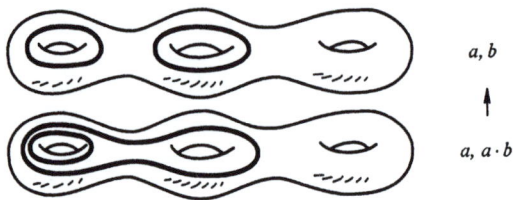

Figure 60

(c) $\left(x^2 - y^2\right)^2 - 2x = 0$;

(d) $\left(x^2 - y^2\right)(x - y) + 1 = 0$;

(e) $(2x + y)^2 (x + y) - x = 0$.

§ 8. Tensors

8.1. Find the valence of the following tensors:

(a) $T_i = \dfrac{\partial f}{\partial x^i}$;

(b) $T_{ij} = \dfrac{\partial^2 f}{\partial x^i \partial x^j}$ at points where the gradient of a function f vanishes;

(c) T^i_j, the entries of the matrix of a linear operator on a vector space;

(d) T_{ij}, the entries of the matrix of a bilinear form on a vector space.

Figure 61

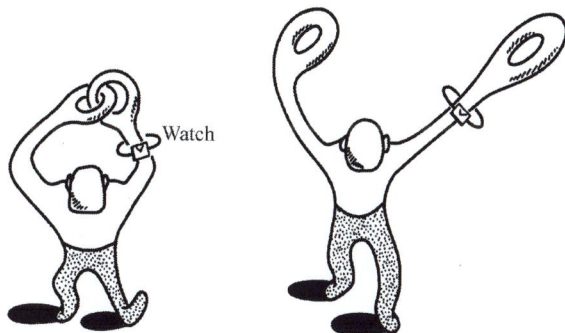

Watch

Figure 62

8.2. Let

$$\delta_j^i = \begin{cases} 0, & \text{if } i \neq j, \\ 1, & \text{if } i = j. \end{cases}$$

Show that $\{\delta_j^i\}$ composes a tensor of valence $(1,1)$. Also, show that the tuple of numbers δ_{ij} given in each coordinate system by the formula

$$\delta_{ij} = \begin{cases} 1, & i = j, \\ 0, & i \neq j. \end{cases}$$

is not a tensor of type $(0,2)$.

8.3. Let $\{\xi^{ij}\}$ be a tensor of valence $(2,0)$. Show that the numbers η_{ij} satisfying the condition $\xi^{ij}\eta_{jk} = \delta_k^i$ compose a tensor of valence $(0,2)$.

8.4. Let the matrix $(g_{ij}(x))$ be a symmetric nonsingular positive-definite matrix in a coordinate system $x = (x^1, \ldots, x^n)$. Prove that all these properties are preserved under any regular change of coordinates.

8.5. Give an example showing that permutation of a subscript and a superscript is not a tensor operation.

8.6. Show that the operation that assigns the tuple of numbers $S^{i_1 \ldots i_p}_{j_1 \ldots j_{q+1}}$ to a tensor $T^{i_1 \ldots i_p}_{j_1 \ldots j_q}$ by the formula

$$S^{i_1 \ldots i_p}_{j_1 \ldots j_{q+1}} = \frac{\partial}{\partial x^{j_1}} T^{i_1 \ldots i_p}_{j_2 \ldots j_{q+1}}$$

is not a tensor operation.

8.7. Let V_n^m be the space of all tensors of type (m, n). Show that if $f : V_n^m \to V_q^p$ is a linear tensor space mapping, the components of this mapping compose a tensor. Find its valence.

8.8. Find the dimension of the tensor space V_n^m.

8.9. Show that any tensor of valence $(2, 0)$ uniquely decomposes into the sum of a symmetric and a skew-symmetric summand. Give an example of a tensor of valence $(3, 0)$ for which this is not true.

8.10. Prove that the alternation and symmetrization operators are projectors in the tensor space. Prove that if $n > 2$, the sum of these operators is not equal to 1. Give an example of a tensor (for $n > 2$) lying in the kernels of the alternation and symmetrization operators.

8.11. Prove that the operation of permutation of indices commutes with the alternation and symmetrization operations.

8.12. Prove that a composition of alternation and symmetrization operators (in any order) yields zero tensor for any tensor of type $(0, k)$, where $k > 1$.

8.13. Let the dimension of the space V be equal to n. Find the dimension of the space $\Lambda^k V$ of skew-symmetric tensors.

8.14. Let the dimension of the space V be equal to n. Find the dimension of the space $S^k V$ of symmetric tensors.

8.15. Describe all invariant tensors of rank 0, 1, 2, 3 and 4. Here, a tensor is said to be invariant if its components remain unchanged under any coordinate changes.

8.16. Show that the mixed and vector products in $\mathbb{R}^3$ are given by tensors of types $(0, 3)$ and $(1, 2)$, respectively. Write their components in an arbitrary basis. Show that the components of these tensors are related to the index lowering and raising operations.

8.17. Express quantities $\det c$, c^i_i, $c^i_j c^j_i$, $c^i_j c^j_k c^k_i$, etc. through the coefficients of the characteristic polynomial of an operator (c^i_j).

8.18. Represent the determinant of a linear operator as a result of performing a sequence of elementary tensor operations.

8.19. Let a tuple of quantities $S^{i_1 \ldots i_p}$ be given. Prove that if, for a fixed q, the tuple of quantities

$$S^{i_1 \ldots i_p} T_{i_1 \ldots i_q}, \quad q < p,$$

is a tensor for any choice of the tensor $T_{i_1 \ldots i_q}$, then $S^{i_1 \ldots i_p}$ is also a tensor.

8.20. If v_{ij} is a tensor, and the equation $av_{ij} + bv_{ji} = 0$ (a and b are numbers) holds in a coordinate system, then this equation holds in any other coordinate system. Besides, if $v_{ij} \neq 0$, then either $a = b$ or $a = -b$.

8.21. Let g_{ij} be a metric tensor. Prove that if tensors $g_{ij}P_k^j$ and $g_{ij}Q_k^j$ are symmetric, then after lowering the upper index the tensor $PQ + QP$ is symmetric, whereas $PQ - QP$ is antisymmetric.

8.22. Let f be a smooth function on a Riemannian manifold M with metric g_{ij}. Prove that the vector field $\mathbf{v} = \operatorname{grad} f$, whose components are given by

$$v^i = g^{ij}\frac{\partial f}{\partial x^j},$$

is orthogonal to the level surfaces of the function f.

8.23. Let A_j^i be a tensor field of type $(1,1)$ on a manifold M. Prove that the formula

$$N(\mathbf{X}, \mathbf{Y}) = A^2[\mathbf{X}, \mathbf{Y}] - A[A\mathbf{X}, \mathbf{Y}] - A[\mathbf{X}, A\mathbf{Y}] + [A\mathbf{X}, A\mathbf{Y}],$$

where $\mathbf{X}$ and $\mathbf{Y}$ are vector fields on M, defines a tensor field N_{jk}^i of type $(1,2)$.

If an operator $A\colon V \to V$ is given, then the following operators are defined:

$$A^{\otimes k}\colon V^{\otimes k} \longrightarrow V^{\otimes k},$$
$$\Lambda^{\otimes k}A\colon \Lambda^{\otimes k}V \longrightarrow \Lambda^{\otimes k},$$
$$S^{\otimes k}A\colon S^{\otimes k}V \longrightarrow S^{\otimes k}.$$

8.24. Find $\operatorname{tr} \Lambda^q A$ through the coefficients of the characteristic polynomial of A in the n-dimensional space V.

8.25. Express the quantities $\operatorname{tr} A \otimes A$, $\operatorname{tr} A^{\otimes k}$ and $\det A \otimes A$ through the trace and the determinant of an operator A.

8.26. Prove that for $\xi \in \Lambda^p(V)$ and $\omega \in \Lambda^1(V)$, the equation $\xi \wedge \omega = 0$ holds if and only if $\xi = \omega \wedge \eta$ for some $\eta \in \Lambda^{p-1}(V)$.

8.27. Prove that if a tensor T_{ijk} is symmetric with respect to first two indices and is skew-symmetric with respect to the second and third indices, then it is equal to zero.

8.28. Prove that if a_{ij} is a symmetric tensor and b^{ij} is a skew-symmetric tensor, then $a_{ij}b^{ij} = 0$.

In the next problems, vectors e_i form a basis of the space V, and e^i is the dual basis of the space V^*, i.e., $e^i(e_j) = \delta_j^i$.

8.29. Find the value of the tensor $e_2 \otimes e^1 + (e_1 + 3e_3) \otimes e^2$ on the pair $e^1 + e^2 + e^3$, $e_1 + 5e_2 + 4e_3$.

8.30. Find the value of the tensor

$$\left(e^1 \otimes e^2 + e^2 \otimes e^3 + e^2 \otimes e^2\right) \otimes \left(e^1 \otimes e^1 \otimes \left(e^1 - e^3\right)\right) -$$
$$\left(e^1 \otimes e^1 \otimes \left(e^1 - e^3\right)\right) \otimes \left(e^1 \otimes e^2 + e^2 \otimes e^3 + e^2 \otimes e^2\right)$$

on the tuple e_1, $e_1 + e_2$, $e_2 + e_3$, e_2, e_2.

8.31. All coordinates of a tensor T of type $(2,3)$ in the basis (e_1, e_2, e_3) are equal to 1. Find the component $\tilde{T}_{123}^{12}$ of this tensor in the basis

$$(\tilde{e}_1, \tilde{e}_2, \tilde{e}_3) = (e_1, e_2, e_3) \begin{pmatrix} 1 & 2 & 3 \\ 0 & 1 & 2 \\ 0 & 0 & 1 \end{pmatrix}.$$

8.32. Find the components:

(a) T_{21}^1 of the tensor $e_1 \otimes e^1 \otimes e^2 + e_2 \otimes e^1 \otimes e^2$ in the basis

$$(\tilde{e}_1, \tilde{e}_2) = (e_1, e_2) \begin{pmatrix} 1 & 1 \\ 2 & 3 \end{pmatrix};$$

(b) $\tilde{T}_{31}^{12}$ of the tensor $e_3 \otimes e_1 \otimes e^2 \otimes e^1 + e_1 \otimes e_2 \otimes e^3 \otimes e^3$ in the basis

$$(\tilde{e}_1, \tilde{e}_2, \tilde{e}_3) = (e_1, e_2, e_3) \begin{pmatrix} 1 & 0 & 0 \\ 2 & 1 & 0 \\ 3 & 2 & 1 \end{pmatrix}.$$

8.33. Find the components of the following tensors:

(a) $(e_1 + e_2) \otimes (e_1 - e_2)$;

(b) $(e_1 + 2e_2) \otimes (e_1 + e_2) - (e_1 + e_2) \otimes (e_1 + 2e_2)$.

8.34. Find the contraction of the following tensors:

(a) $(e_1 + 3e_2 - e_3) \otimes (e^1 - 2e^3 + 3e^4) - (e_1 + e_3) \otimes (e^1 - 3e^3 + e^4)$;

(b) $e_1 \otimes (e^1 + e^2 + e^3 + 3e^4) + e_2 \otimes (e^1 + 2e^2 + 3e^3 + 4e^4) + 2e_3 \otimes (e^1 - e^2 - e^4)$.

8.35. Apply the linear operator given by the tensor $e_3 \otimes e^1$ to the vector $e_1 + e_2 + e_3 + e_4$.

8.36. The tensor $(e_1 + e_2) \otimes (2e^1 - e^3)$ defines a linear operator A. Which tensor is defined by the operator A^2?

8.37. The inner product is given by the matrix

$$\begin{pmatrix} 2 & 1 & 0 & 0 \\ 1 & 1 & 0 & 0 \\ 0 & 0 & 1 & 1 \\ 0 & 0 & 1 & 2 \end{pmatrix}.$$

Raise and lower the index of the following tensors:

(a) $e_3 \otimes e^1 + e_4 \otimes e^2$;

(b) $(e_3 + e_4) \otimes (e^1 + e^2) - e_3 \otimes (e^1 + e^3)$;

(c) $T_j^i = \delta_{2i} + \delta_{4j}$.

§ 9. Vector Fields

9.1. Prove the equivalence of the three definitions of a tangent vector to a manifold at a point P:

(a) a tensor of valence $(1,0)$;

(b) a derivation of smooth functions at the point P;

(c) a class of osculating curves at the point P.

9.2. Find the derivative of the function f at the point P by the direction of the vector ξ, where

(a) $f = \sqrt{x^2 + y^2 + z^2}$; $P = (1,1,1)$, $\xi = (2,1,0)$;

(b) $f = x^2 y + xz^2 - 2$; $P = (1,1,-1)$, $\xi = (1,-2,4)$;

(c) $f = xe^y + ye^x - z^2$; $P = (3,0,2)$, $\xi = (1,1,1)$;

(d) $f = \dfrac{x}{y} - \dfrac{y}{x}$; $P = (1,1)$, $\xi = (3,4)$.

9.3. Find the derivative of the function $f = 1/r$, $r = \sqrt{x^2 + y^2 + z^2}$, by the direction of its gradient.

9.4. Find the derivative of the function $f = yze^x$ by the direction of its gradient.

9.5. Let ∇ be the vector differential operator in $\mathbb{R}^3$ whose components are equal to $\nabla = \left(\dfrac{\partial}{\partial x}, \dfrac{\partial}{\partial y}, \dfrac{\partial}{\partial z} \right)$. Show that

(a) $\operatorname{grad} F = \nabla F$; (b) $\operatorname{div} \mathbf{X} = \langle \nabla, \mathbf{X} \rangle$; (c) $\operatorname{rot} \mathbf{X} = \nabla \times \mathbf{X}$.

9.6. Prove the formula

$$\operatorname{div}(u\mathbf{X}) = u \cdot \operatorname{div} \mathbf{X} + \langle \mathbf{X}, \operatorname{grad} u \rangle,$$

where $\mathbf{X}$ is a vector field and u is a function in $\mathbb{R}^3$.

9.7. Prove the formula

$$\operatorname{rot}(u\mathbf{X}) = u \cdot \operatorname{rot} \mathbf{X} - \mathbf{X} \times \operatorname{grad} u.$$

9.8. Prove that the vector $\mathbf{X} = u \operatorname{grad} v$ is orthogonal to $\operatorname{rot} \mathbf{X}$.

9.9. Show that

(a) $\operatorname{div}(\operatorname{rot} \mathbf{X}) = 0$;

(b) $\operatorname{rot} \operatorname{rot} \mathbf{X} = \operatorname{grad} \operatorname{div} \mathbf{X} - \Delta \mathbf{X}$, where $\Delta = \dfrac{\partial^2}{\partial x^2} + \dfrac{\partial^2}{\partial y^2} + \dfrac{\partial^2}{\partial z^2}$.

9.10. Let $\mathbf{X} = (x,y,z)$. Show that

(a) $\operatorname{div} \mathbf{X} = 3$; (b) $\operatorname{rot} \mathbf{X} = 0$; (c) $\operatorname{div} \left(\dfrac{\mathbf{X}}{|\mathbf{X}|^3} \right) = 0$.

(d) $\operatorname{rot} \left(\dfrac{\mathbf{X}}{|\mathbf{X}|^3} \right) = 0$; (e) $\operatorname{grad} \dfrac{1}{|\mathbf{X}|} = -\dfrac{\mathbf{X}}{|\mathbf{X}|^3}$.

Find a function φ such that $\mathbf{X} = \operatorname{grad} \varphi$.

9.11. Show that the commutator of vector fields considered as operators of differentiation of smooth functions is a vector field.

9.12. In $\mathbb{R}^n$, consider a vector field $\mathbf{V}_A$ such that it assumes the value Ax at a point $x \in \mathbb{R}^n$, where A is a square matrix of order n. Such a vector field is said to be linear. Prove that $[\mathbf{V}_A, \mathbf{V}_B] = -\mathbf{V}_{[A,B]}$. Here, $[A, B] = AB - BA$ is the commutator of matrices.

9.13. Let M be a manifold, and N be its submanifold. Prove that the relation $[\mathbf{X}, \mathbf{Y}]|_N = [\mathbf{X}|_N, \mathbf{Y}|_N]$ holds for any vector fields $\mathbf{X}$ and $\mathbf{Y}$ on M. Here, $\mathbf{X}|_N$ means the restriction of the vector field $\mathbf{X}$ to the submanifold N.

9.14. Calculate the commutator of two vector fields on S^3 $(x^2 + y^2 + z^2 + w^2 = 1)$ one of which is equal to $(y, -x, w, -z)$ at the point $(x, y, z, w) \in S^3$, and the other is equal to $(z, w, -x, -y)$ at this point.

9.15. Let φ_t be the one-parameter group of diffeomorphisms corresponding to a vector field $\boldsymbol{\xi}$. Show that

$$|\boldsymbol{\xi}, \boldsymbol{\eta}| = \frac{d}{dt} \left(\varphi_t^* \left(\boldsymbol{\eta} \right) - \boldsymbol{\eta} \right).$$

9.16. Let $\boldsymbol{\xi}$ and $\boldsymbol{\eta}$ be vector fields, and f and g be smooth functions. Prove the formula

$$[f\,\boldsymbol{\xi}, g\boldsymbol{\eta}] = f\,g\,[\boldsymbol{\xi}, \boldsymbol{\eta}] - g\boldsymbol{\eta}\,(f)\,\boldsymbol{\xi} + f\boldsymbol{\xi}\,(g)\,\boldsymbol{\eta}.$$

9.17. Let $\boldsymbol{\xi}$ and $\boldsymbol{\eta}$ be vector fields, and let φ_t and ψ_t be the one-parameter transformation groups corresponding to them. Show that if $[\boldsymbol{\xi}, \boldsymbol{\eta}] = 0$, then the transformations φ_t commute with the transformations ψ_t.

9.18. Construct the integral trajectories of the following vector fields on the plane:

(a) $\boldsymbol{\xi} = x\dfrac{\partial}{\partial x} + y\dfrac{\partial}{\partial y}$; (b) $\boldsymbol{\xi} = y\dfrac{\partial}{\partial x} - x\dfrac{\partial}{\partial y}$;

(c) $\boldsymbol{\xi} = x\dfrac{\partial}{\partial x} - y\dfrac{\partial}{\partial y}$; (d) $\boldsymbol{\xi} = (x + y)\dfrac{\partial}{\partial x} + y\dfrac{\partial}{\partial y}$;

(e) $\boldsymbol{\xi} = (x - y)\dfrac{\partial}{\partial x} + x\dfrac{\partial}{\partial y}$; (f) $\boldsymbol{\xi} = x^2\dfrac{\partial}{\partial x} + y^2\dfrac{\partial}{\partial y}$.

9.19. Let $\boldsymbol{\xi}$ be a constant vector field in the angular coordinates of the two-dimensional torus T^2. Reveal under which conditions on the coordinates of the field $\boldsymbol{\xi}$ its integral trajectories are closed curves.

9.20. Generalize the previous problem to the case of the torus T^n. Precisely, let $\boldsymbol{\xi} = (\xi^1, \ldots, \xi^n)$ be a constant vector field in the angular coordinates on the torus T^n. Prove that the closure of any its trajectory is homeomorphic to the torus T^k, where k is the number of linearly independent numbers $\xi^1, \ldots, \xi^n$ over the field of rational numbers.

9.21. Let a vector field $\mathbf{Y}$ be fixed. Is it possible to define a tensor field A of type $(1, 1)$ (operator tensor field) such that the equation $A(\mathbf{X}) = [\mathbf{Y}, \mathbf{X}]$ holds for all vector fields $\mathbf{X}$?

§ 10. Connections and Parallel Translation

10.1. Prove the following assertions:

(a) If Γ_{ij}^k and γ_{ij}^k are coefficients of some connections, then $\alpha\Gamma_{ij}^k + \beta\gamma_{ij}^k$ are also coefficients of some connection. Here, α and β are smooth functions, and, moreover, $\alpha + \beta = 1$.

(b) If the connections with coefficients Γ_{ij}^k and γ_{ij}^k have the same geodesics, then the connection $\alpha\Gamma_{ij}^k + \beta\gamma_{ij}^k$ has the same geodesics; here, α and β are smooth functions and $\alpha + \beta = 1$.

(c) Prove that the differences $\Gamma_{ij}^k - \tilde{\Gamma}_{ij}^k$ of two connections ∇ and $\tilde{\nabla}$ compose a tensor of type $(1, 2)$ and that any tensor of type $(1, 2)$ can be represented in such a form. Prove that if we add a tensor of type $(1, 2)$ to the coefficients of a connection, then the coefficients of a connection are obtained.

10.2. Prove that the tensor field δ_j^i of the Kronecker symbols is parallel along any curve for any connection.

10.3. Show that the covariant derivative of a tensor field along a curve depends on the values of the coefficients of connection and the field only on this curve. In other words, if a tensor field is given only at points of a curve, then, nevertheless, we can calculate the covariant derivative of this tensor field along the velocity vector of the curve.

10.4. Prove that under a simultaneous parallel translation of several tensors along a given curve, the tensors obtained from the initial tensors by operations of tensor algebra are also parallel translated.

10.5. Let $f: M \to M'$ be an isometry of Riemannian manifolds, and let $x' = f(x)$, $x \in M$, $x' \in M'$. Equip M and M' with the corresponding Riemannian connections. Prove that the differential $df: TM \to TM'$ is interchangeable with the parallel translations τ and τ' on M and M', respectively.

10.6. Equip a Riemannian manifold with the corresponding symmetric Riemannian connection. Prove that the operation of covariant differentiation commutes with the operations of rising and lowering of indices.

10.7. Let a vector field whose vectors are of the same length be given. Let the manifold considered be equipped with the symmetric Riemannian connection. Prove that the covariant derivative of this vector field in an arbitrary direction is orthogonal to the vectors of the field.

10.8. Let N be a smooth submanifold in M, let γ be a smooth curve on N, and let $\boldsymbol{\xi}$ be a vector field along the curve γ tangent to N. Prove the formula

$$\tilde{\nabla}_\gamma \boldsymbol{\xi} = \mathrm{pr}\,(\nabla_\gamma \boldsymbol{\xi}),$$

where ∇ and $\tilde{\nabla}$ are symmetric Riemannian connections on the manifolds M and N, respectively (the induced metric is considered on N), and pr is the orthogonal projection on the tangent space of N.

10.9. Show that if two submanifolds

(a) of the Euclidean space or

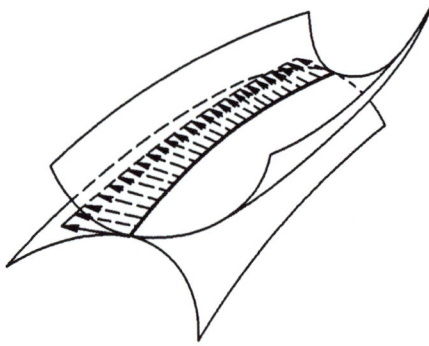

Figure 63

(b) of an arbitrary Riemannian manifold
osculate along a curve γ, then the results of the parallel translation of a vector
along this curve on one submanifold and those of the same curve on the other
submanifold coincide (see Fig. 63).

10.10. Let a surface with metric g_{ij} be given. By definition, we set
$\nabla\varphi = g^{ij}\nabla_i\varphi\nabla_j\varphi$ and $\nabla(\varphi, \psi) = g^{ij}\nabla_i\varphi\nabla_j\psi$. Show that the angle θ between
the curves $\varphi = \text{const}$ and $\psi = \text{const}$ on the surface is defined by

$$\cos\theta = \frac{\nabla(\varphi, \psi)}{\sqrt{\nabla\varphi\nabla\psi}}.$$

10.11. Let a manifold M with a connection ∇ and its submanifold N be
given. Prove the equivalence of the following conditions:

(a) as a result of parallel translation of a vector tangent to N along a curve
lying in N, a vector tangent to N is obtained;

(b) for any two vector fields $\mathbf{X}$ and $\mathbf{Y}$ tangent to N, the vector $\nabla_{\mathbf{X}}\mathbf{Y}$ is
also tangent to N.

10.12. Calculate the Christoffel symbols of the Euclidean metric of the
plane in the polar coordinates.

10.13. Calculate the Christoffel symbols of the metric

$$ds^2 = \lambda(u, v)\left(du^2 + dv^2\right).$$

10.14. Explicitly calculate the Christoffel symbols on the sphere if the
metric is given in the following form:

(a) $ds^2 = d\theta^2 + \sin^2\theta d\varphi^2$; (b) $ds^2 = \dfrac{4\left(dx^2 + dy^2\right)}{\left(1 + x^2 + y^2\right)^2}$;

(c) $ds^2 = \dfrac{4\left(dr^2 + r^2 d\varphi^2\right)}{\left(1 + r^2\right)^2}$.

10.15. Explicitly calculate the Christoffel symbols on the Lobachevskii
plane if the metric has the following form:

(a) $ds^2 = \dfrac{dx^2 + dy^2}{y^2}$; (b) $ds^2 = \dfrac{4\left(dx^2 + dy^2\right)}{\left(1 - x^2 - y^2\right)^2}$;

(c) $ds^2 = \dfrac{4\left(dr^2 + r^2 d\varphi^2\right)}{\left(1 - r^2\right)^2}$.

10.16. Explicitly calculate the Christoffel symbols in the spherical coordinate system in $\mathbb{R}^3$.

10.17. Calculate the Christoffel symbols on the surface of revolution in $\mathbb{R}^3$ given in the form

$$\mathbf{r}\left(u, v\right) = \left(f\left(u\right)\cos v,\ f\left(u\right)\sin v,\ g\left(u\right)\right).$$

10.18. Calculate the Christoffel symbols of the Beltrami pseudosphere given in the form

$$\mathbf{r}\left(u, v\right) = \left(a\sin u \cos v,\ a\sin u \sin v,\ a\left(\ln\tan\frac{u}{2} + \cos u\right)\right).$$

10.19. Find the Christoffel symbols of the metric

$$ds^2 = du^2 + \sinh^2 u\, dv^2.$$

10.20. Calculate the Christoffel symbols on the catenoid

$$\left(x\left(u, v\right), y\left(u, v\right), z\left(u, v\right)\right) = \left(a\cosh\frac{u}{a}\cos v,\ a\cosh\frac{u}{a}\sin v,\ u\right).$$

10.21. Calculate the Christoffel symbols on the helicoid

$$\left(x\left(u, v\right), y\left(u, v\right), z\left(u, v\right)\right) = \left(u\cos v,\ u\sin v,\ hv\right).$$

10.22. Calculate the angle by which the tangent vector to a right circular cylinder in $\mathbb{R}^3$ turns after its parallel translation along a closed curve. Does the result depend on the form of a curve?

10.23. Calculate the angle by which the tangent vector to a right circular cone in $\mathbb{R}^3$ turns after its parallel translation along a closed curve. Find the dependence on the form of the curve.

10.24. Explicitly write and solve the equations of the parallel translation on the sphere with the metric $ds^2 = d\theta^2 + \sin^2\theta\, d\varphi^2$.

(a) along the curve $\theta = \theta_0 = \text{const}$, i.e., along a parallel;

(b) along the curve $\varphi = \varphi_0 = \text{const}$, i.e., along a meridian.

10.25. Reveal the angle by which the tangent vector to the sphere turns after its parallel translation along a parallel.

10.26. Consider the sphere S^2 with the standard metric. Let a vector $\mathbf{X}$ be tangent to S^2 at a point P, let the vector $\mathbf{Y}$ be tangent to S^2 at a point Q, and, moreover, let $\|\mathbf{X}\| = \|\mathbf{Y}\|$. Prove that there exists a smooth regular curve on the sphere S^2 with endpoints at the points P and Q such that the

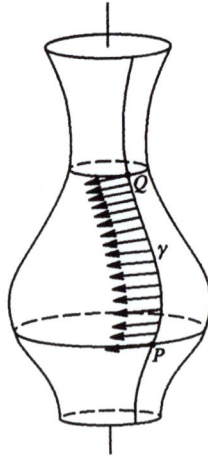

Figure 64

vector **Y** is obtained from the vector **X** as a result of the parallel translation along it.

10.27. On a surface of revolution, translate a tangent vector along a parallel. Find the angle between the initial and final position of the vector.

10.28. (a) Prove that under the parallel translation of a vector tangent to a parallel along a meridian of a surface of revolution, the result of translation is parallel to the initial vector in the ambient Euclidean space (see Fig. 64).

(b) Along a surface of revolution, translate a vector tangent to a meridian along this meridian.

10.29. Along the right helicoid

$$(u \cos v,\ u \sin v,\ hv)$$

translate a vector tangent to the surface in a parallel way. Perform the translation along an arc of the curve

$$(a \cos v,\ a \sin v,\ hv),\quad a = \text{const},\ 0 \le v \le 2\pi.$$

Find the angle (in the space) between the initial and final vectors.

10.30. Write in explicit form and solve the equations of parallel translation in the Lobachevskii plane in the model on the upper half-plane

(a) along the curve $x = x_0 = \text{const}$;

(b) along the curve $y = y_0 = \text{const}$.

10.31. Calculate the angle by which a tangent vector to the sphere turns under its parallel translation along a closed curve γ if

(a) γ is a parallel;

(b) γ is composed of two meridians and a part of the equator lying between them;

(c) γ is composed of two meridians and a part of the parallel lying between them.

10.32. Find the covariant derivatives of the following vector fields given in the polar coordinate system along the coordinate lines $r = \text{const}$ and $\varphi = \text{const}$:

$$\mathbf{v}_1 = \left(\cos \varphi, -\frac{1}{r} \sin \varphi \right), \quad \mathbf{v}_2 = (0, 1), \quad \mathbf{v}_3 = (r, 1).$$

Perform all the calculations in the polar coordinate system.

10.33. The polar coordinates (r, φ) are given on the plane. Translate the vector $\mathbf{v}_0 = (v_0^1, v_0^2)$ in a parallel way along the curve $r = 2$ from the point $\varphi = 0$ to the point $\varphi = \pi/2$. Perform all the calculations in the polar coordinate system.

10.34. (a) Write and solve the equations of parallel translation of a vector along the curve $\theta = \text{const}$ for the metric $ds^2 = d\theta^2 + \sinh^2 \theta \, d\varphi^2$.

(b) Calculate the angle by which a vector turns under its one-fold translation along the curve $\theta = \text{const}$ for the metric $ds^2 = d\theta^2 + \sinh^2 \theta \, d\varphi^2$.

10.35. Translate a vector in a parallel way along a spatial curve on the surface composed of

(a) tangents to this curve;

(b) normals to this curve;

(c) binormals to this curve.

10.36. Find the formula stating the dependence between the angle of turn α of the vector tangent to the sphere of unit radius being a result of its parallel translation along a closed curve γ and the area S of the domain bounded by the curve γ.

10.37. Find the formula stating the dependence between the angle of turn α of the vector tangent to the Lobachevskii plane being a result of its parallel translation along a closed curve γ and the area S of the domain bounded by the curve γ.

10.38. Find all connections on the circle. Establish the formula for the parallel translation of an arbitrary connection on the circle.

10.39. Find an affine connection on the plane with coordinates u^1, u^2 such that the vector fields $\boldsymbol{\xi} = \left(e^{u^1}, 1 \right)$ and $\boldsymbol{\eta} = \left(0, e^{u^2} \right)$ are covariantly constant with respect to it.

10.40. The Laplace–Beltrami operator on a Riemannian manifold is the operator $\Delta f = \nabla^i \nabla_i f$, where ∇ is the symmetric Riemannian connection and $\nabla^i = g^{ij} \nabla_j$. Find an explicit formula for this operator on a surface of revolution.

10.41. Consider a Riemannian manifold with the corresponding symmetric Riemannian connection. Define the gradient of a function f as the vector field $\text{grad} \, f = g^{ij} \nabla_j f$ and the divergence $\text{div} \, \mathbf{v}$ of a vector field $\mathbf{v}$ as the contraction of the covariant derivative of the vector field $\mathbf{v}$ (it has the valence $(1, 1)$): $\text{div} \, \mathbf{v} = \nabla_i v^i$.

(a) Prove that in the Euclidean space the operations of gradient and divergence defined in this way coincide with usual operations.

(b) Prove that on a Riemannian manifold the Laplace operator div grad f coincides with the Laplace–Beltrami operator Δf.

10.42. On a manifold equipped with a torsion-free connection, let two covariantly constant vector fields be given. Prove that they commute.

10.43. Prove the following relations (here, $g = \det \|g_{ij}\|$):

(a) $\Gamma^i_{ij} = \dfrac{1}{2g} \dfrac{\partial}{\partial x^j} \ln g = \dfrac{\partial}{\partial x^j} \ln \sqrt{g}$;

(b) $\nabla_i T^i = \dfrac{1}{\sqrt{g}} \dfrac{\partial}{\partial x^j} \left(\sqrt{g} T^j \right)$;

(c) $\nabla_i T^{ij} = \dfrac{1}{\sqrt{g}} \dfrac{\partial}{\partial x^i} \left(\sqrt{g} T^{ij} \right) + \Gamma^j_{kl} T^{kl}$;

(d) if $A^{ij} = -A^{ji}$, then $\nabla_i A^{ij} = \dfrac{1}{\sqrt{g}} \dfrac{\partial}{\partial x^i} \left(\sqrt{g} A^{ij} \right)$.

10.44. Two linearly independent vector fields with zero commutator are given on the plane. Prove that there exists a unique connection with respect to which these two vector fields are covariantly constant. Reveal whether or not it is symmetric. Present a counterexample in the case of non-commuting vector fields.

10.45. Find a domain in $\mathbb{R}^2$ with coordinates (x, y) in which the following vector fields are linearly independent:

(a) $\mathbf{u} = (-y, x)$, $\mathbf{v} = (1, 0)$;

(b) $\mathbf{u} = (\cos x, \sin x)$, $\mathbf{v} = (-\sin x, \cos x)$;

(c) $\mathbf{u} = \left(\dfrac{x}{r}, \dfrac{y}{r} \right)$, $\mathbf{v} = \left(-\dfrac{y}{r}, \dfrac{x}{r} \right)$, where $r = \sqrt{x^2 + y^2}$.

Find the coefficients of a connection with respect to which these fields are covariantly constant.

10.46. Consider the space $\mathbb{R}^3$ with coordinates x^1, x^2, x^3. Find a domain in $\mathbb{R}^3$ in which the vector fields

$$\mathbf{u} = (1, 0, 0), \ \mathbf{v} = (0, 1, 0), \ \mathbf{w} = \left(0, x^1, 1 \right)$$

are linearly independent. Find the coefficients of a connection with respect to which these fields are covariantly constant.

§11. Geodesics on Two-Dimensional Surfaces

11.1. Prove that a geodesic curve on a two-dimensional surface in the Euclidean space $\mathbb{R}^3$ is completely characterized by one of the following properties:

(a) at each point of the curve where its curvature is different from zero, the normal to the surface is the principal normal of the curve;

(b) at each point of the curve, its geodesic curvature vanishes;

(c) at each point of the curve, its curvature is equal to the absolute value of the normal curvature in the direction of the tangent to this curve.

11.2. Let a surface in $\mathbb{R}^3$ be such that a straight line lies on it. Prove that this line is a geodesic curve on the surface.

11.3. Two surfaces in $\mathbb{R}^3$ are tangent to each other along a curve l. Prove that if l is a geodesic on one of the surfaces, it should be a geodesic on the other surface.

11.4. Let two surfaces transversally intersect each other along a curve l, and, moreover, let l be a geodesic on each of the surfaces. Prove that this curve is a straight line.

11.5. On a surface in $\mathbb{R}^3$, there is a material point that can freely move along the surface. We force the point to move. Prove that the point will move along a geodesic curve of the surface.

11.6. A weightless thread lies on a surface in $\mathbb{R}^3$; this thread cannot leave the surface, but can freely slide on it. Prove that if we stretch the thread between two points of the surface, it goes along a geodesic. (Only the forces of tension and reaction of the surface act on the thread.)

11.7. Prove that the geodesic curvature of a curve $u = u(s)$, $v = v(s)$ on a surface $\mathbf{r} = \mathbf{r}(u, v)$ lying in $\mathbb{R}^3$ can be calculated by the formula

$$k_g = \left| \left(\mathbf{m}, \frac{d}{ds}\mathbf{r}, \frac{d^2}{ds^2}\mathbf{r} \right) \right|,$$

where $\mathbf{m}$ is the unit normal vector to the surface.

11.8. (a) Prove that the differential equation of geodesic curves $u = u(s)$, $v = v(s)$ of a surface $\mathbf{r} = \mathbf{r}(u, v)$ lying in $\mathbb{R}^3$ can be represented in the form $\left(\mathbf{m}, \frac{d}{ds}\mathbf{r}, \frac{d^2}{ds^2}\mathbf{r} \right) = 0$, where $\mathbf{m}$ is the normal vector to the surface.

(b) Deduce from this that exactly one geodesic passes through each point in each direction.

11.9. Prove that under a parallel translation of a vector along a geodesic on a Riemannian manifold, the angle between the vector and the tangent vector of the geodesic is constant.

11.10. Prove that straight lines and only they are geodesic curves on the plane.

11.11. Find the conditions under which the coordinate lines are geodesics on a two-dimensional surface M with coordinates x^1, x^2 and Christoffel symbols Γ_{ij}^k.

11.12. Show that for any connection ∇ there exists a torsion-free connection $\tilde{\nabla}$ with the same geodesics.

11.13. Prove that the meridians of a surface of revolution are geodesics.

11.14. Prove that a parallel of a surface of revolution is a geodesic if and only if the tangent to the meridian at its points is parallel to the axis of

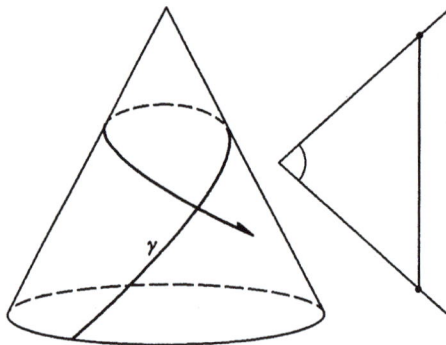

Figure 65 Geodesic on a cone

revolution.

11.15. Find the geodesic curves of a two-dimensional sphere.

11.16. Find the geodesic curves of a cylindrical surface in $\mathbb{R}^3$ (see Problem 5.2a).

11.17. Prove that only one closed geodesic passes through each point of a (not necessary right circular) cylinder.

11.18. Find the geodesic curves of the circular cone $x^2 + y^2 = z^2$.

11.19. Find the geodesic curves of an arbitrary conical surface in $\mathbb{R}^3$ (see Problem 5.2b).

11.20. Prove that on an infinite cone (without boundary) with the angle of development at the vertex $\alpha < \pi$, a unique self-intersecting geodesic emanating from a given point on its ruling is a geodesic that intersects this ruling at the angle $\dfrac{\pi - \alpha}{2}$ (see Fig. 65). Find the angle at which this geodesic intersects itself. If $\alpha \geq \pi$, then there are no self-intersecting geodesics on such a cone.

11.21. Show that the geodesic curves of a surface with the first quadratic form
$$ds^2 = v \left(du^2 + dv^2\right)$$
are depicted by parabolas on the plane with the Cartesian coordinates u, v.

11.22. Find the geodesic curves of the helicoid
$$\mathbf{r}\left(u, v\right) = \left(u \cos v, u \sin v, hv\right).$$

11.23. Show that on a surface with the first quadratic form
$$ds^2 = \left(\varphi\left(u\right) + \psi\left(v\right)\right)\left(du^2 + dv^2\right)$$
(Liouville surface) the geodesics are defined by the equation
$$\frac{du}{\sqrt{\varphi\left(u\right) + a}} \pm \frac{dv}{\sqrt{\psi\left(v\right) - a}} = 0,$$

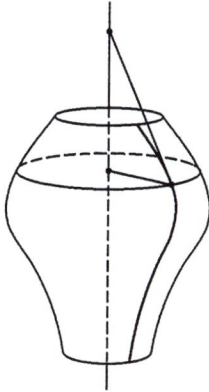

Figure 66

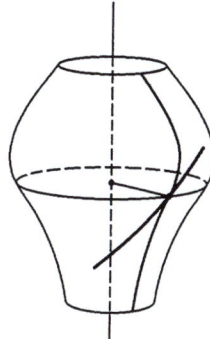

Figure 67

where a is an arbitrary constant.

11.24. (a) Reduce the metric of the surface of revolution

$$(\varphi(u)\cos v, \varphi(u)\sin v, \psi(u))$$

to the form indicated in Problem 11.23.

(b) Find the geodesics on this surface of revolution.

11.25. Prove that on the surface with the metric

$$ds^2 = \left(u^2 + \cos v + 2\right)\left(du^2 + dv^2\right),$$

the curve $v = \pi$ is a geodesic.

11.26. Show that on the surface of revolution

$$\mathbf{r}(t,\varphi) = (x(t)\cos\varphi, x(t)\sin\varphi, z(t))$$

it is possible to choose a parameter $\tau(t)$ along a meridian so that the first quadratic form of the surface becomes $ds^2 = \rho(\tau)\left(d\varphi^2 + d\tau^2\right)$. Show that $\rho(\tau)$ is the distance from a point (τ, φ) to the axis of revolution. Write the equation of a geodesic curve $(\tau(s), \varphi(s))$.

11.27. *Clairaut theorem.* Prove that the geodesic curvature radius of a parallel of a surface of revolution (the quantity inverse to the geodesic curvature) is equal to the segment of the tangent to the meridian contained between the tangent point and the axis of the surface (see Fig. 66).

11.28. (Another formulation of the Clairaut theorem). Prove that on a surface of revolution, along each of the geodesics, the product of the radius of a parallel by the sine of the angle between a geodesic curve and the meridian is constant (see Fig. 67).

11.29. Prove that if a geodesic on a surface of revolution intersects all the parallels meeting at a constant angle, then it is either a meridian or a parallel or a right circular cylinder.

11.30. Prove that all geodesics on the surface of revolution

$$16a^2 \left(x^2 + y^2\right) = z^2 \left(2a^2 - z^2\right)$$

are closed. In the cylindrical coordinates (r, φ, z), this surface can be parameterized as follows:

$$r = \frac{a}{4} \cos u, \quad z = a \left(\sin \frac{u}{2} - \cos \frac{u}{2}\right), \quad \varphi = v.$$

11.31. Present an example of a surface of revolution on which there exists a closed geodesic different from the meridian and the parallel. The surface of revolution should be different from the sphere and the surface of the previous problem.

Note that in reality an example of a surface with a closed geodesic different from the parallel and the meridian is not exceptional; many surfaces of revolution have this property.

11.32. Present an example of a surface of revolution having no closed geodesics.

11.33. Is the visible contour of the ellipsoid a geodesic on it? Here, it is assumed that the "observer" is at infinity, so that the rays of vision are parallel to one another.

11.34. On a Riemannian manifold, consider the beam of geodesics emanating from a fixed point O, i.e., draw a geodesic from the point O in each direction. On each of the geodesics, we distinguish the point distant from the point O by the same distance R. The obtained set is called the geodesic sphere of radius R. Prove that the geodesic sphere is orthogonal to all geodesic radii, i.e., to all geodesics of the above beam.

11.35. Find the geodesic curvature of the spiral

$$\mathbf{r}\left(t\right) = \left(a \cos t, a \sin t, hv\right):$$

(a) on the helicoid $\left(x\left(u, v\right), y\left(u, v\right), z\left(u, v\right)\right) = \left(u \cos v, u \sin v, hv\right)$;

(b) on the cylinder $\left(x\left(u, v\right), y\left(u, v\right), z\left(u, v\right)\right) = \left(a \cos v, a \sin v, u\right)$.

11.36. Calculate the geodesic curvature of the curve $u = \sinh v$, $0 \le v \le v_0$, on the helicoid

$$x = u \cos v, \quad y = u \sin v, \quad z = v.$$

11.37. Find the geodesic curvature of the curve $v = \ln \cosh u$ on a surface with the metric $ds^2 = du^2 + \cosh^2 u \, dv^2$.

11.38. Prove that if curves of a one-parametric family of curves on a surface are such that the arclengths of orthogonal trajectories between any two curves of the family are equal, then the orthogonal trajectories are geodesics.

11.39. Prove that the torsion $\varkappa$ of a geodesic curve passing in the direction of the vector $du \cdot \mathbf{r}_u + dv \cdot \mathbf{r}_v$ is found by the formula

$$-\varkappa = \frac{(LF - ME)\, du^2 + (LG - NE)\, du\, dv + (MG - NF)\, dv^2}{(EG - F^2)\left(E\, du^2 + 2F\, du\, dv + G\, dv^2\right)}.$$

11.40. Find the geodesic curvature of parallels and meridians of the surface of revolution

$$(f(u)\cos v,\ f(u)\sin v,\ u).$$

11.41. Find the geodesic curvatures of the coordinate lines of a surface with the metric $ds^2 = E\,du^2 + G\,dv^2$.

11.42. Let the surface be composed of tangents to a curve of curvature $k(s)$. Draw segments of length l from the tangent point on the tangents. Find the geodesic curvature of the curves composed of the endpoints of these segments on the surface of tangents (see Problem 6.7).

11.43. Show that on any compact Riemannian manifold, any two sufficiently close points can be connected by a geodesic, and, moreover, there exists a unique geodesic of the least length.

11.44. Prove that a smooth curve coinciding with a connected component of the fixed point set of an isometry of Riemannian manifold is a geodesic.

11.45. (a) Describe the geodesics of the torus T^2 in the flat metric.

(b) Prove that the tori obtained by gluing from the rectangle with the vertices $\{(0,0),(2,0),(1,0),(2,1)\}$ and the parallelogram with the vertices $\{(0,0),(2,0),(1,1),(3,1)\}$ are not isometric.

(c) Consider parallelograms of unit area with a common base. Reveal in which case the tori obtained by gluing from these parallelograms are isometric.

(d) Describe all flat metrics on the torus T^2.

11.46. Prove that the geodesics on an open disk with the Lobachevskii metric (Poincaré model) are diameters of the disk and arcs of circles intersecting the boundary of the disk at a right angle.

11.47. Show that the geodesics on the upper half-plane in the Lobachevskii metric are vertical half-lines and arcs of the circles perpendicular to the absolute, i.e., abscissa axes.

11.48. Find the Riemannian metrics of the form

$$ds^2 = e^{2\lambda(u,v,w)}\left(du^2 + dv^2 + dw^2\right),$$

such that the curves $v = \text{const}$ and $w = \text{const}$ are geodesics (under the corresponding parameterization of these curves).

11.49. Present an example

(a) of a metric on the Euclidean plane being not geodesically complete;

(b) of a metric in an open disk being geodesically complete.

11.50. Given a surface of constant Gaussian curvature, find the formula that states a dependence between the angle of turn α of the tangent vector to the surface resulting from the parallel translation along a closed curve γ and the area S of the domain bounded by the curve γ.

11.51. (a) Find all closed geodesics on an infinite cylinder with closed directrix.

(b) For two given points, find the shortest path connecting them.

(c) In which case any arc of a geodesic on the cylinder is a shortest path between its endpoints?

11.52. Study the behavior of shortest paths on a conical surface depending on the value of the complete angle at the vertex of the cone. Do shortest paths passing through the vertex of the cone exist?

11.53. Let the closed surface be a right circular cone with a circular base. This surface has a singular line being the circle at which the cone and its plane base are glued together. Describe shortest paths connecting two given points depending on the location of these points: both of them lie on the lateral surface of the cone, or both lie on the base, or one of them lies on the lateral surface and the other on the base.

11.54. Describe the shortest paths connecting two points on the cube and on the tetrahedron. Study the dependence on the location of the points on the surface.

Consider a point P on a Riemannian manifold M. For a vector $\mathbf{v} \in T_P M$, by $\gamma_{\mathbf{v}}(t)$ we denote the geodesic such that $\gamma_{\mathbf{v}}(0) = P$ and $\gamma_{\mathbf{v}}'(0) = \mathbf{v}$. Define the *geodesic exponential* by the formula $\exp_P(\mathbf{v}) = \gamma_{\mathbf{v}}(1)$ for those $\mathbf{v} \in T_P M$ for which the point $\gamma_{\mathbf{v}}(1)$ is defined.

11.55. (a) Prove that the mapping $\exp_P$ defines a diffeomorphism of a neighborhood V of zero of the space $T_P M$ onto a neighborhood U of the point P in M. Thus, $\exp_P$ defines a coordinate system in the neighborhood $U \subset M$, which is said to be *normal*.

(b) Show that in the normal coordinates centred at a point P, the matrix of the metric is the identity matrix and the Christoffel symbols vanish at the point P.

(c) Consider the normal coordinates on the standard two-dimensional sphere. What is the structure of the maximal domain of the normal coordinate system?

§12. Curvature Tensor

12.1. Calculate the curvature tensor of a one-dimensional manifold with an arbitrary metric.

12.2. Calculate the scalar curvature of the following Riemannian manifolds:

(a) the sphere S^2 of radius R embedded in $\mathbb{R}^3$;

(b) the torus T^2 embedded in $\mathbb{R}^3$ as a surface of revolution (see Problem 6.1c);

(c) the torus T^2 embedded in $\mathbb{R}^4 = \mathbb{C}^2(z, w)$ and given by the equations $|z|^2 + |w|^2 = 1$ and $|z| = |w|$;

(d) the Lobachevskii plane (see Problems 3.2 and 3.3);

(e) the right circular cone in $\mathbb{R}^3$;

(f) the cylindrical surface in $\mathbb{R}^3$;

(g) the sphere S^n of radius R in $\mathbb{R}^{n+1}$;

(h) the Beltrami surface (see Problem 6.4c).

12.3. Prove that a Riemannian metric of a two-dimensional manifold is locally Euclidean if and only if its curvature tensor vanishes identically.

12.4. Calculate the curvature tensor on the sphere S^2 in the spherical coordinates.

12.5. Let $ds^2 = \lambda(x,y)(dx^2 + dy^2)$. Express the scalar curvature of this metric through the function $\lambda(x,y)$ and its derivatives in explicit form.

12.6. Let M be a manifold in $\mathbb{R}^n$ $(n > 2)$. Let $x \in M$, and let P be a two-dimensional subspace of $T_x M$. Define the number $\sigma(P)$ by the formula

$$\sigma(P) = \langle R(\mathbf{e}_1, \mathbf{e}_2)\,\mathbf{e}_2, \mathbf{e}_1 \rangle,$$

where $\mathbf{e}_1$, $\mathbf{e}_2$ is an orthonormal basis of the plane P.

The number $\sigma(P)$ is called the *sectional curvature* of the surface M in the direction of the two-dimensional plane P or, as one says, in the two-dimensional direction P.

(a) Prove that $\sigma(P)$ is independent of the choice of the orthonormal basis of P.

(b) Prove that if $n = 2$, the sectional curvature $\sigma(P)$ coincides with the Gaussian curvature of the surface M^2 at the point x.

(c) Let P be a two-dimensional subspace of $T_x S$. Prove the formula

$$\sigma(P) = R(V),$$

where $R(V)$ is the Gaussian curvature of the two-dimensional surface V composed of the geodesics whose tangent vectors lie in the plane P.

12.7. Let S^n be the n-dimensional sphere $x_1^2 + \cdots + x_{n+1}^2 = r^2$ with the induced metric.

(a) Show that the curvature tensor of the sphere S^n is calculated by the formula

$$R(\mathbf{X}, \mathbf{Y})\,\mathbf{Z} = \frac{1}{r^2}\left(\langle \mathbf{Y}, \mathbf{Z}\rangle\,\mathbf{X} - \langle \mathbf{X}, \mathbf{Z}\rangle\,\mathbf{Y}\right),$$

where $\mathbf{X}$, $\mathbf{Y}$, and $\mathbf{Z}$ are vectors tangent to the sphere.

(b) Show that the sectional curvature of the sphere S^n is constant: $\sigma(P) = \dfrac{1}{r^2}$ for all points x.

(c) Show that the metric

$$ds^2 = \frac{4}{1 + K\left(y_1^2 + \cdots + y_n^2\right)^2}\left(dy_1^2 + \cdots + dy_n^2\right)$$

has a constant sectional curvature.

12.8. (a) Find the Christoffel symbols for a surface with the metric $ds^2 = du^2 + G(u, v) dv^2$.

(b) Find the Gaussian curvature of a surface with the metric

$$ds^2 = du^2 + G(u, v) dv^2.$$

12.9. Find the Gaussian curvature of a surface with the metric

$$ds^2 = du^2 + 2 \cos \omega (u, v) du \, dv + dv^2.$$

12.10. (a) Prove that the Ricci tensor of a two-dimensional surface is proportional to the metric tensor. Find the proportionality coefficient.

(b) Prove that the following equation holds on a three-dimensional manifold:

$$R_{lmnk} = g_{ln} R_{mk} - g_{lk} R_{mn} - g_{mn} R_{lk} + g_{mk} R_{ln} - \frac{1}{2} (g_{ln} g_{mk} - g_{lk} g_{mn}) R.$$

Here, R_{lm} is the Ricci tensor and R is the scalar curvature.

12.11. Prove the following relations:

(a) $\nabla^i \left(R_{ij} - \frac{1}{2} R g_{ij} \right) = 0$;

(b) for a two-dimensional manifold $R_{ij} - \frac{1}{2} R g_{ij} = 0$.

12.12. Let ∇ be a canonical connection in $\mathbb{R}^3$ with the standard Euclidean metric. Consider the following new operation:

$$\tilde{\nabla}_{\mathbf{X}} \mathbf{Y} = \nabla_{\mathbf{X}} \mathbf{Y} + \frac{1}{2} \mathbf{X} \times \mathbf{Y}.$$

Prove that this is a connection. Find its torsion and curvature tensors.

12.13. Let M be a manifold equipped with a connection, and let x be a fixed point of M. Let $Q = \{-1 < u, v < 1\}$ be the square, and let $\mathbf{X}, \mathbf{Y}, \mathbf{Z} \in T_x M$. Further, let $f \colon Q \to M$ be a smooth mapping such that $f(0,0) = x$, $df(\partial/\partial u) = \mathbf{X}$, and $df(\partial/\partial v) = \mathbf{Y}$. For $|t| < 1$, consider the parallel translation τ along the path $x \to f(t, 0) \to f(t, t) \to f(0, t) \to x$. Prove that

$$R(\mathbf{X}, \mathbf{Y}) \mathbf{Z} = \lim_{t \to 0} \frac{1}{t^2} \left(\tau^{-1} \mathbf{Z} - \mathbf{Z} \right).$$

12.14. Find the component R_{1212} of the curvature tensor of the following metrics:

(a) $ds^2 = du^2 + u^2 dv^2$; (b) $ds^2 = du^2 - u^2 dv^2$;

(c) $ds^2 = a^2 (d\theta^2 + \cos^2 \theta \, d\varphi^2)$, $a = $ const.

12.15. A Riemannian manifold in which $R_{ij} = \lambda(x) g_{ij}$ is called the Einstein space. Prove that

(a) the relation $\lambda(x) = R(x)/n$, where R is the scalar curvature and n is the dimension of the manifold, holds for an Einstein space;

(b) any two-dimensional Riemannian manifold is an Einstein space.

12.16. Prove that the scalar curvature of an Einstein space is constant for $n > 2$.

12.17. The Einstein tensor on a Riemannian manifold is defined by the formula

$$G_{ij} = R_{ij} - \frac{R}{2} g_{ij}.$$

Prove that $\nabla_k G_i^k = 0$, where $G_i^k = g^{kl} G_{li}$.

12.18. Prove that a connection on an n-dimensional manifold admitting n covariantly constant linearly independent vector fields at each point has zero curvature tensor.

12.19. At each point, let the sectional curvature of a Riemannian manifold of dimension $n > 2$ be constant, i.e., be independent of a two-dimensional direction. Such manifolds are usually called spaces of *constant curvature*. Show that in a space of constant curvature, the sectional curvature K is connected with the scalar curvature by the relation $K = \dfrac{R}{n(n-1)}$.

12.20. Show that a space of constant curvature is an Einstein space. Taking into account Problem 12.16, we obtain that in a space of constant curvature, for $n > 2$, the scalar curvature is constant and the sectional curvature is independent either of the point or of the two-dimensional direction.

12.21. Show that in the space of constant curvature, any symmetric nondegenerate covariantly constant tensor of type $(0,2)$ has the form $a_{ij} = \lambda g_{ij}$, $\lambda = $ const.

12.22. Prove that on a four-dimensional manifold with the metric

$$ds^2 = 2du^1 du^4 + \left(u^4\right)^2 \left(du^2\right)^2 + 2du^2 du^3$$

the curvature tensor is covariantly constant and the Ricci tensor vanishes.

12.23. Prove that if the curvature tensor of a Riemannian manifold is covariantly constant and the dimension of the manifold is greater than 2, the space is of constant curvature.

12.24. Show that on a two-dimensional surface in $\mathbb{R}^3$ the components of the curvature tensor are represented in the form

$$R_{ijkl} = K \left(g_{il} g_{jk} - g_{ik} g_{jl}\right),$$

where K is the Gaussian curvature.

12.25. Prove that the Christoffel symbols in a coordinate system vanish if and only if the torsion tensor S_{ij}^k and the curvature tensor R_{ijk}^l vanish identically.

§13. Differential Forms and de Rham Cohomologies

In this section, the exterior product of vectors $\mathbf{v}_1, \ldots, \mathbf{v}_p \in V$, $\dim V = n$, is the tensor

$$\mathbf{v}_1 \wedge \cdots \wedge \mathbf{v}_p = \sum_{\sigma \in S_p} (-1)^\sigma \, \mathbf{v}_{\sigma(1)} \otimes \cdots \otimes \mathbf{v}_{\sigma(p)},$$

where S_p is the group of permutations acting on p elements and $(-1)^\sigma$ is the sign of a permutation σ.

With such a definition, the value of an exterior form $\mathbf{e}^1 \wedge \cdots \wedge \mathbf{e}^n$ on a tuple of vectors $\mathbf{v}_1, \ldots, \mathbf{v}_n$ is equal to the volume of the parallelepiped spanned by the vectors $\mathbf{v}_1, \ldots, \mathbf{v}_n$. Here, $\mathbf{e}^1, \ldots, \mathbf{e}^n$ is a basis of the dual space V^*.

13.1. Prove that if vectors $\mathbf{v}_1, \ldots, \mathbf{v}_p \in V$ are linearly dependent, then

$$T(\mathbf{v}_1, \ldots, \mathbf{v}_p) = 0$$

for any form $T \in \Lambda^p(V^*)$.

13.2. Prove that if the forms $\varphi_1, \ldots, \varphi_p \in V^*$ are linearly dependent, then $\varphi_1 \wedge \cdots \wedge \varphi_p = 0$.

13.3. (a) Write the linear form $\omega = f\left(x^2 + y^2\right)(x\,dx + y\,dy)$ in the polar coordinates.

(b) Write the form $\omega = \sqrt{x^2 + y^2}\,dx \wedge dy$ in the polar coordinates.

(c) Write the form $\omega = x\,dx + y\,dy + z\,dz$ in the spherical coordinates.

(d) Write the form

$$\omega = \frac{x\,dy \wedge dz + y\,dz \wedge dx + z\,dx \wedge dy}{\left(x^2 + y^2 + z^2\right)^{3/2}}$$

in the spherical coordinates.

(e) Write the form of the previous item in the cylindrical coordinates.

(f) Write the form $\omega = dx \wedge dy \wedge dz$ in the spherical coordinates.

13.4. Let the mapping $\varphi \colon \mathbb{R}^4 \to \mathbb{R}^4$ be given by the formulas

$$y^1 = x^1 - x^2 x^3 x^4, \quad y^2 = x^2 - x^1 x^3 x^4,$$

$$y^3 = x^3 - x^1 x^2 x^4, \quad y^4 = x^4 - x^1 x^2 x^3.$$

Calculate the form $\varphi^* \omega$ if

(a) $\omega = y^1 dy^1 + y^2 dy^2 + y^3 dy^3 + y^4 dy^4$;

(b) $\omega = dy^1 \wedge dy^2 + dy^3 \wedge dy^4$;

(c) $\omega = dy^1 \wedge dy^2 \wedge dy^3 \wedge dy^4$.

13.5. Let the mapping $\varphi \colon \mathbb{R}^2 \setminus \{0\} \to \mathbb{R}^2 \setminus \{0\}$ be given by the formula

$$(x, y) \mapsto \left(\frac{x}{\sqrt{x^2 + y^2}}, \frac{y}{\sqrt{x^2 + y^2}} \right).$$

Calculate $\varphi^*\omega$ if

(a) $\omega = x\,dx + y\,dy$;

(b) $\omega = x\,dy - y\,dx$;

(c) $\omega = \left(x^2 - y^2\right)dy - 2xy\,dx$.

13.6. Let the mapping $\varphi \colon \mathbb{R}^n \setminus \{0\} \to \mathbb{R}^n \setminus \{0\}$ be given by

$$\varphi\left(x\right) = \frac{x}{\|x\|},$$

where $x = \left(x^1, \ldots, x^n\right)$. For the form

$$\omega = \sum_{k=1}^{n} (-1)^{k+1}\, x^k dx^1 \wedge \cdots \wedge \widehat{dx}^{\,k} \wedge \cdots \wedge dx^n,$$

find $\varphi^*\omega$ and prove that $d\left(\varphi^*\omega\right) = 0$.

13.7. Let $\omega = \dfrac{x\,dy - y\,dx}{x^2 + y^2}$. Find $\varphi^*\omega$ if the mapping φ is given by

(a) $\varphi \colon (x,y) \mapsto \left(e^x \cos y,\, e^x \sin y\right)$;

(b) $\varphi \colon (x,y) \mapsto \left(\cosh y \cos x,\, -\sinh y \sin x\right)$;

(c) $\varphi \colon (x,y) \mapsto \left(x^2 - y^2,\, 2xy\right)$.

13.8. Let $\varphi_1, \ldots, \varphi_n \in V^*$, and let $\mathbf{v}_1, \ldots, \mathbf{v}_n \in V$. Prove that

$$\left(\varphi_1 \wedge \cdots \wedge \varphi_n\right)\left(\mathbf{v}_1, \ldots, \mathbf{v}_n\right) = \det \|\varphi_i\left(\mathbf{v}_j\right)\|\,.$$

13.9. Prove that on an oriented manifold with a Riemannian metric given by a tuple of functions g_{ij} in local coordinates $\left(x^1, \ldots, x^n\right)$, the expression of the form $\sqrt{\det\left(g_{ij}\right)}dx^1 \wedge \cdots \wedge dx^n$ is a differential form correctly defined on the whole manifold. This form is called the *volume form*.

13.10. Show that the operation of exterior differentiation of a differential form can be represented as a composition of the operations of covariant gradient and alternation for an arbitrary symmetric connection on the manifold.

13.11. Calculate the exterior differential of the following differential forms:

(a) $z^2 dx \wedge dy + \left(z^2 + 2y\right) dx \wedge dz$;

(b) $13x\,dx + y^2 dy + xyz\,dz$;

(c) $\left(x + 2y^3\right)\left(dz \wedge dx + \dfrac{1}{2}dy \wedge dx\right)$;

(d) $(x\,dx + y\,dy) / \left(x^2 + y^2\right)$;

(e) $(y\,dx - x\,dy) / \left(x^2 + y^2\right)$;

(f) $f\left(x^2 + y^2\right)(x\,dx + y\,dy)$;

(g) $f\,dg$, where f and g are smooth functions;

(h) $f\left(g\left(x^1, \ldots, x^n\right)\right) dg\left(x^1, \ldots, x^n\right)$.

13.12. Prove the Cartan formula

$$(dw)\left(\mathbf{X}, \mathbf{Y}\right) = \mathbf{X}\left(w\left(\mathbf{Y}\right)\right) - \mathbf{Y}\left(w\left(\mathbf{X}\right)\right) - w\left(\left[\mathbf{X}, \mathbf{Y}\right]\right),$$

where w is a one-dimensional differential form; $\mathbf{X}$ and $\mathbf{Y}$ are vector fields. How does the formula change if the exterior product is defined by

$$\mathbf{v_1} \wedge \cdots \wedge \mathbf{v}_p = \frac{1}{p!} \sum_{\sigma \in S_p} (-1)^\sigma \, \mathbf{v}_{\sigma(1)} \wedge \cdots \wedge \mathbf{v}_{\sigma(p)}?$$

13.13. Prove the following generalizations of the formula from the previous problem (they are also called the Cartan formulas):

(a) If ω is a form of degree 2, then

$$d\omega\left(\mathbf{X}, \mathbf{Y}, \mathbf{Z}\right) = \mathbf{X}\omega\left(\mathbf{Y}, \mathbf{Z}\right) - \mathbf{Y}\omega\left(\mathbf{X}, \mathbf{Z}\right) + \mathbf{Z}\omega\left(\mathbf{X}, \mathbf{Y}\right) -$$
$$\omega\left(\left[\mathbf{X}, \mathbf{Y}\right], \mathbf{Z}\right) + \omega\left(\left[\mathbf{X}, \mathbf{Z}\right], \mathbf{Y}\right) - \omega\left(\left[\mathbf{Y}, \mathbf{Z}\right], \mathbf{X}\right).$$

(b) If ω is a form of degree p, then

$$d\omega\left(\mathbf{X}_1, \ldots, \mathbf{X}_{p+1}\right) = \sum_{i=1}^{p+1} (-1)^{i+1} \, \mathbf{X}_i\omega\left(\mathbf{X}_1, \ldots, \widehat{\mathbf{X}}_i, \ldots, \mathbf{X}_{p+1}\right) +$$
$$\sum_{1 \le i < j \le p+1} (-1)^{i+j} \, \omega\left(\left[\mathbf{X}_i, \mathbf{X}_j\right], \mathbf{X}_1, \ldots, \widehat{\mathbf{X}}_i, \ldots, \widehat{\mathbf{X}}_j, \ldots, \mathbf{X}_{p+1}\right).$$

Let ω be a differential p-form. Denote by $i_\mathbf{X}\omega$, where $\mathbf{X}$ is a vector field, the differential $(p-1)$-form whose value on vectors $\mathbf{X}_1, \ldots, \mathbf{X}_{p-1}$ is defined by the formula

$$i_\mathbf{X}\omega\left(\mathbf{X}_1, \ldots, \mathbf{X}_{p-1}\right) = \omega\left(\mathbf{X}, \mathbf{X}_1, \ldots, \mathbf{X}_{p-1}\right).$$

The form $i_\mathbf{X}\omega$ is called the *inner product* of the form ω and the field $\mathbf{X}$.

13.14. Let ω be a p-form, and let $\mathbf{X}_0, \ldots, \mathbf{X}_p$ be vector fields. Prove that

$$(i_{\mathbf{X}_0}d\omega)\left(\mathbf{X}_1, \ldots, \mathbf{X}_p\right) + (di_{\mathbf{X}_0}\omega)\left(\mathbf{X}_1, \ldots, \mathbf{X}_p\right) =$$
$$\mathbf{X}_0\omega\left(\mathbf{X}_1, \ldots, \mathbf{X}_k\right) - \sum_i \omega\left(\mathbf{X}_1, \ldots, \left[\mathbf{X}_0, \mathbf{X}_i\right], \ldots, \mathbf{X}_p\right).$$

13.15. Let $\mathbf{X}$ be a vector field, and let ω_1 and ω_2 be differential forms. Prove that

$$i_\mathbf{X}\left(\omega_1 \wedge \omega_2\right) = (i_\mathbf{X}\omega_1) \wedge \omega_2 + (-1)^r \, \omega_1 \wedge i_\mathbf{X}\omega_2,$$

where r is the degree of the form ω_1.

In the vector space $\mathbb{R}^n$, let an inner product be given. Introduce the following two operations. The first operation assigns a linear form $\omega = V\left(\mathbf{X}\right)$ to each vector $\mathbf{X}$ such that $\langle \mathbf{X}, \mathbf{Y} \rangle = V\left(\mathbf{X}\right)\left(\mathbf{Y}\right)$. The second operation puts in correspondence the form $*\left(\omega\right)$ of degree $n - p$ to each multilinear skew-symmetric form ω of degree p as follows. Let $\omega_1, \ldots, \omega_n$ be an orthonormal basis of linear forms, and let $\omega = f\omega_{i_1} \wedge \cdots \wedge \omega_{i_p}$. Then $*\left(\omega\right) = (-1)^\sigma f\omega_{j_1} \wedge \cdots \wedge \omega_{j_{n-p}}$, where σ is the parity of the permutation

$$\begin{pmatrix} 1 & \cdots & p & p+1 & \cdots & n \\ i_1 & \cdots & i_p & j_1 & \cdots & j_{n-p} \end{pmatrix}.$$

This operation is usually called the "star operation".

Note that the former operation defines the well-known linear isomorphism between the spaces V and V^*. The latter operation sets an isomorphism between the space of exterior forms of degree p and the space of exterior forms of degree $n - p$.

13.16. Write in explicit form the result of application of the operation $*$ to differential k-forms $(k = 0, 1, 2, 3)$ given in $\mathbb{R}^3$ with the metric $ds^2 = \lambda_1 dx^2 + \lambda_2 dy^2 + \lambda_3 dz^2$, where λ_i are smooth functions.

13.17. Prove that $* (*T) = (-1)^{k(n-k)} T$.

13.18. Show that the following equations hold in the space $\mathbb{R}^3$ for vector fields:

(a) $\operatorname{grad} F = V^{-1} (dF)$; (b) $\operatorname{div} \mathbf{X} = * d * V (\mathbf{X})$;
(c) $\operatorname{rot} \mathbf{X} = V^{-1} * dV (\mathbf{X})$.

13.19. Consider $\mathbb{R}^3$ with the standard Euclidean metric. Prove that the value of the 2-form $V (\mathbf{X}_1) \wedge V (\mathbf{X}_2)$ on a pair of vectors $\mathbf{Y}_1$ and $\mathbf{Y}_2$ can be calculated by the formula

$$V (\mathbf{X}_1) \wedge V (\mathbf{X}_2) (\mathbf{Y}_1, \mathbf{Y}_2) = \det \| \langle \mathbf{X}_i, \mathbf{Y}_j \rangle \| .$$

13.20. Show that the Green, Stokes, and Gauss–Ostrogradskii formulas are particular cases of the general Stokes formula for differential forms.

13.21. Deduce the following integration formulas over the volume V bounded by a closed surface Σ:

(a) $\displaystyle\iiint_V (\varphi \Delta \psi + \langle \operatorname{grad} \varphi, \operatorname{grad} \psi \rangle) \, dV = \iint_\Sigma \varphi \frac{\partial \psi}{\partial \mathbf{n}} \, d\sigma;$

(b) $\displaystyle\iiint_V (\varphi \Delta \psi - \psi \Delta \varphi) \, dV = \iint_\Sigma \left(\varphi \frac{\partial \psi}{\partial \mathbf{n}} - \psi \frac{\partial \varphi}{\partial \mathbf{n}} \right) d\sigma.$

Here, $\partial / \partial \mathbf{n}$ is the derivative along the normal to the surface Σ.

13.22. Find the gradients of the following functions in the spherical coordinates:

(a) $u (r, \theta, \varphi) = r^2 \cos \theta;$
(b) $u (r, \theta, \varphi) = 3r^2 \sin \theta + e^r \cos \varphi - r;$
(c) $u (r, \theta, \varphi) = \cos \theta / r^2.$

13.23. Prove that

$$\operatorname{div} \mathbf{v} = \partial_i v^i + v^i \partial_i \ln \sqrt{|g|} = \frac{1}{\sqrt{|g|}} \partial_i \left(\sqrt{|g|} v^i \right).$$

13.24. Calculate the divergence of a vector field on the plane in: (a) the polar coordinates; (b) the elliptic coordinates; (c) the cylindrical coordinates (see Problems 1.1, 1.2, and 1.3).

13.25. Calculate the divergence of a vector field in the three-dimensional space in: (a) the spherical coordinates; (b) the cylindrical coordinates.

13.26. Find the transformation law of the connection forms

$$\omega_k^i = \Gamma_{jk}^i du^j$$

under the passage from one chart to another. Note that in general the connection forms do not compose a globally defined differential form.

13.27. Let a connection be given on a manifold M. Then the 1-form $\gamma = \Gamma_{ki}^i dx^k$ is defined in each coordinate system.

(a) Prove that the 2-form $d\gamma$ is correctly defined on the whole manifold, i.e., is independent of the choice of a coordinate system.

(b) Prove that under a change of coordinates with Jacobian J the summand $d(\ln J)$ is added to the form γ.

(c) Show that for the Riemannian connection

$$\gamma = d\ln\sqrt{\det(g_{ij})}.$$

13.28. Let M^n be a Riemannian manifold with metric g_{ij}. Prove that the covariant derivative of the volume form

$$\sqrt{\det(g_{ij})}dx^1 \wedge \cdots \wedge dx^n$$

with respect to the Riemannian connection on M^n vanishes.

13.29. Let $\mathbf{F}$ be a vector field in a three-dimensional domain W with smooth boundary ∂W, and let $\mathbf{n}$ be the unit normal vector to ∂W. Prove that

$$\int_W (\operatorname{div}\mathbf{F})\, dx\, dy\, dz = \int_{\partial W} \langle \mathbf{n}, \mathbf{F}\rangle\, d\sigma,$$

where $d\sigma$ is the area element on ∂W and $\langle \mathbf{n}, \mathbf{F}\rangle$ is the inner product.

13.30. Under the conditions of the previous problem, prove that

$$\int_S \langle \operatorname{rot}\mathbf{F}, \mathbf{n}\rangle\, d\sigma = \int_{\partial S} (f_1 dx_1 + f_2 dx_2 + f_3 dx_3),$$

where S is a smooth surface with smooth boundary ∂S and $\mathbf{F} = (f_1, f_2, f_3)$.

13.31. Deduce the Cauchy residue theorem from the Stokes formula.

13.32. Let p and q be arbitrary polynomials in variables $z^1, \ldots, z^n$, and let k and l be real numbers. Let there exist a differential form w such that $dp \wedge w = p\, dz$, $dw = l\, dz$, and $dq \wedge w = kq\, dz$. Prove that $d\left(p^{-k-l}qw\right) = 0$. Here, $dz = dz^1 \wedge \cdots \wedge dz^n$.

13.33. Let $\omega = \omega_{ij} dx^i \wedge dx^j$ be a nondegenerate 2-form on a manifold, and let a^{kl} be the inverse tensor, i.e., $a^{kl}\omega_{li} = \delta_i^k$. Prove that the following two conditions are equivalent:

(a) the form ω is closed;

(b) the operation

$$\{f, g\} = a^{kl} \frac{\partial f}{\partial x^k} \frac{\partial g}{\partial x^l},$$

defined on the set of smooth functions by the tensor a^{kl}, satisfies the Jacobi identity.

13.34. Let $\mathbf{X}_1, \ldots, \mathbf{X}_n$ be linearly independent vector fields on an n-dimensional manifold, and let $\omega^1, \ldots, \omega^n$ be dual 1-forms, i.e., $\omega^i(\mathbf{X}_j) = \delta^i_j$. Prove the formula

$$d\omega^k = -\frac{1}{2} c^k_{ij} \omega^i \wedge \omega^j,$$

where the smooth functions c^k_{ij} are found from the relation $[\mathbf{X}_i, \mathbf{X}_j] = c^k_{ij} \mathbf{X}_k$.

13.35. Prove that if, on a manifold, there exist a nondegenerate form of maximum rank, then the manifold is orientable.

13.36. Let $\omega = \omega_{ij} dx^i \wedge dx^j$ be a nondegenerate 2-form on a manifold M. Prove that the dimension of the manifold M is even and the following equation holds:

$$\underbrace{\omega \wedge \cdots \wedge \omega}_{n \text{ times}} = \pm \frac{1}{n!} \sqrt{\det(\omega_{ij})} dx^1 \wedge \cdots \wedge dx^{2n}.$$

Here, $\dim M = 2n$.

13.37. Let on a manifold there exist a nondegenerate 2-form. Prove that the manifold is orientable.

13.38. Let θ, φ be the standard coordinates on the sphere. Are the differential forms

$$d\theta, \ d\varphi, \ \cos\theta \, d\theta, \ \sin\theta \, d\theta, \ \cos\theta \, d\varphi, \ \sin\theta \, d\varphi, \ d\theta \wedge d\varphi$$

smooth on the whole sphere? In which cases the Stokes formula can be applied to these forms?

13.39. Let Ω be a differential p-form, and let ω be a differential 1-form not equal to zero. Show that Ω is represented in the form $\Omega = \theta \wedge \omega$ if and only if $\Omega \wedge \omega = 0$.

13.40. On a manifold M, let a nondegenerate 2-form be given. Prove that on M, there exists a tensor field J of type $(1,1)$ satisfying the condition $J^2 = -E$, i.e., $J^i_k J^k_m = -\delta^i_m$. Such a field J is called the *almost complex structure* on M.

13.41. Show that the restriction of the form

$$\omega = \frac{x \, dy - y \, dx}{z^2}$$

to any cone centred at the origin is a closed form.

13.42. Calculate the integral of the form $\Omega = x^2 dy \wedge dz + y^2 dz \wedge dx + z^2 dx \wedge dy$ over the domain $D = \{-1 < u < 1, -1 < v < 1\}$ on the surface

$x = u + v$, $y = u - v$, $z = uv$. The coordinate system u, v is assumed to be positively oriented.

13.43. Write the exterior differential form

$$\Omega = \frac{x\,dy \wedge dz + y\,dz \wedge dx + z\,dx \wedge dy}{\left(x^2 + y^2 + z^2\right)^{3/2}}$$

on $\mathbb{R}^3 \setminus \{0\}$ in the cylindrical and spherical coordinates. Show that the integral of this form over a smooth surface that is smoothly projected on the unit sphere from the origin in a one-to-one way is equal to the area of the projection.

13.44. Consider the form

$$\omega = \frac{x\,dy - y\,dx}{x^2 + y^2}.$$

Calculate $d\omega$ and $\displaystyle\int_{x^2+y^2=1} \omega$. Deduce from this that the form ω is closed but is not exact in $\mathbb{R} \setminus \{0\}$.

13.45. Let

$$\omega = \frac{x\,dy \wedge dz + y\,dz \wedge dx + z\,dx \wedge dy}{\left(x^2 + y^2 + z^2\right)^{3/2}}.$$

Calculate $d\omega$. Show that in the half-space $z > 0$, there exists a 1-form θ such that $d\theta = \omega$. Show that in $\mathbb{R}^3 \setminus \{0\}$, there is no 1-form θ such that $d\theta = \omega$.

13.46. Prove that the restriction of a closed (exact) form on a manifold M to a submanifold $N \subset M$ is a closed (exact) form.

13.47. Prove that $H^0(\mathbb{R}) = \mathbb{R}$ and $H^i(\mathbb{R}) = 0$ for $i \neq 0$.

13.48. Calculate the 0-dimensional de Rham cohomologies of an arbitrary smooth manifold.

13.49. Prove that any form of degree n on $\mathbb{R}^n$ is exact.

13.50. P o i n c a r é   l e m m a. Show that

$$H^*\left(\mathbb{R}^{n+1}\right) = H^*\left(\mathbb{R}^n\right).$$

The proof can be performed according to the following scheme. Let i: $\mathbb{R}^n \to \mathbb{R}^n \times \mathbb{R}^1$ be defined by i: $x \mapsto (x, 0)$, and let p: $\mathbb{R}^n \times \mathbb{R}^1 \to \mathbb{R}^n$ be defined by p: $(x, t) \mapsto x$. Since $p \circ i = 1$: $\mathbb{R}^n \to \mathbb{R}^n$, it follows that $i^* \circ p^* = 1$: $H^*(\mathbb{R}^n) \to H^*(\mathbb{R}^n)$. If we show that $p^* \circ i^*$: $H^*\left(\mathbb{R}^{n+1}\right) \to H^*\left(\mathbb{R}^{n+1}\right)$ is also identical, then this implies that p^*: $H^*(\mathbb{R}^n) \to H^*\left(\mathbb{R}^{n+1}\right)$ and i^*: $H^*\left(\mathbb{R}^{n+1}\right) \to H^*(\mathbb{R}^n)$ are mutually inverse isomorphisms. Define the mapping S: $\Omega^p\left(\mathbb{R}^n \times \mathbb{R}^1\right) \to \Omega^{p-1}\left(\mathbb{R}^n \times \mathbb{R}^1\right)$ as follows. The form $\omega \in \Omega^p\left(\mathbb{R}^n \times \mathbb{R}^1\right)$ is uniquely represented as the sum $\omega = \omega_1 + \omega_2$, where the form ω_1 does not contain dt and ω_2 contains it. The form ω_2 has the form $p^*\theta \cdot f(x, t)\,dt$, where $\theta \in \Omega^{p-1}(\mathbb{R}^n)$. We set $S(\omega) = p^*\theta \displaystyle\int_0^t f(x, t)\,dt$.

(a) Show that $(dS - Sd)\,\omega = (-1)^{p-1}\,\omega$, where p is the degree of the form ω.

(b) Show that the mapping

$$1 - p^* \circ i^* : \Omega^* \left(\mathbb{R}^n \times \mathbb{R}^1\right) \to \Omega^* \left(\mathbb{R}^n \times \mathbb{R}^1\right)$$

transforms closed forms into exact forms. Deduce from this that $p^* \circ i^*$: $H^* \left(\mathbb{R}^n \times \mathbb{R}^1\right) \to H^* \left(\mathbb{R}^n \times \mathbb{R}^1\right)$ is the identical mapping.

13.51. Let $f_t \colon X \times [0,1] \to Y$ be a smooth mapping, ω be a differential form on Y, and let $d\omega = 0$. Prove that

$$f_0^* \left(\omega\right) - f_1^* \left(\omega\right) = d\Omega$$

for an appropriate form Ω on X.

13.52. Prove that if $f \colon X \to Y$ is a homotopy equivalence, i.e., there exists a mapping $g \colon Y \to X$ such that $f \circ g \sim \mathrm{id}_Y$ and $g \circ f \sim \mathrm{id}_X$, then f^*: $H^* \left(Y\right) \to H^* \left(X\right)$ is an isomorphism.

13.53. Show that if a manifold X is contractible, then for any closed form ω (i.e., $d\omega = 0$), the equation $d\Omega = \omega$ is solvable.

13.54. Prove that the n-dimensional cohomologies of a smooth closed n-dimensional orientable manifold are nontrivial.

13.55. Let M be a closed compact symplectic manifold, i.e., on this manifold there exists a nondegenerate closed 2-form $\omega = w_{ij}dx^i \wedge dx^j$. Prove that the second cohomology group of the manifold M is nontrivial.

13.56. Let $M = M_1 \cup M_2$ be a disconnected sum of two manifolds of the same dimension. Prove that for all p, there exists a decomposition $H^p \left(M\right) = H^p \left(M_1\right) \oplus H^p \left(M_2\right)$ into a direct sum.

13.57. Let $M = M_1 \cup M_2$, where M_1 and M_2 are open submanifolds, and let the intersection $M_1 \cap M_2$ be diffeomorphic to $\mathbb{R}^n$. Prove that $H^p \left(M\right) = H^p \left(M_1\right) \oplus H^p \left(M_2\right)$ for $p > 0$.

13.58. (a) Prove that a closed form ω on $M^n \times \mathbb{R}^m$ is exact if and only if for a point $t \in \mathbb{R}^m$, the restriction of the form ω to $M^n \times \{t\}$ is an exact form.

(b) Present an example of a manifold M^n and a differential form $\omega \in \Omega^p \left(M^n \times \mathbb{R}^m\right)$ such that for all $t \in \mathbb{R}^m$, the restriction of ω to $M^n \times \{t\}$ is closed, but the form ω itself is not closed.

13.59. (a) Show that the form $\omega \in \Omega^1 \left(S^1\right)$ is exact if and only if $\int_{S^1} \omega = 0$.

(b) Show that the form $\omega \in \Omega^n \left(S^n\right)$ is exact if and only if $\int_{S^n} \omega = 0$.

13.60. Prove that the formula $\omega \mapsto \int_{S^n} \omega$ defines an isomorphism of $H^n \left(S^n\right) = \mathbb{R}$.

13.61. Calculate the cohomologies of S^n.

13.62. Calculate the cohomologies

(a) of the torus T^2;

(b) of the plane $\mathbb{R}^2$ without k points;

(c) of the Klein bottle;

(d) of the projective plane $\mathbb{R}P^2$;

(e) of the space $\mathbb{R}^n$ without k points.

13.63. Prove that on the sphere S^n, where $n > 1$, there are no closed 1-forms that are nonzero at each of its points.

13.64. Let M be a closed manifold on which there exists a nondegenerate closed 2-form. Prove that $\dim H^2(M) > 0$.

13.65. (a) Let $\mathbf{X}_1$, $\mathbf{X}_2$, and $\mathbf{X}_3$ be pairwise commuting vector fields on the sphere S^3. Prove that they cannot be linearly independent at all points of the sphere.

(b) Present an example of two vector fields on the sphere S^3 that are linearly independent at each of its point and commute with each other.

13.66. B r o u w e r t h e o r e m.

(a) Prove that there is no continuous mapping $r\colon \overline{D}^n \to \partial\overline{D}^n = S^{n-1}$ leaving the points of the sphere S^{n-1} fixed. For this purpose, consider the composition of r and $i\colon S^{n-1} \to \overline{D}^n$ (embedding of S^{n-1} in the boundary sphere of the ball D^n) and the mapping induced by it in the cohomologies of the sphere. Here, $\overline{D}^n$ is a closed ball.

(b) Prove that any continuous self-mapping of the ball $\overline{D}^n$ necessarily has a fixed point.

13.67. Consider the sphere S^2 and the equator S^1 on it. Prove that there is no continuous mapping $S^2 \to S^1$ such that it leaves the points of S^1 fixed.

13.68. (a) Prove that there is no continuous mapping $f\colon S^2 \to S^1$ such that $f(-x) = -f(x)$ for all $x \in S^2$.

(b) Let $f\colon S^2 \to \mathbb{R}^2$ be a continuous mapping such that $f(-x) = -f(x)$ for all $x \in S^2$. Prove that there exists a point $x \in S^2$ at which $f(x) = 0$.

13.69. Let M be an orientable manifold with boundary ∂M. Prove that there exists no continuous mapping $M \to \partial M$ that leaves the points of the boundary fixed. The proof can be performed analogously to that of the Brouwer theorem.

§ 14. Topology

14.1. Prove the equivalence of the following two definitions of the interior $\operatorname{Int} Y$ of a subset Y of a topological space X:

(a) $\operatorname{Int} Y$ is the set of all interior points of the space Y, i.e., points that are contained in Y, together with their neighborhoods;

(b) $\operatorname{Int} Y$ is the maximal open set contained in Y, i.e., $\operatorname{Int} Y = \bigcup_{\substack{U \in \tau(X),\\ U \subseteq Y}} U$.

14.2. Prove the equivalence of the following two definitions of the closure $\overline{Y}$ of a subset Y of a topological space X:

(a) $\overline{Y}$ is the set of adherent points of the set Y, i.e., points such that in any neighborhood of each of which there are points of Y;

(b) $\overline{Y}$ is the minimal closed set containing Y, i.e.,

$$\overline{Y} = \bigcap_{\substack{F \text{ is closed,} \\ F \supseteq Y}} F.$$

14.3. Prove that Int Y is an open set and $\overline{Y}$ is a closed set.

14.4. Prove that

(a) Y is open if and only if $Y = \text{Int } Y$;

(b) Y is closed if and only if $Y = \overline{Y}$;

14.5. Prove that $\overline{A \cup B} = \overline{A} \cup \overline{B}$. Show that $\overline{A \cap B}$ may not coincide with $\overline{A} \cap \overline{B}$.

14.6. Show that Int $(A \cup B)$ may not coincide with Int $A \cup$ Int B. Prove that Int $(A \cap B) = $ Int $A \cap$ Int B.

14.7. Let $G \subset \mathbb{R}^1$ be an open set on the real line. Prove that G is a union of disjoint intervals.

14.8. Prove that the Cantor set on the closed interval $[0, 1]$ is closed.

14.9. Let a set X consist of three points. Describe all possible topologies on it. How many of them are Hausdorff? How many of them are connected? How many of them are simply connected?

14.10. Prove that all singletons are closed in a Hausdorff topological space.

14.11. Prove that the topological product of two Hausdorff topological spaces is Hausdorff.

14.12. Prove that a topological space X is Hausdorff if and only if the diagonal $\Delta = \{(x_1, x_2) \,|\, x_1 = x_2\}$ is closed in $X \times X$.

14.13. Let X be an infinite set. Its finite subsets will be called closed. Verify that the resulting topology is not Hausdorff.

14.14. Let X be a metric space. Prove that every singleton is closed.

14.15. Prove that a metric topological space satisfies the Hausdorff separability axiom.

14.16. Is it true that the distance between two disjoint closed sets on the plane (real line) is always greater than 0?

A self-mapping $f\colon X \to Y$ of a metric space X is said to be *contractive* if there exists a constant $\lambda < 1$ such that $\rho(f(x), f(y)) \leq \lambda \rho(x, y)$ for any two points $x, y \in X$.

14.17. Prove that any contractive mapping of a metric space is continuous.

14.18. Prove that any contractive self-mapping of a complete metric space always has a fixed point, and this point is unique.

14.19. Present an example showing that it is not possible to reject the completeness condition of a metric space in Problem 14.18.

14.20. Prove that for any compact set $K \subset \mathbb{R}^n$, there exists a smooth real-valued function f such that $K = f^{-1}(0)$.

14.21. Present an example of a metric space in which there exist two balls such that the ball of greater radius is strictly contained in the ball of lesser radius.

14.22. Let (X, ρ_X) and (Y, ρ_Y) be two metric spaces. On $X \times Y$, define the following distances:

$$\rho_{\max} \left((x_1, y_1), (x_2, y_2) \right) = \max \left(\rho_X (x_1, x_2), \rho_Y (y_1, y_2) \right),$$

$$\rho_2 \left((x_1, y_1), (x_2, y_2) \right) = \left(\rho_X^2 (x_1, x_2) + \rho_Y^2 (y_1, y_2) \right)^{1/2},$$

$$\rho_+ \left((x_1, y_1), (x_2, y_2) \right) = \rho_X (x_1, x_2) + \rho_Y (y_1, y_2).$$

Prove that

(a) these are metrics on $X \times Y$;

(b) the topologies on $X \times Y$ corresponding to them coincide.

14.23. Prove the equivalence of the following definitions of continuity of a mapping $f\colon X \to Y$ of topological spaces. The mapping f is continuous if and only if

(a) for any open $U \subset Y$, the inverse image $f^{-1}(U) \subset X$ is also open;

(b) for any closed $F \subset Y$, the inverse image $f^{-1}(F) \subset X$ is also closed;

(c) for any point $x \in X$ and any neighborhood U of its image $f(x) \in Y$, there exists a neighborhood V of the point x such that $f(V) \subseteq U$;

(d) the graph of f, i.e., the set $\{(x, y) : f(x) = y\}$, is closed in $X \times Y$ provided that Y is Hausdorff.

14.24. Let $f\colon E \to F$, $E = A \cup B$, $A = \overline{A}$, and let $B = \overline{B}$. Then f is continuous if and only if $f|_A$ and $f|_B$ are continuous. If $A \neq \overline{A}$, then this is not true in general. Present an example.

14.25. Prove that if $f_n\colon X \to Y$ is a sequence of continuous mappings and f_n uniformly converge to f (here, X and Y are metric spaces), then f is continuous.

14.26. Let a mapping $f\colon X \to X$ be continuous, and let the topological space X be Hausdorff. Prove that the set $\{x \in X : x = f(x)\}$ of fixed points of the mapping f is closed. Present an example showing that the Hausdorff requirement cannot be rejected.

14.27. Let a topological space Y be Hausdorff, and let mappings $f\colon X \to Y$ and $g\colon X \to Y$ be continuous. Prove that the set $\{x \in X : f(x) = g(x)\}$ is closed. Show that the Hausdorff requirement cannot be rejected.

14.28. Let A be an everywhere dense set in a topological space X, and let Y be Hausdorff. Let continuous mappings $f\colon X \to Y$ and $g\colon X \to Y$ be such that $f|_A = g|_A$. Prove that $f(x) = g(x)$ for all $x \in X$.

14.29. Prove that any infinite closed set of a real line is the closure of its countable subsets.

14.30. Present an example of a metric space X such that its subset $Y \subset X$ (find such Y) is closed and bounded but is not a compact set.

14.31. Let a mapping $f \colon E \to F$ be a continuous mapping "onto", and let E be compact. Prove that F is compact.

14.32. Prove that the cube I^n is a compact space.

14.33. Prove that the n-dimensional sphere $(n < \infty)$ is compact. Is this true for $n = \infty$?

14.34. Let $X \subset Y$, and let Y be a compact space. Prove that X is a compact space if and only if X is a closed subspace.

14.35. Prove that the group of orthogonal matrices of size 3×3 is a compact topological space.

14.36. Prove that the group of orthogonal transformations of the n-dimensional Euclidean space is a compact topological space.

14.37. Reveal whether or not $GL(n, \mathbb{R})$, $SL(n, \mathbb{R})$, $SL(n, \mathbb{C})$, $U(n)$, $SU(n)$, and $SO(n)$ are compact.

14.38. Let X be a compact space, Y be a metric space, and let $f \colon X \to Y$ be a continuous mapping. Prove that f is a uniformly continuous mapping.

14.39. Let A and B be connected subsets of a topological space, and, moreover, let $A \cap \overline{B} \neq \varnothing$. Prove that $A \cup B$ is connected.

11.40. Prove that if E and F are connected, then $E \times F$ is connected.

14.41. Let $f \colon X \to Y$ be a continuous mapping, and let its image coincide with Y. Prove that if X is connected (simply connected) then Y is also connected (simply connected).

14.42. Prove that

(a) the intervals $0 < x < 1$, $0 \le x \le 1$, and $0 \le x < 1$ are connected and simply connected;

(b) if a set $A \in \mathbb{R}^1$ is connected, then A has one of the following forms: $a < x < b$, $a \le x \le b$, $a < x \le b$, or $a \le x < b$, where a and b can take the values $\pm\infty$.

14.43. Prove that the cube I^n and the sphere S^n are connected.

14.44. Let X be a compact connected metric space. Is it possible to connect two of its points by a continuous path?

14.45. Prove that $SO(n)$ is a connected topological space and $O(n)$ consists of two connected components. Prove that $U(n)$ and $SU(n)$ are connected topological spaces.

14.46. Prove that the group $GL(n, \mathbb{C})$ considered as a subset in the space of all complex matrices of size $n \times n$ is an open and connected subset.

14.47. Prove that the group $GL^+(n, \mathbb{R})$ of real matrices of size $n \times n$ with positive determinant is a connected topological space.

14.48. Prove that the group $GL(n, \mathbb{R})$ of real nonsingular matrices of size

$n \times n$ is a topological space consisting of two connected components.

14.49. Prove that the closure of a connected subset of a topological space is connected.

14.50. Let A and B be subsets of a topological space X. Is it true that if $A \cup B$ and $A \cap B$ are connected, then A and B are also connected? A similar question is about the simple connectedness.

14.51. Let A and B be simultaneously open or closed subsets of a topological space X. Prove that if $A \cup B$ and $A \cap B$ are connected, then A and B are also connected.

14.52. (a) Prove that using the operations of closure and passing to the interior of a set, one can obtain no more than seven different subsets from a given subset of a topological space. Present an example of a subset of $\mathbb{R}$ from which exactly seven different subsets are obtained.

(b) Prove that using the operations of closure and passing to the complement, one can obtain no more than 14 different subsets from a given subset of a topological space.

14.53. Let Y be a connected (simply connected) subspace of X. Study the connectedness (simple connectedness) of the interior $\operatorname{Int} Y$ and the boundary $\partial Y = \bar{Y} \setminus \operatorname{Int} Y$.

14.54. Prove that if X is a connected compact space and $f: X \to \mathbb{R}^1$ is a continuous mapping, then the image of f is a closed interval.

14.55. Let A be an open subset of $\mathbb{R}^n$. Prove that the following conditions are equivalent:

(a) A is connected;

(b) A is simply connected;

(c) any two points of A can be connected by a broken line consisting of finitely many segments and entirely lying in A.

14.56. Prove that the following set A is connected but not simply connected.

(a) A consists of the graph of the function $y = \sin \dfrac{1}{x}, x \neq 0$, and the segment $\{(0, y) : y \in [-1, 1]\}$ (the coordinates are Cartesian);

(b) A consists of the circle $r = 1$ and the spiral $r = e^{-1/\varphi}$, $\varphi > 0$.

14.57. Present an example of two metric spaces X and Y and two mappings $f: X \to Y$ and $g: Y \to X$ such that f and g are mutually one-to-one and continuous, but, nevertheless, the spaces X and Y are not homeomorphic.

Recall that the mutual one-to-one property means that the corresponding mapping is surjective and injective, simultaneously.

14.58. Present an example of a continuous bijective mapping $f: X \to Y$ of two homeomorphic topological spaces that is not a homeomorphism.

14.59. Prove that the open disk $x^2 + y^2 < 1$ and the plane $\mathbb{R}^2$ (x, y) are homeomorphic. Show that the open square $\{|x| < 1, |y| < 1\}$ and the plane

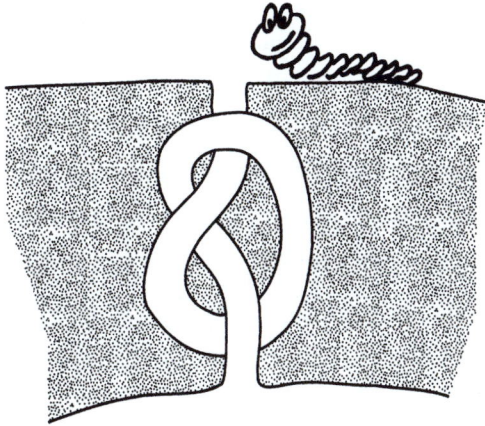

Figure 68

$\mathbb{R}^2 \, (x, y)$ are homeomorphic. Prove that the interval $0 < x < 1$ and the open square $\{|x| < 1, |y| < 1\}$ are not homeomorphic.

14.60. Prove that the cube $\{|x_i| \leq 1, i = 1, 2, \ldots, n\}$ and the ball $\{\sum_{i=1}^{n} x_i^2 \leq 1\}$ are homeomorphic. Prove that the open cube and the open ball are diffeomorphic.

14.61. Prove that the ball $\{\sum_{i=1}^{n} x_i^2 \leq 1\}$ and the upper hemisphere $\{\sum_{i=1}^{n+1} x_i^2 = 1, x_{n+1} \geq 0\}$ are homeomorphic.

14.62. Prove that the ellipsoid $\left\{\sum_{i=1}^{n+1} \dfrac{x_i^2}{a_i^2} = 1\right\}$ is homeomorphic to the sphere S^n.

14.63. Are the segment $0 \leq x \leq 1$ and the letter T homeomorphic?

14.64. Prove that the interval $(-1, 1)$ is homeomorphic to the line $(-\infty, \infty)$. Prove that any two intervals are homeomorphic.

14.65. Are the ball and the sphere homeomorphic?

14.66. Prove the following homeomorphisms:
(a) $\mathbb{R}^n \setminus \mathbb{R}^k \approx S^{n-k-1} \times \mathbb{R}^{k+1}$;
(b) $S^n \setminus S^k \approx S^{n-k-1} \times \mathbb{R}^{k+1}$.

14.67. Prove that the following spaces are homeomorphic:

$$\mathbb{R}^3 \setminus S^1 \quad \text{and} \quad \mathbb{R}^3 \setminus (\{y = z = 0\} \cup \{(1, 1, 1)\}).$$

14.68. Prove that
(a) the spaces $\mathbb{R}^1$ and $\mathbb{R}^2$ are not homeomorphic;
(b) the spaces $\mathbb{R}^1$ and $\mathbb{R}^n$ are not homeomorphic.

14.69. Prove that for no topological space X its square $X \times X$ is homeomorphic to $\mathbb{R}$.

14.70. Let $f\colon D^2 \to D^2$ be a homeomorphism. Prove that then the boundary sphere is mapped onto the boundary sphere.

14.71. Prove that the sphere S^2 is homeomorphic to no subspace of $\mathbb{R}^2$.

14.72. Find the topological type of the surface obtained by gluing lines of the hexagon in accordance with the following word:

(a) $a\,b\,c\,a^{-1}b^{-1}c^{-1}$;

(b) $a\,b\,c\,a^{-1}b^{-1}c$.

14.73. Consider a spherical layer bounded by two concentric spheres. A topological worm gnaws through a passage in the spherical layer from one boundary sphere to the other, which is knotted the way shown in Fig. 68. Find the topological type of the surface bounding the obtained spatial body.

14.74. Give a classification of the letters of the Russian alphabet with accuracy up to a homeomorphism.

§ 15. Homotopy, Degree of Mapping and Index of Vector Field

15.1. Prove that $f\colon X \to Y$ is homotopic to a mapping into a point if
(a) $X = \mathbb{R}^n$; (b) $Y = \mathbb{R}^n$.

15.2. Let f be a self-mapping of the closed interval $[0,1]$, and, moreover, let $f(0) = 0$ and $f(1) = 1$. Prove that there exists a homotopy fixed at the endpoints of the closed interval and deforming the mapping f into the identity mapping.

15.3. Does a vector space $\mathbb{R}^n$ contract into a point along itself?

15.4. Let a space X contract into a point along itself. Prove that any two paths with the same endpoints are homotopic to each other (the homotopy is fixed at the endpoints).

15.5. Prove that on the sphere S^n, $n > 1$, any two paths are homotopic (the endpoints are the same, and the homotopy is fixed at the endpoints).

15.6. Give a classification of letters of the Russian alphabet with accuracy up to a homotopy equivalence.

15.7. (a) Prove that a two-dimensional plane annulus is homotopy equivalent to the circle.

(b) Prove that the Möbius band is homotopy equivalent to the circle.

15.8. (a) Prove that the space $\mathbb{R}^2 \setminus \{0\}$ obtained by removing a point from the plane is homotopy equivalent to the circle.

(b) Prove that the space $\mathbb{R}^n \setminus \{x_1, \ldots, x_k\}$, where $x_1, \ldots, x_k$ are distinct points of $\mathbb{R}^n$, is homotopy equivalent to the union of k spheres of dimension $n-1$.

Recall the definition of the *union* of connected spaces $X_1, \ldots, X_k$. Chose a point $p_i \in X_i$ in each of them. In the disconnected union $X_1 \cup X_2 \cup \cdots \cup X_k$, identify all

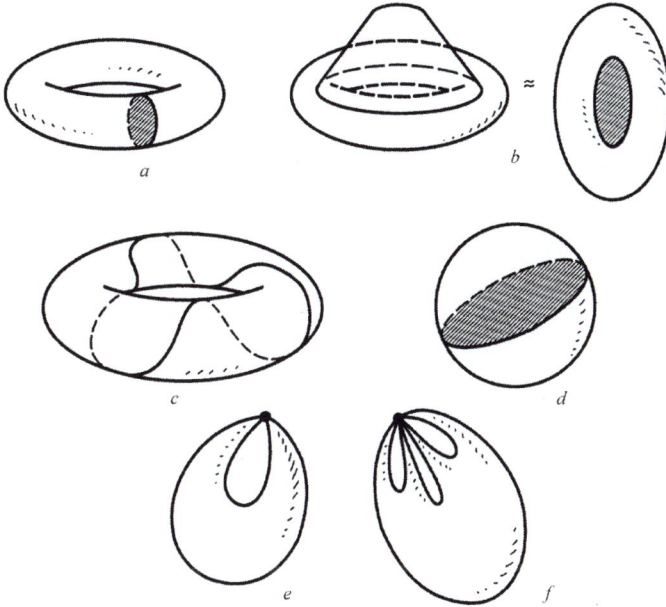

Figure 69

the points $p_1, \ldots, p_k$ (and only them). The obtained space is called the union of the spaces $X_1, \ldots, X_k$ and is denoted by $X_1 \vee X_2 \vee \cdots \vee X_k$.

15.9. Prove that the two-dimensional torus without a point is homotopy equivalent to the union $S^1 \vee S^1$.

15.10. Reveal to which the n-dimensional torus without a point is homotopy equivalent.

15.11. Reveal to which the projective space $\mathbb{R}P^n$ without a point is homotopy equivalent.

15.12. Reveal to which the following spaces are homotopy equivalent (see Fig. 69):

(a) the torus whose meridian is glued up by the disk;

(b) the torus whose parallel is glued up by the disk;

(c) the torus $\left(e^{i\varphi}, e^{i\psi}\right)$ on which the trajectory $\left(e^{2\pi it/p}, e^{2\pi it/q}\right)$, where p and q are coprime integers, is glued up by the disk;

(d) the sphere whose equator is glued up by the disk;

(e) the sphere two of whose points are identified;

(f) the sphere p of whose points are identified.

15.13. Consider simply connected sufficiently good spaces X and Y (for example, manifolds). In each of them, choose k pairwise different points $x_1, \ldots, x_k \in X$ and $y_1, \ldots, y_k \in Y$. Identify the points x_i and y_i for $i =$

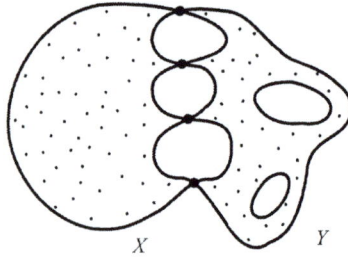

Figure 70

$1, \ldots, k$. Prove that the obtained space (see Fig. 70) is homotopy equivalent to $X \vee Y \vee S^1 \vee \cdots \vee S^1$, where the circle S^1 occurs $k - 1$ times.

15.14. Prove the following homotopy equivalencies:

(a) $\mathbb{R}^n \setminus \mathbb{R}^k \sim S^{n-k-1}$;

(b) $S^n \setminus S^k \sim S^{n-k-1}$;

(c) $\mathbb{R}^n \setminus S^1 \sim S^{n-k-1} \vee S^{n-1}$.

(d) In $\mathbb{R}^n$, consider pairwise disjoint affine subspaces $\Pi^{k_1}, \ldots, \Pi^{k_j}$ of dimensions $k_1, \ldots, k_j$, respectively. To the union of which spheres the space $\mathbb{R}^n \setminus (\Pi^{k_1} \cup \cdots \cup \Pi^{k_j})$ is homotopy equivalent?

(e)* In S^n, consider the tuple $S^{k_1}, \ldots, S^{k_j}$ of pairwise disjoint spheres. To which the space $S^n \setminus (S^{k_1} \cup \cdots \cup S^{k_j})$ is homotopy equivalent?

15.15. Prove that any vector field is homotopic to zero field.

15.16. Prove that on a connected manifold, a vector field with isolated singular points is homotopic to a vector field with a single singular point in the class of vector fields with isolated singularities.

15.17. Prove that if under a homotopy of a vector field $\mathbf{v}$ (in the class of fields with isolated singularities), two singular points X_1 and X_2 merge into one point, then its index is equal to the sum of indices $\operatorname{ind}_{X_1} \mathbf{v} + \operatorname{ind}_{X_2} \mathbf{v}$.

15.18. On the plane $\mathbb{R}^2$, depict the qualitative picture of distribution of integral trajectories of the flows

$$\mathbf{v}_1 = \operatorname{grad} \operatorname{Re} f(z), \quad \mathbf{v}_2 = \operatorname{grad} \operatorname{Im} f(z)$$

for the complex-analytic functions $f(z)$ listed below.

Find singular points of the flows $\mathbf{v}_1$ and $\mathbf{v}_2$.

Study the stability of singular points.

Depict the qualitative picture of behaviour of the trajectories of the flows $\mathbf{v}_1$ and $\mathbf{v}_2$ on the sphere S^2 (extended plane $\mathbb{R}^2 : S^2 = \mathbb{R}^2 \cup (\infty)$).

Depict the disintegration process of the singularity $z = 0$ of these vector fields under a small perturbation of the initial function $f(z)$ under which one obtains a function $g(z)$ for which all singular points of the flows $\mathbf{v}_1$ and $\mathbf{v}_2$ are nondegenerate:

(a) $f(z) = z^n$ (n is an integer);

(b) $f(z) = z + \dfrac{1}{z}$ (Zhukovskii function);

(c) $f(z) = z + \dfrac{1}{z^2}$;

(d) $f(z) = \dfrac{z+1}{z-2}$;

(e) $f(z) = \ln z$;

(f) $f(z) = \ln \dfrac{z-a}{z-b}$;

(g) $f(z) = z^4 \left(2\,(7-5)^2 + 12z^6\right)$ (study it in a neighborhood of $z = 0$);

(h) $f(z) = z^3\,(z-1)^{100}\,(z-2)^{900}$;

(i) $f(z) = 2z - \ln z$;

(j) $f(z) = 1 + z^4\,(z^4 - 4)^{44} \cdot (z^{44} - 44)^{444}$ (study it in a neighborhood of the point $z = 0$);

(k) $f(z) = \dfrac{1}{100} \ln \left(\dfrac{z - 2i}{z - 4}\right)^3$;

(l) $f(z) = \dfrac{1}{z^2 + 2z - 1}$;

(m) $f(z) = \dfrac{2}{z} + 21 \ln\left(z^2\right)$;

(n) $f(z) = z^5 + 2\ln z$;

(o) $f(z) = 2\ln(z-1)^2 - \dfrac{4}{3}\ln(z+10i)^3$;

(p) $f(z) = \dfrac{1}{z^3} - \dfrac{1}{(z-i)^3}$;

(q) $f(z) = \left(2 + \dfrac{5i}{2}\right) \ln \dfrac{4z - 2}{64z + i}$;

(r) $f(z) = \left(1 - \dfrac{i}{2}\right)^4 \ln \left(\dfrac{18z - i}{10z + 1}\right)^2$;

15.19. Find the index of singular points of the vector fields $\operatorname{grad}\operatorname{Re} f(z)$ and $\operatorname{grad}\operatorname{Im} f(z)$, where

(a) $f(z) = z^n$;

(b) $f(z) = z^{-n}$;

(c) $f(z) = \ln z$;

(d) $f(z) = z + \dfrac{1}{z}$;

(e) $f(z) = \ln \dfrac{z-a}{z-b}$.

15.20. Prove that on every connected compact closed manifold, there exists a smooth vector field having exactly one singular point. Find the index of this point as a singular point of a vector field.

15.21. Let a noncompact manifold M be an open domain in a compact closed manifold. Prove that on M, there exists a vector field without singular points.

15.22. Construct a vector field having exactly two singularities on the following oriented surfaces:

(a) on the sphere; (b) on the torus; (c) on the sphere with g handles.

15.23. Construct a smooth vector field with a single singular points on the following surfaces: (a) on the sphere; (b) on the torus; (c) on the pretzel; (d) on the sphere with g handles; (e) on the projective plane; (f) on the Klein bottle; (g) on the sphere with k Möbius bands. Find the indices of these singular points.

15.24. Prove that the index (rotation) of a vector field along a closed curve on the plane does not change under a homotopy of the curve not passing through singular points of the vector field.

15.25. Prove that the index of a vector field along a closed curve is equal to zero if the vector field has no singularities inside the contour.

15.26. Show that the index of a vector field along a closed curve is equal to the sum of indices of singular points of the vector field inside the contour.

15.27. Prove that there is no smooth vector field without singularities

(a) on the sphere S^2;

(b) on the projective plane $\mathbb{R}P^2$.

15.28. Prove that on the Klein bottle, there exists a smooth vector field without singularities. Prove that for any two smooth vector fields on the Klein bottle, there exists a point at which they are linearly dependent.

15.29. Find the degree of the following vector field $\mathbf{X}$ on the hypersurface Γ:

(a) $\xi = A\begin{pmatrix} x \\ y \end{pmatrix}$, where $A \in SO\,(2)$, and Γ: $\dfrac{x^2}{a^2} + \dfrac{y^2}{b^2} = 1$;

(b) $\xi = A_z(\varphi)\begin{pmatrix} x \\ y \\ z \end{pmatrix}$, where $A_z\,(\varphi)$ is the matrix of turn by the angle φ around the axis Oz, and Γ: $\dfrac{x^2 + y^2}{a^2} + \dfrac{z^2}{b^2} = 1$.

15.30. Find the degree of the vector field $\mathbf{X}$ on the hypersurface Γ, where

(a) $\mathbf{X} = \left(x^2 + y^2 - 1,\ \dfrac{x^2}{4} + 4y^2 - 1 \right)$ and Γ: $5x^2 - 6xy + 5y^2 - 3 = 0$;

(b) $\mathbf{X} = \left(2z^2 - y^2 - 4,\ 2x^2 + z^2 - 1,\ x^2 + y^2 + z^2 - 1 \right)$ and Γ: $x^2 + \dfrac{y^2}{9} + z^2 = 1$;

(c) $\mathbf{X} = \left(y^2 + 3z^2 - x^2 - 4,\ x^2 + 2y^2 + z^2 - 1,\ x^2 - y^2 + 2 \right)$ and Γ: $x^2 + y^2 + z^2 = 9$.

15.31. Show that the indices of two vector fields on an arbitrary 2-dimensional closed surface are equal to each other. Is this assertion true for a manifold of any dimension?

15.32. Let $\mathbf{v}(x)$ be a smooth vector field on the plane $\mathbb{R}^2$, L be a smooth self-intersecting contour on the plane $\mathbb{R}^2$, j_L be the index of L in the vector field $\mathbf{v}(x)$, j be the number of internal tangent points of the field $\mathbf{v}$ and the contour L, and let E be the number of external tangent points. Prove that if the number of all tangent points of the field and contour are finite, then $j_L \leq \frac{1}{2}(2 + J - E)$.

15.33. On a Riemannian manifold M^n, consider the vector field $\mathbf{v} = \operatorname{grad} f$, where f is a smooth function. Let $x_0 \in M^n$ be an isolated singular point of the field $\mathbf{v}$, i.e., $\mathbf{v}(x_0) = 0$. Let the Hessian $\left(\dfrac{\partial^2 f}{\partial x^i \partial x^j}\right)$ of the function f be a nonsingular matrix at the point x_0, and, moreover, let its index (the number of negative eigenvalues) be equal to λ. Prove that $\operatorname{ind}_{x_0}\mathbf{v} = (-1)^\lambda$.

15.34. Let $M^{n-1} \subset \mathbb{R}^n$ be a smooth surface, and let $\zeta\colon M^{n-1} \to S^{n-1}$ be the Gaussian mapping. Let l be the line passing through the origin and a point $x \in S^{n-1}$. Consider the height function $f_l\colon M^{n-1} \to \mathbb{R}$ given by the orthogonal projection of M^{n-1} on l.

(a) Prove that all critical points of f_l are inverse images of the points x and $-x$ under the Gaussian mapping ζ.

(b) Prove that the Hessian of the function f_l at these points coincides with the Jacobi matrix of ζ calculated at the same points.

15.35. Let $A \in GL_n(\mathbb{R})$. Find singular points and their indices of a vector field whose value at a point $x \in \mathbb{R}^n$ is equal to

(a) Ax; (b) $Ax + x_0$, where $x_0 \in \mathbb{R}^n$.

15.36. Find critical points and critical values of the mapping $f\colon S^1 \to S^1$, $f(z) = z^k$, $z \in S^1 \subset \mathbb{C}$, and $|z| = 1$. Find its degree.

15.37. (a) Find critical points and critical values of the mapping $f\colon S^2 \to S^2$, where $f(z) = z^3 + 3z^2 + 3z + 1$; here S^2 is the extended complex plane, i.e., $S^2 = \mathbb{C} \cup \{\infty\}$. Find its degree.

(b) Do the same for $f(z) = z^3 - 3z^2 - 9z + 27$.

(c) Consider the general case $f(z) = z^n + a_1 z^{n-1} + \cdots + a_n$, $a_k \in \mathbb{C}$, $k = 1, \ldots, n$.

15.38. (a) Find critical points and critical values of the mapping $f\colon S^2 \to S^2$ defined by the formula $f(z) = \dfrac{z^2 + 4}{z + 1}$. Find its degree.

(b) Do the same for $f(z) = \dfrac{z^2 + 4}{z^3 + 2z^2 + 5}$.

(c) Consider the general case $f(z) = \dfrac{P(z)}{Q(z)}$ for the polynomials $P(z) = z^n + a_1 z^{n-1} + \cdots + a_n$ and $Q(z) = z^m + b_1 z^{m-1} + \cdots + b_m$ with complex coefficients.

15.39. (a) For the mapping $SO(2) \to SO(2)$ given by the formula $A \mapsto A^7$, find critical points and critical values. Find the degree of the mapping.

(b) Do the same for the mapping $O(2) \to O(2)$ defined by the same formula;

(c) The mapping $SU(2) \to SU(2)$ is given by the formula $A \mapsto A^3$. Find its critical points, critical values and the degree.

(d) The mapping $SO(3) \to SO(3)$ is given by the formula $A \mapsto A^3$. Find its critical points, critical values, and degree.

(e) Solve Problems (a)–(d) for the mapping $A \mapsto A^k$.

15.40. Let X, Y, and Z be closed orientable manifolds, and let $f: X \to Y$ and $g: Y \to Z$ be smooth mappings. Prove that $\deg g \circ f = \deg g \cdot \deg f$.

15.41. Let M_1, M_2, N_1, and N_2 be orientable smooth manifolds, and, moreover, $\dim M_1 = \dim N_1$, $\dim M_2 = \dim N_2$. Let $f_1: M_1 \to N_1$ and $f_2: M_2 \to N_2$ be smooth mappings. Prove that $\deg f_1 \times f_2 = \deg f_1 \cdot \deg f_2$, where $f_1 \times f_2: M_1 \times M_2 \to N_1 \times N_2$.

15.42. Let $f: X \to Y$ be a homotopy equivalence, and, moreover, let X and Y be closed orientable manifolds of the same dimension. Prove that $\deg f = \pm 1$.

15.43. Prove that if $f: M \to N$ is homotopic to a mapping into a point, then f necessarily has critical points. Here, M and N are closed orientable manifolds of the same dimension.

15.44. Prove that closed orientable manifolds of different dimensions cannot be homotopy equivalent.

15.45. (a) Prove that a closed orientable manifold is not contractible.

(b) Does the assertion remain valid if we reject the orientability assumption?

15.46. Let $f: S^n \to S^n$ be the central symmetry.

(a) Compute $\deg f$.

(b) Prove that for odd n, the mapping f is homotopic to the identity mapping.

15.47. Let $f: S^n \to S^n$ be a continuous mapping. Let n be even. Prove that there exists a point $x \in S^n$ for which either $x = f(x)$ or $x = -f(x)$.

15.48. On the sphere S^2, let two functions f and g be given, and let $f(x) = -f(\tau x)$ and $g(x) = -g(\tau x)$, where τ is the central symmetry. Prove that f and g have a common zero.

15.49. Let two continuous functions $f(x)$ and $g(x)$ be given on S^2. Prove that there necessarily exists a point $x \in S^2$ such that $f(x) = f(\tau x)$ and $g(x) = g(\tau x)$, where τ is the symmetry of the sphere with respect to its centre.

15.50. Let f be a continuous mapping, $f: S^n \to S^n$, such that $f(x) \neq -x$ for any $x \in S^n$. Prove that

(a) $\deg f = 1$;

(b) if n is even, then there exists a point $x \in S^n$ such that $f(x) = x$.

15.51. Prove that a mapping $f: S^n \to S^n$ having no fixed points is homotopic to the central symmetry $\tau\colon x \mapsto -x$ of the sphere.

15.52. Let the degree of a mapping $f\colon S^n \to S^n$ be equal to $2k + 1$. Show that there exists a pair of diametrically opposite points of the sphere that pass to the diametrically opposite points under the mapping f, i.e., $f(-x) = -f(x)$.

15.53. Let $f\colon X \to S^n$ and $g\colon X \to S^n$ be two continuous mappings. Prove that if for all x, the points $f(x)$ and $g(x)$ are not diametrically opposite, then the mappings f and g are homotopic.

15.54. Recall that the manifold $\mathbb{R}P^n$ is orientable for $n = 2k + 1$. Find the degree of the canonical mapping $f\colon S^{2k+1} \to \mathbb{R}P^{2k+1}$.

15.55. Prove that the degree of any smooth mapping of S^{2k+1} into $\mathbb{R}P^{2k+1}$ is even.

15.56. Present an example of an explicit mapping $f\colon S^n \to S^n$ of degree k.

15.57. Prove that on the sphere S^{2k+1}, there is no even tangent vector field (i.e., a field such that $\mathbf{v}(x) = \mathbf{v}(-x)$) without singularities.

15.58. (a) Prove that there is no continuous self-mapping A of the vector space $\mathbb{R}^3$ such that the vectors x and $A(x)$ are orthogonal for all $x \in \mathbb{R}^3$.

(b) The same problem for $\mathbb{R}^4$.

15.59. Let M^2 be a closed embedded surface in $\mathbb{R}^3$. Let $f\colon M^2 \to S^2$ be the mapping of normals (Gaussian mapping). Let w and w' be area forms of the sphere S^2 and the surface M, respectively. Prove that $f^*(w) = Kw'$, where K is the Gaussian curvature of the surface M. Also, prove that

$$2 \deg f = \int_M K w'$$

and is equal to the Euler characteristic M.

15.60. (a) Construct a smooth mapping of degree 1 of the torus T^2 onto the sphere S^2. Geometrically describe this mapping for the torus and the sphere standardly embedded in $\mathbb{R}^3$.

(b) Construct a smooth mapping of the torus T^2 onto the sphere S^2 of degree k.

(c)* Reveal whether or not there exists a mapping of degree 1 of the sphere S^2 onto the torus T^2.

(d)* For which g_1 and g_2 there exists a smooth mapping of M_{g_1} onto M_{g_2} having degree 1, where M_g is a closed orientable surface of genus g (sphere with g handles)?

(e)* For which g_1 and g_2 there exists a smooth mapping of M_{g_1} onto M_{g_2} having a nonzero degree?

15.61. Prove that two smooth mappings $S^1 \to S^1$ are homotopic if and only if they have the same degree.

15.62. *Hopf theorem.* Let M^n be a closed orientable manifold. Prove that two mappings $M^n \to S^n$ are homotopic if and only if they have the same degree.

15.63. Present an example of a mapping of degree 0 not homotopic to a mapping into a point.

15.64. Let $P_n(z) = z^n + a_1 z^{n-1} + \cdots + a_n$ be a polynomial of degree n with complex coefficients. It defines the self-mapping of the sphere S^2, where the extended complex plane $\mathbb{C} \cup \{\infty\}$ is considered as S^2.

(a) Prove that the formula

$$H_1(z,t) = z^n + (1-t)\left(a_1 z^{n-1} + \ldots a_n\right),$$

where $t \in [0,1]$ defines the homotopy of the initial mapping to the mapping $z \mapsto z^n$.

(b) Let $a_n \neq 0$. Consider the mapping

$$H_2(z,t) = (1-t)(z^n + \ldots a_{n-1}z) + a_n.$$

This is a homotopy of the initial mapping to the mapping $z \mapsto a_n$, which obviously is of zero degree. Hence, the polynomial $P_n(z)$ has no complex roots. Find an error in these arguments.

Part 2

§ 16. Coordinate Systems (Supplementary Problems)

In Problems 16.1–16.4,
(a) find coordinate surfaces and coordinate lines;
(b) calculate the determinants

$$\left| \frac{\partial x_i}{\partial u_j} \right|, \quad \left| \frac{\partial u_i}{\partial x_j} \right|$$

and find at which points of the space $\mathbb{R}^3$ the one-to-one correspondence between the curvilinear and rectilinear Cartesian coordinates is violated for the below enumerated curvilinear coordinates u_1, u_2, u_3 of the space $\mathbb{R}^3$;
(c) are these coordinates orthogonal?

16.1. For the bipolar coordinate system defined by the equations

$$x_1 = \frac{a \sinh u_1}{\cosh u_1 - \cos u_2}, \quad x_2 = \frac{a \sin u_2}{\cosh u_1 - \cos u_2}, \quad x_3 = u_3,$$

where a is a constant factor.

16.2. For the bispherical coordinate system defined by the equations

$$x_1 = \frac{c \sin u_1 \cos u_3}{\cosh u_2 - \cos u_1}, \quad x_2 = \frac{c \sin u_1 \sin u_3}{\cosh u_2 - \cos u_1}, \quad x_3 = \frac{c \sinh u_2}{\cosh u_2 - \cos u_1},$$

where c is a constant factor, $0 \le u_1 < u_2$, $-\infty < u_2 < \infty$, $-\pi < u_3 \le \pi$.

16.3. For the system of elongated spheroidal coordinates defined by the equations

$$x_1 = c u_1 u_2, \quad x_2 = c\sqrt{(u_1^2 - 1)(1 - u_2^2)} \cos u_3,$$

$$x_3 = c\sqrt{(u_1^2 - 1)(1 - u_2^2)} \sin u_3,$$

where $|u_1| \ge 1$, $-1 \le u_2 \le 1$, $0 \le u_3 < 2\pi$, and c is a constant factor.

16.4. For the system of flattened spheroidal coordinates defined by the equations

$$x_1 = cu_1 u_2 \sin u_3, \quad x_2 = c\sqrt{(u_1^2 - 1)(1 - u_2^2)}, \quad x_3 = cu_1 u_2 \cos u_3,$$

where $|u_1| \geq 1$, $-1 \leq u_2 \leq 1$, $0 \leq u_3 < 2\pi$.

16.5. Transform the expression

$$y^2 \frac{\partial^2 V}{\partial x^2} - 2xy \frac{\partial^2 V}{\partial x \partial y} + x^2 \frac{\partial^2 V}{\partial y^2} - x \frac{\partial V}{\partial x} - y \frac{\partial V}{\partial y} + V$$

to the polar coordinates r, φ, i.e., $x = r \cos \varphi$, $y = r \sin \varphi$.

16.6. Transform the expression

$$\frac{\partial^2 V}{\partial x^2} + 2xy^2 \frac{\partial V}{\partial x} + 2(y - y^3) \frac{\partial V}{\partial y} + x^2 y^2 V = 0$$

to new coordinates u, v such that $x = uv$, $y = \dfrac{1}{v}$. Find the domain and the range of this coordinate system.

16.7. Calculate the Laplace operator $\dfrac{\partial^2 V}{\partial x^2} + \dfrac{\partial^2 V}{\partial y^2}$ in the coordinate system (u, v) such that $w = \ln \ln z$, where $w = x + iy$ and $z = u + iv$.

16.8. Calculate the Laplace operator $\dfrac{\partial^2 V}{\partial x^2} + \dfrac{\partial^2 V}{\partial y^2}$ in the coordinate system (u, v) such that $w = \ln \ln \dfrac{z - z_1}{z - z_2}$, where $w = x + iy$ and $z = u + iv$.

16.9. Calculate the Laplace operator $\dfrac{\partial^2 V}{\partial x^2} + \dfrac{\partial^2 V}{\partial y^2}$ in the coordinate system (u, v) such that $w = z^3 - 3z^2 + 1$, where $w = x + iy$ and $z = u + iv$.

§17. Curves and Surfaces: Equations and Parameterizations

Usually, in the present collection of problems a curve $\mathbf{r} = \mathbf{r}(t)$ is said to be smooth if there exist continuous derivatives of all orders of the vector function $\mathbf{r}(t)$. However, in some problems the class of smoothness is made more precise. Namely,

(a) a curve is said to be *smooth of class* C^n if it admits a parameterization $\mathbf{r} = \mathbf{r}(t)$, where the vector function $\mathbf{r}(t)$ has continuous derivatives up to the order n inclusive.

(b) a curve $\mathbf{r} = \mathbf{r}(t)$ is said to be *analytic* if the vector function $\mathbf{r}(t)$ is real-analytic, i.e., if it admits a convergent power series expansion.

17.1. (a) Present examples of irregular curves that admit an analytic parameterization.

(b) Prove that a curve admitting an analytic parameterization can have only those break points at which the velocity vector turns by the angle π.

(c) Prove that if a smooth curve does not admit an analytic parameterization, then at its singular points the angle of turn of the velocity vector can be arbitrary. Construct examples.

17.2. Prove that if a curve is defined by an analytic mapping of the circle into $\mathbb{R}^n$, $n \geq 2$, then its regularity can be violated only at finitely many points.

17.3. Represent the graph of the function $y = |x|$ as a curve of the following type:

(a) with a C^n-smooth parameterization for any natural n being not C^{n+1}-smooth;

(b) with a C^∞-smooth parameterization.

(c) show that for this curve, there is no analytic parameterization.

17.4. Let a plane curve be given by the mapping $x = t^k$, $y = t^n$, $k \leq n$, where k and n are natural numbers and $t \in (-1, 1)$. Study for which k and n this curve is

(a) regular but not analytic in any regular parameterization;

(b) regular analytic;

(c) irregular in any smooth parameterization.

In items (a) and (b), present corresponding regular parameterizations. In all three cases, draw the qualitative form of the curves in a neighborhood of the origin.

17.5. Let a spatial piecewise-regular curve consist of two plane arcs. Which maximal smoothness of the curve as a whole is possible?

17.6. Find a regular analytic parameterization of the Bernoulli lemniscate $\left(x^2 + y^2\right)^2 = a^2 \left(x^2 - y^2\right)$.

17.7. Is it possible to deform the lemniscate into a non-self-intersecting curve in the class of regular curves?

17.8. Let γ be a plane curve, M be a point of the curve γ, and let xOy be the rectangular coordinate system given in the plane of the curve. Denote by T and N the intersection points of the tangent and the normal to this curve with the axis Ox; let P be the projection of the point M on the axis Ox (see Fig. 71).

(a) Find the equation of the curve γ if the length of the segment PN is constant and equal to a.

(b) Find the equation of the curve γ if the length of the segment PT is constant and equal to a.

(c) Find the equation of the curve γ if the length of the segment MN is constant and equal to a.

17.9. Find the equation of the curve γ with the tangent of constant length $MT = a$ (see Fig. 72). Such a curve is called the *tractrix*.

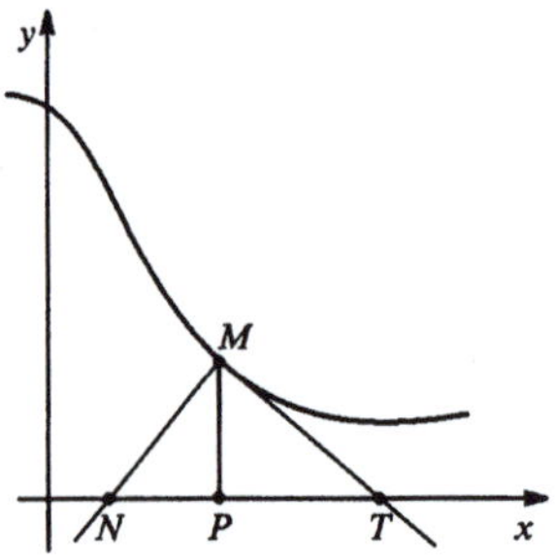

Figure 71 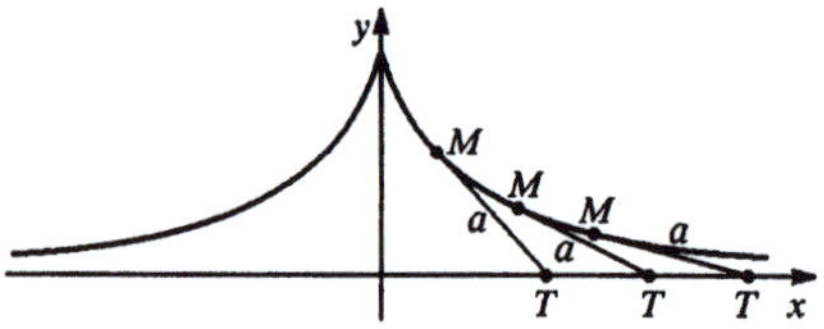   Figure 72

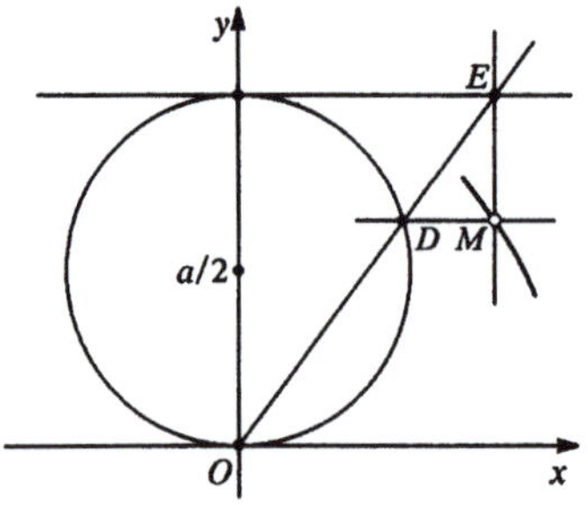

Figure 73 Witch of Agnesi

17.10. An arbitrary ray OE intersects the circle

$$x^2 + \left(y - \frac{a}{2}\right)^2 = \frac{a^2}{4}$$

at a point D, and it intersects its tangent passing through the point of γ diametrically opposite to O at a point E. The lines parallel to the axes Ox and Oy are drawn through the points D and E up to their intersection at a point M, respectively. Compose the equation of the curve formed by points M (*witch of Agnesi*) (see Fig. 73). Draw this curve.

17.11. Let γ be a closed smooth curve. Prove that for any vector $\mathbf{a}$, there exists a point $x \in \gamma$ at which the tangent to γ is orthogonal to $\mathbf{a}$.

17.12. Let two points move in space so that the distance between them remains constant. Prove that the projections of their velocities on the direction of the line connecting these points are equal to each other.

17.13. Prove that if on a segment $[a, b]$, a vector function $\mathbf{r}(t)$, together with its derivative $\mathbf{r}'$, is continuous, and, moreover, the vector $\mathbf{r}$ is parallel to $\mathbf{r}'$, and, in addition, $\mathbf{r}' \neq 0$ and $\mathbf{r} \neq 0$, then the hodograph of the vector function $\mathbf{r} = \mathbf{r}(t)$ is a line segment.

17.14. On a closed interval $[a, b]$, let a smooth vector function $\mathbf{r}(t)$ be given, the derivatives $\mathbf{r}'$ and $\mathbf{r}''$ be different from zero for all $t \in [a, b]$, and be

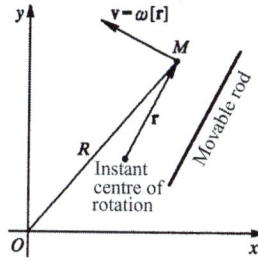

Figure 74 Motion of a rigid rod

collinear, i.e., the vector $\mathbf{r}'$ be parallel to the vector $\mathbf{r}''$ for all $t \in [a, b]$. Prove that the hodograph of this vector function $\mathbf{r} = \mathbf{r}\,(t)$ is a line segment.

17.15. Let a rigid rod move on the plane; moreover, let its endpoints circumscribe two curves $\mathbf{r}_1\,(t)$ and $\mathbf{r}_2\,(t)$. Find the equation of the fixed centroid. The *fixed centroid* is a set of intersection points of lines passing through the endpoints of the rod perpendicular to the velocity directions of its endpoints.

17.16. Consider the motion of a rod described in the previous problem. The *movable centroid* is the set of instant centres of rotation with respect to the movable rod. Compose the equation of the movable centroid if the laws of motion of the endpoints $\mathbf{r}_1\,(t)$ and $\mathbf{r}_2\,(t)$ of the rod are given.

17.17. Let the motion (isometry) of the plane be given by the motion of a rigid rod on the plane. Prove that the linear velocity $\mathbf{v}$ of a point is defined by the equation $\mathbf{v} = \omega\,[\mathbf{r}]$, where $\mathbf{r}$ is the radius-vector of the point $M\,(R)$ with respect to the instant centre of rotation (see **17.15**, **17.16**) and $[\mathbf{r}]$ is the vector obtained from $\mathbf{r}$ by turning by the angle $\pi/2$ counterclockwise. Express ω through $\mathbf{r}_1$ and $\mathbf{r}_2$; also find the velocity $\mathbf{v}$ of the point $M\,(R)$ (see Fig. 74).

17.18. Compose the equations of the normal and the tangent to the following curves:

(a) $\mathbf{r}\,(t) = (a\cos t, b\sin t)$ (ellipse);

(b) $\mathbf{r}\,(t) = \left(\dfrac{a}{2}\left(t + \dfrac{1}{t}\right), \dfrac{b}{2}\left(t - \dfrac{1}{t}\right)\right)$ (hyperbola);

(c) $\mathbf{r}\,(t) = \left(a\cos^3 t, a\sin^3 t\right)$ (astroid);

(d) $\mathbf{r}\,(t) = (a\,(t - \sin t), a\,(1 - \cos t))$ (cycloid);

(e) $\mathbf{r}\,(t) = \left(\dfrac{1}{2}t^2 - \dfrac{1}{4}t^4, \dfrac{1}{2}t^2 + \dfrac{1}{3}t^3\right)$ at the point $t = 0$;

(f) $\mathbf{r}\,(t)\,(a\varphi\cos\varphi, a\varphi\sin\varphi)$ (Archimedean spiral).

17.19. Prove that the length of the segment of the tangent to the astroid

$$x^{2/3} + y^{2/3} = a^{2/3}$$

contained between the coordinate axes is equal to a (see Fig. 75).

17.20. Prove that the cardioids given by the equations

$$r = a\,(1 + \cos\varphi), \quad r = a\,(1 - \cos\varphi)$$

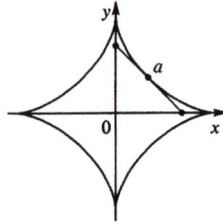

Figure 75 Astroid

in the polar coordinates are orthogonal at the intersection points different from the origin.

17.21. Find the envelope of the family of lines connecting the endpoints of pairs of the conjugate diameters of the ellipse.

17.22. Find the envelope of the family of lines that cut off a triangle of constant area from the sides of a right angle.

17.23. Find the envelope of the family of lines that cut off segments of a given area from a given parabola.

17.24. Find the envelope of the family of lines that cut off a triangle of a given perimeter from the sides of a given angle.

17.25. Find the envelope of the family of circles constructed on parallel chords of a circle as on diameters.

17.26. Find the envelope of the family of ellipses having common principal axes and a given sum of semiaxes.

17.27. On the plane, a mirror having the form of a circle is placed. A beam of parallel rays falls on it. Find the envelope of the reflected rays. This curve is called the *caustic*.

17.28. Find the envelope of the family of ellipses having a given area and common principal axes.

17.29. Find the envelope of the family of circles centred at an ellipse and passing through one of its foci.

17.30. Find the envelope of the family of circles of radius a centred at a curve $\mathbf{r} = \mathbf{r}(t)$.

17.31. Let a vector function $\mathbf{r}(t)$ be defined, continuous, and twice differentiable on a closed interval $[a, b]$. Let the vectors $\mathbf{r}'$ and $\mathbf{r}''$ be not collinear at each point of this closed interval. Find the envelope of the normals of the curve $\mathbf{r} = \mathbf{r}(t)$.

17.32. Find the envelope of rays reflected from a circle if a luminous point lies on the circle.

17.33. For the curve

$$\mathbf{r}(t) = \left(t^3 - t^2 - 5, \ 3t^2 + 1, \ 2t^3 - 16\right),$$

write the equation of the tangent line of the equation of the normal plane at the point for which $t = 2$.

17.34. Find the tangent line and the normal plane of the curve

$$\mathbf{r}(t) = \left(t^4 + t^2 + 1, \ 4t^3 + 5t + 2, \ t^4 - t^3\right)$$

at the point $A(3, -7, 2)$.

17.35. Find the tangent line and the normal plane to the curve

$$\mathbf{r}(t) = \left(t^2 - 2t + 3, \ t^3 - 2t^2 + t, \ 2t^3 - 6t + 2\right)$$

at the point $A(2, 0, -2)$.

17.36. Compose the equations of the tangent line and normal plane of the curve in $\mathbb{R}^3$ given by the transversal intersection of two surfaces

$$F_1(x, y, z) = 0, \quad F_2(x, y, z) = 0.$$

17.37. Find the length of the arc of the helical line

$$x = 3a\cos t, \quad y = 3a\sin t, \quad z = 4at,$$

going from the intersection point with the plane xOy up to an arbitrary point $M(t)$.

17.38. Find the arclength of one loop between two points of intersection of the curve

$$x = a(t - \sin t), \quad y = a(1 - \cos t), \quad z = 4a\cos\frac{t}{2}$$

with the plane xOy.

17.39. Find the length of the arc of the curve

$$x^3 = 3a^2 y, \quad 2xz = a^2$$

between the planes $y = a/3$ and $y = 9a$.

17.40. Find the length of the closed curve

$$x = \cos^3 t, \quad y = \sin^3 t, \quad z = \cos 2t.$$

17.41. At each point of a curve $\mathbf{r} = \mathbf{r}(t)$, a tangent vector $\mathbf{T} = \mathbf{T}(t) \neq 0$ is given. The function $\mathbf{r}(t)$ is defined and continuous and has a continuous derivative $\mathbf{r}'(t)$ on a closed interval $[a, b]$. The function $\mathbf{T}(t)$ is continuous on the closed interval $[a, b]$. Prove that it is possible to introduce a parameterization on this curve such that $\dfrac{d\mathbf{r}}{dt} = \mathbf{T}$. In other words, each sufficiently good vector field tangent to a curve can be considered as its velocity field under an appropriate choice of the parameter along the curve.

17.42. Find a necessary and sufficient condition under which the family of lines

$$\mathbf{r} = \rho(u) + \lambda \mathbf{e}(u) \quad (|\mathbf{e}| = 1)$$

in the space $\mathbb{R}^3$ admits an envelope. Find the envelope.

17.43. Compose the parametric equation of a cylinder for which the curve $\rho = \rho(u)$ is the directrix and the rulings are parallel to a vector **e**.

17.44. Compose the parametric equation of a cone with the vertex at the origin of a radius-vector $\rho(t)$ for which the curve $\rho = \rho(t)$ is the directrix.

17.45. Compose the parametric equation of a surface composed of tangents to a given curve $\rho = \rho(t)$. Such a surface is called a *developable surface*.

17.46. A circle of radius a moves in such a way that its centre moves along a given curve $\rho = \rho(s)$, and the plane in which it is located is the normal plane to this curve at each moment. Here, for simplicity of writing an answer, the parameter s is natural. Compose the parametric equation of the surface circumscribed by the circle.

17.47. A plane curve $x = \varphi(v)$, $z = \psi(v)$ rotates around the axis Oz. Compose the parametric equations of the surface of revolution. Consider a particular case where the meridian is given by the equation $x = f(z)$.

18.48. Compose the equation of the surface composed of the principal normals of the helical line.

17.49. Compose the equation of the surface composed of the family of principal normals to a given curve $\rho = \rho(s)$.

17.50. A line moves in such a way that the intersection point M of this line with a circle moves along the circle, and, moreover, the line remains in the normal plane to the circle at the corresponding point and turns by the angle equal to the angle $\widehat{MOM_0}$ which was past by a point moving along the circle. Compose the equation of the surface circumscribed by the moving line assuming that the initial position of the line is the axis Ox and the circle is given by the equations $x^2 + y^2 = a^2$, $z = 0$.

17.51. Two curves $\mathbf{r} = \mathbf{r}(u)$ and $\rho = \rho(v)$ are given. Compose the equation of the surface circumscribed by the midpoint of the segment whose endpoints lie on these curves. Such a surface is called a *translation surface*.

17.52. Assume that a plane (directing plane), a line parallel to it (directing line), and a curve (directing curve) are given. Draw the line l through each point of the directing curve parallel to the directing plane and intersecting the line. Consider the surface swept by the line l at its motion. Such a surface is called the conoid (Fig. 76). Compose the equation of the conoid if the following data are given: the directing plane yOz, the directing line $y = 0$, $z = h$, and the directing curve $\dfrac{x^2}{a^2} + \dfrac{y^2}{b^2} = 1$, $z = 0$ (ellipse).

17.53. Compose the equation of the conoid for which the directing line, directing plane and directing curve are, respectively, given by the equations

(a) $x = a$, $y = 0$; (b) $z = 0$; (c) $y^2 = 2pz$, $x = 0$.

17.54. The *cylindroid* is a surface composed of lines parallel to a plane (see Fig. 77). The cylindroid can be given by two directing curves (lying on it) and the directing plane (to which the rulings of the cylindroid are parallel). A

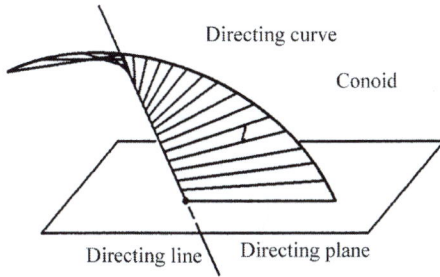

Figure 76 Conoid

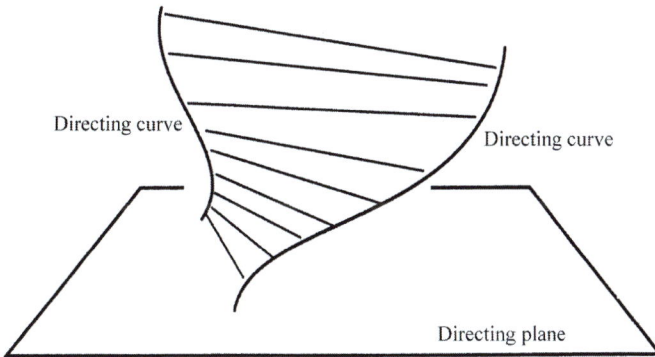

Figure 77 Cylindroid

particular case of the cylindroid is the conoid described in Problems 17.52 and 17.53. The cylindroid transforms into the conoid if one of the two directrices is a straight line. Compose the equation of the cylindroid if its directrices are the circles $x^2 + z^2 - 2ax = 0$, $y = 0$ and $y^2 + z^2 - 2ay = 0$, $x = 0$ and the directing plane is the plane xOy.

17.55. In $\mathbb{R}^3$, let a curve $\rho = \rho(u)$ be given, and at each of its points, let a vector $\mathbf{a}(u)$ be given. A *ruled surface* is a surface given by the parametric equation

$$\mathbf{r}(u, v) = \rho(u) + v\mathbf{a}(u).$$

The lines passing through the points of the curve $\rho(u)$ in the direction of the vector $\mathbf{a}(u)$ are called *rectilinear rulings* of the ruled surface. Note that the conoid and the cylindroid are ruled surfaces.

Compose the equation of a ruled surface whose rulings are parallel to the plane $y - z = 0$ and intersect the parabolas $y^2 = 2px$, $z = 0$ and $z^2 = -2px$, $y = 0$.

17.56. Compose the equation of the ruled surface whose rulings intersect the axis Oz, are parallel to the plane xOy, and intersect the curve $xyz = a^3$, $x^2 + y^2 = b^2$.

17.57. Compose the equation of the ruled surface whose rulings intersect the line $\mathbf{r} = \mathbf{a} + u\mathbf{b}$ and the curve $\rho = \rho(v)$ and are perpendicular to the vector $\mathbf{n}$.

17.58. Compose the equation of the ruled surface whose rulings are parallel to the plane xOy and intersect two ellipses

$$\frac{y^2}{b^2} + \frac{z^2}{c^2} = 1, \quad x = a;$$

$$\frac{y^2}{c^2} + \frac{z^2}{b^2} = 1, \quad x = -a.$$

17.59. Compose the equation of a ruled surface composed of lines intersecting the curve $\rho = \left(u, u^2, u^3\right)$, parallel to the plane xOy, and intersecting the axis Oz.

17.60. Compose the equation of a surface composed of lines parallel to the plane $x + y + z = 0$ and intersecting the axis Oz and the circle $\rho = (b, a \cos u, a \sin u)$.

17.61. Compose the parametric equations of a surface composed of lines intersecting the circle $x^2 + z^2 = 1$, $y = 0$ and the lines $y = 1$, $z = 1$ and $x = 1$, $z = 0$.

17.62. Compose the equation of a surface composed of tangent lines to the helical line $\rho(v) = (a \cos v, a \sin v, bv)$. Such a surface is called the *developable helicoid*.

17.63. Compose the equation of a conical surface with the vertex at the point $(0, 0, -c)$ and the ruling, the curve $\left(x^2 + y^2\right)^2 = a^2 \left(x^2 - y^2\right)$, lying in the plane $z = 0$.

17.64. In a plane π, a line AB and a curve $\rho = \rho(u)$ are given. The curve ρ uniformly moves in the plane π so that each of its point moves parallel to AB. At the same time, the plane π uniformly rotates around AB. Compose the equation of the surface circumscribed by the curve ρ. This surface is called the *helical surface*. The right helicoid is a particular case of the helical surface. In this case, $\rho = \rho(u)$ is the line orthogonal to AB.

17.65. Let $\mathbf{r} = \mathbf{r}(u)$ be a curve whose curvature k is different from zero. The normal plane is drawn through each of its points, and in this plane the circle of a given radius a, $a > 0$, $ak < 1$, centred at the curve $\mathbf{r} = \mathbf{r}(u)$ is constructed. These circles sweep a tube-type surface S in the space. Such surfaces are called *tubes* or *canal surfaces*.

(a) Compose the equation of the surface S.

(b) Prove that any normal of the surface S intersects the curve $\mathbf{r} = \mathbf{r}(u)$ and is perpendicular to the velocity vector of this curve.

17.66. Find the surface S knowing that all its normals intersect at one point O.

17.67. Show that the volume of the tetrahedron composed by the intersection of the coordinate planes and the tangent plane of the surface

$$x = u, \quad y = v, \quad z = \frac{a^3}{uv}$$

is independent of the choice of the tangent point on the surface.

17.68. Show that the sum of squares of lengths of the segments cut off by the tangent plane of the surface

$$x = u^3 \sin^3 v, \quad y = u^3 \cos^3 v, \quad z = \left(a^2 - u^2\right)^{3/2}$$

from the coordinate axes is constant.

17.69. Show that the tangent space to the conoid

$$x = u \cos v, \quad y = u \sin v, \quad z = a \sin 2v$$

intersects the conoid along the ellipse.

17.70. Prove that the plane tangent to the surface $z = xf\left(y/x\right)$ passes through the same point.

17.71. Compose the equation of a tangent plane and that of the normal to the helicoid

$$\mathbf{r}\left(u, v\right) = \left(v \cos u, v \sin u, ku\right).$$

17.72. Compose the equation of the tangent plane to the surface

$$xyz = a^3.$$

17.73. Let the surface be composed of tangents to a curve γ. Prove that this surface has the same tangent plane at all points of the same tangent to the curve γ.

17.74. Let the surface be composed of principal normals of a curve γ. Compose the equations of the tangent plane and normal at an arbitrary point of this surface.

17.75. Compose the equation of the tangent plane and normal to the surface composed of binormals of a curve γ.

17.76. Prove that the normal of a surface of revolution coincides with the principal normal of the meridian and intersects the axis of revolution.

17.77. Prove that if all normals of a surface intersect the same line, then the surface is a surface of revolution.

17.78. A ruled surface (for the definition, see Problem 17.55) is said to be *developable* if at all points of an arbitrary rectilinear ruling, the tangent plane to the surface is the same. Prove that the ruled surface

$$\mathbf{r}\left(u, v\right) = \rho\left(u\right) + v\mathbf{a}\left(u\right)$$

is developable if and only if

$$\left(\mathbf{r}', \mathbf{a}, \mathbf{a}'\right) = 0.$$

17.79. Prove that any developable surface can be partitioned into the following parts: (1) a part of a plane; (2) a part of a cylinder; (3) a part of a cone; (4) a part of a figure consisting of tangents to a non-plane curve. In the latter case, the curve is called the cuspidal edge.

17.80. Find the envelope and the cuspidal edge of the family of ellipsoids

$$\alpha^2 \left(\frac{x^2}{a^2} + \frac{y^2}{b^2} \right) + \frac{z^2}{c^2} = 1,$$

where α is the parameter of the family.

17.81. Find the envelope of a family of spheres constructed on chords parallel to the large axis of the ellipse

$$\frac{x^2}{a^2} + \frac{y^2}{b^2} = 1, \quad z = 0$$

as on diameters.

17.82. Find the envelope and the cuspidal edge of a family of spheres whose diameters are chords of the circle

$$x^2 + y^2 - 2x = 0, z = 0$$

passing through the origin.

17.83. Two parabolas are located in perpendicular planes and have a common vertex and a common tangent at the vertex. Find the envelope of a family of planes tangent to both parabolas.

17.84. Find the envelope of a family of spheres of constant radius centred at a given curve $\rho = \rho(s)$ (canal surface).

17.85. Find the cuspidal edge of a family of spheres of radius a centred at the curve $\rho = \rho(s)$.

17.86. Find the envelope and the cuspidal edge of a family of spheres of constant radius a centred at the circle

$$x^2 + y^2 = b^2, \quad z = 0.$$

17.87. Find the envelope and the cuspidal edge of a family of spheres passing through the origin and centred at the curve

$$\mathbf{r}(u) = \left(u^3, u^2, u \right).$$

17.88. Find the envelope of the family of ellipsoids

$$\frac{x^2}{a^2} + \frac{y^2}{b^2} + \frac{z^2}{c^2} = 1$$

with the given sum

$$a + b + c = l$$

of the semiaxes.

17.89. Find the surface whose tangent planes cut off segments from the coordinate axes such that the sum of the squares of their lengths is equal to a^2.

17.90. Find the surface whose tangent planes cut off a tetrahedron of constant volume a^3 from the coordinate angle.

17.91. Find the envelope and the cuspidal edge of a family of planes

$$xa^2 + y\alpha + z = 0,$$

where α is the parameter of the family.

17.92. Find the envelope and the cuspidal edge of a family of planes

$$x \sin \alpha - y \cos \alpha + z = \alpha C,$$

where α is the parameter of the family.

17.93. Find the envelope, the characteristics, and the cuspidal edge of a family of osculating planes of a given curve.

17.94. Find the envelope, the characteristics, and the cuspidal edge of a family of normal planes of a given curve.

17.95. Find the characteristics, the envelope, and the cuspidal edge of a family of planes

$$\langle \mathbf{r}, \mathbf{n} \rangle + D = 0, \quad \mathbf{n} = \mathbf{n}(u), \quad D = D(u), \quad |\mathbf{n}| = 1,$$

where u is the parameter of the family.

17.96. Find the developable surface passing through two parabolas

$$y^2 = 4ax, \quad z = 0$$

and

$$x^2 = 4ay, \quad z = b.$$

17.97. Show that the surface

$$x = \cos v - (u + v) \sin v, \quad y = \sin v + (u + v) \cos v, \quad z = u + 2v$$

is developable.

17.98. Show that the surface

$$x = u^2 + \frac{v}{3}, \quad y = 2u^3 + uv, \quad z = u^4 + \frac{2u^2 v}{3}$$

is developable.

17.99. The paraboloid

$$x = 2au \cos v, \quad y = 2bu \sin v, \quad z = 2u^2 \left(a \cos^2 v + b \sin^2 v \right),$$

where a and b are constants, is given. Define the equation of curves on the surface such that the tangent planes to the surface make an angle with the plane xOy constant along the curve.

Show that the characteristics of this family of tangent planes make a constant angle with the axis z. Find the cuspidal edge of the envelope.

17.100. Find the cuspidal edge of a developable surface tangent to the surface $az = xy$ along its intersection line with the cylinder $x^2 = by$.

17.101. Show that a developable surface passing through the circles $x^2 + y^2 = a^2$, $z = 0$ and $x^2 + z^2 = b^2$, $y = 0$ intersects the plane $x = 0$ along the equilateral hyperbola.

§18. Theory of Curves (Supplementary Problems)

18.1. Find the curvature of the curves:

(a) $\begin{cases} x\,(t) = a\,(1+m)\cos mt - am\cos(1+m)\,t, \\ y\,(t) = a\,(1+m)\sin mt - am\sin(1+m)\,t \end{cases}$ (epicycloid);

(b) $x^2 y^2 = \left(a^2 - y^2\right)(b+y)^2$ (conchoid).

18.2. Calculate the curvature of the following curves:

(a) $y = -\ln\cos x$;

(b) $x = 3t^2$, $y = 3t - t^3$ for $t = 1$;

(c) $x = a\,(\cos t + t\sin t)$, $y = a\,(\sin t - t\cos t)$ for $t = \pi/2$;

(d) $x = a\,(2\cos t - \cos 2t)$, $y = a\,(2\sin t - \sin 2t)$.

18.3. Find the curvature of the following curves given in the polar coordinates:

(a) $r = a\varphi$;

(b) $r = a\varphi^k$;

(c) $r = a^\varphi$ at the point $\varphi = 0$.

18.4. Find the curvature of the following curves:

(a) $y = \sin x$;

(b) $y = \sinh x$;

(c) $y = \cosh x$;

(d) $y = \tan x$;

(e) $y = \tanh x$;

(f) $y = \cot x$.

18.5. Find the curvature of the curve

$$x = t - \sin t, \quad y = 1 - \cos t, \quad z = 4\sin\frac{t}{2}.$$

18.6. Find the parametric equation of the curve

(a) $s^2 + 9R^2 = 16a^2$;

(b) $s^2 + R^2 = 16a^2$;

(c) $R^2 + a^2 = a^2 e^{-2s-a}$.

18.7. Prove that if coordinates of points of a curve satisfy the equation

$$\frac{\left(x^2 + y^2 + z^2 - a^2\right)\left(dx^2 + dy^2 + dz^2\right)}{\left(x\,dx + y\,dy + z\,dz\right)^2} = 1,$$

then the tangents to the curve also are tangent to the sphere $x^2 + y^2 + z^2 = a^2$.

18.8. Prove that tangents to the curve

$$\mathbf{r}\left(t\right) = \left(\frac{1}{2}\left(\sin t + \cos t\right),\ \frac{1}{2}\left(\sin t - \cos t\right),\ e^{-t}\right)$$

intersect the plane xOy along the circle $x^2 + y^2 = 1$.

18.9. Let O be the centre of an ellipse, A be the endpoint of one of the semiaxes $(OA = a)$, and let B be the endpoint of the other semiaxis $(OB = b)$. Consider a point C such that $CAOB$ is a rectangle. The perpendicular dropped from C on AB is extended up to the intersection with AO at a point P and with BO at a point Q. Prove that P is the curvature centre of the ellipse at the vertex A and Q is the curvature centre at the point B.

18.10. Write the equation of an osculating plane to the curve $\mathbf{r}\left(u\right) = \left(u^2,\ u,\ u^3 - 20\right)$ at the point $A\left(9, 3, 7\right)$.

18.11. Show that the curve

$$\mathbf{r}\left(u\right) = \left(au + b,\ cu + d,\ u^2\right)$$

has the same osculating plane at all points.

18.12. Compose the equations of the osculating plane, the principal normal, and the binormal of the curve

$$y^2 = x,\quad x^2 = z$$

at the point $(1, 1, 1)$.

18.13. At each point of the curve

$$x\left(t\right) = t - \sin t,\quad y\left(t\right) = 1 - \cos t,\quad z\left(t\right) = 4\sin\frac{t}{2}$$

a segment of length equal to the quadruple curvature of the curve at this point is taken in the positive direction. Find the equation of the osculating plane of the curve circumscribed by the endpoint of the segment.

18.14. Given the helical line

$$\mathbf{r}\left(t\right) = \left(a\cos t,\ a\sin t,\ bt\right),$$

compose the equations of the tangent, the normal plane, the binormal, the osculating plane, and the principal normal.

18.15. Given the curve

$$\mathbf{r}(t) = \left(t^2, 1 - t, t^3\right),$$

compose the equations of the tangent, the normal plane, the binormal, the osculating plane, and the principal normal at the point $t = 1$.

18.16. Prove that

(a) if all osculating planes of a regular smooth curve are parallel to one another, then the curve is a plane curve;

(b) if all osculating planes of a regular smooth curve have a common point, then the curve is a plane curve;

(c) if all tangents to a curve are perpendicular to one direction, then the curve is a plane curve;

(d) if all normal planes are parallel to the same direction, then the curve is a plane curve.

18.17. Prove that in a neighborhood of a point at which the torsion of a curve is different from zero, the properties presented below hold. In both cases, find the principal terms of the deviation d of the curve from these planes as compared with the arclength calculated from the point considered.

(a) A curve intersects its osculating plane lying to its different sides.

(b) A curve lies to one side of its rectifying plane.

18.18. Calculate the curvature and torsion radii of the curve

$$x^3 = 3a^2 y, \quad 2xz = a^2.$$

18.19. Let a curve

$$\mathbf{r}(u) = (v \cos u, \, v \sin u, \, kv),$$

where $v = v(u)$, be given. Prove that this curve is located on a cone. Define the function $v(u)$ so that this curve intersects the rulings of the cone at a constant angle θ.

18.20. The *generalized helix* or the *slope curve* is a spatial curve whose tangents make a constant angle with a fixed direction. Prove that a curve is a generalized helix if and only if one of the following conditions holds:

(a) the principal normals are perpendicular to a fixed direction;

(b) the binormals make a constant angle with a fixed direction;

(c) the ratio between the curvature and the torsion is constant;

(d) all rectifying planes of the curve are parallel to a line (recall that the rectifying plane at a given point of the curve passes through this point and is perpendicular to the principal normal);

(e) $\left(\dfrac{d}{ds}\mathbf{v}, \dfrac{d^2}{ds^2}\mathbf{v}, \dfrac{d^3}{ds^3}\mathbf{v} \right) = 0;$

(f) $\left(\dfrac{d}{ds}\mathbf{b}, \dfrac{d^2}{ds^2}\mathbf{b}, \dfrac{d^3}{ds^3}\mathbf{b} \right) = 0;$

(g) the vector

$$\frac{\mathbf{v}/k + \mathbf{b}/\varkappa}{\sqrt{1/k + 1/\varkappa}} = \frac{\varkappa\mathbf{v} + k\mathbf{b}}{\sqrt{k^2 + \varkappa^2}}$$

is constant.

18.21. Prove that the curve

$$x = a \int \sin \alpha\,(t)\,dt, \quad y = a \int \cos \alpha\,(t)\,dt, \quad z = ht$$

is a generalized helix.

18.22. Prove that the curve $x^2 = 3y$, $2xy = 9z$ is a generalized helix. Find the vector with which the tangents to the curve make a constant angle.

18.23. For which values of parameters a, b, and c the curve $\mathbf{r}\,(t) = (at, bt^2, ct^3)$ is the generalized helical line?

Let a curve $\mathbf{r} = \mathbf{r}\,(s)$, where s is the natural parameter, be given. The curve $\rho = \dot{\mathbf{r}}\,(s)$ lies on the sphere of unit radius centred at the origin and is called the *tangent spherical image* of the curve. The normal and binormal spherical images are defined in a similar way. Thus, for example, it is clear from the definition that the curve $\mathbf{r} = \mathbf{r}\,(s)$ is a generalized helix if and only if its tangent spherical image is an arc of the circle.

18.24. Prove that the binormal spherical image of the generalized helix is a circle.

18.25. Find the tangent, normal, and binormal spherical images of the helical line

$$\mathbf{r}\,(t) = (a \cos t, \, a \sin t, \, bt).$$

18.26. Let $\mathbf{r} = \mathbf{r}\,(s)$ be a curve parameterized by the natural parameter.

(a) Prove that the tangent spherical image of the curve $\mathbf{r} = \mathbf{r}\,(s)$ degenerates into a point if and only if $\mathbf{r} = \mathbf{r}\,(s)$ is a straight line.

(b) Prove that the binormal spherical image of the curve $\mathbf{r} = \mathbf{r}\,(s)$ degenerates into a point if and only if $\mathbf{r} = \mathbf{r}\,(s)$ is a plane curve.

(c) Prove that the normal spherical image of the curve cannot be a point.

18.27. Let $\mathbf{r} = \mathbf{r}\,(s)$ be a curve parameterized by the natural parameter, and, moreover, let $k\varkappa \neq 0$. Prove that the tangent to the tangent spherical image is parallel to the tangent to the binormal spherical image at the points corresponding to the same value of the parameter s.

18.28. Let $\mathbf{r} = \mathbf{r}\,(s)$ be a curve parameterized by the natural parameter. Prove that if the tangent spherical image of this curve lies in the plane passing through the origin, then the curve $\mathbf{r} = \mathbf{r}\,(s)$ is a plane curve.

18.29. Show that at the inversion the osculating circle of a given curve passes to the osculating circle of the image of this curve. In this case, it is assumed that the centre of inversion does not coincide with the osculating point of the curve with the circle.

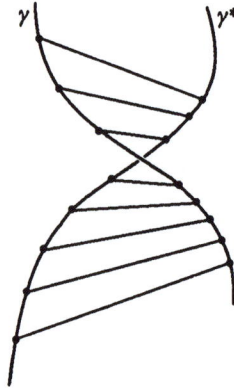

Figure 78 Bertrand curves

Recall that the *spherical curve* is a curve $\mathbf{r} = \mathbf{r}(t)$ for which there exist a constant vector $\mathbf{m}$ and a constant real number R such that $|\mathbf{r}(t) - \mathbf{m}| = R$.

18.30. Let $\mathbf{r} = \mathbf{r}(t)$ be a regular curve, and let there exist a point $\mathbf{a}$ lying in all normal planes of the curve $\mathbf{r} = \mathbf{r}(t)$. Prove that $\mathbf{r} = \mathbf{r}(t)$ is a spherical curve.

18.31. Prove that

$$\mathbf{r}(t) = (-\cos 2t, \; -2\cos t, \; \sin 2t)$$

is a spherical curve.

Hint. Show that the point $(-1, 0, 0)$ lies in each normal plane of the curve considered. See Problem 18.30.

18.32. Using the results of the previous problems, prove that the curve $\mathbf{r} = \mathbf{r}(s)$ lies on the sphere if and only if there exist constant real numbers A and B such that

$$k \left(A \cos \int_0^s \varkappa ds + B \sin \int_0^s \varkappa ds \right) \equiv 1.$$

18.33. The curves γ and γ^* are called the *Bertrand curves* if it is possible to state a one-to-one correspondence between the points of these curves under which the principal normals (as affine lines) coincide at the corresponding points (see Fig. 78).

This means that the segment joining the corresponding points of the curves is a segment of the principal normal for both curves.

Prove the following properties of the Bertrand curves γ and γ^*:

(a) the distance between the corresponding points of the curve is constant;

(b) the curvature and the torsion of each of the Bertrand curves are connected by the relation

$$ak + b\varkappa = 1,$$

where a and b are constant numbers. Herewith, each of the curves has its own numbers a and b;

(c) the tangents to the curves γ and γ^* at the corresponding points make a constant angle.

18.34. (a) Prove that two arbitrary concentric surfaces in the plane compose a pair of Bertrand curves.

(b) Let

$$\mathbf{r}_1(t) = \frac{1}{2}\left(\frac{1}{\cos t} - t\sqrt{1 - t^2}, 1 - t^2, 0\right),$$

$$\mathbf{r}_2(t) = \frac{1}{2}\left(\frac{1}{\cos t} - t\sqrt{1 - t^2} - t, 1 - t^2 + t\sqrt{1 - t^2}, 0\right).$$

Prove that $\mathbf{r}_1(t)$ and $\mathbf{r}_2(t)$ compose a pair of Bertrand curves.

18.35. Prove that if the curvature and torsion of a curve γ are connected by the linear dependence $ak + b\varkappa = 1$, where a and b are nonzero numbers, then there exists a curve γ^* such that the curves γ and γ^* compose a pair of Bertrand curves.

18.36. Prove that for a constant curvature curve, there necessarily exists a curve such that both these curves compose a pair of Bertrand curves.

18.37. Prove that a curve $\mathbf{r}(t)$ is a Bertrand curve if and only if it can be given by the vector equation

$$\mathbf{r}(t) = a \int \mathbf{e}(t)\, dt + c \int \mathbf{e}(t) \times \mathbf{e}'(t)\, dt,$$

where $\mathbf{e}(t)$ is a vector function such that $|\mathbf{e}(t)| = 1$ and $|\mathbf{e}'(t)| = 1$.

18.38. Let $\mathbf{r} = \mathbf{r}(t)$ be a regular smooth curve, and let $\varkappa \neq 0$. Prove that $\mathbf{r}(t)$ is an ordinary helical line if and only if there exist at least two different curves each of which composes a pair of Bertrand curves with $\mathbf{r}(t)$.

18.39. Prove that if there is a one-to-one correspondence between the points of two different curves γ and γ^* under which the binormals of the curves coincide (as affine lines) at the corresponding points, then these curves are plane curves.

18.40. For a smooth curve γ, let there exist a curve γ^* such that the principal normals of the curve γ are binormals of γ^* at the corresponding points. Prove that the curvature and torsion of the curve γ satisfy the relation $k = \lambda\left(k^2 + \varkappa^2\right)$, where λ is a fixed number.

18.41. Assume that it is possible to state a correspondence between the points of curves γ_1 and γ_2 under which the tangents at the corresponding points are parallel. Prove that the modules of the ratios of the torsion and curvature at the corresponding points of the curves are equal. Also, prove that the normals (binormals) are parallel at the corresponding points.

18.42. Let a correspondence be stated between the points of curves γ_1 and γ_2 such that at the corresponding points, the tangents of γ_1 are parallel to the binormals of γ_2. Prove that at the corresponding points, the binormals of γ_1 are parallel to the tangents of γ_2 and the principal normals of both curves are parallel to each other. Moreover, show that their curvatures and absolute values of the torsions are inversely proportional, i.e., $\dfrac{k_1}{k_2} = \dfrac{|\varkappa_2|}{|\varkappa_1|}$.

An *oval* is a regular simple closed plane curve with $k > 0$. A *vertex* of a regular plane curve is a point at which the curvature k has a relative maximum or minimum.

Let $\mathbf{r}(s)$ be an oval, and let P be a point on $\mathbf{r}(s)$. Then there exists a point P' at which the tangent vector $\mathbf{v}$ to the oval is opposite to the tangent vector at the point P, i.e., $\mathbf{v}(P') = -\mathbf{v}(P)$. The tangents at the points P and P' are parallel. Clearly, such a point P' is unique. It is called the *opposite point* to the point P.

The *width* $w(s)$ of the oval at the point $P = r(s)$ is the distance between the tangent lines to the oval at the opposite points P and P'.

The oval is said to be of *constant width* if its width at the point P is constant, i.e., it is independent of the choice of P.

18.43. Prove that the concept of vertex of a curve is independent of the choice of its parameterization.

18.44. Prove that if $\mathbf{r} = \mathbf{r}(s)$ is an oval, then the vector $\mathbf{v}''$ is parallel to the vector $\mathbf{v}$ at least at four points.

18.45. *Theorem on four vertices.* Prove that any oval has at least four vertices.

18.46. Show that the theorem on four vertices (see Problem 18.45) is not true if the closedness condition is omitted.

18.47. Prove that if $\mathbf{r}(s)$ is an oval of constant width w, then its length is equal to πw.

18.48. Let $\mathbf{r} = \mathbf{r}(s)$ be an oval of constant width. Prove that the line connecting a pair of opposite points P and P' of the oval is orthogonal to the tangents at the points P and P'.

18.49. Let $\mathbf{r}(s)$ be a plane curve of constant width. Show that the sum of curvature radii $1/k$ at opposite points is a constant independent of the choice of the points.

18.50. Let $\mathbf{r}(s)$ be an oval of length L given in the natural parameter. Denote by θ the angle between the horizontal line and the tangent vector $\mathbf{v}(s)$.

(a) Show that the mapping $\theta\colon [0, L] \to [0, 2\pi]$ is one-to-one. Show that the mapping $\mathbf{r} \circ \theta^{-1}$ is a smooth regular parameterization of the oval.

(b) Let $\rho(\theta)$ be a parameterization of the oval such that $\mathbf{r}(s) = \rho(\theta(s))$ (see the previous item). Prove that the point opposite to $\mathbf{r}(s)$ is $\mathbf{R}(s) = \rho(\theta(s) + \pi)$.

(c) Prove that the curve $\mathbf{R}(s)$ is regular.

18.51. Let $\rho(\theta)$ be an oval parameterized by the angle θ as in the previous problem. Let $w(\theta)$ be the width of the oval at the point $\rho(\theta)$. Prove that

$$\int_0^{2\pi} w\,d\theta = 2L,$$

where L is the width of the oval.

18.52. Let $\rho(\theta)$ be an oval parameterized by the angle θ, and let $k(\theta)$ and $w(\theta)$ be its curvature and width, respectively. Prove that

$$\frac{d^2 w}{d\theta^2} + w = \frac{1}{k(\theta)} + \frac{1}{k(\theta + \pi)}.$$

18.53. Prove that the curvature of a curve γ at a point M is equal to the curvature of the projection of γ on the osculating plane at the point M.

18.54. Express the curvature and torsion of the evolvent of a curve through the curvature and torsion of the initial curve.

18.55. Prove that the evolvent of the slope curve is a plane curve. For the definition of the slope curve, see above.

The *pedal* of a spatial curve with respect to a point O is the set of bases of perpendiculars dropped from the point O on the tangents of the given curve.

18.56. Find the equation of the pedal of the curve $\mathbf{r} = \mathbf{r}(t)$.

18.57. Find the pedal of the helical line $\mathbf{r}(t) = (a\cos t, a\sin t, ht)$ with respect to the origin. Prove that it lies on the one-sheet hyperboloid

$$\frac{x^2}{a^2} + \frac{y^2}{a^2} - \frac{h^2 z^2}{a^4} = 1.$$

18.58. Let a curve $\mathbf{r} = \mathbf{r}(s)$ have a constant curvature. Prove that the osculating sphere and osculating circle are of the same radius.

18.59. Prove that the curvature radius of a curve on the sphere of radius R cannot be greater than R.

18.60. Prove that a non-plane regular smooth curve on the sphere cannot have a constant curvature.

18.61. Find the general form of the curvature of a spherical curve with a constant torsion.

18.62. Prove that a curve on the sphere is defined by assigning its curvature or its torsion as a function of arclength.

18.63. (a) Prove that on the sphere there exist curves with a constant nonzero torsion.

(b) Show that on the sphere there is no closed regular curve with torsion of a constant sign different from 0.

(c) The previous items imply that on the sphere a curve with a constant nonzero torsion cannot be closed. The question is: how "far" a spherical curve with a constant torsion can go, i.e., is it necessarily of finite length or it has infinite length and tightly wraps around the sphere?

18.64. Prove that if a curve is given by the equation $\mathbf{r} = \mathbf{r}\,(s)$ in the natural parameter s, then the radius-vector of the centre of the osculating sphere is given by $\rho = \mathbf{r} + R\mathbf{n} + \dfrac{\dot{R}}{\varkappa}\mathbf{b}$, and the radius of the osculating sphere is equal to

$$R_s = \sqrt{R^2 + \frac{\dot{R}^2}{\varkappa^2}}.$$

18.65. Find the osculating circle of the curve:

(a) $\mathbf{r}\,(t) = (t - \sin t, 1 - \cos t, \sin t)$ for $t = 0$;

(b) $\mathbf{r}\,(t) = \left(2t, \ln t, t^2\right)$ for $t = 1$.

18.66. Find the radius of the osculating sphere of the curve $\mathbf{r}\,(t) = \left(t, t^2, t^3\right)$ at the point $t = 0$.

18.67. Find the radius of the osculating sphere of the following curves at an arbitrary point:

(a) $\mathbf{r}\,(t) = (a \cos t, a \sin t, ht)$;

(b) $\mathbf{r}\,(t) = \left(e^t, e^{-t}, t\sqrt{2}\right)$;

(c) $\mathbf{r}\,(t) = (e^t \sin t, e^t \cos t, e^t)$.

18.68. Prove that the osculating plane of a curve intersects the osculating sphere along the osculating circle at the same point.

18.69. Find the locus of centres of the osculating spheres of the helical line

$$\mathbf{r}\,(t) = (a \cos t, a \sin t, ht).$$

18.70. Prove that if the radius of an osculating surface is constant, then the curve either lies on the sphere or has a constant curvature.

18.71. Let a curve γ_1 be the locus of centres of osculating spheres of a curve γ_2, and vice versa. Prove that the curvatures of the curves γ_1 and γ_2 are constant and equal, their osculating planes are mutually perpendicular and the product of their torsions at the corresponding points is equal to the square of the curvature.

18.72. Prove that if at each point a curve has a tangency of the third order with its osculating plane, then this curve is plane.

18.73. Consider all possible second-order curves passing through a given point of a plane regular curve γ. Show that among them there exists a unique curve whose order of tangency with the curve γ is equal to 4. Find the type and parameters of this second-order curve in terms of the curvature $k\,(s)$ of the curve γ.

18.74. A beam of rays falls onto a plane curve $\mathbf{r} = \mathbf{r}\,(s)$ from the origin. Compose the equation of the envelope of the reflected rays (caustic).

18.75. Which form takes the caustic equation of a plane curve with respect to the origin if the equation of the curve is given in the form $\mathbf{r} = \mathbf{r}(t)$ in an arbitrary parameter?

18.76. A beam of parallel rays having the direction of a vector $\mathbf{e}$ ($|\mathbf{e}| = 1$) falls onto a plane curve given by the equation $\mathbf{r} = \mathbf{r}(s)$. Compose the equation of the envelope of the rays reflected from the given curve (caustic). Consider the case where the curve is given by the equation $\mathbf{r} = \mathbf{r}(t)$ in an arbitrary parameter, and also the case where it is given by the equation $y = f(x)$.

18.77. Let $\mathbf{r}_1\colon [0, a] \to \mathbb{R}^2$ be a segment of a curve parameterized by the natural parameter, and let $\mathbf{r}_2(s)$ be the curve

$$\mathbf{r}_2(s) = \mathbf{r}_1(s) + (a_0 - s)\,\mathbf{v}(s),$$

where $\mathbf{v}(s)$ is the tangent vector to $\mathbf{r}_1(s)$ and $a_0 > a$ is a constant. Show that the unit tangent to $\mathbf{r}_2(s)$ is orthogonal to $\mathbf{v}(s)$ at each point.

Note that the curve $\mathbf{r}_1$ is the *evolute* of the curve $\mathbf{r}_2$ and the curve $\mathbf{r}_2$ is the *evolvent* of the curve $\mathbf{r}_1$.

18.78. Find the evolutes, evolvents and curvature radii of the curves:

(a) $\dfrac{x^2}{a^2} + \dfrac{y^2}{b^2} = 1$;

(b) $\dfrac{x^2}{a^2} - \dfrac{y^2}{b^2} = 1$;

(c) $x^2 - 2py = 0$;

(d) $\begin{cases} x = a\,(t - \sin t), \\ y = a\,(1 - \cos t); \end{cases}$

(e) $r = a\,(1 + \cos \varphi)$ (in the polar coordinate system);

(f) $y = a \cosh \dfrac{x}{a}$;

(g) $r = \lambda \varphi$ (in the polar coordinate system);

(h) $r^2 = 2a^2 \cos 2\varphi$ (in the polar coordinate system);

(i) $r^m = a^m \cos m\varphi$ (in the polar coordinate system), where m is an integer positive constant;

(j) $r = a \cot \varphi$.

18.79. Find the evolute of the logarithmic spiral $r = a^\varphi$. Prove that this is the same spiral but turned around the origin by an angle. Find this angle.

18.80. A ray located on the plane Oxy uniformly rotates in the plane around its initial point O. A point M on the ray uniformly moves along it with a speed proportional to the distance OM. Under these conditions,

(a) deduce the equation of the trajectory of motion of the point in the polar coordinates with the pole at the point O;

(b) show that on the curve obtained (called the Bernoulli logarithmic spiral), the tangents make a constant angle with the radius-vector;

(c) this property of the logarithmic spirals are characteristic of them;

(d) the spiral tends to its *pole* $r = 0$, having a constant length;

(e) write the natural equations of the logarithmic spiral;

(f) the curvature radius of the spiral is proportional to its arclength counted off from the pole, and this property is also a characteristic property of the logarithmic spirals;

(g) the evolute of the logarithmic spiral is congruent to the spiral itself;

(h) find the value of the parameter a for which the evolute of the spiral coincides with the spiral itself, i.e., the spiral is the *envelope* of its normals;

(i) each logarithmic spiral is an orbit of action of the group G of the plane's linear transformations, which consists in the similarity transformation with the similarity coefficient $t > 0$, followed by the turn of the plane by the angle $\alpha = \ln t$ (the similarity centre and the centre of rotation are at the common pole of the spirals);

(j) for the arc $r = e^{\varphi}$, $\varphi_1 \leq \varphi \leq \varphi_2$ of the spiral, find an element of the group G that transforms it into the arc of the same spiral starting from the point with the polar coordinates $\varphi = \pi + \varphi_1$, $r = e^{\pi + \varphi_1}$; does the image of the arc have the same angular opening $\varphi_2 - \varphi_1$ as the initial arc?

18.81. Find singular points of the evolute of the ellipse. How many normals to the ellipse can be drawn through an arbitrary point of the plane?

18.82. Find the evolvent of the circle.

18.83. Find the curvature centre of the Steiner curve

$$x = 2r \cos \frac{t}{3} + r \cos \frac{2t}{3}, \quad y = 2r \sin \frac{t}{3} - r \sin \frac{2t}{3}$$

at an arbitrary point. The Steiner curve is the hypercycloid composed as a result of rolling the circle of radius r along the circle of radius $3r$.

18.84. Prove that the evolute of a Steiner curve is the Steiner curve similar to the initial curve, the similarity coefficient being equal to 3. Herewith, the evolute is turned by the angle $\frac{\pi}{3}$ relative to the initial curve.

18.85. Find the evolute of an astroid. Prove that its evolute is an astroid similar to the initial astroid with the similarity coefficient equal to 2, which is turned by the angle $\frac{\pi}{2}$ relative to the initial one. Note that the astroid is the hypercycloid composed as a result of rolling a circle of radius r along a circle of radius $4r$. Furthermore, the properties of the evolute of the astroid described in this problem are similar to the properties of the evolute of the Steiner curve (see the previous problem).

18.86. Study the evolutes of hypercycloids obtained as a result of rolling a circle of radius r along a circle of radius R for various ratios R/r (see Problems 18.84 and 18.85).

18.87. Find the evolute of the folium of Descartes

$$x = \frac{3at}{1 + t^3}, \quad y = \frac{3at^2}{1 + t^3}.$$

18.88. Find the curvature of a plane curve whose curvature is equal to the curvature of its evolute (at the corresponding points).

18.89. By the natural equation of a curve, $R = \varphi(s)$, compose the natural equation of the evolute of this curve.

18.90. Write the natural equation of the evolvent from the natural equation of a curve.

18.91. Prove the following assertions:

(a) the tangents to the evolute are the normals of the initial curve;

(b) any orthogonal curve of a family of tangents to a curve is the evolvent of the initial curve;

(c) the arclength of the evolute is equal to the module of the difference of the curvature radii at the corresponding points of the initial curve;

(d) the curvature radius of the evolute is equal to

$$\frac{1}{2}\frac{d\left(R^2\right)}{ds},$$

where R is the curvature radius of the curve itself.

18.92. Let a non-plane curve have a constant nonzero curvature, and let its torsion be different from zero. Consider the set of curvature centres of this curve. Denote the curve obtained as γ^*. Prove that the curvature of the curve γ^* is constant. Find the torsion of γ^*.

18.93. Let a smooth curve C be given by the equation $\mathbf{r} = \mathbf{r}(t)$, where the function $\mathbf{r}(t)$ is defined on a closed interval $[a, b]$. At a point M, let the derivatives $\mathbf{r}'$, $\mathbf{r}''$, and $\mathbf{r}'''$ be non-coplanar. Prove that the osculating plane of the curve C at the point M intersects the curve C.

18.94. Let a smooth curve C be given by the equation $\mathbf{r} = \mathbf{r}(t)$, where the function $\mathbf{r}(t)$ is defined on a closed interval $[a, b]$. At a point $M(t)$, let the vector $\mathbf{r}'$ be not parallel to the vector $\mathbf{r}''$. Calculate the limit

$$\lim_{\Delta t \to 0}\frac{d}{|\Delta t|^3},$$

where d is the distance from the point $M(t + \Delta t)$ to the osculating plane of the curve C at the point $M(t)$. Consider a particular case where the curve is given by the equation $\mathbf{r} = \mathbf{r}(s)$ in the natural parameter.

18.95. Prove that the torsion of an asymptotic line on the surface with $K < 0$ is equal to $\pm\sqrt{-K}$.

18.96. Let the curvature of a surface of revolution be strictly negative. Do there exist closed asymptotic lines on this surface?

Let $\mathbf{r} = \mathbf{r}(s)$, $s \in [0, a]$, be a plane piecewise-regular curve of class C^2 parameterized by the natural parameter. The index of rotation (rotation number) $i_{r(s)}$ of this curve is the number

$$i_{r(s)} = \frac{\int\limits_0^a k\,ds + \sum\limits_{i=0}^{n-1}\Delta\theta_i}{2\pi},$$

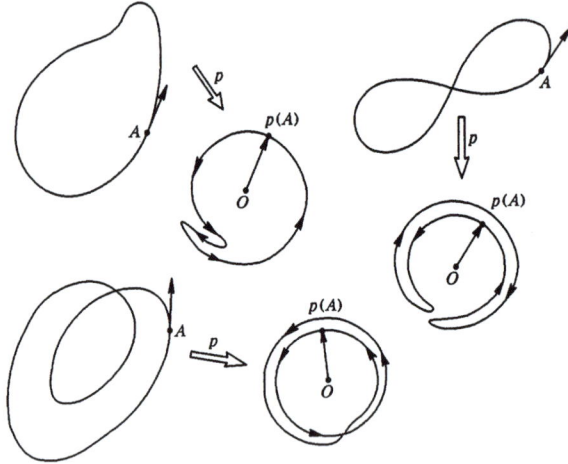

Figure 79 Index of rotation of a curve

where k is the curvature of the curve, s_i $(0 \le i \le n-1)$ are irregularity points, $\mathbf{v}^-(s_i) = \lim\limits_{s \to s_i^-} \mathbf{v}(s)$, $\mathbf{v}^+(s_i) = \lim\limits_{s \to s_i^+} \mathbf{v}(s)$ and $\Delta\theta_i$ is the angle between the vectors $\mathbf{v}^-(s_i)$ and $\mathbf{v}^+(s_i)$ (see Fig. 79).

18.97. Calculate the indices of rotation of the curves given by the following equations:

(a) $\mathbf{r}(t) = (a + \rho\cos t, \rho\sin t)$, $0 \le t \le 2\pi$, $|a| < \rho$;
(b) $\mathbf{r}(t) = (a + \rho\cos t, \rho\sin t)$, $0 \le t \le 2\pi$, $0 < \rho < |a|$;
(c) $\mathbf{r}(t) = (\rho\cos 2t, -\rho\sin 2t)$, $0 \le t \le 2\pi$, $\rho > 0$;
(d) $\mathbf{r}(t) = ((1/2)\cos t, \sin t)$, $0 \le t \le 2\pi$;
(e) $\mathbf{r}(t) = (2\cos t, -\sin t)$, $0 \le t \le 6\pi$;
(f) $\mathbf{r}(t) = (\cos t, \sin^2 t)$, $0 \le t \le 2\pi$.

18.98. Let $\gamma \subset \mathbb{R}^2$ be a closed curve (not necessarily simple, i.e., self-intersections are admitted). Prove that

$$l^2 \ge 4\pi \int_{\mathbb{R}^2} \omega(x)\, ds,$$

where the function $\omega(x)$ is the rotation number of the curve γ around a point $x \in \mathbb{R}^2$.

18.99. Prove that if $\mathbf{r}(s)$ is a simple closed regular plane curve, then the tangent circular image $v\colon [0, L] \to S^1$ of the curve is a mapping "onto".

18.100. Prove that if $\mathbf{r} = \mathbf{r}(s)$ is a regular closed curve, then its tangent spherical image cannot lie on any open hemisphere.

18.101. Prove that the tangent spherical image of a regular closed curve cannot lie in any closed hemisphere, except for the case where it is a large

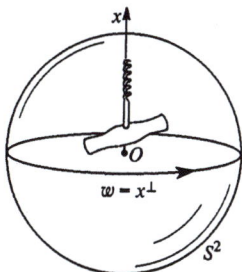

Figure 80

circle bounding a hemisphere.

18.102. Let γ be a closed smooth curve on a unit sphere S^2. Prove that it is contained in an open hemisphere if

(a) the length l of the curve γ is strictly less than 2π;

(b) the length l of the curve γ is equal to 2π, but the curve is not the union of two large semicircles.

The *total torsion* of a regular spatial curve $\mathbf{r} = \mathbf{r}(s)$ of length L parameterized by the natural parameter is the number $\int\limits_0^L \varkappa\, ds$. Note that the total curvature and the torsion of a curve in $\mathbb{R}^3$ defined above for the natural parameter s along the curve can be calculated in an arbitrary parameter $t = t(s)$. In this case, the corresponding expression, for example, for the torsion has the form

$$\int\limits_a^b \varkappa(t)\left|\frac{d\mathbf{r}}{dt}\right| dt,$$

where $a = t(0)$ and $b = t(L)$.

18.103. Using the results of Problems 18.100–18.102, prove the following assertion: the total curvature of a closed spatial curve γ is not less than 2π, and, moreover, it is equal to 2π if and only if γ is a plane convex curve (Fenchel theorem).

18.104. Let γ be a spatial closed curve. Assume that $0 \le k \le 1/R$ for some real number $R > 0$. Prove that the length l of the curve γ satisfies the inequality $l \ge 2\pi R$.

18.105. Calculate the tangent spherical image of the ellipse

$$\mathbf{r}(t) = (2\cos t,\ \sin t,\ 0),\quad 0 \le t \le 2\pi.$$

What we can say about this image taking into account the Fenchel theorem?

Consider a standard unit two-dimensional sphere S^2 in $\mathbb{R}^3$. A set of oriented large circles of the sphere is in one-to-one correspondence with points of the sphere.

This correspondence is defined as follows. An endpoint of the unit positive normal that is perpendicular to the plane of the circle and emanates from the origin is assigned to each large circle. Here, the normal is said to be positive if a right corkscrew moves along it when rotated in the direction indicated on the large circle (see Fig. 80).

The measure of the set of oriented large circles is the measure of the corresponding set of points of S^2.

If $x \in S^2$, then by $w = x^\perp$ we denote the large oriented circle corresponding to the point x. For a regular curve γ on the sphere, by $n_\gamma(x)$ we denote the number of points in $\gamma \cap x^\perp$ (which can be infinite). Note that the number $n_\gamma(x)$ is independent of the parameterization of the curve γ.

18.106. Let γ be a regular curve of length l on the sphere S^2. Prove that the measure of the set of oriented large circles intersecting γ (with account for the multiplicities) is equal to $4l$. In other words, $\int_{S^2} n_\gamma(w)\, d\sigma = 4l$ (Crofton formula).

A closed simple curve γ in $\mathbb{R}^3$ is said to be *non-knotted* if there exists a one-to-one continuous function $g\colon D^2 \to \mathbb{R}^3$ (D^2 is a unit disk) mapping the boundary S^1 of the disk D^2 onto the image of the curve γ. Otherwise, the curve is said to be *knotted*.

18.107. Prove that if γ is a simple knotted regular curve, then its total curvature is greater than or equal to 4π.

18.108. Using the Crofton formula, prove that for any closed spatial regular curve, its total curvature $\int_\gamma k(s)\, ds$ is no less than 2π (a particular case of the Fenchel theorem).

18.109. For a smooth closed regular curve γ, let the spherical indicatrix of its normals have no self-intersections. Prove that in this case, the following inequality holds for the oriented curvature k (see Problem **4.20**) of the curve γ:

$$-2\pi < \int_0^l k(s)\, ds < 2\pi,$$

where l is the length of L. Deduce from this that the curvature of the curve L necessarily changes its sign.

18.110. Prove that for any real number r, there exists a closed curve γ such that its total torsion $\int_\gamma \varkappa\, ds$ is equal to r.

18.111. Prove that for a closed curve γ given in the form $\mathbf{r} = \mathbf{r}(s)$ and lying on the sphere S^2, its total torsion $\int_\gamma \varkappa(s)\, ds$ vanishes.

18.112. Let M be a surface in $\mathbb{R}^3$ such that $\int_\gamma \varkappa ds = 0$ for all closed curves γ lying on M. Prove that M is a part of the plane or a part of the sphere.

18.113. Prove that for any closed spherical curve γ, the integral $\int_\gamma \frac{\varkappa}{k}\, ds$ vanishes.

18.114. Find the geodesic curvature of a curve $u = \text{const}$ on the surface $\mathbf{r} = (u \cos v, u \sin v, av)$.

18.115. Find the geodesic curvature of a curve $ax + by + c = 0$ in the upper half-plane with the metric

$$ds^2 = \frac{dx^2 + dy^2}{y^2}.$$

18.116. Prove that if each normal of a plane closed regular curve divides it in two equal parts, then this curve is a circle.

18.117. Prove the following assertion: a periodic function $k(s)$ with period S is the curvature function of a plane closed curve if and only if the following two equations hold:

$$\int_0^S \cos \left(\int_0^s k(t)\, dt + C \right) ds = 0,$$

$$\int_0^S \sin \left(\int_0^s k(t)\, dt + C \right) ds = 0,$$

where C is a constant.

The *support function* $h(t)$ of a plane curve $\mathbf{r} = \mathbf{r}(t)$ is the distance from a fixed point O to the variable tangent of the curve at the point $\mathbf{r}(t)$.

18.118. Let a plane convex curve be given. Then, as the parameter on the curve, one can take the angle α made by the tangent at a point of the curve with a fixed direction on the plane. Prove that the following formula holds for the curvature radius of this curve:

$$R(\alpha) = h(\alpha) + h''(\alpha),$$

where $h(\alpha)$ is the support function of the curve.

18.119. Prove that the area of the domain bounded by a plane convex regular closed curve γ is equal to $\frac{1}{2} \int_\gamma h(t)\, dt$, where $h(t)$ is the support function of the curve.

18.120. Let a closed plane convex regular curve of length L bound the domain of area S. Let R be the radius of the circumscribed circle, and let r be the radius of the inscribed circle. Prove the following inequalities:

(a) $L^2 - 4\pi S \geq 0$;

(b) $L^2 - 4\pi S \geq \pi^2 (R - r)^2$;

(c) $L^2 - 4\pi S \geq S^2 \left(\dfrac{1}{r} - \dfrac{1}{R} \right)^2$;

(d) $L^2 - 4\pi S \geq L^2 \left(\dfrac{R - r}{R + r} \right)^2$;

(e) $\dfrac{L - \sqrt{L^2 - 4\pi S}}{2\pi} \leq r \leq R \leq \dfrac{L + \sqrt{L^2 - 4\pi S}}{2\pi}$,

where the equalities are attained if and only if the curve is a circle.

18.121. Prove that an oval having four vertices intersects an arbitrary circle at no more than four points.

18.122. Prove that if an oval intersects a circle at $2n$ points, then it has at least $2n$ vertices.

18.123. Let γ be a spatial curve of constant curvature, and let γ^* be the set of its curvature centres. Prove that the curve γ^* has the same curvature as γ. Prove that γ is the set of curvature centres of the curve γ^*.

18.124. Prove that the osculating sphere has a constant radius if and only if the corresponding curve either lies on the sphere or has a constant curvature.

18.125. Prove that a spatial curve given in the form

$$\mathbf{r}(s) = \frac{1}{\varkappa} \int_0^s \left(\mathbf{b}(t) \times \frac{d\mathbf{b}}{dt}(t) \right) dt$$

has a constant torsion $\varkappa$. Here, $\mathbf{b}(t)$ is an arbitrary curve on the unit sphere and $\varkappa$ is an arbitrary nonzero real number.

Conversely, any spatial curve of nonzero constant torsion $\varkappa$ can be represented in this form for an appropriate curve $\mathbf{b}(t)$.

18.126. Prove that any curve of nonconstant curvature, for which the relation

$$\left(\frac{d(1/k)}{ds} \right)^2 = \varkappa^2 \left(R^2 - \frac{1}{k^2} \right)$$

holds, lies on a sphere of radius R.

18.127. Prove that if a curve lies on a sphere, then its curvature and torsion satisfy the relation

$$\frac{\varkappa}{k} + \frac{d}{ds} \left(\frac{1}{\varkappa} \frac{d}{ds} \left(\frac{1}{k} \right) \right) = 0.$$

Conversely, if the curvature and torsion of a curve satisfy this relation, then the curve lies on the sphere.

18.128. If the curvature and torsion of a spatial curve do not vanish, then this curve is spherical if and only if

$$\frac{\varkappa}{k} = \frac{d}{ds} \left(\frac{1}{\varkappa k^2} \frac{d}{ds} k \right).$$

If one set $\rho = \dfrac{1}{k}$ and $\sigma = \displaystyle\int \varkappa \, ds$, then the equation can be rewritten in the form

$$\frac{d^2\rho}{d\sigma^2} + \rho = 0.$$

18.129. Show that for a generic curve lying on a sphere, the relations of the previous three problems are equivalent (see also Problem 4.52).

18.130. Show that for an arbitrary number A, there exists a biregular closed curve such that $\displaystyle\int_\gamma \varkappa \, ds \geq A$. Recall that a curve is said to be biregular if $\dot{\mathbf{r}} \neq 0$ and $\ddot{\mathbf{r}} \neq 0$.

18.131. If a plane curve, together with the chord connecting its endpoints, bounds a convex domain, then when twisting this curve, i.e., when replacing it by a spatial curve of the same curvature, the length of the chord increases.

18.132. Let γ be a closed regular smooth curve. Prove that the set of directions in which the tangent lines tangent the curve at points with zero curvature is of zero angular measure. Prove that there are only finitely many directions in which the tangents tangent the curve more than at one point.

18.133. Prove the Jacobi theorem: if a spherical indicatrix of the principal normals of a closed curve has no self-intersections, then it divides the sphere into two equal parts.

§ 19. Riemannian Metric (Supplementary Problems)

19.1. Prove that in a rectangular triangle on the Lobachevskii plane, the perpendicular dropped from the midpoint of the hypotenuse on one of the legs is less that half of the other leg.

19.2. Let three points A, B, and C belong to one line, and let a point D not belong to it. Prove that the midpoints of the segments DA, DB, and DC do not lie on one line:

(a) on the sphere;

(b) on the Lobachevskii plane.

Recall that this is not true on the Euclidean plane (see Fig. 81).

19.3. Consider the Lobachevskii plane in the Poincaré model in the open disk of radius 1 with the metric

$$ds^2 = \frac{4}{\left(1 - u^2 - v^2\right)^2} \left(du^2 + dv^2\right).$$

Consider the diameter of a disk given by the equation $v = 0$. As is known, it is a line of the Lobachevskii plane. Find the equidistant curve of this line.

19.4. For the model of the Lobachevskii plane in the upper half-plane, find the area of triangles bounded by the lines:

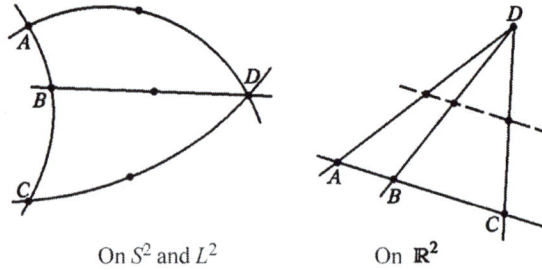

On S^2 and L^2 On $\mathbb{R}^2$

Figure 81

(a) $x = -a$, $x = a$, $x^2 + y^2 = a^2$;

(b) $x = a$, $x = 2a$, $(x-a)^2 + y^2 = a^2$;

(c) $x = 0$, $x = a/2$, $x^2 + y^2 = 4a^2$;

(d) $x = 0$, $x = a/2$, $x^2 + y^2 = a^2$;

(e) $x = -a$, $x = a$, $x^2 + y^2 = 4a^2$;

(f) $x = -a$, $x = a$, $x^2 + y^2 = 2a^2$;

(g) $x = a$, $x = b$, $(x-a)^2 + y^2 = 2b^2$.

19.5. For the model of the Lobachevskii plane in the upper half-plane, find the area of a triangle with the vertices $(-2, 2)$, $(0, 2)$, $(2, 2)$.

19.6. For the model of the Lobachevskii plane in the upper half-plane find the area of a quadrangle bounded by the lines $x = 0$, $x = a$, $y = a/2$, $y = a$.

19.7. For the model of the Lobachevskii plane in the upper half-plane, find the area of a quadrangle with the vertices at the points $(0, a\sqrt{12})$, $(0, 2a)$, (a, a), $(a, a\sqrt{3})$.

19.8. To which domain in the disk with the Lobachevskii plane (Poincaré model) is the Beltrami pseudosphere with a cut along the meridian isometric? To which domain is the universal covering over the Beltrami pseudosphere (without a cut) isometric?

Hint. After cutting along the meridian, the Beltrami pseudosphere proves isometric to the area on the Lobachevskii plane, bounded by the so-called oricycle (see Fig. 82).

19.9. Consider a set G of isometries of the form $z \mapsto \dfrac{az + b}{cz + d}$ of the Lobachevskii plane in the model on the upper half-plane having exactly one fixed point $x_0 \in \mathbb{R}$. Find the orbit of an arbitrary point of the Lobachevskii plane under the action of G.

19.10. Consider a set G of isometries of the form $z \mapsto \dfrac{az + b}{cz + d}$ of the Lobachevskii plane in the model on the upper half-plane having exactly two fixed points x_1 and $x_2 \in \mathbb{R}$. Find the orbit of an arbitrary point of the

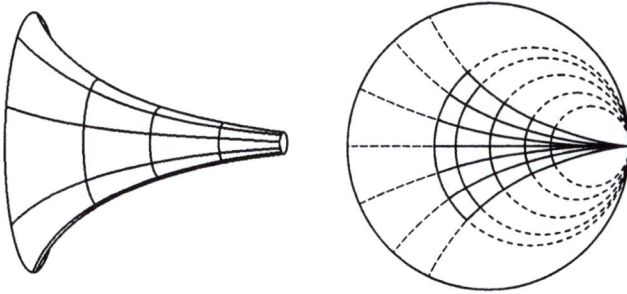

Figure 82

Lobachevskii plane under the action of G.

19.11. Find the linear-fractional transformation $w = \dfrac{az + b}{cz + d}$ under which:

(a) the upper half-plane is mapped onto itself, and, moreover, $w(0) = 1$, $w(1) = 2$ and $w(2) = \infty$;

(b) the upper half-plane is mapped onto the unit disk, and, moreover, $w(i) = 0$ and $\arg w'(i) = -\pi/2$;

(c) the upper half-plane is mapped onto the unit disk, and, moreover, $w(-1) = 1$, $w(0) = i$ and $w(1) = -1$;

(d) the unit disk is mapped onto the lower half-plane, and, moreover, $w(1) = 1$, $w(i) = 0$ and $w(-i) = -1$;

(e) the unit disk is mapped onto the upper half-plane, and, moreover, $w(-1) = \infty$, $w(1) = 0$ and $w(i) = 1$;

(f) the unit disk is mapped onto itself; moreover, $w(1/2) = 0$ and $\arg w'(1/2) = \pi/2$.

19.12. On the Lobachevskii plane, consider a triangle ABC around which a circle is circumscribed such that the side AB is its diameter. Prove that the relation $\angle C = \angle A + \angle B$ holds in the triangle ABC.

19.13. On the Lobachevskii plane, consider a triangle with sides a, b and c. Let r be the radius of the inscribed circle, let l_a and m_a be the bisectrix and median, respectively, drawn to the side a, and let α be the angle opposite to it. Prove the relations:

(a) $\tanh l_a = \dfrac{2 \sinh b \sinh c \cos(\alpha/2)}{\sinh(b + c)}$;

(b) $\cosh m_a = \dfrac{\cosh[(b + c)/2] \cosh[(b - c)/2]}{\cosh(a/2)}$;

(c) $\tanh^2 r = \dfrac{\sinh(p - a) \sinh(p - b) \sinh(p - c)}{\sinh p}$, where $p = \dfrac{1}{2}(a + b + c)$.

19.14. Consider the transformation of the unit disk centred at the origin, given by

$$(x, y) \mapsto \left(\frac{x}{\sqrt{1 - x^2 - y^2 + 1}}, \frac{y}{\sqrt{1 - x^2 - y^2 + 1}} \right).$$

(a) Prove that under this transformation, the unit disk is diffeomorphically mapped onto itself.

(b) Calculate the metric to which the metric of the Lobachevskii plane passes under this transformation.

(c) Find the Christoffel symbols of the metric from item (b).

(d) Calculate the parallel translation along chords of the unit disk in the metric of item (b).

(e) Find the geodesics of the metric of item (b).

19.15. In the unit disk $x^2 + y^2 < 1$, consider the metric having the matrix

$$\frac{1}{1 - x^2 - y^2} \begin{pmatrix} 1 - y^2 & xy \\ xy & 1 - x^2 \end{pmatrix}.$$

Show that the transformation

$$x = x' \cos \alpha - y' \sin \alpha, \quad y = x' \sin \alpha + y' \cos \alpha$$

is the motion of this metric.

19.16. Prove that the angle φ between the circles on the unit sphere S^2 given by the planes $a_i x + b_i y + c_i z = d_i$, $i = 1, 2$, is calculated by the formula

$$\cos \varphi = \frac{a_1 a_2 + b_1 b_2 + c_1 c_2 - d_1 d_2}{\sqrt{a_1^2 + b_1^2 + c_1^2 - d_1^2} \sqrt{a_2^2 + b_2^2 + c_2^2 - d_2^2}}.$$

19.17. Calculate the first quadratic form of the following surfaces in $\mathbb{R}^3$:

(a) $\mathbf{r}(u, v) = \left(\frac{a}{2} \left(v + \frac{1}{v} \right) \cos u, \frac{b}{2} \left(v + \frac{1}{v} \right) \sin u, \frac{c}{2} \left(v - \frac{1}{v} \right) \right)$ (one-sheeted hyperboloid);

(b) $\mathbf{r}(u, v) = \left(a \frac{uv + 1}{v + u}, b \frac{v - u}{v + u}, c \frac{uv - 1}{v + u} \right)$ (one-sheeted hyperboloid);

(c) $\mathbf{r}(u, v) = \left(\frac{a}{2} \left(v - \frac{1}{v} \right) \cos u, \frac{b}{2} \left(v - \frac{1}{v} \right) \sin u, \frac{c}{2} \left(v + \frac{1}{v} \right) \right)$ (two-sheeted hyperboloid);

(d) $\mathbf{r}(u, v) = \left(v\sqrt{p} \cos u, v\sqrt{q} \sin u, \frac{v^2}{2} \right)$ (elliptic paraboloid);

(e) $\mathbf{r}(u, v) = \left((u + v)\sqrt{p}, (u - v)\sqrt{q}, 2uv \right)$ (hyperbolic paraboloid);

(f) $\mathbf{r}(u, v) = (a \cos u, b \sin u, v)$ (elliptic cylinder);

(g) $\mathbf{r}(u, v) = \left(\frac{a}{2} \left(u + \frac{1}{u} \right), \frac{b}{2} \left(u - \frac{1}{u} \right), v \right)$ (hyperbolic cylinder).

In the next Problems 19.18 and 19.19, prove that the surfaces considered are second-order surfaces (sometimes, they are called quadrics) and find their type.

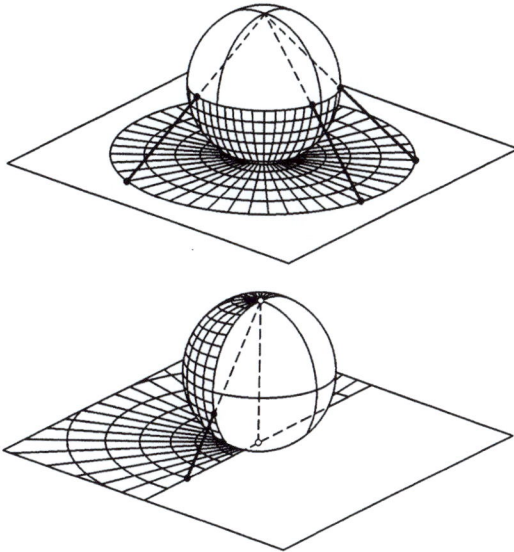

Figure 83 Stereographic projection of the sphere on the plane

19.18. Find the first quadratic form of the surface

$$x = a \cosh u \cos v, \quad y = a \cosh u \sin v, \quad z = c \sinh u, \quad a, c = \text{const}.$$

19.19. Find the first quadratic form of the surface

$$x = a \sinh u \cos v, \quad y = a \sinh u \sin v, \quad z = c \cosh u, \quad a, c = \text{const}.$$

19.20. Prove that under the corresponding choice of curvilinear coordinates on a surface of revolution, its first quadratic form can be reduced to the form

$$ds^2 = du^2 + G(u)\, dv^2.$$

19.21. Reduce the first quadratic form of the sphere, torus, catenoid, and the Beltrami pseudosphere to the form

$$ds^2 = du^2 + G(u)\, dv^2.$$

19.22. A curvilinear coordinate system on a surface is said to be *isothermal* (sometimes, it is said to be *conformal*) if the first quadratic form of the surface in such coordinates has the form

$$ds^2 = \lambda(u, v)\left(du^2 + dv^2\right).$$

Find the isothermal coordinates on the Beltrami pseudosphere.

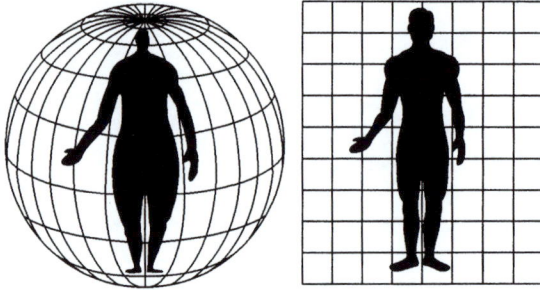

Figure 84 Mercator projection

19.23. The *Liouville surface* is a surface whose first quadratic form can be reduced to the form

$$ds^2 = (f(u) + g(v)) \left(du^2 + dv^2\right).$$

Prove that a surface locally isometric to a surface of revolution is a Liouville surface.

19.24. Prove that any surface of revolution can be locally conformally mapped onto the plane.

19.25. Prove that the metric $ds^2 = dx^2 + f(x)\,dy^2$, $0 < f(x) < \infty$, reduces to the form

$$ds^2 = g(u,v) \left(du^2 + dv^2\right)$$

(isothermal coordinates).

19.26. For a surface of revolution

$$x = f(r)\cos\varphi, \quad y = f(r)\sin\varphi, \quad z = g(r),$$

where $u \in (a,b)$ and $\varphi \in [0, 2\pi]$, find the isothermal coordinates (u,v) in which $\rho = \sqrt{u^2 + v^2}$ is a function of r. In particular, for the catenoid

$$x = \cosh z \cos\varphi, \quad y = \cosh z \sin\varphi, \quad z = z$$

find the representation of its radius-vector in such isothermal coordinates.

19.27. On a plane with coordinates (x,y), a matrix-valued function $F(x,y)$ is given. Verify that the function F defines a Riemannian metric. Find the coordinates in which the metric has the form $du^2 + G(u)\,dv^2$:

(a) $F(x,y) = \begin{pmatrix} 1 + 2x^2 & 4xy \\ 4xy & 1 + 2y^2 \end{pmatrix}$;

(b) $F(x,y) = \begin{pmatrix} 1 + 16x^2\left(x^2 + y^2\right)^2 & 16xy\left(x^2 + y^2\right)^2 \\ 16xy\left(x^2 + y^2\right)^2 & 1 + 16y^2\left(x^2 + y^2\right)^2 \end{pmatrix}$;

(c) in the unit disk $F(x,y) = \dfrac{1}{1 - x^2 - y^2}\begin{pmatrix} 1 - y^2 & xy \\ xy & 1 - x^2 \end{pmatrix}$;

(d) $F(x, y) = \dfrac{1}{\left(x^2 + y^2\right)^4} \begin{pmatrix} 4x^2 + \left(x^2 + y^2\right)^4 & 4xy \\ 4xy & 4y^2 + \left(x^2 + y^2\right)^4 \end{pmatrix}.$

19.28. Prove that on every real-analytic surface M^2, it is possible to introduce local isothermal coordinates.

19.29. Prove that any conformal mapping of a sphere on the plane is a composition of a motion of the sphere along itself and the stereographic projection of the sphere on this plane (see Fig. 83).

19.30. The Mercator projection of the Earth surface is defined as follows. On a chart, we introduce rectangular coordinates (x, y) such that with each line on the chart, we associate a line of constant azimuth (a fixed position of the compass arrow) on the Earth surface (see Fig. 84).

(a) Prove that in the Mercator projection, with a point on the Earth surface with spherical coordinates (θ, φ), we associate a point with the coordinates $x = \varphi$, $y = \ln \cot \theta/2$ on the chart.

(b) Write the metric of the sphere in the coordinates x, y of the Mercator projection.

19.31. Find the metric in the two-dimensional velocity space in special relativity theory.

19.32. In the previous problem, make the change of coordinates $v \to \tanh \chi$ (v is the speed of a moving point).

19.33. Write the metric of the previous problem in the polar coordinates of the unit disk.

19.34. Prove that in the class of the second-order surfaces, any surface of the second order of nonzero curvature is uniquely defined by its metric, even locally. We say that a surface S is uniquely defined by its metric if any surface isometric to it is obtained from S by a motion with a probable mirror reflection.

19.35. On the plane, let two isometric domains be given. Show that they are congruent.

19.36. Prove that two isometric domains on a sphere can always be obtained from each other by a rotation of the space, i.e., they are congruent.

19.37. On a cylindrical surface, let two isometric domains be given. Are they necessarily congruent in $\mathbb{R}^3$? Find the conditions under which any two isometric domains on a cylindrical surface are congruent in $\mathbb{R}^3$.

19.38. On the Beltrami pseudosphere, consider all possible disks of the same internal radius r. Reveal which of these disks are congruent or not in $\mathbb{R}^3$.

19.39. Are the metrics $ds^2 = \dfrac{a}{r^2}\left(dx^2 + dy^2\right)$ and $ds^2 = \dfrac{b}{r^2}\left(dx^2 + dy^2\right)$, $a \neq b$, $0 < r^2 = x^2 + y^2 < \infty$ isometric? Find the realization of each of them in $\mathbb{R}^3$ on the surface of a cylinder.

19.40. Show that any surface of the three-dimensional space is bent to give an "exit" to the four-dimensional space.

19.41. Prove that two complete cylindrical surfaces homeomorphic to the annulus are isometric if and only if their plane sections perpendicular to their rulings have the same length.

19.42. Construct the isometric mapping of a domain $\varphi_1 \le \varphi \le \varphi_2$, $z_1 \le z \le z_2$ on a cylinder $x = R\cos\varphi$, $y = R\sin\varphi$, $z = z$, $0 \le \varphi \le 2\pi$, $-\infty \le z \le +\infty$ onto an isometric domain on the surface of a right circular cone with known parameters.

19.43. Give an example of a complete Euclidean metric in the open unit disk.

19.44. Prove that a compact complete locally convex surface is convex as a whole.

19.45. Prove that a simply connected complete locally Euclidean two-dimen- sional manifold is as a whole isometric to the Euclidean space with the standard metric

$$dx^2 + dy^2.$$

19.46. Prove that it is not possible to define a complete Euclidean metric in the isothermal form.

19.47. In some range of variables (x, y), let the metric

$$ds^2 = E(x)\,dx^2 + G(y)\,dy^2$$

with continuous coefficients $E(x) > 0$ and $G(y) > 0$ be given. Prove that

(a) this metric is locally Euclidean;

(b) if this metric is given in a close simply connected domain, then it is possible to continuously extend it to the interior of the convex hull of the domain.

(c) if this metric is given in a multiply connected domain, then it can be continuously extended to the convex hull of the exterior boundary of the domain;

(d) this metric is Euclidean as a whole, i.e., its domain can be isometrically mapped as a whole into the Euclidean plane with the standard metric $du^2 + dv^2$.

Under which conditions on the domain and the coefficients is this metric complete?

19.48. Given the metric

$$ds^2 = \frac{1}{(1-r)^p}dr^2 + \frac{r^2}{(1-r)^q}d\varphi^2,$$

(a) assuming that r and φ are polar coordinates on the plane x, y, show that this metric is nondegenerate in the whole disk D: $x^2 + y^2 = r^2 < 1$.

(b) find all geodesic curves emanating from the origin and determine the values of p and q under which this metric is complete in the disk D;

(c) verify that this metric is locally Euclidean only if $p = 4$ and $q = 2$ and, for these p and q, find its isometric mapping onto the Euclidean plane with the standard metric.

19.49. Surfaces with the radius-vector $x = u\cos v$, $y = u\sin v$, $z = hv + f(u)$ are called general helical surfaces.

(a) Verify that on a part where v varies in some interval of length 2π, the sections of the general helical surface by the planes passing through the axis Oz consist of curves each of which is congruent to the curve $z = f(x)$, $y = 0$.

(b) Verify that for $h = 0$, the general helical surface transforms into a surface of revolution, and, for $f(u) = \text{const}$, it transforms into a helicoid.

(c) Prove the Boor theorem: the metric of the general helical surface is isometric to the metric of the form

$$ds^2 = dU^2 + G(U)\, d\varphi^2,$$

which is realized on a surface of revolution, i.e., any general helical surface is locally isometric to a surface of revolution, and, conversely, any surface of revolution in a neighborhood of any of its points except for the pole, is locally isometric to a general helical surface.

(d) The Boor theorem can be refined: a general helical surface locally admits a covering over a surface of revolution, i.e., it locally deforms to preserve the metric or, in other words, bends onto a surface of revolution.

(e) A general helical surface S admits a *bend of sliding along itself*; namely, there exists a deformation $S \to S_t$ expressed through the radii-vectors according to the law S_t: $\mathbf{r}(u, v; t) = \mathbf{r}(u, v + t)$ that transforms the points of the surface S into other points of the same surface with preservation of the metric. In this case, the distances between the points of S are equal to those between their images on S_t. Show that this bend is *trivial*, i.e., it is obtained by the motion of the whole surface in the space as a rigid body.

19.50. Show that if a surface admits a bend of sliding along itself, then its metric is isometric to the metric of rotation.

19.51. The ellipsoid $\dfrac{x^2}{a^2} + \dfrac{y^2}{b^2} + \dfrac{z^2}{c^2} = 1$ is represented in the parametric form

$$x = uf(v)\cos v, \quad y = uf(v)\sin v, \quad z = g(u),$$

where

$$f(v) = \frac{ab}{\sqrt{a^2\sin^2 v + b^2\cos^2 v}}, \quad g(u) = c\sqrt{1 - u^2}.$$

Choose the function $\psi(v)$ so that the deformation

$$x = u\sqrt{f^2 - t^2}\cos\psi, \quad y = u\sqrt{f^2 - t^2}\sin\psi, \quad z = \int\sqrt{g'^2 + t}\,du$$

with parameter t is a bend of some part of the ellipsoid. To which part of the ellipsoid does this deformation extend for small t?

§ 20. Gaussian Curvature and Mean Curvature

In Problems **20.1–20.4**, it is required to find the second quadratic form of the following surfaces.

20.1. (a) $\mathbf{r}(u, v) = (a \cosh u \cos v, \ a \cosh u \sin v, \ c \sinh u)$;

(b) $\mathbf{r}(u, v) = (a \sinh u \cos v, \ a \sinh u \sin v, \ c \cosh u)$.

20.2. $\mathbf{r}(u, v) = (u \cos v, \ u \sin v, \ u^2)$.

20.3. $\mathbf{r}(u, v) = (R \cos v, \ R \sin v, \ u)$.

20.4. $\mathbf{r}(u, v) = (u \cos v, \ u \sin v, \ ku)$.

20.5. Show that principal curvatures of the right conoid

$$x = u \cos v, \quad y = u \sin v, \quad z = f(v),$$

where $f(v)$ is an arbitrary smooth function whose derivative does not vanish, have different signs.

20.6. Find the Gaussian curvature and mean curvature of a surface composed of binormals to a given curve.

20.7. Find the Gaussian curvature and mean curvature of a surface composed of principal normals to a given curve.

20.8. Verify that for a torus of revolution, the integral Gaussian curvature vanishes and the integral mean curvature is different from zero, i.e.,

$$\iint K dS = 0, \quad \iint H dS \neq 0.$$

Note that the relation $\iint K dS = 0$ holds for any torus.

20.9. Find the curvature lines on the surface

$$x = \frac{a}{2}(u - v), \quad y = \frac{b}{2}(u + v), \quad z = \frac{uv}{2}.$$

20.10. Find the curvature lines of the helicoid

$$x = u \cos v, \quad y = u \sin v, \quad z = av.$$

20.11. Show that under a locally isometric covering of the helicoid

$$x = u \cos v, \quad y = u \sin v, \quad z = av$$

on the catenoid

$$x = \sqrt{u^2 + a^2} \cos v, \quad y = \sqrt{u^2 + a^2} \sin v,$$

$$z = a \ln\left(u + \sqrt{u^2 + a^2}\right),$$

the curvature lines pass to the asymptotic lines.

20.12. Find the curvature lines of the surface

$$\mathbf{r}\left(s, u\right) = \rho\left(s\right) + f\left(u\right)\mathbf{a} + g\left(u\right)\left(\mathbf{v}\left(s\right) \times \mathbf{a}\right),$$

where $\mathbf{v}\left(s\right) = \dot{\rho}\left(s\right)$, $\left|\mathbf{v}\left(s\right)\right| = 1$, $\langle\mathbf{v}\left(s\right), \mathbf{a}\rangle = 0$, $\left|\mathbf{a}\right| = 1$, and $\mathbf{a}$ is a constant vector.

20.13. The plane curve γ is given by the equation $\rho = \rho\left(s\right)$, where s is a natural parameter, $k = k\left(s\right)$ is its curvature, $0 < k < 1/a$, $\mathbf{n}$ is the principal normal to γ and $\mathbf{b}$ is the unit vector of the normal to the plane of the curve γ. The surface S is given by the equation

$$\mathbf{r}\left(s, \varphi\right) = \rho\left(s\right) + a\mathbf{n}\left(s\right)\cos\varphi + a\mathbf{b}\sin\varphi.$$

(a) Verify that the surface S is regular.
(b) Find the Gaussian curvature of the surface S.
(c) Find the mean curvature of the surface S.
(d) Find the curvature lines of the surface S.

20.14. Find the curvature lines of the surface

$$\mathbf{r}\left(s, \varphi\right) = \rho\left(s\right) + a\mathbf{n}\left(s\right)\cos\varphi + a\mathbf{b}\left(s\right)\sin\varphi;$$

here, $\mathbf{n}$ and $\mathbf{b}$ are unit vectors of normal and binormal of a curve $\rho = \rho\left(s\right)$ with natural parameter s, curvature $k\left(s\right) < 1/a$, and torsion $\varkappa\left(s\right)$.

20.15. Find the curvature lines of the surface

$$\mathbf{r}\left(u, v\right) = \left(u\left(3v^2 - u^2 - \frac{1}{3}\right), v\left(3u^2 - v^2 - \frac{1}{3}\right), 2uv\right).$$

20.16. Assume that the first quadratic form of a surface is

$$ds^2 = Edu^2 + Gdv^2.$$

Prove that

$$K = -\frac{1}{2\sqrt{EG}}\left\{\frac{\partial}{\partial v}\left(\frac{\partial E/\partial v}{\sqrt{EG}}\right) + \frac{\partial}{\partial u}\left(\frac{\partial G/\partial u}{\sqrt{EG}}\right)\right\}.$$

20.17. Find the expression for the Gaussian curvature of a surface with coordinates such that in these coordinates, its first quadratic form looks like

$$ds^2 = du^2 + G\left(u, v\right)dv^2.$$

20.18. Find the Gaussian curvature of a surface whose first quadratic form is

$$ds^2 = B\left(u, v\right)\left(du^2 + dv^2\right).$$

20.19. Find the curvature of a surface whose first quadratic form is

$$ds^2 = du^2 + e^{2u}dv^2.$$

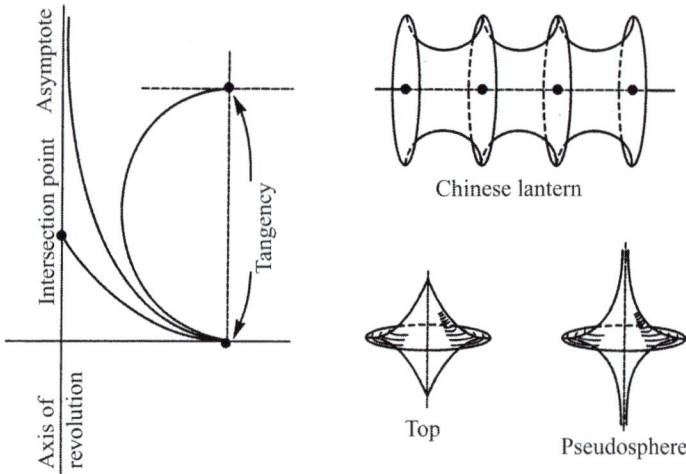

Figure 85

20.20. Prove that the Gaussian curvature K of a two-dimensional surface is expressed only through its metric, i.e., through the coefficients of the first quadratic form and their derivatives. This implies that the Gaussian curvature does not change under isometries of the surface.

20.21. Is it possible to choose λ, φ, and ψ in such a way that the surface

$$\mathbf{r}\,(u, \theta) = (\varphi\,(u) \cos \lambda\theta,\ \varphi\,(u) \sin \lambda\theta,\ \psi\,(u))$$

has the Gaussian curvature equal to 1?

(a) Find an example different from the sphere.

(b) Describe all such surfaces (see Problem 6.4).

20.22. Is it possible to choose λ, φ, and ψ in such a way that the surface

$$\mathbf{r}\,(u, \theta) = (\varphi\,(u) \cos \lambda\theta,\ \varphi\,(u) \sin \lambda\theta,\ \psi\,(u))$$

has the Gaussian curvature equal to -1? Find an example different from the pseudosphere.

20.23. Prove that any surface of revolution of constant negative curvature refers to one of the following three types:

(1) the Minding coil (looks like a Chinese lantern);

(2) the Minding top, i.e., the surface with a peak on the axis of revolution;

(3) Beltrami pseudosphere.

The graphs of the rulings corresponding to these three types of surfaces are presented in Figs. 85 and 86.

20.24. Prove that each surface of revolution (closed or with the boundary along the parallel) admits a bending in the class of surfaces of revolution with mapping of parallels into parallels and preservation of the form of the area.

Figure 86 Minding coil, Minding top and Beltrami pseudosphere

20.25. Prove that under the projective transformation of the space $\mathbb{R}^3$, the property of a point of a surface to be elliptic, parabolic, or planar is preserved.

20.26. Prove that all points of a surface of the second order have the same type (elliptic, hyperbolic or parabolic).

20.27. In the isothermal coordinates (u, v), let a surface be given as

$$x = \operatorname{Re} F_1(w), \quad y = \operatorname{Re} F_2(w), \quad z = \operatorname{Re} F_3(w),$$

where $F_k(w)$, $k = 1, 2, 3$, are analytic functions of the complex argument $w = u + iv$. Such a representation is called the *Weierstrass representation*.

(a) Which necessary and sufficient condition must these functions satisfy in order for the surface to be minimal?

(b) Calculate the metric form of this minimal surface.

20.28. Let a minimal surface M given by the Weierstrass representation (see Problem 20.27) be subject to the deformation M_t of the form

$$x = \operatorname{Re}\left(F_1(w)\,e^{it}\right), \quad y = \operatorname{Re}\left(F_2(w)\,e^{it}\right), \quad z = \operatorname{Re}\left(F_3(w)\,e^{it}\right)$$

with deformation parameter t. Prove the fulfillment of the following assertions:

(a) this deformation is a bending of the surface M. In particular, all surfaces M_t are isometric to one another;

(b) the bendings are executed in the class of minimal surfaces;

(c) all surfaces M_t have parallel normals at the points corresponding to one another under the isometry.

20.29. The minimal surfaces defined by formulas of Problem 20.28 are said to be *associated*. Prove that the bending of a minimal surface is executed only in the class of associated minimal surfaces.

20.30. Consider the surface M given by

$$x = u - \frac{4}{3}u^3 + 4uv^2, \quad y = v - \frac{4}{3}v^3 + 4u^2v, \quad z = 2\left(u^2 - v^2\right),$$

and its deformation M_t,

$$x = u - \left(\frac{4}{3}u^3 - 4uv^2\right)\cos 2t + \left(4u^2v - \frac{4}{3}v^3\right)\sin 2t,$$

$$y = v + \left(\frac{4}{3}u^3 - 4uv^2\right)\sin 2t + \left(4u^2v - \frac{4}{3}v^3\right)\cos 2t$$

$$z = 2\left(u^2 - v^2\right)\cos t - 4uv\sin t.$$

Prove that this deformation is a bending and that the initial surface M and all its deformations M_t are minimal surfaces. For which value of the parameter is the surface M_t a mirror image of the surface $M_{t'}$?

20.31. Let a surface M be such that there exists its continuous bending on its central-symmetric image M^*. In these cases, we say that M is *coverable* on M^*. Prove that M is also coverable on its mirror image (reflection in the plane).

20.32. Prove that a minimal surface is coverable on its mirror image.

20.33. Prove that on a surface of constant curvature, the neighborhoods of any two of its points have a three-parameter family of isometric mappings onto one another. Note that in this case, the centres of the neighborhoods need not pass one to another.

20.34. Prove that there exists no surface on which neighborhoods of any two of its points have exactly a two-parameter family of isometric mappings one onto another (see the previous problem). Note that there exist surfaces having exactly a one-parameter family of isometries. These are surfaces of revolution, which are not surfaces of constant curvature.

20.35. Prove that if two isometric surfaces have parallel normals at the points corresponding to each other under the isometry, then their mean curvatures are either equal or differ one from the other by the sign.

20.36. Prove that if two isometric surfaces have parallel normals at the points corresponding to each other under the isometry and the mean curvature of at least one of these surface is different from zero, then these surfaces are congruent. If both surfaces are minimal, then they can be not congruent. Moreover, any minimal surface admits a bending with preservation of directions of normals at the points corresponding to each other under the isometry.

20.37. Assume that it is known that two metrics $ds^2 = g_{ij}du^i du^j$ and $d\sigma^2 = \gamma_{ij}d\xi^i d\xi^j$, $1 \le i, j \le n$, are locally isometric. Deduce the set of equation to which the isometry $f\colon (u) \to (\xi)$, where (u) and (ξ) denote the neighborhoods of the corresponding points in the domains of the metrics, must satisfy. Separately, consider the case $n = 2$ and show that in this case, the search for the desired mapping can be reduced to the solution of a quasilinear system.

20.38. On the plane of variables (u, v), the metric

$$ds^2 = e^{-2u^2}\left(du^2 + dv^2\right)$$

is given. Verify that it is a metric of positive curvature, that it is not complete and that it cannot be realized on any convex surface in $\mathbb{R}^3$.

20.39. (a) Verify that the metric

$$ds^2 = \frac{du^2 + dv^2}{-Kr^2 \left(A + \ln\left(1/r\right)\right)^2}$$

where $r^2 = u^2 + v^2$, has a constant curvature $K < 0$.

(b) Is this metric complete in the annulus $0 < r < e^A$?

20.40. (a) Verify that the metric

$$ds^2 = \frac{a^2 r^{2a-2}}{\left(1 - Ar^{2a}\right)^2} \left(du^2 + dv^2\right),$$

where $r^2 = u^2 + v^2$ and $A = \dfrac{-K}{4}$, has a constant curvature $K < 0$.

(b) Let $a > 0$. Is the metric of item (a) complete in the annulus $0 < r < \left(\dfrac{1}{A}\right)^{1/2a}$?

(c) Consider two metrics from item (a): the first for $a > 0$ and the second for $a < 0$. Find the maximal domains of existence of both metrics and verify whether these metric are isometric in these domains.

20.41. (a) Verify that the metric

$$ds^2 = \frac{a^2}{\left(r\sqrt{-K}\left(A\sin\left(a\ln r\right) + B\cos\left(a\ln r\right)\right)\right)^2} \left(du^2 + dv^2\right),$$

where $r^2 = u^2 + v^2$ and $A^2 + B^2 = 1$, has a constant curvature $K < 0$.

(b) Find the maximal domains of existence of the metric of item (a) and verify whether it is complete in these domains.

20.42. For the metrics of the previous three problems, find their isometric immersions in $\mathbb{R}^3$ in the form of surfaces of revolution of the type $x = f(r)\cos n\varphi$, $y = f(r)\sin n\varphi$, $z = g(r)$, where $u = r\cos\varphi$ and $v = r\sin\varphi$. Describe the domains of immersion of these metrics depending on the values of the integer parameter n.

20.43. Consider a family of surfaces of revolution obtained one from another by a parallel translation along the axis of revolution. Consider a surface of revolution M having the same axis and intersecting the surfaces of the family at a right angle. Prove that at the intersection point, the surfaces of the family and the surface M has Gaussian curvatures of opposite sign and equal modules.

§ 21. Parameterization of the Known Two-Dimensional Surfaces

In this section, it is required to verify that the formulas given in the problems define parameterizations of embeddings or immersions of the known surfaces.

21.1. The surface with singularities parametrically given in $\mathbb{R}^3$ by the equations

$$\mathbf{r}\,(\theta, \varphi) = ((1 + \cos 2\theta) \cos 2\varphi,\ (1 + \cos 2\theta) \sin 2\varphi,\ \sin 2\theta \sin \varphi)\,,$$

$$-\frac{\pi}{2} \le \theta \le \frac{\pi}{2},\quad 0 \le \varphi \le 2\pi,$$

is one of the models (see Fig. 87) of the projective plane $\mathbb{R}P^2$. Describe the singular points of this surface.

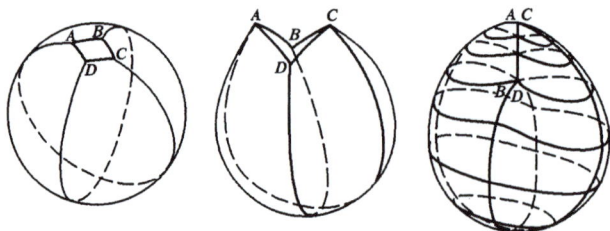

Figure 87 Gluing of the projective plane from the square

Prove that the sphere

$$\mathbf{r}\,(\theta, \varphi) = (\cos \theta \cos \varphi,\ \cos \theta \sin \varphi,\ \sin \theta)$$

is a two-sheeted covering of this surface. Find the correspondence between the pairs of points of the sphere and the points of $\mathbb{R}P^2$ in terms of the parameters θ and φ.

Note that if from the described model of the projective plane, we cut off a small disk by a plane, then the remaining part is a model of the Möbius band. It is called the *crossed cap* (see Fig. 88). The equation $\varphi = 0$ corresponds to the line of self-intersection of the cap. The pinch points (Fig. 89) are $(\theta, \varphi) = (0, 0)$ and $\theta = \pi/2$.

21.2. Recall that one of the models of the projective plane is the Boy surface in $\mathbb{R}^3$ (see Figs. 90 and 91).

The Aperi parameterization of the Boy surface is

$$r\,(\theta, \varphi) = A \begin{pmatrix} r_1 \cos 2\varphi \\ r_1 \sin 2\varphi \\ 1 \end{pmatrix} + B \begin{pmatrix} r_2 \cos \varphi \\ -r_2 \sin \varphi \\ 0 \end{pmatrix},$$

$$A = \frac{\cos^2 \theta}{1 - b \sin 3\varphi \sin 2\theta},\quad B = \frac{\sin \theta \cos \theta}{1 - b \sin 3\varphi \sin 2\theta}.$$

Here, $r_1 = \sqrt{2}/3$ and $r_2 = 2/3$. For $1/\sqrt{6} < b < 1$, there are no pinch points on the surface.

Prove that the standard sphere is a two-sheeted covering of this parameterized surface.

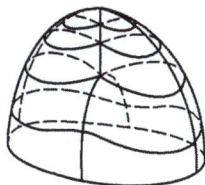

Figure 88 Crossed cap

Figure 89 Neighborhood of a pinch point

Figure 90 Boy surface, the immersion of the projective plane in $\mathbb{R}^3$

Figure 91 "Transparent" Boy surface. Here, the structure of the set of its self-intersection points is clearly seen

The structure of the Boy surface is shown in Figs. 92 and 93.

21.3. Let θ and φ be the spherical coordinates on S^2 (the angle θ is measured from the horizontal plane). Prove that the formulas

$$\mathbf{r}(\theta, \varphi) = (w, x, y, z),$$

$$w(\theta, \varphi) = (\cos\theta)^2 \cos 2\varphi, \quad x(\theta, \varphi) = \sin 2\theta \cos\varphi,$$

$$y(\theta, \varphi) = \sin 2\theta \sin\varphi, \quad z(\theta, \varphi) = (\cos\theta)^2 \sin 2\varphi$$

define the embedding of the projective plane in $\mathbb{R}^4$.

21.4. Prove that the following vector functions define the immersions of the Klein bottle in $\mathbb{R}^3$ (see Fig. 94).

(a) $\mathbf{r}(u, v) = (R_x, R_y, R_z^2)$, where R_x, R_y, and R_z are components of the following vector function in the Cartesian coordinates:

$$\mathbf{R}(u, v) = \mathbf{R}_0(u) + \rho(u)(\mathbf{e}_1(u)\cos v + \mathbf{e}_2 \sin v).$$

Here,

$$\mathbf{R}_0(u) = \begin{pmatrix} a \sin 2u \\ 0 \\ b \cos u \end{pmatrix},$$

Figure 92 Start of the process of constructing the Boy surface from the Möbius band

Figure 93 Finish of the process of constructing the Boy surface from the Möbius band

$$\mathbf{e}_1\left(u\right)=\begin{pmatrix} b\sin u \\ 0 \\ a\cos 2u \end{pmatrix}\frac{1}{|R_0'|}, \quad \mathbf{e}_2=\begin{pmatrix} 0 \\ 1 \\ 0 \end{pmatrix}.$$

In other words, $\mathbf{e}_1$ and $\mathbf{e}_2$ are unit vectors orthogonal to $\mathbf{R}_0'$,

$$\rho=\rho_0+\rho_1\sin^{2n}\left(u-u_0\right).$$

Here, for example, it is possible to set

$$a=\frac{1}{4}, \quad b=1, \quad \rho_0=0.1, \quad \rho_1=0.6, \quad u_0=0.3, \quad n=4.$$

(b) One more parameterization in which a is a real number is the following:

$$\mathbf{r}_a\left(u,v\right)=\begin{pmatrix} \left(a+\cos\dfrac{u}{2}\sin v-\sin\dfrac{u}{2}\sin 2v\right)\cos u \\[2mm] \left(a+\cos\dfrac{u}{2}\sin v-\sin\dfrac{u}{2}\sin 2v\right)\sin u \\[2mm] \sin\dfrac{u}{2}\sin v+\cos\dfrac{u}{2}\sin 2v \end{pmatrix}.$$

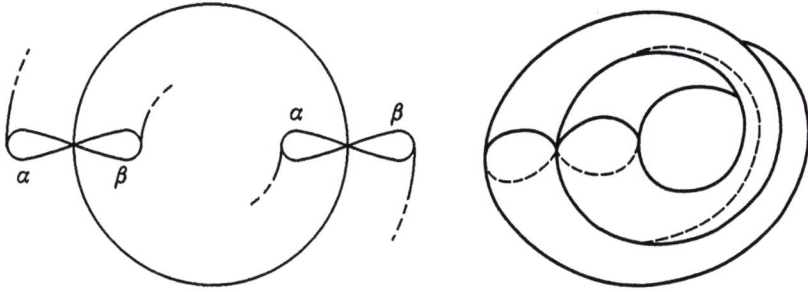

Figure 94 Immersion of the Klein bottle

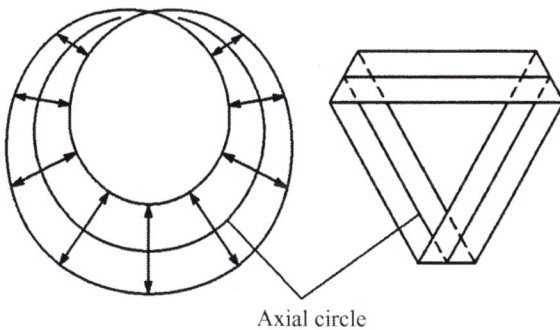

Axial circle

Figure 95 Möbius band in $\mathbb{R}^3$

21.5. Prove that the following vector function defines the parameterization of the Möbius band in $\mathbb{R}^3$:

$$\mathbf{r}\left(t, \varphi\right) = \left(\left(1 + t\sin\frac{\varphi}{2}\right)\cos\varphi, \left(1 + t\sin\frac{\varphi}{2}\right)\sin\varphi, t\cos\frac{\varphi}{2}\right).$$

Prove that for $t \in \left(\dfrac{1}{2}, \dfrac{1}{2}\right)$ this mapping is an embedding.

21.6. Define an open Möbius band as the skew (twisted) product of a circle by an interval. The circle is said to be axial, and the length of the interval is called the width of the Möbius band (see Fig. 95).

(a) Prove that the Möbius band of infinite width with plane metric is not embedded in $\mathbb{R}^3$.

(b) Find the maximum ratio between the length of the axial circle and the width of the Möbius band under which there exists an isometric embedding of the plane Möbius band in $\mathbb{R}^3$.

21.7. Prove that the following algebraic surface in $\mathbb{R}^3$ is the crossed cap, i.e., the model of the Möbius band:

$$\left(k_1 x^2 + k_2 y^2\right)\left(x^2 + y^2 + z^2\right) - 2z\left(x^2 + y^2\right) = 0,$$

where $k_1 \neq k_2$. Prove that this surface can be parameterized as follows:

$$x = \frac{\cos\theta\cos\varphi}{k_1\cos^2\varphi + k_2\sin^2\varphi}, \qquad y = \frac{\cos\theta\sin\varphi}{k_1\cos^2\varphi + k_2\sin^2\varphi},$$

$$z = \frac{1+\sin\theta}{k_1\cos^2\varphi + k_2\sin^2\varphi}.$$

(a) In this case, it is required to prove that the first (algebraic) equation defines a *surface*, i.e., the parameterization presented above indeed defines the set of *all* solutions of this equation.

(b) Prove that the presented formulas define a smooth regular parameterization outside singular points. Find the singular points.

(c) Verify that under this parameterization, the square $-\pi/2 \le \theta \le \pi/2$, $0 \le \varphi \le \pi$ is glued into the Möbius band.

21.8. Prove that the formulas $x = u^2 - v^2$, $y = uv$, $z = uw$, and $t = vw$ define a regular immersion of the sphere $u^2 + v^2 + w^2 = 1$ in $\mathbb{R}^4$. Note that under this mapping, opposite points of the sphere pass into a single point, and, therefore, there arises a regular mapping of $\mathbb{R}P^2$ into $\mathbb{R}^4$. Prove that in this case, one obtains an embedding of the projective plane in $\mathbb{R}^4$.

§ 22. Surfaces in $\mathbb{R}^3$

22.1. Prove that if at some point of a surface in $\mathbb{R}^3$ the Gaussian curvature $K > 0$, then the surface locally lies to one side of the tangent plane at this point.

22.2. Let two surfaces M_1 and M_2 be tangent to each other at some common point. Moreover, in some neighborhood of this point, let both surfaces lie to one side of the plane, and let the surface M_1 entirely lie inside the surface M_2 (see Fig. 96). Prove that the nonstrict inequality $K_1 \ge K_2$ holds for their Gaussian curvatures. This fact is local.

22.3. Is it possible to realize the Möbius band in $\mathbb{R}^3$ as a smooth surface whose curvature is everywhere positive?

22.4. Is it possible to embed the Möbius band in $\mathbb{R}^3$ as a smooth surface that is projected on a plane in a single-valued way?

22.5. (a) Prove that a nonorientable surface (closed or with boundary) cannot be immersed in $\mathbb{R}^3$ in such a way that at each of its points, either the Gaussian curvature K is strictly positive or the Gaussian curvature K vanishes, but the mean curvature H is different from zero.

(b) Construct an embedding of the plane (i.e., with locally Euclidean metric) Möbius band in $\mathbb{R}^3$. Explain why the existence of such an embedding does not contradict the previous item.

22.6. Prove that a closed smooth surface in $\mathbb{R}^3$ always has a point at which

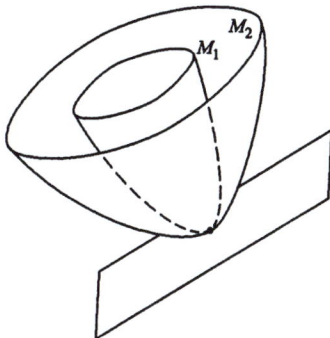

Figure 96

(a) the Gaussian curvature is no less than 0;

(b) the Gaussian curvature is strictly greater than 0.

22.7. (a) Prove that a smooth closed surface of negative Gaussian curvature cannot be immersed in $\mathbb{R}^3$.

(b) Prove that a smooth closed surface of nonpositive Gaussian curvature cannot be immersed in $\mathbb{R}^3$.

(c) Prove that a plane torus cannot be smoothly and isometrically immersed in $\mathbb{R}^3$.

22.8. (a) Prove that on a smooth oriented surface immersed in $\mathbb{R}^3$ and having the genus less than 2, there necessarily exists a point at which the Gaussian curvature K is strictly less than 0.

(b) Let the torus (or Klein bottle) be smoothly immersed in $\mathbb{R}^3$. Prove that there necessarily exists a point at which $K < 0$.

22.9. Prove that the spherical mapping of a minimal surface is conformal in a neighborhood of every point being not planar.

22.10. Prove the following assertions:

(a) on the Möbius band, there exists a metric of positive constant curvature. This is an example of a metric that is not isometrically immersed in $\mathbb{R}^3$ (see Problem 22.3);

(b) on the Möbius band, it is possible to introduce a metric of constant negative curvature.

22.11. Is it possible to isometrically embed the Möbius band with metric of constant negative curvature in $\mathbb{R}^3$?

22.12. Is it possible to realize the Möbius band in $\mathbb{R}^3$ as a domain on a surface of revolution?

22.13. Consider a surface of revolution with respect to the axis z and introduce the action of the group $\mathbb{Z}/2$ on it as follows: $(x, y, z) \to (-x, -y, z)$. Take the quotient of the surface by this action. We obtain a new surface with metric.

Consider the example of a one-sheeted hyperboloid of revolution and iso-metrically realize the quotient surface obtained by the method described above in $\mathbb{R}^3$ as a surface of revolution.

22.14. Find the area of the spherical image of a surface

$$r = r\left(u, v\right), \quad u_1 < u < u_2, \quad v_1 < v < v_2,$$

with the first quadratic form

$$ds^2 = du^2 + G dv^2.$$

22.15. Find the area of the spherical image of an elliptic paraboloid.

22.16. Find the image of the Gaussian mapping for

(a) a one-sheeted hyperboloid;

(b) a two-sheeted hyperboloid;

(c) a catenoid.

22.17. Find the spherical image of a torus realized in $\mathbb{R}^3$ in the standard way as a surface of revolution.

22.18. Prove that the area of the spherical image of one sheet of the two–sheeted hyperboloid is less than 2π.

22.19. Study the spherical images of the following convex surfaces:

(a) an ovaloid, i.e., a smooth closed convex surface. Prove that its spherical image covers the whole sphere, and if the ovaloid is strictly convex, then the spherical image is one-to-one;

(b) the paraboloid $z = x^2 + y^2$;

(c) the surface obtained by rotating the curve $z = x^2$, $z \leq a$ smoothly extended by the tangent up to infinity around the axis Oz. Show that choosing $a > 0$, we can obtain that the spherical image of such a surface can fill in a domain on the sphere having an arbitrary area $S < 2\pi$.

22.20. Construct an example of a complete surface of strictly positive curvature whose spherical image has the area less than 2π.

22.21. Prove that there exists a conformal mapping of a surface of revo-lution onto the plane under which the meridians and parallels pass to straight lines on the plane.

22.22. Prove that there exists a conformal mapping of a surface of revo-lution onto the plane under which the meridians pass to lines passing through the origin and the parallels pass to circles centred at the origin.

22.23. Calculate the integral $\iint\limits_M |K|\, dS$, where K is the Gaussian curva-ture of a surface M if M is

(a) the ellipsoid;

(b) the elliptic paraboloid;

(c) the torus.

22.24. A rectangle P bounded by two parallels and two meridians is cut from the sphere. The boundaries corresponding to meridians are identified according to the equality of the length of segment being glued so that we obtain a manifold M homeomorphic to the annulus. Verify that the metric on M is analytic and obtain an embedding of M in $\mathbb{R}^3$ as a surface of revolution.

22.25. Under the conditions of the previous problem, the identification of the arcs of the meridians is performed in such a way that topologically we obtain the Möbius band. Verify that the obtained metric of positive curvature on the Möbius band is smooth.

22.26. Find all surfaces of revolution of constant negative Gaussian curvature $K = -1$. They are called the Minding surfaces. In the unit disk with the Lobachevskii metric (Poincaré model), find domains isometric to the Minding surfaces (after a cut along the meridian) and their universal coverings.

22.27. Find all surfaces of revolution of constant mean curvature.

§ 23. Topology of Two-Dimensional Surfaces

In this section, we consider partitions of a surface into finitely many closed polygons which intersect one another along common edges and vertices. A partition in which any two polygons are either disjoint or intersect along *one* common edge or *one* common vertex is said to be *regular*. A partition of a surface into triangles is called a *triangulation*. Further in this section, by V we denote the number of vertices, by E the number of edges and by F the number of faces (polygons) of the partition of the surface. The number $V - E + F$ is called the *Euler characteristic* $\chi(M)$ of the surface M. This number is independent of the choice of a partition of the surface, and hence it is a topological invariant of the surface.

23.1. Construct some triangulations of the sphere, of the torus, and of the projective plane and find the Euler characteristics of these surfaces.

23.2. From the fact that the Euler characteristic of the sphere is $\chi\left(S^2\right) = 2$ deduce the existence of exactly five combinatorially right polygons (see Fig. 97). Recall that a partition of a surface is said to be combinatorially right if all faces have the same number of edges and the same number of edges meet at each vertex.

23.3. Prove that the number of vertices of a right triangulation of a two-dimensional closed surface satisfies the estimate

$$V \geq \frac{7 + \sqrt{49 - 24\chi(M)}}{2}.$$

In particular, a right triangulation of the torus contains no less than seven vertices, and a right triangulation of $\mathbb{R}P^2$ contains not less than six vertices.

23.4. Present an example of a right triangulation:

(a) of the torus with seven vertices;

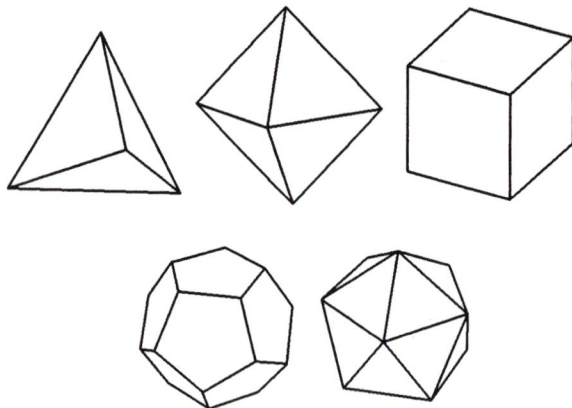

Figure 97 Tetrahedron, octahedron, cube, dodecahedron and icosahedron

(b) of the projective plane with six vertices.

23.5. Prove that the minimum number of triangles of a right triangulation of the torus is equal to 14. Construct an example of such a triangulation. Compare with the example from the previous problem.

23.6. (a) Reveal how many right triangulations of the torus with the minimum number of vertices exist. How many right triangulations of the torus with the least number of triangles exist? Do right triangulations of the torus minimal in one sense but not minimal in another exist?

(b) Answer these questions for the projective plane.

23.7. Prove the Heawood inequality

$$\mathrm{col}\,(M) \leq \frac{7 + \sqrt{49 - 24\chi\,(M)}}{2}$$

for all closed two-dimensional surfaces, except for the sphere and the Klein bottle. Here, $\mathrm{col}\,(M)$ denotes the chromatic number of the surface M, i.e., the minimal number of colors needed for coloring the partition of the surface in such a way that polygons with a common edge have different colors. What can we say about this inequality for the Klein bottle and the sphere?

23.8. Prove that $\mathrm{col}\,(M_g) = \mathrm{col}\,(M_g^k)$ and $\mathrm{col}\,(N_\alpha) = \mathrm{col}\,(N_\alpha^k)$. Here, M_g is a sphere with g handles, N_α is a sphere with α Möbius bands, M_g^k is a sphere with g handles from which k disjoint disks are removed, and N_α^k is a sphere with α Möbius bands from which k disjoint disks are removed (see Fig. 98).

23.9. Present an example of a chart on the projective plane which does not admit coloring into five colors.

23.10. Prove that the operation of the connected sum of surfaces is independent of the method for gluing the boundary circles, i.e., on the choice of their orientation in gluing.

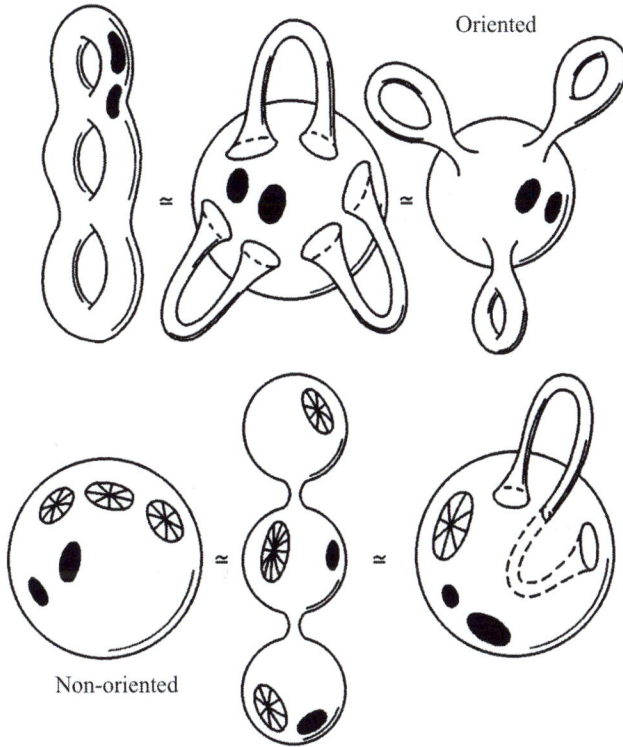

Oriented

Non-oriented

Figure 98 Two-dimensional surfaces with boundary.

23.11. Prove that $T^2 \sharp \mathbb{R}P^2 = \mathbb{R}P^2 \sharp \mathbb{R}P^2 \sharp \mathbb{R}P^2$. Here, $X \sharp Y$ is the connected sum of the surfaces X and Y.

23.12. Cut the Klein bottle into two Möbius bands.

23.13. Cut a pretzel (a sphere with two handles) in such a way that a plane connected dodecagon is obtained.

23.14. Cut the sphere with three handles in such a way that a plane connected 12-gon whose vertices represent one point of the surface is obtained.

23.15. Prove that the connected sum $X \sharp Y$ of two surfaces X and Y is orientable if and only if both surfaces are orientable.

23.16. Prove that

$$\chi(X \sharp Y) = \chi(X) + \chi(Y) - 2.$$

23.17. Prove that

$$\chi(M_g) = 2 - 2g, \quad \chi(N_\alpha) = 2 - \alpha.$$

23.18. Prove that

$$\chi(S^2 \setminus kD^2) = 2 - k.$$

Here, k disjoint disks are removed from the sphere.

23.19. If the sphere is partitioned into n-gons and k edges occur at each of the vertices, then
$$\frac{1}{n} + \frac{1}{k} = \frac{1}{2} + \frac{1}{E}.$$

The graph "n houses and m wells" is the graph whose $n + m$ vertices are divided into two disjoint groups of n and m elements, respectively, and, moreover, each vertex from one group is connected with each vertex of the other group by a unique edge. There are no other edges in this graph.

23.20. Prove that on the sphere (plane), it is not possible to arrange the following graphs so that any two different edges intersect one another by no more than a common vertex:

(a) the graph "three houses and three wells";

(b) the graph with five vertices each pair of whose different vertices is connected by a unique edge.

23.21. Prove that the graph "three houses and three wells" can be arranged on the projective plane (Möbius band) without self-intersections.

23.22. Prove that the combinatorially right partition of a torus consists of triangles, quadrangles or hexagons (for the definition of a combinatorially right partition, see Problem 23.2).

23.23. Let a closed surface Q be partitioned into hexagons in a combinatorially right way, and, moreover, let four faces meet at each vertex. Prove that if the number of vertices is odd, then Q is non-orientable.

23.24. On a closed surface, let three curves p, q and r with common endpoints be drawn so that they have no common interior points pairwise. If a cut along one of the curves $p \cup q$, $q \cup r$, or $r \cup p$ leaves the surface Q connected, then at least one of the remaining two curves has this property.

23.25. Show that if we cut off a hole from a closed non-orientable surface N_α, then the obtained surface can be placed in the space $\mathbb{R}^3$ without self-intersections.

23.26. Prove that the graph "four houses and four wells" cannot be arranged on the projective plane without self-intersections, but this is possible for the torus.

23.27. Prove that if the graph "m houses and n walls" can be arranged on a surface Q without self-intersections, then
$$\chi(Q) \leq m + n - \frac{mn}{2}.$$

§ 24. Curves on Surfaces

In $\mathbb{R}^3$, consider a two-dimensional surface M^2 with parameterization $\mathbf{r}(u, v)$ and the second quadratic form $L\,du^2 + 2M\,du\,dv + N\,dv^2$. Consider two families of curves on M^2. These families are said to be *conjugate* if at each point P the directions of curves of the first and second families passing through P are conjugate with respect to the second quadratic form of the surface. More exactly, at P, let a curve of the first family have the direction (ξ_1, η_1), and the line of the second family have the direction (ξ_2, η_2). Then the conjugacy condition can be written as

$$L\xi_1\xi_2 + M(\xi_1\eta_2 + \xi_2\eta_1) + N\eta_1\eta_2 = 0.$$

If the curves of the first family are known, then this condition gives the differential equation of curves of the conjugate family.

A curve $(u(t), v(t))$ on the surface M^2 is said to be *asymptotic* if at each point the velocity vector of the curve has an asymptotic direction with respect to the second form of the surface. The respective differential equation can be written as

$$L(u')^2 + 2Mu'v' + N(v')^2 = 0.$$

The asymptotic curve is defined by the condition that the normal curvature of the surface vanishes along the asymptotic curve.

A curve $(u(t), v(t))$ is called a *curvature line* of the surface M^2 if at each of its points, the velocity vector has a principal direction. At each point of the surface that is not umbilical, there exist two principal directions; they are conjugated and orthogonal. If (ξ_1, η_1) and (ξ_2, η_2) are vectors having principal directions, then they satisfy the following two equations:

$$E\xi_1\xi_2 + F(\xi_1\eta_2 + \xi_2\eta_1) + G\eta_1\eta_2 = 0,$$

$$L\xi_1\xi_2 + M(\xi_1\eta_2 + \xi_2\eta_1) + N\eta_1\eta_2 = 0.$$

This implies that

$$\begin{pmatrix} \xi_1^2 & -\xi_1\eta_1 & \eta_1^2 \\ E & F & G \\ L & M & N \end{pmatrix} = 0.$$

In particular, a curve $(u(t), v(t))$ is a curvature line if and only if it satisfies the differential equation

$$\begin{pmatrix} (u')^2 & -u'v' & (v')^2 \\ E & F & G \\ L & M & N \end{pmatrix} = 0.$$

24.1. Let the surface

$$x = u + v, \quad y = u^2 + v^2, \quad z = 2uv$$

and the family of lines $u^2 - cv^2 = c$ on it be given. Find the equation of the family conjugated to it.

24.2. Let σ be a surface, l be a line, $\{\gamma_1\}$ be a family of sections of the surface by planes passing through the straight line l, and let $\{\gamma_2\}$ be a family of curves at which the cones with vertices on l tangent the surface. Show that the families $\{\gamma_1\}$ and $\{\gamma_2\}$ compose a conjugate net.

24.3. On the surface

$$x = u \cos v, \quad y = u^2 \sin v, \quad z = hv,$$

a family of curves v is given. Find the conjugated family.

24.4. Find surfaces $z = f(x, y)$ with a conjugated net such that it projects on the net of coordinate lines on the plane xOy.

24.5. Show that a mapping of two surfaces onto each other under which each conjugated net passes to the orthogonal net of the other transforms the second quadratic form of one surface into the form proportional to the first quadratic form of the other.

24.6. Find the asymptotic curves of the one-sheeted hyperboloid

$$\frac{x^2}{a^2} + \frac{y^2}{b^2} - \frac{z^2}{c^2} = 1.$$

24.7. Find the asymptotic lines of the surface

$$z = f(x) - f(y)$$

and define the function f in such a way that the curves of one asymptotic family are orthogonal to the asymptotic curves of the other.

24.8. Find the asymptotic curves of the surface $z = xy^3 - yx^3$ passing through the point $(1, 2, 6)$.

24.9. Find the asymptotic curves of the catenoid.

24.10. Show that on the right helicoid, one family of asymptotic curves consists of lines and the other consists of helices.

24.11. Prove that

(a) the coordinate lines $u = \text{const}$ are asymptotic if and only if the coefficient N of the second quadratic form is equal to 0;

(b) the coordinate net consists of asymptotic curves if and only if the coefficients L and N of the second quadratic form are equal to 0.

24.12. Prove that if a surface is minimal, then the asymptotic curves on it are orthogonal.

24.13. Prove that on the plane any curve is asymptotic, and, vice versa, a surface on which any curve is asymptotic is a part of the plane.

24.14. Prove that if an asymptotic curve on a surface is plane, then it is either a parabolic or straight line.

24.15. Find asymptotic lines of a surface composed of midpoints of chords of the helical line.

24.16. Let the curvature of a surface of revolution be strictly less than 0. Prove that on this surface, there are no closed asymptotic curves.

24.17. Let the surface be composed of principal normals of a curve γ. Prove that the curve γ is an asymptotic curve of this surface.

24.18. Prove that if asymptotic curves of a surface intersect by a constant angle, then the Gaussian curvature of the surface is proportional to the square of mean curvature.

24.19. Find the asymptotic curves of the surface

$$(u \cos v, \ u \sin v, \ a \cos \lambda v),$$

where a and λ are constants.

24.20. Find the asymptotic curves of the surface

$$\left(3u + 3v, \ 3u^2 + 3v^2, \ 2u^3 + 2v^3\right).$$

24.21. Find the asymptotic curves of the surface $z = y \cos x$.

24.22. Parameterize the hyperbolic paraboloid

$$\frac{x^2}{a^2} - \frac{y^2}{b^2} = 2z,$$

so that the coordinate lines become asymptotic curves.

24.23. Show that the curves

$$\gamma_1(t) = \left(t, a, a^2 t\right) \quad \text{and} \quad \gamma_2(t) = \left(t, \frac{b}{t^2}, \frac{b^2}{t^3}\right)$$

are asymptotic on the surface $z = xy^2$.

24.24. Show that if the coordinate net on a surface is asymptotic, then the following equations hold:

$$\frac{\partial \ln |K|}{\partial u} = 2 \frac{FE_v - EG_u}{EG - F^2}, \qquad \frac{\partial \ln |K|}{\partial v} = 2 \frac{FG_u - GE_v}{EG - F^2},$$

where K is the Gaussian curvature of the surface.

24.25. Prove that the asymptotic curves on a surface of constant negative curvature compose a Chebyshev net and, conversely, if an asymptotic net on a surface is Chebyshev, then the Gaussian curvature of the surface is constant. A net is said to be *Chebyshev* if for any quadrangle composed of the curves of the net, the lengths of opposite sides are equal.

24.26. B e l t r a m i – E n n e p e r   t h e o r e m. Prove that if asymptotic curves of different families have nonzero curvatures at their common point, then their torsions have the same modules and opposite signs. Moreover, the square of the torsion of an asymptotic curve is equal to the module of the Gaussian curvature at this point.

24.27. Two surfaces intersect at a right angle. Prove that if on one surface the intersection curve is asymptotic, then it is geodesic on the other, and vice versa.

24.28. Find the curvature curves of a right helicoid.

24.29. Find the curvature curves on an arbitrary cylindrical surface.

24.30. Find the curvature curves on an arbitrary conical surface.

24.31. Prove that on a plane and on a sphere, any curve is a curvature curve. Vice versa, if on a surface any curve is a curvature curve, then such a surface is a plane or a sphere (or their part).

24.32. L i o u v i l l e t h e o r e m. Prove that under a conformal self-mapping of the space, the sphere (plane) passes to the sphere or to the plane.

24.33. Prove that if the coordinate net consists of curvature curves, then the principal curvatures of the surface are given by the formulas:

$$k_1 = \frac{L}{E}, \quad k_2 = \frac{N}{G}.$$

24.34. R o d r i g u e s t h e o r e m. Prove that a curve γ on a surface $\mathbf{r} = \mathbf{r}(u, v)$ is a curvature curve if and only if the following equation holds along the curve:

$$d\mathbf{m} = -k\, d\mathbf{r}.$$

where $\mathbf{m}$ is the unit normal vector to the surface and k is the normal curvature of the surface along the curve γ.

24.35. Prove that

(a) if two surfaces intersect along a curve at a constant angle and this curve is a curvature curve on one of the surfaces, then it is a curvature curve on the other surface;

(b) if two surfaces intersect along a curve being a curvature curve, then the surfaces intersect at a right angle.

24.36. D u p i n t h e o r e m. Let three families of surfaces $f_1(x, y, z) =$ const, $f_2(x, y, z) =$ const and $f_3(x, y, z) =$ const be given, and, moreover, let the Jacobian satisfy

$$\frac{D(f_1, f_2, f_3)}{D(x, y, z)} \neq 0.$$

Assume that the surfaces intersect at a right angle pairwise. This is the so-called triorthogonal system of surfaces. Prove that the intersection lines of the surfaces from different families are curvature curves on each of the surfaces.

24.37. Prove that each surface in $\mathbb{R}^3$ can be included into a triorthogonal family of surfaces.

24.38. Let two families of surfaces orthogonally intersecting along curves being curvature curves for both surfaces be given. Prove that there exists a third family of surface that composes a triorthogonal system of surfaces with these surfaces.

24.39. Surfaces M_1 and M_2 are said to be parallel if the normals of one surface are the normals of the other as affine lines. Points of the surfaces M_1 and M_2 lying on common normals are said to be corresponding. Prove that under such a correspondence, the curvature curves pass to the curvature curves.

24.40. Prove that under a conformal self-mapping of the space, the curvature curves of the initial surface transform into curvature curves of the transformed surface.

24.41. Let γ be a curvature curve on a surface M, and let the normal curvature k_n of γ be constant and different from zero. Prove that the surface M is tangent to a sphere of radius $1/k_n$ along γ.

24.42. Prove that under a locally isometric covering of the helicoid on the catenoid (see Problem 5.27), its asymptotic curves pass to the curvature curves and its curvature curves pass to the asymptotic curves.

24.43. Prove that a geodesic curve on a surface is a curvature curve if and only if this curve is plane.

24.44. Prove that if on a surface all geodesics are plane curves, then the surface is either a plane or a sphere.

24.45. (a) Let two domains on two surfaces of the 2nd order be glued along a common arc of the boundary so that the result is a C^1-smooth surface. Show that the gluing line is a plane curve. Do every two surfaces of the 2nd order admit such a gluing?

(b) Let domains on two surfaces S_1 and S_2 of the 2nd order be glued along a common arc of the boundary composing a C^2-smooth surface S as a whole. Then the surfaces S_1 and S_2 coincide as second-order surfaces. In other words, the domains being glued are adjacent domains on the same surface of the 2nd order, so that, in fact, S is analytic.

24.46. If a non-developable ruled surface is smooth of class C^n, $n \geq 2$, then it is possible to introduce the asymptotic parameterization on it (when the rulings are one of the families of the coordinate lines) in which the surface is smooth of class C^{n-1}.

24.47. Let a ruled surface be C^1-smooth with respect to an asymptotic parameterization. Show that it is possible to introduce a C^1-smooth asymptotic parameterization on it in which the directing line is orthogonal to the rulings intersecting it.

24.48. Describe the metrics of surfaces in which the internal coordinates u and v are natural parameters on the coordinate lines.

24.49. Describe the surfaces on which the coordinates in the curvature curves are simultaneously the natural parameters of the curvature curves.

24.50. Describe the surfaces on which curvature curves are geodesics.

24.51. If on a smooth non-developable ruled surface M, there exists a rectilinear segment intersecting all the rulings, then on M it is possible to introduce the internal coordinates in which the asymptotic curves can be found in quadratures.

24.52. Prove that a non-developable ruled smooth regular surface cannot have a constant Gaussian curvature.

24.53. Show that among the ruled surfaces, the right helicoid is a unique minimal surface besides the plane.

24.54. Show that ruled surfaces, except for the plane and the helicoid, cannot have a constant mean curvature.

24.55. Show that the plane is a unique minimal surface of constant Gaussian curvature.

24.56. We say that a surface in $\mathbb{R}^n$, $n \geq 3$, has a *constant exterior geometry* if any two of its points have neighborhoods congruent in $\mathbb{R}^n$.

(a) Show that in $\mathbb{R}^3$ the plane, sphere and right circular cylinder (or domains on them) are the only surfaces with constant exterior geometry.

(b) Prove that the surface

$$x_1 = R_1 \cos u, \quad x_2 = R_1 \sin u, \quad x_3 = R_2 \cos v, \quad x_4 = R_2 \sin v$$

in $\mathbb{R}^4$, where $0 \leq u, v \leq 2\pi$ and R_1 and R_2 are constant numbers, is homeomorphic to the torus, has zero Gaussian curvature and its exterior geometry in $\mathbb{R}^4$ is constant. This surface is called the *generalized Clifford torus*. It can be given by one more method. Identify $\mathbb{R}^4$ with the complex space $\mathbb{C}^2$ with coordinates z, w. Then the Clifford torus is obtained as the intersection of the three-dimensional sphere $|z|^2 + |w|^2 = 1$ with the cone $|z| = |w|$. Note that the induced metric on the Clifford torus is plane.

(c) Show that the following surface in $\mathbb{R}^5$

$$x_1 = \frac{xy}{\sqrt{3}}, \quad x_2 = \frac{yz}{\sqrt{3}}, \quad x_3 = \frac{xz}{\sqrt{3}},$$

$$x_4 = \frac{x^2 - y^2}{2\sqrt{3}}, \quad x_5 = \frac{x^2 + y^2 - 2z^2}{6},$$

with the condition $x^2 + y^2 + z^2 = 1$ is a homeomorphic embedding of the projective plane in $\mathbb{R}^5$ having a constant exterior geometry. This surface is called the *Veronese surface*.

24.57. Prove that if an asymptotic curve on a surface is simultaneously a geodesic, then it is a straight line.

24.58. Prove that the property of a curve on a surface to be asymptotic is invariant under projective transformations of the ambient space.

24.59. Prove that if a surface contains a straight line, then this line is an asymptotic curve of the surface.

24.60. Prove that asymptotic curves of a surface compose an orthogonal net if and only if the surface is minimal.

24.61. Prove that if an asymptotic curve on a surface consists of parabolic points, then this curve is plane and its plane is the tangent plane to the surface at all points of the curve. The converse is also true: if the plane of a plane curve on a surface is the tangent plane to the surface at all points of the curve, then this curve is asymptotic and consists of parabolic points.

24.62. Prove that if a curvature curve L is asymptotic, then the Gaussian curvature of the surface vanishes along L, and the curve L is plane.

24.63. Show that on a surface of negative curvature, the binormal to an asymptotic curve coincides with the normal to the surface.

24.64. Prove that if under a bending of a surface different from a ruled surface, all asymptotic curves of the same family pass to asymptotic curves, then the surface remains congruent to itself.

24.65. Show that a non-developable ruled surface does not admit isometric transformations under which all the asymptotic curves of the initial surface pass to the asymptotic curves of the transformed surface.

24.66. Prove that if on a surface of constant Gaussian curvature $K = -1$, the coordinate lines $u = \text{const}$ and $v = \text{const}$ are asymptotic curves of the surface, then in these coordinates, the first form looks like $ds^2 = du^2 + 2\cos w \, du \, dv + dv^2$, where w is the angle between the asymptotic curves.

24.67. Show that under the conditions of the previous problem, the angle $w(u, v)$ satisfies the sine-Gordon equation $w''_{uv} = \sin w$.

24.68. Show that if on a surface of negative curvature the isometric coordinates (u, v) are simultaneously asymptotic, then the surface is minimal with the condition $M = \text{const}$.

24.69. Prove that on a simply connected surface of negative curvature, there is no closed asymptotic curve.

§ 25. Manifolds (Supplementary Problems)

Recall that a manifold is said to be *closed* if it is compact and has no boundary.

25.1. Prove that the groups $SL(n, \mathbb{R})$ and $SL(n, \mathbb{C})$ are smooth submanifolds in the spaces of real and complex square matrices of order n, respectively. Find their dimensions. Find the number of connected components of these matrix groups.

25.2. Prove that the group $SO(n)$ is a smooth submanifold in the space $\mathbb{R}^{n^2}$ of all square matrices of order n. Find the dimension and the number of connected components of this group.

25.3. Show that the groups $U(n)$ and $SU(n)$ are smooth submanifolds in the space $\mathbb{C}^{n^2}$ of complex square matrices of order n. Find the dimension and the number of connected components of these groups.

25.4. Show that the matrix mapping $A \mapsto e^A$ is a smooth homeomorphism in a neighborhood of zero matrix in the inverse image and in a neighborhood of the identity matrix in the image. Show that the inverse mapping is given by the correspondence $B \to \ln(B)$.

25.5. Show that on each of the groups listed in Problems 25.1–25.3, as local coordinate systems in a neighborhood U_A of a matrix A, it is possible to take some of the Cartesian coordinates of the matrix $\ln(A^{-1}X)$. Show that in these coordinates, the functions of coordinate change are smooth functions of class C^∞.

25.6. Show that the projective space $\mathbb{R}P^n$ is the quotient space $S^n/\mathbb{Z}_2$ under some action of the group $\mathbb{Z}_2$ on the sphere S^n. Find this action.

25.7. Show that the complex projective space $\mathbb{C}P^n$ is the quotient space S^{2n+1}/S^1 under some action of the group S^1 on the sphere S^{2n+1}. Find this action.

25.8. Construct a smooth (of class C^∞) function f in $\mathbb{R}^n$ such that $f = 1$ on a ball of radius 1, $f = 0$ outside a ball of radius 2, and $0 \le f \le 1$.

25.9. Let M be a manifold, and let $p \in U \subset M$ be a neighborhood of p. Prove that there exists a smooth function f such that $0 \le f \le 1$, $f(p) = 1$, and $f(x) = 0$ on $M \setminus U$.

25.10. Let M be a manifold, $A = \overline{A}$ be a closed set, and let $U \supset A$ be an open domain. Prove that there exists a smooth function f such that $0 \le f \le 1$, $f|_A = 1$, and $f|_{M\setminus U} = 0$.

25.11. (Weak variant of the Whitney theorem). Prove that a compact smooth manifold M^n is embedded in the Euclidean space $\mathbb{R}^N$ of an appropriate dimension $N < \infty$.

25.12. Prove that a smooth function on a smooth compact manifold M can be represented as a coordinate for some embedding $M \subset \mathbb{R}^N$.

25.13. Prove that the product of spheres is embedded in $\mathbb{R}^N$ as a submanifold of codimension one.

25.14. Prove that if $\dim X < \dim Y$ and $f\colon X \to Y$ is a smooth mapping, then the image of the mapping f does not coincide with Y.

25.15. (a) Prove that a 2-dimensional smooth closed manifold is smoothly immersed in $\mathbb{R}^3$.

(b) Prove that a 2-dimensional closed smooth orientable manifold is embedded in $\mathbb{R}^3$.

(c) Prove that a 2-dimensional smooth closed non-orientable manifold is embedded in $\mathbb{R}^4$ but not embedded in $\mathbb{R}^3$.

25.16. Whitney theorem. Prove that a smooth closed manifold M^n can be embedded in the Euclidean space $\mathbb{R}^{2n+1}$ and can be immersed in $\mathbb{R}^{2n}$.

Remark. Use Problem 25.11. After that, choose the direction of projection in order to avoid self-intersections and singular points on the projection. See Fig. 99.

25.17. Present an example of an immersion of a manifold in $\mathbb{R}^n$ that is in one-to-one correspondence with the image but is not an embedding.

25.18. Prove that for compact manifolds, an embedding is always a homeomorphism onto the image.

25.19. Present an example of an embedding whose image is not a submanifold.

25.20. Let M be a manifold with boundary ∂M. Prove that the manifold M can be embedded in the half-space $(x_{N+1} \ge 0)$ of the Euclidean space $\mathbb{R}^{N+1}$ so that ∂M lies in the half-space $(x_{N+1} = 0)$.

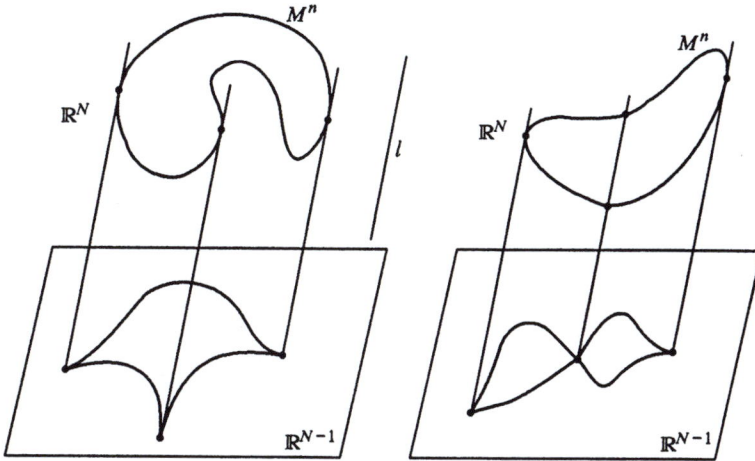

Figure 99

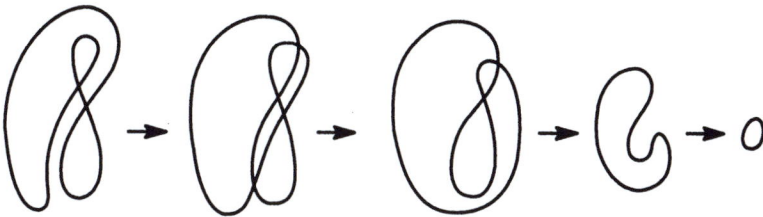

Figure 100

25.21. Let the boundary ∂M consist of two connected components: $\partial M = M_1 \cup M_2$, $M_1 \cap M_2 = \varnothing$. Prove that M can be embedded in $\mathbb{R}^N \times [0, 1]$ so that M_1 lies in $\mathbb{R}^N \times 0$ and M_2 lies in $\mathbb{R}^N \times 1$.

25.22. Prove that a noncompact smooth manifold M^n (without boundary) can be embedded in the Euclidean space $\mathbb{R}^{2n+1}$ and immersed in $\mathbb{R}^{2n}$.

25.23. Consider a half of the Klein bottle depicted in Fig. 161 (see Answers). Let us raise the plane containing the boundary Γ of this surface and simultaneously deform Γ in the way shown in Fig. 100. Glue up the resulting plane circle with a disk. Prove that the obtained surface M (composed of the initial part of the Klein bottle, the trace of the deformation of Γ, and the disk) is an immersion of $\mathbb{R}P^2$ in $\mathbb{R}^3$. Prove that M is a Boy surface (see Problem 21.2).

25.24. Describe the set of self-intersection points of the immersion of $\mathbb{R}P^2$ in $\mathbb{R}^3$ constructed in the previous problem. Find the multiplicities of these points, i.e., find how many sheets of the surface intersect at this self-intersection point.

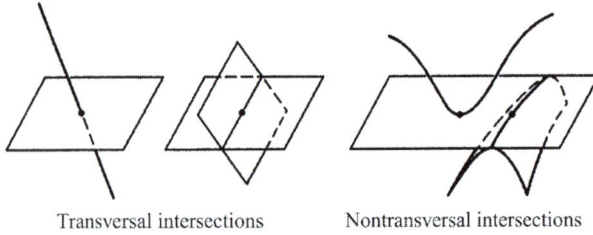

Transversal intersections   Nontransversal intersections

Figure 101

25.25. Consider the immersion of $\mathbb{R}P^2$ in $\mathbb{R}^3$ described in the previous two problems. Denote by $i\left(\mathbb{R}P^2\right)$ the image of $\mathbb{R}P^2$ in $\mathbb{R}^3$. At each point $x \in i\left(\mathbb{R}P^2\right)$, which is not a self-intersection point of the surface, consider a segment of length 2ε orthogonal to $i\left(\mathbb{R}P^2\right)$ centred at x, where ε is sufficiently small. As i is a smooth mapping, the obtained beam of orthogonal segments can be also extended to each self-intersection point. In this case, at each point, we obtain exactly that number of segments which is equal to the multiplicity of this point. In $\mathbb{R}^3$, consider a set consisting of the endpoints of all orthogonal segments. Prove that it is the image of the two-dimensional sphere under a smooth immersion in $\mathbb{R}^3$.

Let $f\colon X \to Y$ be a smooth mapping of smooth manifolds, and let $M \subset Y$ be a smooth submanifold. *The mapping f is said* to be *transversal along the submanifold M* if for any point $x \in f^{-1}(M)$, the tangent space $T_{f(x)}(Y)$ to the manifold Y is a sum (in general, not direct) of the tangent space $T_{f(x)}(M)$ to the manifold M and the image $df(T_x(X))$ of the tangent space to the manifold X. Two submanifolds M_1 and M_2 of the manifold X transversally intersect if the embedding of one of them is transversal along the other submanifold (see Fig. 101).

25.26. Let $f\colon X^n \to Y^n$ be a smooth mapping of a compact closed manifold X into a manifold Y of the same dimension n. Let y_0 be a regular value of the mapping f. Prove that $f^{-1}(y_0)$ consists of finitely many points.

25.27. Prove that if $y \in Y$ is a regular value of the mapping $f\colon X \to Y$, then f is a mapping transversal along y.

25.28. Prove that the definition of the transversal intersection is independent of the choice of the order in the pair M_1 and M_2.

25.29. Prove that if $f\colon X \to Y$ is a mapping transversal along a submanifold $M \subset Y$, then the inverse image $f^{-1}(M)$ is a submanifold in the manifold X. Calculate the dimension of $f^{-1}(M)$.

25.30. Prove that any simple closed arc with self-intersections in $\mathbb{R}^2$ can be transformed into a closed simple arc in $\mathbb{R}^3$ without self-intersections by a small motion.

25.31. Reveal whether or not the following submanifolds intersect transversally:

(a) the plane xy and the axis z in $\mathbb{R}^3$;

(b) the plane xy and the plane spanned by the vectors $(3,2,0)$ and $(0,4,-1)$ in $\mathbb{R}^3$;

(c) the subspace $V \times \{0\}$ and the diagonal in the product $V \times V$;

(d) the spaces of symmetric and skew-symmetric matrices.

25.32. Reveal for which values of a, the surfaces $x^2 + y^2 - z^2 = 1$ and $x^2 + y^2 + z^2 = a$ intersect transversally.

25.33. Show that the set $V_{n.k}$ of all orthonormal systems of k vectors in the Euclidean space $\mathbb{R}^n$ admits a smooth manifold structure. Find its dimension. Show that $V_{n,1} = S^{n-1}$ and $V_{n,n} = O(n)$.

25.34. Show that the set $G_{n,k}$ of all k-dimensional subspaces in the Euclidean space $\mathbb{R}^n$ admits a smooth manifold structure. Find its dimension. Show that $G_{n,1} = \mathbb{R}P^{n-1}$.

25.35. Let $f\colon S^n \to \mathbb{R}P^n$ be a mapping such that it assigns the line passing through the point x and the origin in $\mathbb{R}^{n+1}$ to a point $x \in S^n$. Show that f is smooth and all its points are regular.

25.36. Let $f\colon SO(n) \to S^{n-1}$ be a mapping such that it puts its first column in correspondence to each orthogonal matrix. Prove that the mapping f is regular and all its points are regular. Find the inverse image $f^{-1}(y)$.

25.37. Let $f\colon U(n) \to S^{2n-1}$ be a mapping such that to each unitary matrix, it puts in correspondence its first column. Prove that the mapping f is smooth and all points of f are regular. Find the inverse image $f^{-1}(y)$.

25.38. Let $f\colon V_{n,k} \to V_{n,s}$, $s \le k$, be a mapping such that to an orthonormal system of k vectors, it puts in correspondence its first s vectors. Show that each point is regular for the mapping f. Show that the inverse image $f^{-1}(y)$ is homeomorphic to the manifold $V_{n-s,k-s}$.

25.39. Let $f\colon O(n) \to G_{n,k}$ be a mapping such that to each orthogonal matrix, it puts in correspondence the subspace spanned by the first k columns. Show that all points of the image f are regular. Show that the inverse image $f^{-1}(y)$ is homeomorphic to the manifold $O(n-k) \times O(k)$.

25.40. Let $f\colon X \times Y \to M$ be a smooth mapping, and let $m_0 \in M$ be a regular point. Consider a family of mappings $f_y\colon X \to M$, $f_y(x) = f(x,y)$. Prove that the point m_0 is regular for the mapping f_y for almost all values of the parameter y, i.e., when y runs over an open everywhere dense subset in Y.

25.41. Solve Problem 25.40 replacing the point m_0 by a submanifold $N \subset M$ and the regularity condition by the transversality condition of the mapping along the submanifold N.

25.42. Let a surface M in $\mathbb{R}^n$, $n \ge 3$ (for example, $n = 3$) be given by the equation $f(x_1, \ldots, x_n) = c$, where c is a regular value of the smooth function f. Prove that M is an orientable surface.

25.43. Verify whether or not the following manifolds are orientable: (a) the group $GL(n, \mathbb{R})$; (b) the group $U(n)$; (c) the group $SO(n)$.

25.44. Prove that the projective space $\mathbb{R}P^n$ is orientable exactly for odd n.

25.45. Show that the complex projective space $\mathbb{C}P^n$ is orientable.

25.46. Prove that an arbitrary complex-analytic manifold is orientable.

25.47. Prove that the space of the tangent bundle of a manifold is a manifold, and, moreover, it is always orientable.

25.48. Prove that T^*M is a manifold. Is it orientable?

25.49. Prove that the union of unit spheres of the tangent bundle of a Riemannian n-dimensional manifold M is a smooth manifold of dimension $(2n - 1)$. Prove that this manifold fibers over M with fiber S^{n-1}.

25.50. Prove that the plane annulus, Klein bottle and torus are fiber spaces.

25.51. Prove that a fiber bundle over a contractible base is trivial, i.e., it is a direct product of the base by the fiber.

25.52. Prove that the circle, Euclidean space and torus are parallelizable manifolds.

25.53. Show that parallelizable manifolds are orientable. In particular, the tangent bundles of the projective plane and the Klein bottle are nontrivial,

25.54. Prove that if a point is removed from a two-dimensional closed surface, then the obtained manifold is parallelizable.

25.55. Prove that a manifold of dimension n is parallelizable if and only if there are n linearly independent smooth vector fields on it.

25.56. Prove that a manifold on which a Lie group structure is defined is parallelizable.

25.57. Prove that the sphere S^2 is not parallelizable.

25.58. Present an example of a smooth mapping of manifolds under which the image of a smooth regular curve is no longer regular at some points.

25.59. Prove that linearly independent vectors $\mathbf{a}_1, \ldots, \mathbf{a}_k$ at a point $A \in M^n$, $k \leq n$, can be assumed to be basis vectors of some chart of the manifold, i.e., there exist coordinates $(x^1, \ldots, x^n)$ in a neighborhood of the point A such that

$$\left. \frac{\partial}{\partial x^i} \right|_A = \mathbf{a}_i, \quad 1 \leq i \leq k.$$

25.60. Prove that a vector field $\mathbf{a}$ on a manifold not vanishing at a point A is a basis field for some chart containing the point A.

25.61. Consider vector fields $\mathbf{a}_1, \ldots, \mathbf{a}_k$ on a manifold M^n, where $k \leq n$, linearly independent in a neighborhood of some point $A \in M^n$ and such that $[\mathbf{a}_i, \mathbf{a}_j] = 0$ for all i and j. Prove that in some neighborhood of the point A, it is possible to define coordinates $(x^1, \ldots, x^n)$ such that

$$\mathbf{a}_i = \frac{\partial}{\partial x^i}, 1 \leq i \leq k.$$

25.62. Prove that an orientable 2-dimensional surface admits a complex structure.

25.63. Prove that the manifolds $S^1 \times S^{2n-1}$ and $S^{2n-1} \times S^{2n-1}$ admit a complex structure.

25.64. Prove that a closed odd-dimensional Riemannian manifold of positive curvature is orientable.

25.65. Let $f \colon X \to Y$ is a smooth mapping of closed manifolds. Let f be a covering, and, moreover, let all points of Y be regular. Prove that for any point $y \in Y$, the inverse image of its sufficiently small neighborhood $U(y)$ is homeomorphic to the direct product $U(y) \times f^{-1}(y)$, and, thus, f is a fiber bundle. In particular, if Y is a connected manifold, then the fibers $f^{-1}(y)$ are pairwise diffeomorphic.

A function $w = f(z^1, \ldots, z^n)$, $z^k = x^k + iy^k$, is said to be *holomorphic*, if it is continuously differentiable and its differential is a complex linear form at each point $(z^1, \ldots, z^n)$.

25.66. Prove that if $f(z^1, \ldots, z^n)$ is a holomorphic function, then

$$\frac{\partial \operatorname{Re} f}{\partial x^k} = \frac{\partial \operatorname{Im} f}{\partial y^k}, \qquad \frac{\partial \operatorname{Im} f}{\partial x^k} = -\frac{\partial \operatorname{Re} f}{\partial y^k}.$$

25.67. Let $w^j = f^j(z^1, \ldots, z^n)$ be a holomorphic vector function mapping $\mathbb{C}^n$ into $\mathbb{C}^m$. Find the relation between the real Jacobi matrix of this mapping and its complex Jacobi matrix.

25.68. Prove that a holomorphic vector function $f \colon \mathbb{C}^n \to \mathbb{C}^n$ composes a local coordinate system if and only if its complex Jacobian is different from zero.

25.69. Show that S^2 admits a complex-analytic structure. Describe the simplest atlas of charts in explicit form.

25.70. Show that complex projective spaces $\mathbb{C}P^n$ admit complex-analytic structure. Describe the simplest atlas of charts in explicit form.

§ 26. Tensor Analysis

26.1. Let V_1 and V_2 be vector spaces. Prove the isomorphisms

$$\Lambda^k(V_1 \oplus V_2) = \bigoplus_{i+j=k} \Lambda^i(V_1) \otimes \Lambda^j(V_2),$$

$$S^k(V_1 \oplus V_2) = \bigoplus_{i+j=k} S^i(V_1) \otimes S^j(V_2).$$

26.2. Prove that

$$T^{[i_1}_{j_1} \cdots T^{i_p]}_{j_p} = T^{[i_1}_{[j_1} \cdots T^{i_p]}_{j_p]} = T^{i_1}_{[j_1} \cdots T^{i_p}_{j_p]}.$$

26.3. Prove that if $v^{[i_1 i_2 i_3} v^{j_1 j_2] j_3} = 0$, then $v^{[i_1 i_2 i_3} v^{j_1] j_2 j_3} = 0$.

26.4. Is it true that if a covariant tensor of valency 3 satisfies the conditions

$$T_{ijk} = T_{jik}, \quad T_{ijk}u^i u^j u^k = 0$$

for any choice of contravariant vector u, then the components of the tensor satisfy the condition

$$T_{ijk} + T_{jki} + T_{kij} = 0.$$

26.5. Prove that if the relation

$$u^\gamma_{\alpha\beta} v^\alpha v^\beta w_\gamma = 0$$

holds for any choice of tensors v^α and w_α satisfying the condition $v^\alpha w_\alpha = 0$, then

$$u^\gamma_{(\alpha\beta)} = s_{(\alpha} \delta^\gamma_{\beta)},$$

where s_α is a tensor.

26.6. Prove that if the dimension of a space V is greater than 2, then $\Lambda^2 \left(\Lambda^2 V \right) \neq \Lambda^4 V$.

26.7. Prove that if $\operatorname{tr} \Lambda^q A = 0$ for all $q > 0$, then the operator A is nilpotent.

26.8. On a space V of dimension n, let a linear operator A be given. If the operator $\Lambda^{n-1}A$ on the space $\Lambda^{n-1}V$ is nonzero, then it is either nonsingular or has the rank 1.

26.9. Prove that if an operator A is diagonalizable, then the operator $A^{\otimes k}$ is also diagonalizable.

26.10. Prove that the rank of the tensor $A_{ij} = a_i b_j$ is equal to unity and the rank of the tensor $a_i b_j + a_j b_i$ is equal to 2.

26.11. To which condition the tensor A_{ij} must satisfy to be of the form $A_{ij} = a_i b_j$?

26.12. Prove that if a tensor T_{ijk} is symmetric with respect to the first two indices, then

$$T_{(ijk)} = \frac{1}{3} \left(T_{ijk} + T_{jki} + T_{kij} \right).$$

26.13. Prove that if a tensor T_{ijk} is skew-symmetric with respect to the first two indices, then

$$T_{[ijk]} = \frac{1}{3} \left(T_{ijk} + T_{jki} + T_{kij} \right).$$

26.14. Let two tensors a_{ij} and b^{klm} be given. Construct tensors of the first, third and fifth valency from them by using one multiplication and the contraction. How many are there?

26.15. Let a tensor T_{ijkl} have the properties

$$T_{hijk} + T_{hikj} = 0, \quad T_{hijk} + T_{hjki} + T_{hkij} = 0.$$

(a) Prove that if $T_{hijk} - T_{hjik} = 0$, then $T_{hijk} = 0$.

(b) Prove that if $T_{hijk} + T_{hjik} = 0$, then $T_{hijk} = 0$.

26.16. Prove that if a three-valency tensor T_{ij}^h satisfies the conditions

$$T_{ij}^h = T_{ji}^h, \quad T_{ij}^h u^i u^j = 0$$

for any tensor (vector) u^j, then the tensor T_{ij}^h is zero.

26.17. Show that if the tensor $T_{i_1 \dots i_m}$ satisfies the conditions

$$T_{i_1 \dots i_m} u^{i_1} \cdots u^{i_m} = 0$$

for any choice of the vector (tensor) u^i, then $T_{(i_1 \dots i_m)} = 0$.

Let $\varphi \colon u \to V$ be a diffeomorphism of domains of the Euclidean space. Let a tensor field T of type (p, q) on V have the components $T_{j_1, \dots, j_q}^{i_1, \dots, i_p}$ in the coordinates $(x) = (x^1, \dots, x^n)$. Clearly, the composition $(x \circ \varphi)$ defines curvilinear coordinates on U. Define the tensor field $\varphi^* T$ in U in this coordinate system by the formula

$$(\varphi^* T)_{j_1 \dots j_q}^{i_1 \dots i_p} = T_{j_1 \dots j_q}^{i_1 \dots i_p} \circ \varphi.$$

Let $\mathbf{X}$ be a smooth vector field, and let T be some tensor field. The field $\mathbf{X}$ defines the one-parameter group φ_t (recall that φ_t is defined from the solution of the differential equation $\frac{d}{dt} \varphi_t(P)|_{t=0} = \mathbf{X}(P)$). Define the Lie derivative of the field T along the vector field $\mathbf{X}$ by the formula

$$L_{\mathbf{X}} T(P) = \lim_{t \to 0} \frac{(\varphi_t^* T)(P) - T(P)}{t}.$$

26.18. Prove that if a tensor field T is of the type $(0, 0)$, i.e., is a smooth function, then $\varphi^* T = T \circ \varphi$.

26.19. Prove that if a tensor field T is of type $(1, 0)$, i.e., is a vector field, then

$$(\varphi^* T)(P) = (d\varphi|_P)^{-1} T(\varphi(P)).$$

26.20. Prove that if a tensor field ξ is of type $(0, 1)$, i.e., is covector, then

$$(\varphi^* \xi)(P) = (d\varphi|_P)^* \xi(\varphi(P)).$$

Here, $d\varphi|_P \colon T_P U \to T_{\varphi(P)} V$ is the differential of the mapping φ at the point $P \in U$ and $(d\varphi|_P)^* \colon T_{\varphi(P)}^* V \to T_P^* U$ is the conjugate mapping of the cotangent spaces.

26.21. Prove that the operation $T \mapsto \varphi^* T$ preserves the algebraic operations on tensor fields: the sum, the multiplication by a function, the tensor product and contraction.

26.22. Let $\mathbf{X}$ and $\mathbf{Y}$ be vector fields in a domain V, and let $\varphi \colon U \to V$ be a diffeomorphism. Prove that $[\varphi^* \mathbf{X}, \varphi^* \mathbf{Y}] = \varphi^* [\mathbf{X}, \mathbf{Y}]$.

26.23. Prove that the Lie derivative is a linear operation.

26.24. Prove that if T and S are tensor fields and $\mathbf{X}$ is a vector field, then

$$L_{\mathbf{X}}\left(T \otimes S\right) = L_{\mathbf{X}}T \otimes S + T \otimes L_{\mathbf{X}}S.$$

26.25. Prove that the Lie derivative commutes with the contraction.

26.26. Prove that if $\mathbf{X}$ is a vector field and f is a smooth function, then $L_{\mathbf{X}}f = \mathbf{X}f$.

26.27. (a) Let $\left(x^1,\ldots,x^n\right)$ be local coordinates in a domain U, and let $\mathbf{X}$ be a vector field with components $\left(X^1,\ldots,X^n\right)$ in the coordinates (x). Prove that

$$\left(L_{\mathbf{X}}dx\right)^i = \frac{\partial X^i}{\partial x^k}dx^k.$$

(b) Prove that the following formula holds for a covector $\xi = \xi_i dx^i$:

$$L_{\mathbf{X}}\xi = \left(X^k \frac{\partial \xi_i}{\partial x^k} + \xi_k \frac{\partial X^k}{\partial x^i}\right) dx^i.$$

26.28. Prove that if $\mathbf{X}$ and $\mathbf{Y}$ are vector fields, then $L_{\mathbf{X}}\mathbf{Y} = [\mathbf{X}, \mathbf{Y}]$.

26.29. Obtain the formula for the Lie derivative of a tensor field of type (p, q).

26.30. Prove that the equation $dL_{\mathbf{X}}\omega = L_X d\omega$ holds for a differential form ω and a vector field $\mathbf{X}$.

26.31. Prove that
$$L_{\mathbf{X}}\omega = i_{\mathbf{X}}d\omega + di_{\mathbf{X}}\omega,$$

where ω is a differential form and $\mathbf{X}$ is a vector field.

26.32. For vector fields $\mathbf{X}$ and $\mathbf{Y},$ prove the following formulas:
(a) $[L_{\mathbf{X}}, i_{\mathbf{Y}}] = i_{[\mathbf{X},\mathbf{Y}]}$; (b) $[L_{\mathbf{X}}, L_{\mathbf{Y}}] = L_{[\mathbf{X},\mathbf{Y}]}.$

§ 27. Geodesics on Manifolds

27.1. Prove that the geodesic torsion of a curve $u = u(s)$, $v = v(s)$ lying on a surface $\mathbf{r} = \mathbf{r}(u, v)$ is calculated by the formula

$$\varkappa_g = (\dot{\mathbf{r}}, \dot{\mathbf{m}}, \mathbf{m}),$$

where $\mathbf{m}$ is a unit vector of the normal of the surface.

27.2. Prove the following assertion: for a curve on a surface to be a curvature line, it is necessary and sufficient that the geodesic torsion vanishes at each its point.

27.3. Find geodesic curves on a developable surface.

27.4. Prove that the torsion of a geodesic curves tangent to a curvature curve of a surface vanishes.

27.5. Prove that each plane geodesic curve different from a line is a curvature curve of the surface considered.

27.6. Prove that geodesic equations on a manifold M with metric (g_{ij}) can be written in the cotangent bundle T^*M in the Hamiltonian form

$$\frac{dp_i}{dt} = \frac{\partial H}{\partial x^i}, \quad \frac{dx^i}{dt} = \frac{\partial H}{\partial p_i},$$

where $(x^1, \ldots, x^n, p_1, \ldots, p_n)$ are the standard coordinates on the cotangent bundle and $H(p,x) = \frac{1}{2}g^{ij}(x)p_ip_j$. Besides, $p_i = g_{ij}\dot{x}^j$, where $\dot{x}^j$ is the velocity vector of a geodesic.

27.7. Consider a surface with Riemannian metric. There are two types of curves on it: Darboux circles and Gaussian circles. A *Darboux circle* is a curve on the surface having a constant geodesic curvature. A *Gaussian circle* is the set of points distant from a fixed point by the same distance measured along geodesic radii. Prove that the set of Darboux circles coincides with the set of Gaussian circles if and only if the surface considered is a surface of constant curvature.

27.8. Prove that a two-dimensional surface such that the length of any Gaussian circle on it is equal to $2\pi R$, where R is the radius measured along the geodesics, is locally isometric to the plane.

27.9. (a) Prove that on a compact simply connected manifold M^n, there always exists a pair of conjugated points.

(b) What happens if M^n is not simply connected?

(c) On a manifold, consider a beam of geodesics emanating from the same point. Does there exist a conjugate point on each of them?

27.10. Describe all surfaces of revolution on which all geodesics are closed.

27.11. (a) Prove that on a ruled surface, any segment of a ruling is the shortest path between its endpoints.

(b) Prove the following more general fact: on any complete simply connected surface of negative curvature, an arc of any geodesic is the shortest path between its endpoints.

27.12. Present an example of two non-isometric Riemannian manifolds with common geodesics. Two manifolds have common geodesics if there exists a diffeomorphism of one manifold onto the other transforming geodesics into geodesics.

27.13. Prove that all compact closed surfaces are geodesic complete.

The following problems are devoted to matrix Lie groups. In one of the following sections, Lie groups will be considered in full generality, i.e., as smooth manifolds equipped with a group structure in the smooth category.

In this section, in order to simplify the presentation, we restrict ourselves to the consideration of only those Lie groups which are realized in the form of some matrix groups. Such Lie groups are said to be *matrix*. Consider the general linear group

$GL\left(n,\mathbb{R}\right)$, the group of nonsingular matrices of size $n \times n$ with real entries. It is an open domain in the Euclidean space $\mathbb{R}^{n^2}$ of all real matrices of size $n \times n$. On $\mathbb{R}^{n^2}$, consider the ordinary Euclidean metric given in the form

$$\langle X, Y \rangle = \text{tr}\, X \cdot Y^{\top},$$

where $Y^{\top}$ is the transposed matrix. This metric induces the Riemannian metric on $GL\left(n,\mathbb{R}\right)$ and all its subgroups. Such a metric on a matrix group G is called the *Killing metric*. Also, we can consider the case of $GL\left(n,\mathbb{C}\right)$ where the metric is given by

$$\langle X, Y \rangle = \text{Re}\,\text{tr}\, X \cdot \overline{Y}^{\top}.$$

The one-parametric (one-dimensional) subgroups of a matrix group G are exactly subgroups having the form $\{e^{tX} : t \in \mathbb{R}\}$, where X is an arbitrary matrix from the tangent space of the group G at its identity matrix E.

Recall that the tangent space L of a matrix group G at the identity equipped with the bilinear skew-symmetric operation $[X,Y] = XY - YX$, is a Lie algebra; in particular, the following Jacobi identity holds:

$$[[X,Y],Z] + [[Z,X],Y] + [[Y,Z],X] = 0.$$

Here, XY is the ordinary matrix product. The operation $[X,Y]$ is called the *commutator*. On the Lie algebra L of a matrix group G, the linear operators Ad_g and ad_X are defined by

$$\text{Ad}_g\left(X\right) = gXg^{-1}, \quad \text{ad}_X\left(Y\right) = [X,Y],$$

where $g \in G$ and $X,Y \in L$.

A vector field $\mathbf{V}$ on a group G is said to be left-invariant (right-invariant) if it transforms into itself under all left (right) shifts of the group. Such a field is uniquely defined by its value at the identity of the group, i.e., by a vector X from the Lie algebra L. Such a field is denoted by L_X (resp. R_X).

27.14. (a) Prove that the operator ad_X is a derivation of the Lie algebra L, i.e.,

$$\text{ad}_X\left([Y,Z]\right) = [\text{ad}_X\left(Y\right), Z] + [Y, \text{ad}_X\left(Z\right)].$$

Prove that this formula is equivalent to the Jacobi identity.

(b) Prove that the Riemannian metric defined above on an arbitrary subgroup G of the group $GL\left(n,\mathbb{R}\right)$ is bi-invariant, i.e., the left and right shifts on G are its isometries.

(c) Prove the assertion analogous to that of the previous item for subgroups of the group $GL\left(n,\mathbb{C}\right)$.

(d) Prove that for any $X \in L$, the operator ad_X is skew-symmetric with respect to the Killing metric:

$$\langle \text{ad}_X\left(Y\right), Z \rangle = -\langle Y, \text{ad}_X\left(Z\right) \rangle.$$

(e) Prove that for any $g \in G$, the operator Ad_g preserves the Killing metric, i.e., $\langle \text{Ad}_g\left(X\right), \text{Ad}_g\left(Y\right) \rangle = \langle X, Y \rangle$.

27.15. Let G be a matrix group, and let L be its Lie algebra. Introduce the connection ∇ on the group G by the following formula for left-invariant fields L_X and L_Y:

$$\nabla_{L_X} L_Y = \frac{1}{2} L_{[X,Y]} = \frac{1}{2} \left[L_X, L_Y \right].$$

Note that, as follows from item (a) of this problem, this formula uniquely defines the connection not only on left-invariant but also on arbitrary vector fields on the group G.

(a) Prove that the formula presented above uniquely defines the Christoffel symbols Γ^i_{jk}.

(b) Prove that this connection on G is symmetric and is in concordance with the Killing metric.

(c) Prove that geodesics of a matrix group G with the Riemannian metric described above passing through the identity are all its one-parameter subgroups and only they. Prove that all other geodesics on G are obtained by right (left) shifts of the one-parameter subgroups.

(d) Prove that the curvature tensor R of the connection described in this problem on a matrix group G is given by the following formulas on left-invariant vector fields L_X, L_Y, and L_Z:

$$R\left(L_X, L_Y \right) L_Z = -\frac{1}{4} L_{[[X,Y],Z]},$$

$$\left\langle R\left(L_X, L_Y \right) L_Z, L_W \right\rangle = -\frac{1}{4} \left\langle [X,Y], [Z,W] \right\rangle.$$

Note that these formulas are sufficient for defining the curvature tensor R on arbitrary vector fields on the group G.

(e) Show that the components R^i_{jkl} are uniquely reconstructed from these formulas.

27.16. Show that the exponential mapping $\exp : L \to G$ is a diffeomorphism of some neighborhood of zero matrix in the Lie algebra L onto some neighborhood of the identity matrix in the group G.

27.17. (a) For the group $SL\left(2, \mathbb{R} \right)$, deduce the explicit formula for the exponential mapping.

(b) For the group $SL\left(2, \mathbb{R} \right)$, the exponential mapping is not a mapping "onto".

(c) Study the case of the group $SL\left(2, \mathbb{C} \right)$.

(d) Prove that for a connected compact group G, the exponential mapping is always a mapping "onto".

§ 28. Curvature Tensor

28.1. Calculate the scalar curvature of the group $SO\left(n \right)$ with two-sided invariant metric.

28.2. In the n-dimensional Riemannian space, let linear independent vector fields $\mathbf{X}_1, \ldots, \mathbf{X}_n$ be chosen. Let $g_{ij} = \langle \mathbf{X}_i, \mathbf{X}_j \rangle$ and $[\mathbf{X}_i, \mathbf{X}_j] = c_{ij}^k \mathbf{X}_k$. Express Γ_{ij}^k through g_{ij} and c_{ij}^k if $\nabla_{\mathbf{X}_j} \mathbf{X}_i = \Gamma_{ij}^k \mathbf{X}_k$.

28.3. Prove that the curvature tensor of an n-dimensional manifold has $\dfrac{n^2 \left(n^2 - 1\right)}{12}$ algebraically independent components. In other words, the curvature tensor of general form has no other symmetries, except for the "known" ones.

28.4. Under what condition a symmetric connection ∇ is Riemannian, i.e., there exists a Riemannian metric g_{ij} such that $\nabla_k g_{ij} \equiv 0$?

28.5. (a) Prove that on an N-dimensional manifold, the set of quantities

$$
C_{lmnk} = R_{lmnk} - \frac{1}{N-2} \left(g_{ln} R_{mk} - g_{lk} R_{mn} - g_{mn} R_{lk} + g_{mk} R_{ln} \right) +
$$

$$
+ \frac{R}{(N-1)(N-2)} \left(g_{ln} g_{mk} - g_{lk} g_{mn} \right)
$$

is a tensor. The tensor C_{mlk}^n is called the Weyl tensor.

(b) Prove that it has the same algebraic properties as the curvature tensor and, moreover, satisfies $\dfrac{N(N+1)}{2}$ conditions $C_{mlk}^l = 0$.

(c) Prove that two Riemannian metrics with the same Weyl tensors are conformally equivalent.

28.6. (a) Show that in an n-dimensional manifold ($n \geq 3$), it is possible to construct $\dfrac{n(n-1)(n-2)(n-3)}{12}$ algebraically independent scalars from the metric tensor g_{ij} and its curvature tensor R_{ijkl}.

(b) Show that for $n = 3$, these quantities are the scalar curvature R, the contraction $R_{ij} R^{ij}$ and $\dfrac{\det R_{ij}}{\det g_{ij}}$, where R_{ij} is the Ricci tensor.

28.7. Prove that if the curvature tensor of a Riemannian manifold is identically equal to 0, then the result of the parallel translation along a curve γ is independent of a homotopy of the path γ provided that the endpoints of the path are fixed.

28.8. (a) If for a Riemannian simply connected manifold $R_{jkl}^i = 0$, then $TM^n = M^n \times \mathbb{R}^n$, where TM is the tangent bundle of the manifold M. In other words, in this case the tangent bundle is trivial, and the manifold itself is parallelizable.

(b) Reveal what happens in the non-simply connected case.

28.9. Prove that any direction in $T_x M$ at an arbitrary point of a Riemannian manifold M is a principal direction of the Ricci tensor if and only if M is an Einstein space.

28.10. Show that the following metrics are of constant curvature. Find this curvature.

(a) $ds^2 = du^2 + dv^2$;

(b) $ds^2 = du^2 + \cos^2 \dfrac{u}{a} dv^2$;

(c) $ds^2 = du^2 + \cosh^2 \dfrac{u}{a} dv^2$, $a = \text{const} \neq 0$.

28.11. Prove that an affine connection of the general form satisfies the Bianchi identity $\nabla_{[l} R^i_{km]j} = S^p_{[lk} R^i_{m]pj}$.

28.12. Let $\omega^i_j = \Gamma^i_{kj} du^k$ be connection forms, and let

$$\Omega^i_j = \frac{1}{2} R^i_{kmj} du^k \wedge du^m$$

be curvature connection forms. Prove that the structural Cartan equations

$$\Omega^i_j = d\omega^i_j + \omega^i_k \wedge \omega^k_j$$

hold.

28.13. Show that the Bianchi formula can be written in the form

$$d\Omega^i_j = \Omega^i_k \wedge \omega^k_j - \omega^i_k \wedge \Omega^k_j.$$

28.14. Let an arbitrary torsion-free affine connection be fixed. In a neighborhood U of some point O, let coordinates $\left(x^1, \ldots, x^n\right)$ be chosen, and, moreover, let the coordinates of the point O be equal to $\left(x_0^1, \ldots, x_0^n\right)$. Introduce new coordinates $\left(x^{1'}, \ldots, x^{n'}\right)$ in U, setting

$$x^{k'} = \left(x^k - x_0^k\right) + \frac{1}{2} \Gamma^k_{ij} \left(x^i - x_0^i\right) \left(x^j - x_0^j\right).$$

Prove that in the new coordinate system, the point O has zero coordinates, and the Christoffel symbols vanish at O. Such a coordinate system is said to be *geodesic*.

28.15. Verify that in the geodesic coordinate system at a point O, the values of components of the curvature tensor and its covariant derivatives are given by the formulas

$$R^h_{ij\cdot k}(O) = \frac{\partial \Gamma^h_{ki}(O)}{\partial x^j} - \frac{\partial \Gamma^h_{ji}(O)}{\partial x^k},$$

$$\nabla_l R^h_{ij\cdot k} = \frac{\partial^2 \Gamma^h_{ki}(O)}{\partial x^j \partial x^l} - \frac{\partial^2 \Gamma^h_{ji}(O)}{\partial x^k \partial x^l}.$$

28.16. (a) Taking into account the previous problem, verify the fulfillment of the Bianchi formula

$$\nabla_m R^n_{ikl} + \nabla_l R^n_{imk} + \nabla_k R^n_{ilm} = 0.$$

(b) Prove the following equation for the Ricci tensor R^l_m and the scalar curvature R:

$$\nabla_l R^l_m = \frac{1}{2} \frac{\partial R}{\partial x_m}.$$

28.17. Let H_1, H_2, and H_3 be the Lamé coefficients of some curvilinear coordinates in $\mathbb{R}^3$. Prove the following relations:

$$\frac{\partial}{\partial s}q^1\left(\frac{1}{H_1}\frac{\partial H_2}{\partial s}q^1\right)+\frac{\partial}{\partial s}q^2\left(\frac{1}{H_2}\frac{\partial H_1}{\partial s}q^2\right)+\frac{1}{H_3^2}\frac{\partial H_1}{ds}q^3\frac{\partial H_2}{ds}q^3=0;$$

$$\frac{\partial}{\partial s}q^2\left(\frac{1}{H_2}\frac{\partial H_3}{\partial s}q^2\right)+\frac{\partial}{\partial s}q^3\left(\frac{1}{H_3}\frac{\partial H_2}{\partial s}q^3\right)+\frac{1}{H_1^2}\frac{\partial H_2}{ds}q^1\frac{\partial H_3}{ds}q^1=0;$$

$$\frac{\partial}{\partial s}q^3\left(\frac{1}{H_3}\frac{\partial H_1}{\partial s}q^3\right)+\frac{\partial}{\partial s}q^1\left(\frac{1}{H_1}\frac{\partial H_3}{\partial s}q^1\right)+\frac{1}{H_2^2}\frac{\partial H_3}{ds}q^2\frac{\partial H_1}{ds}q^2=0;$$

$$\frac{\partial^2 H_1}{\partial q^2\partial q^3}=\frac{1}{H_3}\frac{\partial H_3}{\partial s}q^2\frac{\partial H_1}{\partial s}q^3+\frac{1}{H_2}\frac{\partial H_1}{\partial s}q^2\frac{\partial H_2}{ds}q^3;$$

$$\frac{\partial^2 H_2}{\partial q^3\partial q^1}=\frac{1}{H_1}\frac{\partial H_1}{\partial s}q^3\frac{\partial H_2}{\partial s}q^1+\frac{1}{H_3}\frac{\partial H_2}{\partial s}q^3\frac{\partial H_3}{ds}q^1;$$

$$\frac{\partial^2 H_3}{\partial q^1\partial q^2}=\frac{1}{H_2}\frac{\partial H_2}{\partial s}q^1\frac{\partial H_3}{\partial s}q^2+\frac{1}{H_1}\frac{\partial H_3}{\partial s}q^1\frac{\partial H_1}{ds}q^2.$$

28.18. Prove that the smooth functions

$$H_1\left(q^1,q^2,q^3\right),\quad H_2\left(q^1,q^2,q^3\right),\quad H_3\left(q^1,q^2,q^3\right)$$

satisfying the relations of the previous problem are the Lamé coefficients of some transformation

$$x_s=x_s\left(q^1,q^2,q^3\right),\quad s=1,2,3.$$

28.19. Prove that the fundamental group of a complete Riemannian manifold of nonpositive curvature contains no elements of finite order. Prove that $\pi_1(M)$, where M is a complete Riemannian manifold of strictly negative curvature, has the following property: if two elements commute (i.e., $ab=ba$, where $a,b\in\pi_1(M)$), then a and b belong to the same cyclic subgroup.

28.20. Prove that a closed orientable Riemannian manifold M^n of strictly positive curvature and even dimension is simply connected.

28.21. (a) Prove that any compact closed Riemannian manifold of constant positive curvature γ is isometric either to the sphere S^n or to $\mathbb{R}P^n$ (of radius $1/\sqrt{\gamma}$).

(b) Let M^n be a compact closed simply connected complete Riemannian manifold, and let $C(l)$ be the set of first conjugated points for some point $l\in M^n$. Prove that if M^n is a symmetric space, then the complement $M^n\setminus C(l)$ is homeomorphic to an open disk.

28.22. Prove that a complete noncompact Riemannian manifold of positive curvature and dimension m, where $m=2$ or $m\geq 5$, is diffeomorphic to $\mathbb{R}^m$.

28.23. Let x and y be two close points on a standard sphere S^2, and let $f(z)$ be the area of the geodesic triangle with vertices at the points x, y, and z.

(a) Is the function $f(z)$ harmonic on the sphere S^2?

(b) Study the case of the n-dimensional sphere (here, $f(z)$ is the volume of the geodesic simplex one of whose faces is fixed and z is a free vertex).

(c) Study the same problem on the Lobachevskii plane.

28.24. Prove that if M^n is a complete simply connected Riemannian manifold such that n is odd and on M^n there exists a point p such that the set of first points conjugated to p is regular and of order k, then $k = n - 1$ and M^n is homeomorphic to the sphere S^n (by the order of a point we mean its multiplicity).

§ 29. Vector Fields

29.1. Find the derivative of the function $f = \ln(x^2 + y^2)$ at the point $P = (1, 2)$ in the direction of the curve $y^2 = 4x$.

29.2. Find the derivative of the function $f = \arctan(y/x)$ at the point $P = (2, -2)$ in the direction of the curve $x^2 + y^2 - 4x = 0$.

29.3. Find the derivative of the function f at the point P in the direction of the curve γ, where:

(a) $f = x^2 + y^2$, $P = (1, 2)$, γ: $x^2 + y^2 = 5$;

(b) $f = 2xy + y^2$, $P = (\sqrt{2}, 1)$, γ: $\dfrac{x^2}{4} + \dfrac{y^2}{2} = 1$;

(c) $f = x^2 - y^2$, $P = (5, 4)$, γ: $x^2 - y^2 = 9$;

(d) $f = \ln(xy + yz + xz)$, $P = (0, 1, 1)$, γ: $x = \cos t, y = \sin t, z = 1$;

(e) $f = \ln(x^2 + y^2 + z^2)$, $P = \left(0, R, \dfrac{\pi a}{2}\right)$, γ: $x = R\cos t$, $y = R\sin t$, $z = at$.

29.4. Find the derivative of the function $f = \dfrac{x^2}{a^2} + \dfrac{y^2}{b^2} + \dfrac{z^2}{c^2}$ at an arbitrary point $P = (x, y, z)$ in the direction of the radius-vector of this point.

29.5. Find the derivative of a function $f = f(x, y, z)$ in the direction of the gradient of a function g.

29.6. Let $\mathbf{v}(x, y, z)$ be the velocity field of a rigid body rotating around some axis. Show that

(a) $\operatorname{div}(\mathbf{v}) = 0$;

(b) $\operatorname{rot}(\mathbf{v}) = 2\mathbf{w}$, where $\mathbf{w}$ is the angular velocity vector.

29.7. Let $\mathbf{X} = (x, y, z)$, and let $\mathbf{Y}$ be a constant vector field. Show that $\operatorname{rot}(\mathbf{Y} \times \mathbf{X}) = 2\mathbf{Y}$.

29.8. Show that $\operatorname{rot}\operatorname{grad} F = 0$.

29.9. Prove the formula

$$\Delta(FG) = F\Delta G + G\Delta F + 2\langle \text{grad } F, \text{grad } G\rangle.$$

29.10. Solve the equation rot $\mathbf{X} = \mathbf{Y}$, where
(a) $\mathbf{Y} = (1, 1, 1)$; (b) $\mathbf{Y} = (2y, 2z, 0)$;
(c) $\mathbf{Y} = (0, 0, e^x - e^y)$; (d) $\mathbf{Y} = (6y^2, 6z, 6x)$;
(e) $\mathbf{Y} = (3y^2, -3x^2, -(y^2 + 2x))$;
(f) $\mathbf{Y} = (0, 2\cos xz, 0)$;

(g) $\mathbf{Y} = \left(-\dfrac{y}{x^2 + y^2}, \dfrac{x}{x^2 + y^2}, 0\right)$;

(h) $\mathbf{Y} = \left(ye^{x^2}, 2yz, -\left(2xyze^{x^2} + z^2\right)\right)$.

29.11. Prove that with each vector field we associate the one-parameter group of diffeomorphisms φ_t whose trajectory are tangent to this field.

29.12. Let V be a linear finite-dimensional space of vector fields closed with respect to the commutator, i.e., $[\xi, \eta] \in V$ for $\xi, \eta \in V$. Prove that V is a Lie algebra.

29.13. In the previous problem, show that the simply connected Lie group G corresponding to the algebra V acts on a compact manifold, and, moreover, each field $\xi \in V$ defines a one-dimensional subgroup in G, orbits of whose action are tangent to the vector field ξ.

29.14. Let P and Q be two arbitrary points in the disk $D^n \subset \mathbb{R}^n$. Find a diffeomorphism φ of the space $\mathbb{R}^n$ such that $\varphi(P) = Q$ and $\varphi(x) = x$ for $x \notin D^n$.

29.15. (a) On the standard sphere S^3, construct three smooth vector fields linearly independent at each point.

(b) In explicit form, find the integral trajectories of fields obtained by multiplying the radius-vector of a point on the sphere by the imaginary quaternions i, j and k. Here, the sphere S^3 is realized as the set of quaternions of unit length.

29.16. Prove that on a manifold, with each one-parametric group of smooth homeomorphisms, one associates the smooth velocity vector field of points.

29.17. Present an example of a vector field on a noncompact manifold whose trajectory is not generated by some one-parametric transformation group.

29.18. (a) Show that for any two points x_1 and x_2 of the open ball $\overset{\circ}{D}{}^n$, there exists a diffeomorphism φ of the ball onto itself such that it interchanges the points x_1 and x_2, i.e., $\varphi(x_1) = x_2$ and $\varphi(x_2) = x_1$.

(b) Prove that for any two points $x_1, x_2 \in \overset{\circ}{D}{}^n$ there exists a one-parameter group of diffeomorphisms φ_t such that φ_0 is the identity mapping and $\varphi_1(x_1) = x_2$.

(c) Let X be a smooth connected manifold, and let x_0 and x_1 be two of its arbitrary points. Find a one-parameter group of diffeomorphisms φ_t such that φ_0 is the identity mapping and $\varphi_1(x_0) = x_1$. Show that without loss of generality, one can assume that all transformations φ_t are identical outside a certain compact set.

29.19. Let a finite group G smoothly act on a smooth manifold X. Show that if the action of G is free (i.e., each point $x \in X$ is fixed only under the action of the identity of the group G), then the quotient space X/G is a manifold.

29.20. Let $f(z)$ be c complex-analytic function of one variable. Prove that singular points (zeros) of the vector fields grad Re $f(z)$ and grad Im $f(z)$ coincide with zeros of the derivative $f'(z)$.

29.21. Find the integral trajectories of the flow $\mathbf{v}_1(x)$ orthogonal to the flow $\mathbf{v}_2(x)$, where $\mathbf{v}_2(x) = \operatorname{grad} f(x)$, $x \in \mathbb{R}^2$, and $f(x)$ is the value of the angle AxB. Here, A and B are fixed points of the plane $\mathbb{R}^2$ and x is a variable point.

29.22. Prove that an irrotational flow $\mathbf{v} = (P, Q)$, where P and Q are components of the flow on the plane $\mathbb{R}^2(x, y)$ is potential, i.e., $\mathbf{v} = \operatorname{grad} f(x, y)$ for some smooth function f. What can be said about the potential f if, in addition, the flow is incompressible, i.e., div $(\mathbf{v}) = 0$?

29.23. Let a vector field ξ satisfy the condition div $(\xi) = 0$. Show that the shift operator along the integral trajectories is unitary.

29.24. Find all homotopy classes of vector fields without singularities on the torus T^2. The homotopy must be realized in the class of vector fields without singular points.

29.25. Prove that if a vector field $\mathbf{X}$ on the two-dimensional torus is homotopic to the vector field $\dfrac{\partial}{\partial \varphi_1}$ in the class of vector fields without singular points, then it has a periodic trajectory.

29.26. Find the maximum number of linearly independent tangent vector fields on a smooth closed surface M^2.

29.27. Let m and n be the rotation numbers of a vector field on the torus T^2, and let $\lambda = (m, n)$. Prove that this field has λ periodic solutions (closed trajectories).

29.28. P o i n c a r é – B e n d i x o n   t h e o r e m. In a certain domain on a plane, let a vector field be given. In this domain, fix a certain compact set K containing no singular points of the vector field. If K contains a point such that the integral trajectory of the vector field emanating from this point does not leave K, then K necessarily contains a periodic trajectory of the vector field.

29.29. Let a point P on the plane be a limit point of a certain trajectory of a vector field, i.e., it belongs to the closure of this trajectory. Prove that the trajectory passing through P is limit for the initial trajectory.

29.30. Prove that the set of vector fields having only isolated singularities is simply connected.

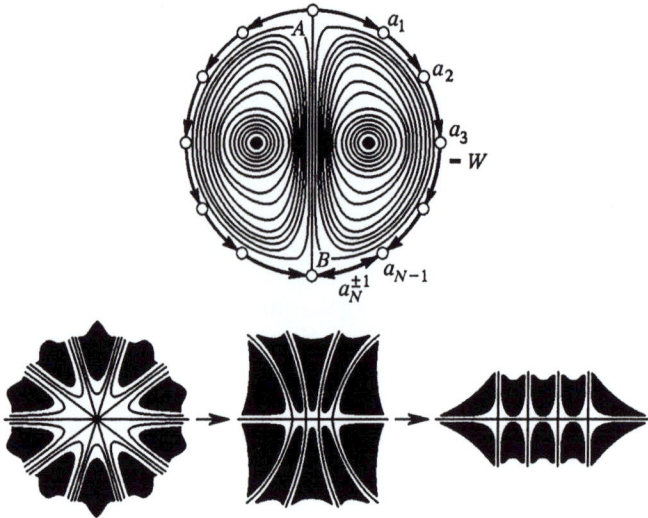

Figure 102

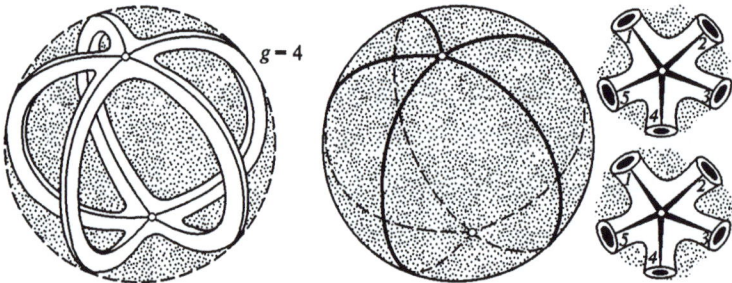

Figure 103

29.31. Prove that the sum of indices of singularities of a vector field on a closed manifold does not change under smooth deformations.

29.32. Prove that the set of all integral trajectories of the vector field $\mathbf{v}(x) = (x^1, -x^0, x^3, -x^2)$, where

$$x = (x^0, x^1, x^2, x^3) \in S^3 : (|x| = 1) \subset \mathbb{R}^4,$$

is homeomorphic to the sphere S^2. Find the connection with the Hopf bundle $S^3 \to S^2$. How this field is connected with quaternions?

29.33. We set

$$\frac{\partial}{\partial z} = \frac{1}{2}\left(\frac{\partial}{\partial x} - i\frac{\partial}{\partial y}\right), \quad \frac{\partial}{\partial \bar{z}} = \frac{1}{2}\left(\frac{\partial}{\partial x} + i\frac{\partial}{\partial y}\right).$$

Show that a function f is holomorphic if and only if $\dfrac{\partial}{\partial \bar{z}^k}(f) \equiv 0$ for all k (see also Problem 25.66).

29.34. (a) Prove that an arbitrary two-dimensional oriented manifold becomes parallelizable after removing a single point.

(b) Is this assertion true for non-orientable surfaces?

(c) Is it possible to make a non-orientable surface (with or without boundary) parallelizable removing points and embedded circles from it?

29.35. (a) On a closed two-dimensional surface, construct a smooth function having exactly three critical points.

(b) For which surface such a function can be a Morse function (this means that all its critical points are nondegenerate) (see Fig. 102)?

29.36. (a) On a closed two-dimensional surface, construct a smooth function having exactly four critical points.

(b) For which surface such a function can be a Morse function (this means that all its critical points are nondegenerate) (see Fig. 103)?

§ 30. Transformation Groups

30.1. Let a finite group G smoothly act on a manifold X and let $x_0 \in X$ be a fixed point for the action of any element of the group G. Prove that in a neighborhood of the point x_0, there exists a local coordinate system in which the action of the group G is linear.

30.2. Generalize the previous problem to the case of an arbitrary compact Lie group.

30.3. Prove that the set of all fixed points of the action of a finite group G on a smooth manifold is a union of smooth submanifolds (in general, of different dimensions).

30.4. Let G be a Lie group. Show that the action of the group G on itself by left (right) shifts is smooth.

30.5. Let a Lie group G act on itself by inner automorphisms. Show that the set of fixed points coincides with the centre of the group G.

30.6. Show that the isometry group of the Euclidean space is generated by orthogonal transformations and parallel translations (see Fig. 104). In this figure, the frame $Oxyz$ is fixed, and the frame $O'x'y'z'$ is obtained from the frame $Oxyz$ by an isometry of the Euclidean space.

30.7. Show that the isometry group of the standard n-dimensional sphere is isomorphic to the group of orthogonal transformations of the $(n+1)$-dimensional Euclidean space.

30.8. Show that the Lie groups $Sp(1)$ and $SU(2)$ are isomorphic. Show that they are diffeomorphic to the sphere S^3. Establish the connection with quaternions.

30.9. (a) In the space $\mathbb{H}$ of quaternions, define the linear transformation

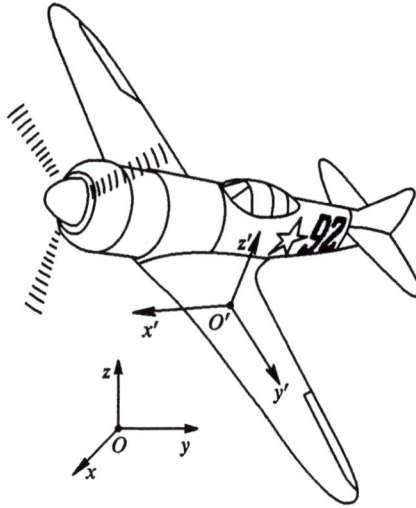

Figure 104

L_A by the formula $L_A\colon x \mapsto Ax$, where $x, A \in \mathbb{H}$ and the module of A is equal to 1. Prove that the set of all linear transformations of the form L_A composes a group isomorphic to $SU(2)$.

(b) In the space $\mathbb{H}$ of quaternions, consider the linear transformations defined by the formula $L_{A,B}\colon x \mapsto AxB$, where $x, A, B \in \mathbb{H}$ and the modules of A and B are equal to 1. Prove that the set of all such transformations forms a group isomorphic to $SO(4)$.

(c) Prove that $SO(4)$ is isomorphic to the quotient group $S^3 \times S^3/\mathbb{Z}_2$, where S^3 is equipped with a structure of the group $SU(2) \cong Sp(1)$

(d) Find the fundamental group $SO(n)$ for any n. Consider first the cases $n = 3$ and $n = 4$.

30.10. Denote by $A_z(\varphi)$, $A_y(\varphi)$, and $A_x(\varphi)$, respectively, the following matrices

$$\left\{\begin{array}{ccc} \cos\varphi & \sin\varphi & 0 \\ -\sin\varphi & \cos\varphi & 0 \\ 0 & 0 & 1 \end{array}\right\}, \quad \left\{\begin{array}{ccc} \cos\varphi & 0 & \sin\varphi \\ 0 & 1 & 0 \\ -\sin\varphi & 0 & \cos\varphi \end{array}\right\},$$

and

$$\left\{\begin{array}{ccc} 1 & 0 & 0 \\ 0 & \cos\varphi & \sin\varphi \\ 0 & -\sin\varphi & \cos\varphi \end{array}\right\}.$$

Clearly, these are rotation matrices with respect to the corresponding coordinate axes. Any matrix $A \in SO(3)$ can be represented in the form

$$A = A_i(\varphi) A_j(\theta) A_i(\psi)$$

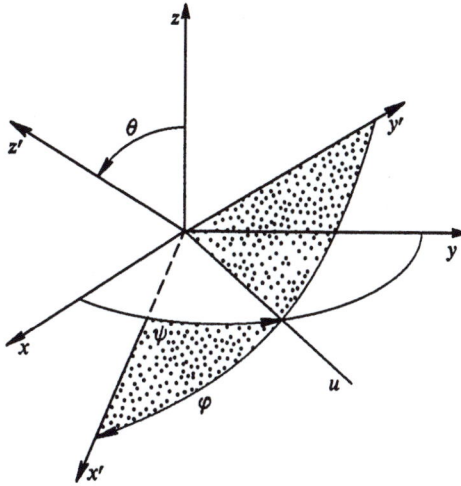

Figure 105 Euler angles; the axis u is the intersection of the planes Oxy and $Ox'y'$

of three rotations with respect to two coordinate axes. Here, the subscripts i and j assume the values in the set $\{x, y, z\}$, and, moreover, $i \neq j$. Figure 105 corresponds to the representation of the matrix A in the form of the product $A_z(\varphi) A_x(\theta) A_z(\psi)$.

The angles φ, θ, and ψ are called the *Euler angles*.

(a) How many variants are there for choosing the Euler angles?

(b) Prove that the Euler angles are local regular coordinates on almost the whole group $SO(3)$. Find the set of matrices in $SO(3)$ for which the Euler angles are not regular coordinates.

(c) Find the expression for the Killing metric on $SO(3)$ in the Euler angles.

(d) Find the volume of the group $SO(3)$.

30.11. Not only Euler angles can serve as coordinates on the group $SO(3)$. Indeed, an arbitrary matrix $A \in SO(3)$ is represented in the form

$$A = A_i(\varphi) A_j(\theta) A_k(\psi),$$

where the subscripts i, j, and k assume different values in the set $\{x, y, z\}$. In this case, the angles φ, θ and ψ are called the *navigation angles* (see Fig. 106).

(a) How many are variants for choosing navigation angles?

(b) Prove that the navigation angles are local regular coordinates on almost whole group $SO(3)$. Find the set of matrices from $SO(3)$ for which the navigation angles are not regular coordinates.

(c) Find the expression for the Killing metric on $SO(3)$ in the navigation angles.

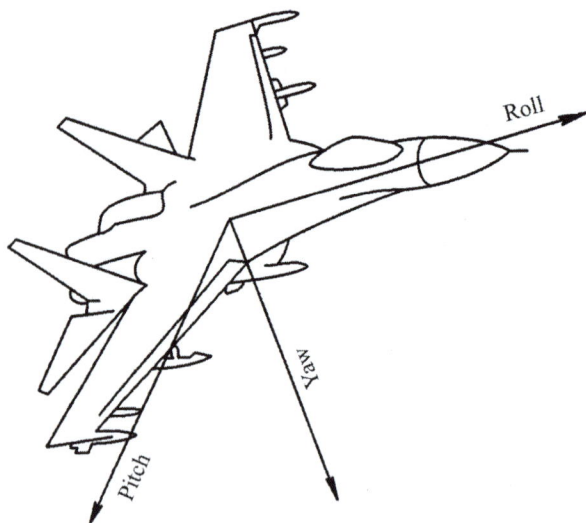

Figure 106 Navigation angles, the angles of turn around the indicated axes

30.12. Prove that $O = \dfrac{1+A}{1-A}$ is an orthogonal transformation of $\mathbb{R}^3$. Find its rotation axis, the angle of turn, and the Euler angles if

$$\text{(a) } A = \begin{pmatrix} 0 & 1 & 2 \\ -1 & 0 & -3 \\ -2 & 3 & 0 \end{pmatrix}, \quad \text{(b) } A = \begin{pmatrix} 0 & -1 & 3 \\ 1 & 0 & 1 \\ -3 & -1 & 0 \end{pmatrix},$$

$$\text{(c) } A = \begin{pmatrix} 0 & 2 & 2 \\ -2 & 0 & 2 \\ -2 & -2 & 0 \end{pmatrix}, \quad \text{(d) } A = \begin{pmatrix} 0 & 3 & -1 \\ -3 & 0 & 2 \\ 1 & -2 & 0 \end{pmatrix}.$$

30.13. Find the rotation axis and the angle of turn if all Euler angles of the orthogonal transformation are equal to

(a) $\pi/3$, (b) $\pi/4$.

30.14. (a) Prove that the Lie groups $SO(n)$, $SU(n)$, $U(n)$, and $Sp(n)$ are connected.

(b) Prove that the group $O(n)$ has two connected components.

(c) Find the number of connected components of the group of motions of the pseudo-Euclidean space $\mathbb{R}^n_1$.

(d) Prove that the group $SL(2, \mathbb{R})/\{\pm E\}$ is connected.

30.15. Realize the group $U(n)$ and its Lie algebra $u(n)$ by submanifolds in the Euclidean space of all square complex matrices of size $n \times n$ (a natural embedding of unitary and skew-Hermitian matrices in this space).

(a) Prove that $U(n) \subset S^{2n^2-1}$, where the sphere S^{2n^2-1} is standardly embedded in $\mathbb{R}^{2n^2} = \mathbb{C}^{n^2}$ and is of radius $\sqrt{n}$.

(b) Prove that the Riemannian metric induced on the group $SU(n)$ considered as a submanifold in S^{2n^2-1} coincides with two-sided invariant Killing metric on the group $SU(n)$.

(c) Find the intersection $U(n) \cap u(n)$ as a submanifold of the space $\mathbb{C}^{n^2}$.

30.16. Formulate and solve analogous problems for the groups $O(n)$ and $Sp(n)$.

30.17. Prove that the group of all isometries of a Riemannian manifold is a smooth manifold, i.e., a Lie group. Estimate from above its dimension through the dimension of the Riemannian manifold.

30.18. Enumerate all finite-dimensional Lie transformation groups of the real line $\mathbb{R}^1$.

30.19. Find the group of all linear-fractional transformations preserving the disk $|z| \leq 1$ in the complex plane. Prove that this group is isomorphic to the group $SL(2,\mathbb{R})/\mathbb{Z}_2$ and also to the group of all transformations preserving the form $dx^2 + dy^2 - dt^2$ in $\mathbb{R}^3$ (x, y, t). Establish the connection with the Lobachevskii geometry.

30.20. Prove that the connected component of the identity of the isometry group of the Lobachevskii plane (in the standard metric of constant curvature) is isomorphic to $SL(2,\mathbb{R})/\mathbb{Z}_2$. Find the total number of components in the group of motions of the Lobachevskii plane.

30.21. Prove that the phase space of the system consisting of a single material point sliding along the two-dimensional sphere with velocity whose module is constant is diffeomorphic to $\mathbb{R}P^3$.

30.22. A material ball is clamped between two parallel planes tangent to it. Under the motion of the planes preserving the parallelity of the planes and the distance between them, the ball rotates without sliding at contact points. Consider all such motions of the ball induced by the motion of the upper plane under which the lower contact point circumscribes a closed trajectory on the lower plane, i.e., the contact point returns to the initial place. Which part of the group $SO(3)$ can be obtained by such rotations of the ball (the turns of the ball are fixed after its return to the initial point)?

30.23. Find the quotient group G/G_0, where G is the group of motions of the Lobachevskii plane and G_0 is the connected component of the identity. Find all conformal transformations of this standard metric.

30.24. Find all discrete subgroups in the group G of affine transformations of the real line $\mathbb{R}^1$.

30.25. Prove that all left-invariant vector fields on a Lie group G are in a one-to-one correspondence with the vectors of the tangent space $T_e(g)$ at the identity of the group G.

30.26. Prove that the commutator of two left-invariant vector fields on a Lie group G is again a left-invariant vector field, i.e., the operation of commutation transforms the space $T_e G$ into a Lie group.

30.27. Let $\boldsymbol{\xi}$ be a left-invariant vector field, and let φ_t be the one-parameter transformation group corresponding to it. Prove that φ_t is a right shift for any t, i.e., $\varphi_t\left(g\right) = gh_t$, where h_t is certain element of the group G.

Let G be a Lie group, and let $x^1, \ldots, x^n$ be a local coordinate system in a neighborhood of the identity,. Then the multiplication operation induces the vector-valued function $\mathbf{q} = \mathbf{q}\left(x, y\right) = xyx^{-1}y^{-1}$, $x = \left(x^1, \ldots, x^n\right)$, $y = \left(y^1, \ldots, y^n\right)$. The Taylor series expansion of the function $\mathbf{q} = \mathbf{q}\left(x, y\right)$ has the form

$$q^i = \sum_{j,k} c^i_{jk} x^j y^k + \varepsilon^i_3,$$

where ε^i_3 is the quantity of the third order of smallness with respect to the coordinates x^i, y^i. The bilinear expression

$$\zeta^i = \sum_{j,k} c^i_{jk} \xi^j \eta^k$$

defines a certain operation on tangent vectors at the identity of the group G. This operation is denoted by $\zeta = [\xi, \eta]$, and ζ is called the *commutation of the vectors* ξ *and* η. Therefore, the tangent space $T_e\left(G\right)$ transforms into an algebra called the *Lie algebra of the Lie group* G.

30.28. Prove that the following properties hold on the Lie algebra L:
(a) $[\xi, \eta] = -[\eta, \xi]$;
(b) the Jacobi identity

$$[[\xi, \eta], \zeta] + [[\eta, \zeta], \xi] + [[\zeta, \xi], \eta] = 0.$$

30.29. Verify that the operation in the Lie algebra L passes to the commutator of vector fields on the Lie group G if to a vector ξ, we put in correspondence the (right-) left-invariant vector field,

30.30. Let $\mathbf{x}\left(t\right)$ and $\mathbf{y}\left(t\right)$ be two curves passing through the identity of the group G, and, moreover, let $\xi = \dfrac{d\mathbf{x}}{dt}\left(0\right)$, $\eta = \dfrac{d\mathbf{y}}{dt}\left(0\right)$. Show that

$$[\xi, \eta] = \frac{d}{dt}\left(\mathbf{x}\left(\sqrt{t}\right) \mathbf{y}\left(\sqrt{t}\right) \mathbf{x}^{-1}\left(\sqrt{t}\right) \mathbf{y}^{-1}\left(\sqrt{t}\right)\right)\Big|_{t=0}.$$

30.31. Let $\gamma\left(t\right)$ be a one-parameter subgroup of a Lie group. Assume that γ has self-intersections. Show that there exists a number $L > 0$ such that $\gamma\left(t + L\right) = \gamma\left(t\right)$ for all $t \in \mathbb{R}$. In particular, this implies that any one-parameter subgroup of a Lie group considered as a one-dimensional Lie group is homeomorphic either to a line or to a circle.

Remark. A noncompact one-parameter subgroup can be embedded in a Lie group in a rather complicated way, for example, in the form of the so-called irrational or dense winding of the torus of a certain dimension.

30.32. Let G be a compact connected Lie group. Show that each point $x \in G$ belongs to a certain one-parameter subgroup.

30.33. Let G be a compact group smoothly acting on a manifold M. Show that on M, there exists a Riemannian metric for which G is the isometry group.

30.34. Show that a commutative connected Lie group is locally isomorphic to a finite-dimensional vector space.

30.35. Show that a compact commutative connected Lie group is isomorphic to the torus.

30.36. Show that a commutative connected Lie group is isomorphic to the Cartesian product of the torus to the vector space.

30.37. Let a Lie group G be a subgroup of the matrix group $GL\,(n, \mathbb{C}) \subset \mathbb{C}^{n^2} = \mathrm{End}\,(n, \mathbb{C})$. Show that in the Lie algebra L of the group G understanding as a subspace in $\mathrm{End}\,(n, \mathbb{C})$, the commutation operation coincides with the ordinary commutation of matrices., i.e. $[\boldsymbol{\xi}, \boldsymbol{\eta}] = \boldsymbol{\xi}\boldsymbol{\eta} - \boldsymbol{\eta}\boldsymbol{\xi}$, where $\boldsymbol{\xi}, \boldsymbol{\eta} \in L$.

30.38. Describe the Lie algebras of the following matrix Lie groups:

$$SL\,(n, \mathbb{C}), \quad SL\,(n, \mathbb{R}), \quad U\,(n), \quad O\,(n), \quad O\,(n, m), \quad Sp\,(n).$$

30.39. Prove that a finite group cannot effectively act on $\mathbb{R}^n$.

§ 31. Differential Forms

31.1. Calculate the geodesic curvature k_g of a curve $x = x\,(s)$, $y = y\,(s)$, $z = z\,(s)$ lying on the sphere of radius R and deduce from this the following formula for finding the area S of a domain on the sphere bounded by a closed line L:

$$S = \left(2\pi \pm \oint_L \sqrt{k^2 - \frac{1}{R^2}}\, ds \right) R^2.$$

Here, k is the curvature of the curve. We call attention that in integrating, it is necessary to take into account the sign of the geodesic curvature $k_g = \sqrt{k^2 - \frac{1}{R^2}}$. Precisely, if the vector of the geodesic curvature is directed inside the domain, then we take the sign "$-$"; otherwise, we take the sign "$+$". However, this rule is automatically taken into account by the formula for the geodesic curvature $k_g = \dfrac{(\mathbf{r}', \mathbf{r}'', \mathbf{n})}{|\mathbf{r}'|^3}$ (see Problem **11.7**).

31.2. Find $\displaystyle\int_M \omega$ in the following cases:

(a) $\omega = (x + y)\,dx + (x - y)\,dy$, M: $\dfrac{x^2}{a^2} + \dfrac{y^2}{b^2} = 1$;

(b) $\omega = (2a - y)\,dx + x\,dy$, M: $x = a\,(t - \sin t)$, $y = a\,(1 - \cos t)$, $0 \le t \le 2\pi$;

(c) $\omega = \left(y^2 - z^2\right) dx + 2yz\,dy - x^2 dz$, M: $x = t$, $y = t^2$, $z = t^3$, $0 \le t \le 1$;

(d) $\omega = y\,dx + z\,dy + x\,dz$, M: $x = a\cos t$, $y = a\sin t$, $z = bt$, $0 \le t \le 2\pi$;

(e) $\omega = (y+z)\,dx + (x+z)\,dy + (x+y)\,dz$, M: $x = a\sin^2 t$, $y = 2a\cos t \cdot$
$\sin t$, $z = a\cos^2 t$, $0 \le t \le 2\pi$;

(f) $\omega = (y-z)\,dx + (z-x)\,dy + (x-y)\,dz$, M: $x^2 + y^2 = a^2$, $\dfrac{x}{a} + \dfrac{z}{h} = 1$;

(g) $\omega = (y^2 + z^2)\,dx + (x^2 + z^2)\,dy + (x^2 + y^2)\,dz$, M: $x^2 + y^2 + z^2 = 2Rx$,
$x^2 + y^2 = Rx$;

(h) $\omega = (y-z)\,dy \wedge dz + (z-x)\,dz \wedge dx + (x-y)\,dx \wedge dy$, M: $x^2 + y^2 + z^2 = 1$;

(i) $\omega = x\,dy \wedge dz + y\,dz \wedge dx + z\,dx \wedge dy$, M: $x^2 + y^2 + z^2 = a^2$;

(j) $\omega = \dfrac{dy \wedge dz}{x} + \dfrac{dz \wedge dx}{y} + \dfrac{dx \wedge dy}{z}$, M: $\dfrac{x^2}{a^2} + \dfrac{y^2}{b^2} + \dfrac{z^2}{c^2} = 1$.

We will orient surfaces by the exterior normal. In item (a) the curve is oriented counterclockwise, and in items (f) and (g) counterclockwise if we see from the positive direction of the axis Ox.

31.3. Calculate the surface integral $\displaystyle\iint \varphi \frac{\partial \psi}{\partial n}\,d\sigma$ for the following closed surface Σ:

(a) for $\varphi = z^2$, $\psi = x^2 + y^2 - z^2$ if Σ bounds the domain $x^2 + y^2 + z^2 \le 1$ and $y \ge 0$;

(b) for $\varphi = 2x^2$, $\psi = x^2 + z^2$ if Σ bounds the domain $x^2 + y^2 \le 1$ and $0 \le z \le 1$;

(c) for $\varphi = \psi = \dfrac{x+y+z}{\sqrt{3}}$ if Σ is the sphere $x^2 + y^2 + z^2 = r^2$;

(d) for $\varphi = 1$, $\psi = e^x \sin y + e^y \sin x + z$ if Σ is the triaxial ellipsoid $\dfrac{x^2}{a^2} + \dfrac{y^2}{b^2} + \dfrac{z^2}{c^2} = 1$.

Here, $\dfrac{\partial}{\partial n}$ denotes the derivative in direction of the exterior normal to the surface.

31.4. Find the gradients of the following functions in the cylindrical coordinates:

(a) $u = \rho^2 + 2\rho\cos\varphi - e^z \sin\varphi$;

(b) $u = \rho\cos\varphi + z\sin^2\varphi - e^\rho$.

31.5. Find $\operatorname{div} \mathbf{X}$ in the cylindrical coordinates for the following vector fields:

(a) $\mathbf{X} = (\rho, \ z\sin\varphi, \ e\varphi\cos z)$;

(b) $\mathbf{X} = (\varphi \arctan \rho, \ 2, \ -z^2 e^z)$.

31.6. Find the divergence of the vector field

$$\mathbf{X} = \left(r^2, \ -2\cos^2\varphi, \ \frac{\varphi}{r^2 + 1} \right)$$

in the spherical coordinates.

31.7. Find the vorticity of the following vector fields in the spherical coordinates:

(a) $\mathbf{X} = (2r + a\cos\varphi, -a\sin\theta, r\cos\theta)$, $\alpha = \text{const}$;

(b) $\mathbf{X} = (r^2, 2\cos\theta, -\varphi)$.

31.8. Verify that the following vector fields in the spherical coordinates (r, θ, φ) are potential:

(a) $\mathbf{X} = (2\cos\theta/r^3, \sin\theta/r^3, 0)$;

(b) $\mathbf{X} = (f(r), 0, 0)$.

31.9. Find the potentials of the following vector fields in the cylindrical coordinates (ρ, φ, z):

(a) $\mathbf{X} = \left(1, \dfrac{1}{\rho}, 1\right)$; (b) $\mathbf{X} = \left(\rho, \dfrac{\varphi}{\rho}, z\right)$; (c) $\mathbf{X} = (\varphi z, z, \rho\varphi)$;

(d) $\mathbf{X} = \left(e^\rho \sin\varphi, \dfrac{e^\rho \cos\varphi}{\rho}, 2z\right)$;

(e) $\mathbf{X} = (\varphi\cos z, \cos z, -\rho\varphi\sin z)$.

31.10. Find the potential of the following vector fields in the spherical coordinates (r, θ, φ):

(a) $\mathbf{X} = (\theta, 1, 0)$; (b) $\mathbf{X} = \left(2r, \dfrac{1}{r}, \dfrac{1}{r}\sin\theta\right)$;

(c) $\mathbf{X} = \left(\dfrac{\varphi^2}{2}, \dfrac{\theta}{r}, \dfrac{\varphi}{\sin\theta}\right)$;

(d) $\mathbf{X} = (\cos\varphi\sin\theta, \cos\varphi\cos\theta, -\sin\varphi)$;

(e) $\mathbf{X} = \left(e^r \sin\theta, \dfrac{e^r \cos\theta}{r}, \dfrac{2\varphi}{(1+\varphi^2)r\sin\theta}\right)$.

31.11. Find the circulation of the vector field

$$\mathbf{X} = (r, 0, (R+r)\sin\theta)$$

in the spherical coordinates on the circle $\{r = R, \theta = \pi/2\}$.

31.12. Calculate the linear integral of the vector field $\mathbf{X}$ along the line L given in the cylindrical coordinates if

(a) $\mathbf{X} = (z, \rho\varphi, \cos\varphi)$, L is a line segment: $\{\rho = a, \varphi = 0, \ 0 \le z \le 1\}$;

(b) $\mathbf{X} = (\rho, 2\rho\varphi, z)$, L is a semicircle: $\{\rho = 1, z = 0, \ 0 \le \varphi \le \pi\}$;

(c) $\mathbf{X} = (e^\rho \cos\varphi, \rho\sin\varphi, \rho)$, L is a spiral: $\{\rho = R, z = \varphi, 0 \le \varphi \le 2\pi\}$;

(d) $\mathbf{X} = (z, \rho z, \rho)$, L is a circle: $\{\rho = 1, z = 0\}$;

(e) $\mathbf{X} = (\rho\sin\varphi, -\rho^2 z, \rho^2)$, L is a circle: $\{\rho = R, z = R\}$;

(f) $\mathbf{X} = (z\cos\varphi, \rho, \varphi^2)$, L: $\{\rho = \sin\varphi, z = 1\}$.

31.13. Calculate the linear integral of the vector field $\mathbf{X}$ along the line L given in the spherical coordinates if

(a) $\mathbf{X} = (4r^3, \tan\varphi/2, \theta\varphi, \cos^2\varphi)$, L: $\{\varphi = \pi/2, \theta = \pi/4, \ 0 \le r \le 1\}$;

(b) $\mathbf{X} = (\sin^2\theta, \sin\theta, r\varphi\theta)$, L: $\{\varphi = \pi/2, r = 1/\sin\theta, \ \pi/4 \le \theta \le \pi/2\}$;

(c) $\mathbf{X} = (r\theta, 0, r\sin\theta)$, L: $\{r = 1, \theta = \pi/4, 0 \le \varphi \le 2\pi\}$;

(d) $\mathbf{X} = (r\sin\theta, \theta e^\theta, 0)$, L: $\{r = \sin\varphi, \theta = \pi/2, \ 0 \le \varphi \le \pi\}$;

(e) $\mathbf{X} = (0, 0, r\varphi\theta)$, L is the contour bounding the half-disk $\{r \leq R, \varphi = \pi/4\}$;

(f) $\mathbf{X} = (r, 0, (R + r)\sin\theta)$, L: $\{r = R, \theta = \pi/2\}$;

(g) $\mathbf{X} = (e^r \cos\theta, 2\theta\cos\varphi, \varphi)$, L: $\{r = 1, \theta = 0, 0 \leq \varphi \leq \pi\}$.

31.14. Find the flow of the vector field $\mathbf{X}$ given in the cylindrical coordinates through the surface S if

(a) $\mathbf{X} = (\rho, -\cos\varphi, z)$, S bounds the domain $\{\rho \leq 2, 0 \leq z \leq 2\}$;

(b) $\mathbf{X} = (\rho, \rho\varphi, -2z)$, S bounds the domain $\{\rho \leq 1, 0 \leq \varphi \leq \pi/2, -1 \leq z \leq 1\}$.

31.15. Find the flow of the vector field $\mathbf{X}$ given in the spherical coordinates through the surface S if

(a) $\mathbf{X} = (1/r^2, 0, 0)$, S surrounds the origin;

(b) $\mathbf{X} = (r, r\sin\theta, -3r\varphi\sin\theta)$, S bounds the domain $\{r \leq R, \theta \leq \pi/2\}$;

(c) $\mathbf{X} = (r^2, 0, R^2\cos\varphi)$, S: $\{r = R\}$;

(d) $\mathbf{X} = (r, 0, 0)$, S bounds the domain $\{r \leq R, \theta \leq \pi/2\}$;

(e) $\mathbf{X} = (r^2, 0, R^2 r\sin\theta\cos\varphi)$, S bounds the domain $\{r \leq R, 0 \leq \varphi \leq \pi/2, \theta \leq \pi/2\}$.

31.16. Prove that bi-invariant forms on a Lie group are closed.

§ 32. Homotopy Theory

32.1. Represent as a cell complex the following:

(a) the torus;

(b) the Klein bottle;

(c) the suspension over a cell complex K.

32.2. Prove that the sphere S^∞ and the ball D^∞ are cell spaces.

32.3. Prove that the topology of a CW-complex is the weakest among the topologies in which all characteristic mappings are continuous.

32.4. Prove that on a CW-complex, a function is continuous if and only if it is continuous on every finite subcomplex.

32.5. Prove that the torus with a disk spanned on the meridian is homotopy equivalent to the union $S^1 \vee S^2$.

32.6. Prove that the torus with a disk spanned on the meridian and the parallel are homotopy equivalent to the sphere S^2.

32.7. Generalize Problems 32.5 and 32.6 to the case of the product $S^k \times S^{n-k}$.

32.8. Prove the homotopy equivalence of the spaces

$$(X \times S^n)(X \vee S^n) = \Sigma^n X.$$

32.9. Prove the following homotopy equivalences:

(a) $\Sigma\left(X \vee Y\right) \sim \Sigma X \vee \Sigma Y$;

(b) $\Sigma\left(X \wedge Y\right) \sim \Sigma\left(X \times Y\right) / \left(\Sigma X \vee \Sigma Y\right)$.

32.10. Let $P\left(X; A, B\right)$ be the path spaces with origin in A and end in B. Let $A \subset B$. Prove that $P\left(X; A, B\right)$ contains a subspace homeomorphic to A.

32.11. Let $f\colon X \to \Sigma$ be a continuous mapping of simplexes, and let $Y \subset X$ be the subcomplex on which f is simplicial. Prove that there exists a subdivision of the complex X identical on Y such that the mapping f is homotopic to a certain simplicial mapping q, and, moreover the homotopy is constant on Y.

32.12. Let X be a simplicial complex, and let S_x be the star of a vertex $x \in X$. Prove that any two simplexes of the star S_x intersect along a certain face.

32.13. Prove that a simplicial mapping of simplicial complexes is continuous.

32.14. Let X be a simplicial complex, and let $\varepsilon > 0$. Prove that there exists a subdivision of the complex X such that the diameter of each new complex is less than ε.

32.15. Prove that a one-dimensional cell complex is the space of type $K\left(\pi, 1\right)$, where π is a free group.

32.16. Prove that a contractible space is homotopy equivalent to a point.

32.17. Prove that any two space of type $K\left(\pi, 1\right)$ are weakly homotopy equivalent.

32.18. Prove that

(a) $S^n \wedge S^k = S^{n+k}$;

(b) S^n / S^k is homotopy equivalent to $S^n \vee S^{k+1}$;

(c) $S^n \backslash S^k$ is homotopy equivalent to S^{n-k-1};

(d) $S^n \backslash S^k$ is homotopy equivalent to $S^{n-k-1} \times \mathbb{R}^{k+1}$.

Here $S^k \subset S^n$ is the standard embedding.

32.19. Prove that on a CW-complex, a function is continuous if and only if it is continuous on every finite subcomplex.

32.20. Prove that if the product of two topological spaces is homeomorphic to the suspension over a certain topological space, then either both factors are contractible or one of them reduces to a point.

32.21. Let a space X be contractible to a subspace Y, and, moreover, let the homotopy be fixed (constant) on Y. Prove that any path in X with ends in Y is homotopic to a path entirely lying in Y (the homotopy is fixed at the ends).

32.22. Prove that any connected cell complex is homotopy equivalent to a cell complex with one vertex.

32.23. Prove that the sphere S^{n-1} can be represented as the union $\left(S^r \times D^{n-r}\right) \cup \left(D^{r+1} \times S^{n-r-1}\right)$ with common boundary $S^r \times S^{n-r-1}$.

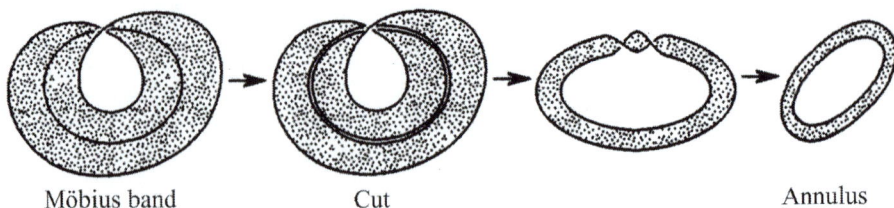

Möbius band   Cut   Annulus

Figure 107

32.24. In the Euclidean space $\mathbb{R}^n$, consider the standard sphere S^{n-1} and two spheres

$$S^{r-1} = \{x_{r+1} = \ldots = x_n = 0\}, \quad S^{n-r-1} = \{x_1 = \ldots = x_r = 0\}.$$

embedded in it. Prove that any pair of points $y \in S^{r-1}$ and $x \in S^{n-r-1}$ is connected by a large arc without common points outside these spheres.

32.25. Find the topological type of the closed one-sheeted hyperboloid $\Gamma = \{x^2 + y^2 - z^2 = 1\}$ in the projective space $\mathbb{R}P^3$.

32.26. Cut the Möbius band embedded in $\mathbb{R}^3$ along the middle line. Is the obtained manifold orientable? Repeat the process of cutting several times. Describe the obtained disconnected manifold and find the index of linking of arbitrary two connected components (see Fig. 107).

32.27. Prove that the space of polynomials of degree 3 is homotopy equivalent to the complement of the trifolium in the sphere S^3. Construct an explicit deformation.

32.28. In $\mathbb{R}^3$, consider two linked circles S_1^1 and S_2^1 (see Fig. 108). Present a geometric proof of the relation $aba^{-1}b^{-1} = 1$ in the group $\pi_1(\mathbb{R}^3 \setminus (A \cup B), x_0)$.

32.29. Consider $\mathbb{R}P^2$ as the sphere with identified opposite points. Denote by a the path in $\mathbb{R}P^2$ which is represented on the sphere by an arc of large disk connected two diametrically opposite points. Present a geometric proof of the relation $a^2 = 1$ in the group $\pi_1\left(\mathbb{R}P^2\right)$.

32.30. Consider the set of points of $\mathbb{C}^n$ with pairwise different coordinates. Prove that the obtained space has the type of Eilenberg–Maclane complex $K(\pi, 1)$.

32.31. Construct an example of two not homotopy equivalent spaces X_1 and X_2 and two one-to-one continuous mappings $f\colon X_1 \to X_2$ and $g\colon X_2 \to X_1$.

32.32. Denote by $\{X, Y\}$ the set of all continuous mappings from the space X into the space Y. Let $h\colon X \to X'$ be a continuous mapping, and let the correspondence $\Phi\colon \{X', Y\} \to \{X, Y\}$ be defined by the formula $\Phi(a) = ah$. Prove that this correspondence Φ transforms homotopic mappings into homotopic mappings.

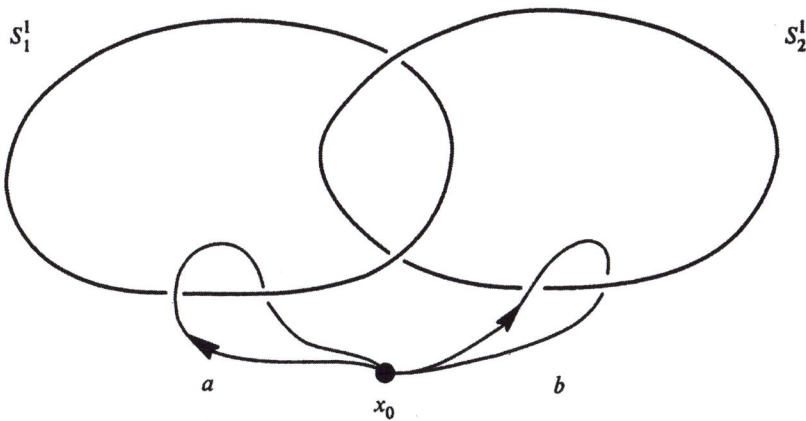

Figure 108

32.33. Prove that the following homotopy equivalence holds:

$$\Sigma \left(S^n \times S^m \right) \sim S^{n+1} \vee S^{m+1} \vee S^{n+m+1}.$$

32.34. Prove that the infinite-dimensional sphere S^∞ is contractible to a point along itself.

32.35. Prove that a connected finite graph is homotopy equivalent to the union of circles $\vee S^1$.

32.36. Let a mapping $p \colon X \to Y$ satisfy the covering homotopy axiom. Prove that the inverse images of points are homotopy equivalent.

32.37. Let a space X be contracted to its linearly connected subspace A. Prove that the space X is linearly connected.

32.38. In a space X, fix two points x_0 and x_1. Let Y be space of paths with origin at the point x_0 and passing through the point x_1. Prove that the space Y is contractible.

32.39. Prove that the path space $P(X; X, X)$ is contractible to $X \subset P(X; X, X)$, being fixed on X.

32.40. Let a sequence of spaces $X_n \subset X_{n+1}$ be given, and let X_{n+1} be contractible to X_n being fixed on X_n. Prove that the space

$$X = \bigcup_n X_n$$

is contractible to X_0 being fixed on X_0.

32.41. Prove that each open n-dimensional manifold is homotopy equivalent to an $(n-1)$-dimensional complex.

32.42. Prove that if a space X is contractible to a subspace A being fixed on A, then A is homotopy equivalent to X.

32.43. Consider the following space X (clearly, it is a cell complex). Denote by M_2 the ordinary Möbius band with boundary ∂M_2 homeomorphic to the circle. The "triple" Möbius band M_3 is constructed as follows. Let T consist of three segments of equal length lying in the same plane at equal angles. In the product $T \times I$, identify the subspaces $T \times 0$ and $T \times 1$ twisting them by the angle $2\pi/3$. The boundary of the obtained space M_3 is also homeomorphic to the circle. The space X is obtained from M_2 and M_3 by identifying the segments $I_1 \subset \partial M_2$ and $I_2 \subset \partial M_3$. Prove that ∂X is a (not deformational) retract of the space X.

32.44. Calculate the sets $\pi\left(S^1 \times S^1, S^2\right)$ and $\pi\left(S^k \times S^{n-k}, S^n\right)$.

32.45. For a topological space X, denote by $\mathrm{Cat}_1(X)$ (resp. by $\mathrm{Cat}_2(X)$) the minimum number of closed subspaces X_i such that $X = \cup_i X$ and the embeddings $X_i \subset X$ are homotopic to constant mappings (resp. X_i are contractible). Find $\mathrm{Cat}_1(\mathbb{R}P^n)$ and $\mathrm{Cat}_2(\mathbb{R}P^n)$.

32.46. Calculate $\mathrm{Cat}_1(K)$ and $\mathrm{Cat}_2(K)$ for the sphere with three identified points.

32.47. Let M^2 be a compact closed oriented 2-dimensional manifold of genus h, i.e., M^2 is the sphere S^2 with h handles. Find $\Sigma^2 M^2$ (two-fold suspension) with accuracy up to the homotopy type.

32.48. In $\mathbb{R}P^n$, consider a certain standard chart with inhomogeneous coordinates $x_1, \dots, x_n$. Find the homotopy type of the following sets:

(a) $\mathbb{R}P^n \setminus \check{S}^k$, where $\check{S}^k = \{x_1^2 + \dots + x_{k+1}^2 = 1, x_{k+2} = \dots = x_n = 0\}$;

(b) $\mathbb{R}P^n \setminus \check{M}_q^k$, where $\check{M}_q^k = \{x_1^2 + \dots + x_k^2 - x_{k+1}^2 - \dots - x_{k+q}^2 - 1 = 0,$ $x_{k+q+1} = \dots = x_n = 0\}$;

(c) $\check{S}^k$; (d) $\check{M}_q^k$.

32.49. In the open manifold $\mathbb{R}^n \times S^{n-k}$, consider a small ball D^n, and instead of it, glue the projective space $\mathbb{R}P^n$, i.e., identify points x and $-x$ on the boundary of the ball $S^{n-1} = \partial D^n$. Prove that the obtained space is homotopy equivalent to $\mathbb{R}P^{n-1} \vee S^{n-k}$.

32.50. Given a topological manifold M^n whose boundary is a topological manifold P^{n-1}. It is known that the boundary P^{n-1} is contractible to a point along the manifold M^n.

(a) Prove that the manifold is contractible to a point.

(b) Prove that if the manifold P^{n-1} is simply connected, then the manifold M^n is homeomorphic to the disk D^n (under the assumption that P^{n-1} is contractible to a point along M^n).

(c) Construct an example of a pair $\left(M^n, P^{n-1}\right)$ such that the manifold P^{n-1} is contractible to a point along M^n but M^n is not homeomorphic to the disk D^n. As a consequence, show that $\pi_1\left(P^{n-1}\right) \neq 0$.

32.51. Find the homotopy type of the space $\mathbb{C}^n \setminus \Delta$, where $\Delta = \cup_{ij}\Delta_{ij}$, $\Delta_{ij} = \{x \in \mathbb{C}^n | x_i = x_j\}$.

32.52. Calculate the number of the following mappings (with accuracy up to a homotopy):

(a) $\mathbb{R}P^n \to \mathbb{R}P^n$; (b) $\mathbb{C}P^n \to \mathbb{C}P^n$; (c) $\mathbb{R}P^{n+1} \to \mathbb{R}P^n$; (d) $\mathbb{C}P^{n+1} \to \mathbb{C}P^n$; (e) $\Sigma\mathbb{R}P^n \to \mathbb{R}P^n$; (f) $\Sigma\mathbb{C}P^n \to \mathbb{C}P^n$; (g) $\Sigma\mathbb{R}P^n \to \mathbb{R}P^{n+1}$; (h) $\Sigma\mathbb{C}P^n \to \mathbb{C}P^{n+1}$.

32.53. (a) Let the spaces X and Y be connected. Prove that

$$\mathrm{Cat}\,(\mathrm{join}\,(X,Y)) = \min\,(\mathrm{Cat}\,(X)\,,\mathrm{Cat}\,(Y))\,,$$

where Cat denotes the Lyusternik–Shnirelman category (Cat_1 in the notation of Problem 32.45).

(b) Find $\mathrm{Cat}\,(S^1 \times S^2)$.

32.54. Let the spaces $X_i, 1 \le i \le N$, be linearly connected and $X = X_1 \times X_2 \times \cdots \times X_N$. Prove that

$$\mathrm{Cat}\,(X_i) \le \mathrm{Cat}\,(X) \le 1 + \sum_{i=1}^{N} (\mathrm{Cat}\,(X_i) - 1)\,.$$

32.55. (a) Calculate $\mathrm{Cat}\,(\mathbb{R}P^n)$, $\mathrm{Cat}\,(T^n)$, $\mathrm{Cat}\,(S^m \times S^n)$.

(b) Prove that if the sphere S^n is covered by q closed sets (not necessarily connected) $V_1, V_2, \ldots, V_q$, where $q \le n$, then there always exists at least one set V_i such that it contains two diametrically opposite points $-x$ and x of the sphere S^n.

32.56. Let $M \subset \mathbb{R}^n$ be an arbitrary subset in the Euclidean space (for example, a smooth submanifold), and let $\mathbb{R}^n \subset \mathbb{R}^{n+1}$ be the standard embedding. Prove that the following homotopy equivalence holds: $\mathbb{R}^{n+1} \setminus M \simeq \Sigma\,(\mathbb{R}^n \setminus M)$.

32.57. *Connection between the Lyusternik–Shnirelman category and the so-called cohomology length of a complex (or manifold).* Let M^n be a smooth compact connected closed manifold. Consider the ring $H^*\,(M^n, G)$, where $G = \mathbb{Z}$ if M^n is oriented and $G = \mathbb{Z}_2$ if M^n is not oriented. Denote by $l\,(M^n; G)$ the maximum integer for which there exists a sequence of elements $x_1, x_2, \ldots, x_l$ of the ring $H^*\,(M^n; G)$ ($\deg x_a > 0, 1 \le a \le l$) such that their product $x_1 \wedge x_2 \wedge \ldots \wedge x_l\,(M^n; G) \ne 0$ in the ring $H^*\,(M^n; G)$. The number $l\,(M^n; G)$ is called the *cohomology length* of the manifold M^n. Prove that

$$\mathrm{Cat}\,(M^n) \ge l\,(M^n; G)\,.$$

32.58. Prove that for any arcwise connected topological space X and for any point x_0 of it, the group $\pi_1\,(\Omega X, x_0)$ is Abelian.

32.59. Prove that every contractible space is simply connected.

32.60. Prove that the group $\pi_1\,(\bigvee_A S^1)$ is a free group with A generators.

32.61. Prove that if Y_1 and Y_2 are homotopy equivalent, then the following isomorphisms hold: $\pi_1\,(Y_1) \cong \pi_1\,(Y_2)$ and $\pi_k\,(Y_1) \cong \pi_k\,(Y_2)$, $k \ge 2$.

32.62. Prove that $\pi_1(X \vee Y) = \pi_1(X) * \pi(Y)$, where $\pi_1(X) * \pi_1(Y)$ is the free product of the groups $\pi_1(X)$ and $\pi_1(Y)$.

32.63. Find the fundamental group of the complement to the knot-trifolium in $\mathbb{R}^3$ (and also in the sphere S^3) and prove that the trifolium "is not untied", i.e., there is no homeomorphism of the Euclidean space (or sphere) onto itself that transforms the trifolium into the standardly embedded untied circle, i.e., a trivial knot.

32.64. Find the fundamental group of the complement to the knot Γ in $\mathbb{R}^3$, which is defined as follows: the circle depicting the knot lies on the two-dimensional standardly embedded torus $T^2 \subset \mathbb{R}^3$ on which it goes around the parallel of the torus p times and its meridian q times; moreover, the numbers p and q are relatively prime. (The knot-trifolium can be represented as such a knot Γ, where $p = 2, q = 3$.) Reveal the role of the condition of mutual primality of the numbers p and q.

32.65. Let $X = Y\bigcup_W Z$, where Y, Z, and W are finite CW-complexes, $W = Y \cap Z$, W is arcwise connected, and $X = Y\bigcup_W Z$ denotes the complex obtained by gluing Y and Z along the common subset W. Calculate the group $\pi_1(X)$ if the groups $\pi_1(Y)$, $\pi_1(Z)$, and $\pi_1(W)$ are known. Separately, consider the case where W is disconnected.

32.66. Given an arbitrary group G with finitely many generators and relations. Prove that there exists a finite complex X whose fundamental group is isomorphic to G. Is it possible to choose a finite-dimensional manifold, for example, even dimensional as such a complex X?

32.67. Calculate the group $\pi_1(X)$, where X is the union of three circles.

32.68. Construct a two-dimensional complex X whose fundamental group is equal to $\mathbb{Z}/p\mathbb{Z}$. For which p is it possible to choose a two-dimensional smooth closed compact manifold as such a complex?

32.69. Calculate the fundamental group of the two-dimensional sphere with three handles. Reveal whether or not this group is commutative and find its commutant. Calculate the fundamental group of the two-dimensional torus.

32.70. Let a simplicial complex X have N one-dimensional simplexes. Prove that its fundamental group has no more than N generators.

32.71. Prove that $\pi_1(X) = \pi_1(X_2)$, where X is a CW-complex and X_2 is its two-dimensional skeleton, i.e., the union of all cells of dimension 1 and 2.

32.72. Find $\pi_2(X)$, where $X = S^1 \vee S^2$. Is this group finitely generated?

32.73. Find the fundamental group of "figure eight curve" (the union of two circles).

32.74. Let f be a path in X, and let $a \in \pi_1(X, x_0)$, $f(0) = x_0$. Prove that there exists a path g such that $g(0) = x_0$, $g(1) = f(1)$, and $fg^{-1} \in a$.

32.75. Let X be an arcwise connected space. Prove that the group $\pi_1(X, x)$ is isomorphic to the group $\pi_1(X, y)$ for any two points $x, y \in X$,

32.76. Calculate $\pi_1(X)$ and $\pi_n(X)$, where X is the union $S^1 \vee S^n$.

32.77. Prove that if X is a one-dimensional CW-complex, then $\pi_1(X)$ is a free group.

32.78. Prove that the group $G = \mathbb{Z} \oplus \mathbb{Z} \oplus \mathbb{Z} \oplus \mathbb{Z}$ cannot be the fundamental group of any 3-dimensional manifold.

32.79. Calculate $\pi_1(P_g)$, where P_g is a 2-dimensional compact closed orientable surface of genus g.

32.80. Calculate $\pi_1(TP_g)$, where TP_g is the manifold of linear elements of a surface of genus g.

32.81. Calculate the fundamental group of the "Klein bottle" constructing the covering space with action of a discrete group.

32.82. Let P be a two-dimensional surface with nonempty boundary (open surface). Prove that $\pi_1(P)$ is a free group.

32.83. Prove that if X is a CW-complex, then $\pi_1(X)$ is a group whose generators are one-dimensional cells, and the complete set of relations is determined by the boundaries of 2-dimensional cells.

32.84. Let G be a continuous groupoid with unit. Prove that G is homotopically simple in all dimensions and, as a consequence, that $\pi_1(G)$ is an Abelian group.

32.85. Let X be a continuous groupoid with unit, and let $G \subset \pi_1(X)$ be a subgroup. Prove that

(a) it is possible to introduce a multiplication in $\hat{X}_G$ so that $p_G \colon \hat{X}_G \to X$ (where p_G is the projection of the covering $\hat{X}_G$ on X) becomes a homomorphism;

(b) if X is a group, then $\hat{X}_G$ (the covering by the subgroup G) is also a group. Examine the example $\mathbb{Z}_2 \to \mathrm{Spin}\,(n) \to SO\,(n)$, $n > 2$.

32.86. Prove that the following isomorphism holds:

$$\pi_n\underbrace{\left(S^n \vee \ldots \vee S^n\right)}_{k \text{ times}} \cong \underbrace{\pi_n\left(S^n\right) \oplus \ldots \oplus \pi_n\left(S^n\right)}_{k \text{ times}}.$$

32.87. Prove that the groups $\pi_i(X)$ are commutative for $i > 1$ for any CW-complex X.

32.88. By examining an example show that the cutting axiom is not true for the group $\pi_i(X, Y)$ (it holds for usual (co)homology theories), i.e., there exist pair (X, Y) such that

$$\pi_i(X, Y) \neq \pi_i(X/Y).$$

32.89. Show that for any arcwise connected space Y and for any point $x_0 \in Y$, the following isomorphism holds: $\pi_k(Y, x_0) \cong \pi_{k-1}(\Omega_{x_0} Y, w_{x_0})$, where w_{x_0} is a constant loop at the point x_0.

32.90. Prove that $\pi_1(\mathbb{R}P^n) = \mathbb{Z}_2$, $n > 1$, and $\pi_k(\mathbb{R}P^n) = \pi_k(S^n)$, $n \geq 1$, $k > 1$, where $\mathbb{R}P^n$ is the real projective space.

32.91. Prove the following assertions:

(a) if A is a contractible subspace in the space X (X and A are CW-complexes) to a point $x_0 \in A$, then for $n \geq 1$, the homomorphism i_*: $\pi_n (A, x_0) \to \pi_n (X, x_0)$ is trivial, and for $n \geq 3$, the following decomposition holds:

$$\pi_n (X, A, x_0) \cong \pi_n (X, x_0) \oplus \pi_{n-1} (A, x_0) ;$$

(b) if $i: X \vee Y \to X \times Y$ is an embedding, then the following exact sequence holds: $\pi_g (X \vee Y) \xrightarrow{i_*} \pi_g (X \times Y) \to 0$.

32.92. Prove that $\pi_1 (\mathbb{C}, P^n) = 0$; $\pi_2 = (\mathbb{C}, P^n) = \mathbb{Z}$, $n > 0$; $\pi_k (\mathbb{C}, P^n) = \pi_k (S^{2n+1})$, $k \geq 2$.

32.93. Prove that if a CW-complex X has no cells of dimension from 1 up to k inclusively, then $\pi_i (X) = 0$ for $i \leq k$.

32.94. Let X and Y be CW-complexes. Prove that

$$\pi_i (X \times Y) = \pi_i (X) \oplus \pi_i (Y) .$$

Calculate the action of $\pi_1 (X \times Y)$ on $\pi_i (X \times Y)$. Construct the universal covering over $X \times Y$.

32.95. Find the homotopy groups $\pi_q (S^n)$, $0 \leq q \leq n$, and prove that $\pi_n (S^n) = \mathbb{Z}$, where S^n is the sphere.

32.96. Prove that $\pi_i (S^3) = \pi_i (S^2)$ for $i \geq 3$, and as a consequence, prove that $\pi_3 (S^2) = \mathbb{Z}$.

32.97. Prove that:

(a) $\pi_1 (SO(3)) = \mathbb{Z}_2$, $\pi_2 (SO(3)) = \pi_2 (SO) = 0$, where $SO = \varinjlim SO(n)$;

(b) $\pi_3 (SO(4)) = \mathbb{Z}$, $\pi_1 (U) = \mathbb{Z}$, $\pi_2 (U) = 0$, where $U = \varinjlim U(n)$ (the embeddings $U(n) \subset U(n+1)$ and $SO(n) \subset SO(n+1)$ are standard;

(c) $\pi_3 (SO(5)) = \mathbb{Z}$.

32.98. Find the groups $\pi_g (S^1 \vee S^1)$, $g \geq 0$.

32.99. Calculate the groups $\pi_1 (X)$ and $\pi_n (X)$ and the action of the group $\pi_1 (X)$ on the group $\pi_n (X)$ in the following cases:

(a) $X = \mathbb{R} P^n$; (b) $X = S^1 \vee S^n$;

(c) $X = \partial B (\xi^{n+1})$, where $B (\xi^{n+1})$ is the space of a nontrivial $O(n+1)$-bundle of disks on S^1.

32.100. If the mapping $f: (X, A) \to (Y, B)$ establishes the following isomorphisms $\pi_k (X) \approx \pi_k (Y)$ and $\pi_k (A) \approx \pi_k (B)$ for all k, then it establishes the following isomorphisms for all k: $\pi_k (X, A) \approx \pi_k (X, B)$.

32.101. Calculate the groups $\pi_{n-k} \left(V_{n,k}^{\mathbb{R}} \right)$, where $V_{n,k}^{\mathbb{R}}$ is the real Stiefel manifold.

32.102. Prove that when k grows, the groups $\pi_k (S^n)$ cannot become trivial starting from a certain number k.

32.103. Prove that $\pi_3 (S^2) = \mathbb{Z}$ and $\pi_{n+1} (S^n) = \mathbb{Z}_2$ for $n \geq 3$.

32.104. Find $\pi_3 (S^2 \vee S^2)$, $\pi_3 (S^1 \vee S^2)$, and $\pi_3 (S^2 \vee S^2 \vee S^2)$.

32.105. Calculate first relative homotopy groups of the pair $\left(\mathbb{C}P^2, S^2\right)$, where the embedding $S^2 \cong \mathbb{C}P^1 \subset \mathbb{C}P^2$ is standard.

32.106. Prove the following assertions:

(a) if a 3-dimensional compact closed manifold M^3 is simply connected, then M is homotopy equivalent to the sphere (i.e., M^3 is the homotopy sphere).

(b) if M^n is a smooth compact closed manifold, and, moreover, $\pi_i\left(M^n\right) = 0$ for $i \leq \left[\dfrac{n}{2}\right]$, then M^n is homotopy equivalent to the sphere S^n.

32.107. Construct an example of a 3-dimensional closed compact manifold M^3 such that M^3 is the homology sphere (i.e., it has the same integral homologies as S^3); however, $\pi_1\left(M^3\right) \neq 0$. Construct an example of a finitely generated group G coinciding with its first commutant.

32.108. Prove that the set of homotopy classes of the mappings $[S^n, X]$ is isomorphic to the set of classes of conjugated elements of the group $\pi^n\left(X, x_0\right)$ under the action of $\pi_1\left(X, x_0\right)$ (X is a connected complex).

32.109. Calculate $\pi_2\left(\mathbb{R}^2, X\right)$, where $\mathbb{R}^2$ is the plane and X is the "figure eight curve" embedded into the 2-dimensional plane.

32.110. Calculate $\pi_i\left(\mathbb{C}P^n\right)$ for $i \leq 2n + 1$.

32.111. Let $\pi_n\left(X\right) = 0$, and let a finite group G act on X and Y without fixed points. Show that there exists (and is unique up to a homotopy) a mapping $f\colon Y \to X$ that commutes with the action of the group G.

32.112. Prove that $\left[\mathbb{C}P^2, S^2\right] = \pi_4\left(S^2\right)$, where $[X, Y]$ is the set of homotopy classes of mappings of X into Y.

32.113. Let (X, A) be a pair of topological spaces, where X is arcwise connected and $X \supset A$. Let Λ be the set of paths in the space X starting from a fixed point x_0 and ending at points of the subspace A. Prove that $\pi_g\left(X, A, a\right) = \pi_{g-1}\left(\Lambda, \lambda_a\right)$, where λ_a is an arbitrary path from x_0 to $a \in A$.

32.114. Prove that the following conditions are equivalent to the n-connectedness:

(a) $\pi_0\left(S^q, X\right)$ consists of a single element for $q \leq n$ (pointed mappings)

(b) any continuous mapping $S^q \to X$ extends to a continuous mapping of the disk $D^{q+1} \to X$, $q \leq n$.

32.115. Prove that $\pi_0\left(X, \Omega\Omega Z\right)$ is an Abelian group, where X and Z are topological spaces and ΩZ is the loop space of the space Z. Prove that ΩZ is an H-space.

32.116. Let A be a retract of X. Prove that for $n \geq 1$, for any point $x_0 \in A$, the homomorphism induced by the embedding

$$i_*\colon \pi_n\left(A, x_0\right) \to \pi_n\left(X, x_0\right),$$

is a monomorphism, and for $n \geq 2$, it defines the following decomposition into the direct sum:

$$\pi_n\left(X, x_0\right) \cong \pi_n\left(A, x_0\right) \oplus \pi_n\left(X, A, x_0\right).$$

32.117. Prove that $\pi_0(\Sigma\Sigma Z, X)$ is an Abelian group. Establish its connection with $\pi_0(Z, \Omega\Omega X)$.

32.118. Let $TS^n \to S^n$ be the standard tangent bundle over the sphere S^n. Calculate the homomorphism $\partial_*: \pi_n(S^n) \to \pi_{n-1}(S^{n-1})$ in the exact homotopy sequence of the associated bundle.

32.119. Let $f: X \to Y$ be a continuous mapping $(f(x_0) = y_0)$. Prove that the induced mapping $f_*: \pi_n(X, x_0) \to \pi_n(Y, y_0)$ is a group homomorphism.

32.120. Let $Y \to X$ be a bundle with fixed points x_0 and y_0 and with fiber F, and, moreover, let F_0 be the fiber over the point x_0 and $y_0 \in F_0$. Prove that $\pi_n(Y, F_0; y_0) \cong \pi_n(X, x_0)$.

32.121. Let E and X be topological spaces, let X be arcwise connected, and let $p: E \to X$ be a continuous mapping such that the following isomorphism holds for any points $x \in X$ and $y \in p^{-1}(x)$:

$$p_*: \pi_i(E, p^{-1}(x), y) \to \pi_i(X, x), \; i \geq 0$$

(for $i = 0, 1$, the isomorphism holds without additional group structure). Prove that for any points x_1 and x_2, the topological spaces $p^{-1}(x_1)$ and $p^{-1}(x_2)$ are weakly homotopy equivalent.

32.122. Prove the exactness of the following sequence of homotopy groups of the pair (X, A):

$$\ldots \to \pi_i(A) \to \pi_i(X) \to \pi_i(X, A) \to \pi_{i-1}(A) \to \ldots.$$

32.123. Prove that if X is a smooth compact closed submanifold of codimension one in the Euclidean space, then X is orientable.

32.124. Prove that if the fundamental group of a compact closed manifold is trivial, then the manifold is orientable. Prove that if a manifold X is not orientable, then $\pi_1(X)$ contains a subgroup of index 2.

32.125. Prove that if X is a non-orientable space, then the suspension ΣX is not a manifold.

32.126. Prove that the Euler characteristic $\chi(X)$ of any odd-dimensional orientable closed manifold is trivial.

32.127. Present examples of

(a) a non-orientable manifold two-sidedly embedded in another manifold (of dimension greater by one);

(b) an oriented manifold one-sidedly embedded into another manifold.

32.128. Let X_1 and X_2 be two solid tori, let $f: \partial X_1 \to \partial X_2$ be a diffeomorphism, and let $M_f^3 = X_1 \cup_f X_2$. Present diffeomorphisms f for which the manifold M_f^3 is diffeomorphic to

(a) S^3; (b) $S^2 \times S^1$; (c) $\mathbb{R}P^3$.

32.129. In terms of the previous problem, consider the mapping $f_*: \pi_1(\partial X_1) \to \pi_1(\partial X_2)$, i.e., $f_*: \mathbb{Z} \oplus \mathbb{Z} \to \mathbb{Z} \oplus \mathbb{Z}$ induced by the diffeomorphism of the solid tori X_1 and X_2. Obviously, the homomorphism f_* is given by the integral matrix

$$\begin{pmatrix} a & b \\ c & d \end{pmatrix}.$$

Prove that this matrix is unimodular and compute the fundamental group of the manifold M_f^3 in terms of the matrix f_*.

32.130. Let X^n be the space of polynomials $f_n(z)$ (in one complex variable) without multiple roots. Find the groups $\pi_k(X_n)$.

32.131. Prove that a finite CW-complex is homotopy equivalent to a manifold with boundary.

§ 33. Coverings and Bundles

33.1. Let $p\colon X \to Y$ be a covering such that $f_*(\pi_1(X, x_0))$ is a normal divisor of the group $\pi_1(Y, y_0)$, $p(x_0) = y_0$. Prove that every element $\alpha \in \pi_1(Y, y_0)$ generates a homeomorphism φ of the covering, i.e., $p\varphi(x) = p(x)$.

33.2. Let $p\colon X \to Y$ be a covering, $p(x_0) = y_0$. Prove that the induced mapping $p_*\colon \pi_1(X, x_0) \to \pi_1(Y, y_0)$ is a homomorphism.

33.3. Let $p\colon X \to Y$ be a covering, $p(x_0) = y_0$. Prove that the induced mapping $p_*\colon \pi_1(X, x_0) \to \pi_1(Y, y_0)$ is a monomorphism.

33.4. Let $p\colon Y \to X$ be a covering, $\pi_1(Y) = 0$. Prove that each element $\alpha \in \pi_1(X)$ is determined by the homeomorphism of the space Y onto itself, $\alpha\colon Y \to Y$, such that the diagram

$$\begin{array}{ccc} Y & \xrightarrow{\alpha} & Y \\ p \searrow & & \swarrow p \\ & X & \end{array}$$

is commutative.

33.5. Let $p\colon X \to Y$ be a connected covering, and let $F = p^{-1}(y_0)$ be the inverse image of a point $y_0 \in Y$, $x_0 \in F$. Prove that there exists a one-to-one correspondence between F and $\pi_1(Y, y_0)$ provided that $\pi_1(X, x_0) = 0$.

33.6. Let $p\colon X \to Y$ be a covering, $F\colon I^2 \to Y$ be a continuous mapping, where I^2 is a square, and Let $f\colon I^1 \to X$ be also continuous, and, moreover $pf(t) = F(t, 0)$. Prove that there exists a unique continuous mapping $G\colon I^2 \to X$ such that $G(t, 0) = f(t)$ and $pG(t, s) = F(t, s)$.

33.7. Let $p\colon X \to Y$ be a covering, let f and g be paths on X, and let $f(0) = g(0)$. Let $pf(1) = pg(1)$, and let the paths pf and pg be homotopic. Prove that $f(1) = g(1)$.

33.8. Let $p\colon X \to Y$ be a covering, and let f and g be paths on X, $f(0) = g(0)$. Is it necessarily that $f(1) = g(1)$ if $pf(1) = pg(1)$?

33.9. Let $p\colon X \to Y$ be a covering, and let f and g be paths on Y and $\bar{f}$ and $\bar{g}$ be paths on X such that $p\bar{f} = f$, $p\bar{g} = g$, $\bar{f}(0) = \bar{g}(0)$. Prove that if f and g are homotopic, then $\bar{f}$ and $\bar{g}$ are also homotopic.

33.10. Let $p\colon X \to Y$ be a covering, let f be a path in Y, and let x_0 be a point in X such that $p(x_0) = f(0)$. Prove that there exists a unique path g in X such that $pg = f$.

33.11. Prove that any covering is a bundle in the sense of Serre.

33.12. Prove that every 2-sheeted covering is regular. Which algebraic fact corresponds to this assertion?

33.13. Construct a 3-sheeted irregular covering of the pretzel (the sphere with two handles).

33.14. Let M^2 be a non-orientable compact smooth closed manifold. Prove the existence of a 2-sheeted covering $p\colon M_+^2 \to M^2$, where M_+^2 is an orientable manifold. Present M_+^2 in explicit form. Which property has the fundamental group of a non-orientable manifold?

33.15. Construct the covering $S^n \to \mathbb{R}P^n$ with fiber $\mathbb{Z}_2$ and prove that

(a) $\mathbb{R}P^n$ is orientable for $n = 2k - 1$ and non-orientable for $n = 2k$.

(b) $\pi_1(\mathbb{R}P^n) = \mathbb{Z}_2$, $\pi_i(\mathbb{R}P^n) = \pi_i(S^n)$ for $n > 1, i > 1$.

33.16. Prove that a covering is regular if and only if all its paths lying over the same path in the basis all are simultaneously closed or non-closed.

33.17. Let $p\colon \hat{X} \to X$ be a covering. Prove that every path in X is uniquely covered in $\hat{X}$ with accuracy up to the choice of the origin of the path in the inverse image, and the multiplicity of the projection p is the same at all points of the base.

33.18. Construct all coverings over the circle and prove that $\pi_1(S^1) = \mathbb{Z}$, $\pi_i(S^1) = 0$ for $i \geq 2$.

33.19. Construct a regular covering $p\colon P_k \to P_2$ with fiber $\mathbb{Z}_{k-1}$, where $k > 2$ and P_k is the sphere with k handles.

33.20. Construct the universal covering over $\bigvee_A S^1$ and prove that $\pi_i(\bigvee_A S^1) = 0$ for $i > 1$. Find $\pi_1(\bigvee_A S^1)$.

33.21. Construct a covering $\varphi\colon \hat{X} \to P_2$ (pretzel) such that $\hat{X}$ is contracted to a graph, and as consequence, prove that

(a) the universal covering over P_2 is contractible, $P_2 \sim K(\pi, 1)$;

(b) if M^2 is a 2-dimensional closed manifold and $\pi_1(M^2)$ is an infinite group, then $M^2 \sim K(\pi, 1)$ (is homotopy equivalent).

33.22. Find the connection between the universal coverings of P_k (the sphere with k handles) and the Lobachevskii plane.

33.23. Prove that all coverings of the torus T^2 are regular and find them. Construct an example of two non-equivalent but homeomorphic coverings of the torus T^2.

33.24. Can the torus T^2 cover

(a) the sphere; (b) the projective plane?

33.25. Can the projective plane $\mathbb{R}P^2$ cover the torus?

33.26. Let X be a finite complex. Find the connection between an arbitrary subgroup $G \subset \pi_1(X)$, the Euler characteristic $\chi(X)$, and $\chi(X_G)$, where $\hat{X}_G$ is the covering constructed according to the subgroup $G \subset \pi_1(X)$.

33.27. Construct the universal covering of the torus $P_1($ (the sphere with one handle) and the Klein bottle N_2 (the sphere with two Möbius band); calculate the homotopy groups of P_1 and N_2. Can the torus P_1 be a 2-sheeted regular covering of the Klein bottle? If yes, then present the covering an calculate the image of $\pi_1(P_1)$ in $\pi_1(N_2)$ under the monomorphism of the covering.

33.28. Prove that if $\pi_1(M^n) = 0$ or $\pi_1(M^n)$ is a simple or finite group of order $p \neq 2$ (p is a prime number), then the manifold M^n is orientable.

33.29. In explicit form, construct seven smooth linearly independent vector fields on the sphere S^7. Use the octave algebra (Gravs–Cayley numbers). Construct the integral trajectories of these vector fields.

33.30. Prove that if k linear operators $A_1, \ldots, A_k$ such that $A_i^2 = -E$ and $A_i A_j + A_j A_i = 0$ (for all i and j) are given in $\mathbb{R}^n$, the it is possible to define k linearly independent vector fields on the sphere $S^{n-1} \subset \mathbb{R}^n$.

33.31. If the homotopy groups of the base and the fiber of a bundle have a finite rank, then the homotopy groups of the bundle space are also of finite rank, and, moreover, the rank of the q-dimensional group of the bundle space does not exceed the sum of ranks of the q-dimensional homotopy groups of the base and the fiber.

33.32. Prove that the universal covering over X is a covering for any other covering space.

33.33. Prove that the standard bundle $EX \xrightarrow{\Omega X} X$ (Serre bundle), where X is a manifold, is a locally trivial bundle.

33.34. Let a bundle $p\colon E \to B$ admit a cross-section surface $\chi\colon B \to E$, and, moreover, $e_0 = \chi(b_0)$. Prove that for $n \geq 1$, the mapping p_* is an epimorphism, and for $n \geq 2$, it defines the decomposition into the direct sum

$$\pi_n(E, e_0) = \pi_n(B, b_0) \oplus \pi_n(F, e_0).$$

33.35. Prove that if all homotopy groups of the base and the fiber are finite, then the homotopy groups of the bundle space are also finite and their orders do not exceed the product of orders of the homotopy groups of the base and the fiber of the same dimension.

33.36. Let EX be the space of all paths on X emanating from a fixed point, and let ΩX be the space of all loops on X. Define the mapping $p\colon EX \to X$ as a mapping that to a path, it puts in correspondence its final point. Prove that $p\colon EX \to X$ satisfies the covering homotopy axiom, i.e., is a bundle in the Serre sense, and, moreover, its fiber is Ω_X. Sometimes, the described bundle is called the Serre bundle.

33.37. Prove that if $p_{1,2}\colon \hat{X}_{1,2} \to X$ are coverings and $\text{Im}\,(p_1)_* = \text{Im}\,(p_2)_*$, then $\left(\hat{X}_1, p_1, X\right)$ and $\left(\hat{X}_2, p_2, X\right)$ are fibrewise homeomorphic. Here, $\text{Im}\,(p_i)_* \subset \pi_1(X)$.

33.38. Prove that over each connected complex X, there exists a covering p: $\hat{X} \to X$ such that $\pi_1\left(\hat{X}\right) = 0$. Such a covering is said to be universal.

33.39. Prove that the set of vector bundles with linearly connected structure group G over the sphere S^n is isomorphic to $\pi_{n-1}(G)$.

33.40. By examining an example, show that there is no "exact homology sequence of a bundle."

33.41. Let $p\colon E \to B$ be a locally trivial bundle with fiber F and base B, and, moreover, let F and B be finite complexes. Then $\chi(E) \leq \chi(B) \cdot \chi(F)$.

33.42. A material point move with velocity whose module is constant (a) along the torus T^n; (b) along the sphere S^n. Find the phase space of this system.

33.43. Let $p\colon E \to B$ be a bundle with arcwise connected base B and fiber F. Let $\widetilde{\mathrm{Cat}} = \mathrm{Cat} - 1$ be the reduced Lyusternik–Shnirelman category, i.e., $\widetilde{\mathrm{Cat}}$ (point) $= 0$. Prove that

$$\widetilde{\mathrm{Cat}}\,(E) \leq \widetilde{\mathrm{Cat}}_E\,(F) \cdot \widetilde{\mathrm{Cat}}\,(B) + \widetilde{\mathrm{Cat}}\,(B) + \mathrm{Cat}_E\,(F),$$

where $\mathrm{Cat}_E\,(F)$ is the relative category of the fiber F with respect to E.

33.44. Prove that if $p\colon X \to Y$ is a bundle in the sense of Serre, then p is a mapping "onto".

33.45. Prove that if $p\colon X \to Y$ is a bundle in the Serre sense, then $p^{-1}(y_1)$ and $p^{-1}(y_2)$ are homotopy equivalent for any $y_1, y_2 \in Y$.

33.46. Prove that the manifold of linear elements of a manifold M is a bundle with base M.

33.47. Prove that any locally trivial bundle (skew product) is a bundle in the Serre sense.

33.48. Prove that the path space EX with fixed initial point of the space X is a bundle in the Serre sense with base X.

33.49. Prove that the direct product $X \times Y$ of topological spaces with projection on one of the factors is a bundle in the Serre sense.

33.50. Let $p\colon X \to Y$ be a covering, $p(x_0) = y_0$, and let f and g be paths such that $f(0) = g(0) = y_0$; $f(1) = g(1)$. Let $fg^{-1} \in p_*\,(\pi_1(X, x_0))$. Let $\widehat{f}$ and $\widehat{g}$ be coverings of these paths. Prove that $\widehat{f}(1) = \widehat{g}(1)$.

33.51. Represent the torus T^2 in the form $T^2 = \{g\}$, where

$$g = \begin{pmatrix} e^{i\varphi_1} & 0 \\ 0 & e^{i\varphi_2} \end{pmatrix}.$$

Consider the following equivalence relations on R:

(a) $\left(e^{i\varphi_1}, e^{i\varphi_2}\right) R \left(-e^{i\varphi_1}, e^{-i\varphi_2}\right)$;

(b) $\left(e^{i\varphi_1}, e^{i\varphi_2}\right) R \left(-e^{i\varphi_1}, -e^{-i\varphi_2}\right)$;

(c) $\left(e^{i\varphi_1}, e^{i\varphi_2}\right) R \left(e^{-i\varphi_1}, e^{-i\varphi_2}\right)$.

Find the space $X = T^2/R$ and calculate the image $f_* = (\pi_1(T^2)) \subset \pi_1(X)$, where $f\colon T^2 \to X = T^2/X$ is the projection related to the relation R. Is f a covering?

33.52. How many there are bundles of the form

(a) $T^3 \to S^1$, where T^3 is the 3-dimensional torus;

(b) $T^n \to S^1$, where T^n is the n-dimensional torus (the bundles are considered with accuracy up to homotopy equivalence)?

33.53. Let $C = A * B$ be the free product of arbitrary groups A and B. Prove that for any subgroup $M \subset C$ the relation $M = A_1 * B_1 * F$ holds, where $A_1 \subset A$, $B_1 \subset B$, and F is a free group. Give a topological proof using coverings.

33.54. Let G be a simply connected compact Lie group, and let $\sigma\colon G \to G$ be an arbitrary involutive automorphism (i.e., $\sigma^2 = 1$). We set $H = \{g \in G | \sigma(g) = g\}$; $V = \{g \in G | \sigma(g) = g^{-1}\}$. Prove that $G = VHV$, i.e., any element $g \in G$ admits the representation in the form

$$g = vhv, \quad v \in V, \quad h \in H.$$

Prove that $V \equiv G/H$ (homogeneous space).

33.55. The following (Cartan) construction is known. Let $\sigma\colon G \to G$ be an arbitrary involutive automorphism of a compact connected Lie group. We set

$$H = \{g \in G | \sigma(g) = g\} : V = \{g \in G | \sigma(g) = g^{-1}\}.$$

Then $V \cong G/H$ and $V \subset G$ is a completely geodesic submanifold, and, therefore, V is a symmetric space. The submanifold V is called the Cartan model of the symmetric space G/H. Any symmetric space admits a (moreover, almost always uniquely defined) Cartan model.

(a) Prove that the projection $p\colon G \to V$ defined by the formula $p(g) = g\sigma(g^{-1})$ defines the principal bundle $0 \to H \to G \to G/H \to 0$.

(b) Let V be a Cartan model, let $\pi_1(V) = 0$, $e \in G$, $e \in V$, and let e be the identity in G. Prove that if a point $x \in V$ is conjugated to e along a geodesic $\gamma(t) \subset V$ in the group G, then the point x is conjugated to e along γ in the manifold $V \subset G$ itself.

33.56. Calculate the homotopy groups $\pi_i(M_g)$, $i \geq 1$ of a two-dimensional manifold M_g of genus g.

33.57. Let M^2 be a compact closed oriented 2-dimensional manifold of genus g. Find the homotopy type of $\Sigma^2 M^2$.

33.58. Let $S^1 \times S^1 \subset \mathbb{R}^3$ be the standard embedding of the torus in the Euclidean space. Prove that there is no homeomorphism of the pair $(\mathbb{R}^3, S^1 \times S^1)$ onto itself whose restriction on the torus is defined by the matrix

$$\begin{pmatrix} 0 & 1 \\ -1 & 0 \end{pmatrix}.$$

33.59. Let π be the fundamental group of a 2-dimensional surface, and let $f\colon \pi \to \pi$ be an epimorphism. Prove that f is an isomorphism.

33.60. Using three essentially different methods, prove that on the sphere S^2, there is no continuous vector field without singular points (i.e., different from zero at each point).

33.61. Let $p\colon X \to Y$ be a two-sheeted covering. Prove that then any path in Y is covered exactly two paths.

33.62. Construct the universal covering space for the orthogonal group $SO\,(n)$.

33.63. Prove that any two-dimensional closed oriented smooth manifold of genus $g \geq 2$ can be locally isometrically covered by the Lobachevskii plane (equipped with the standard metric of constant negative curvature). In other words, prove that the fundamental group of the surface of indicated form can be represented as a discrete subgroup in the isometry group of the Lobachevskii plane acting effectively.

In particular, this implies that a two-dimensional smooth orientable closed manifold can be equipped with a Riemannian metric of constant negative curvature.

33.64. Which spaces can cover the Klein bottle?

33.65. Let S_g be the sphere with g handles. Which S_h can cover S_g?

33.66. Prove that for any compact non-orientable two-dimensional manifold, there exists exactly one compact two-dimensional orientable manifold that two-sheetedly covers it.

33.67. Prove that the Beltrami surface (surface of constant negative curvature standardly embedded in $\mathbb{R}^3$) can be infinitely sheetedly and locally isometrically covered by a certain domain lying in the Lobachevskii plane. Find this domain. Prove that it is homeomorphic to the two-dimensional disk. Find the corresponding group of this covering.

33.68. Can the two-dimensional torus be a two-sheeted covering of the Klein bottle?

33.69. Calculate the permutation group of leaves of the Riemann surface of the algebraic function $w = \sqrt[n]{z}$ arising going around a branch point of this function (zero point).

§ 34. Critical Points, Degree of Mapping, Morse Theory

34.1. Let $f, g\colon S^n \to S^n$ be simplicial mappings. Prove that

(a) the inverse image of each interior point of a simplex of highest dimension consists in the same number of points taken with the sign of orientation;

(b) if f and g are homotopic, then the numbers of points of the inverse images taken with the sign of orientation coincide;

(c) if the numbers of points of the inverse images coincide, then the mapping f and g are homotopic.

34.2. Prove that every continuous self-mapping of the closed ball D^n always has a fixed point.

34.3. Let $f\colon SU(n) \to SU(n)$ be a smooth mapping, and let $f(g) = g^k$ Find $\deg f$.

34.4. Let $f\colon M^n \to \mathbb{R}^p$ be a smooth immersion of a closed compact orientable closed n-dimensional $(n < p)$ manifold in $\mathbb{R}^p$. Let $v(f)$ be the normal beam of this immersion, and let $Sv(f)$ the associated beam of spheres, i.e., $Sv(f) = \partial v(f)$ is the boundary of a sufficiently small tubular neighborhood of the submanifold $f(M^n) \subset \mathbb{R}^p$. Let $T\colon Sv(f) \to S^{p-1}$ be the usual (spherical) Gaussian mapping.

Find the degree $\deg T$ $(\dim Sv(f) = p - 1)$ if the Euler characteristic of the manifold M^n is known. Does $\deg T$ depend on the method of immersion of M^n in $\mathbb{R}^3$. What happens if M^n is not oriented? Especially, examine the case $p = n + 1$.

34.5. It is known that a 2-dimensional orientable closed compact manifold M^2 of genus g can be embedded in the Euclidean space $\mathbb{R}^3$ (x, y, z). Find the minimum number of saddle points (degenerate in general) of the function $f(p) = z$, $p \in i(M^2)$ where i is the embedding and f is the height function.

34.6. Prove that nondegenerate critical points of smooth function on a smooth manifold are isolated.

34.7. Let $f(x)$ be a function on a 2-dimensional compact orientable surface of genus g (sphere with g handles) having finitely many critical points, which all are nondegenerate. Prove that the number of minima minus the number of saddles plus the number of maxima is equal to $2 - 2g$.

34.8. Let $f\colon M^n \to \mathbb{R}$ be a smooth function on a smooth manifold. Prove that almost all values of f are regular.

34.9. Prove that the alternated sum of singular (critical) points of a smooth function $f(x)$ (under the assumption that all its singularities are nondegenerate) given on a smooth compact manifold is independent of the function (by the alternated sum we mean $\Sigma_{\lambda=0}^n (-1)^\lambda m_\lambda$, where $n = \dim M$, λ is the index of the critical point, m_λ is the number of critical points of index λ).

34.10. Let $f(x)$ be a complex-analytic function of one variable x. Prove that the number of critical values of the function $S^2 \to S^2$ is of zero measure.

34.11. Let $M_c^n = \{x | f(x) = c\}$. Prove that if M_c^n contains no critical points of the function f, then M_c^n is a submanifold and codim $M_c^n = 1$ in M^n.

34.12. Prove that the concept of nondegenerate critical point is independent of the choice of a local chart containing this point.

34.13. Show that for the standard embedding of the torus $T^2 \subset \mathbb{R}^3$ (surface of revolution around the axis Oz), the coordinate x orthogonal to the axis of revolution of the torus T^2 has only nondegenerate critical points.

34.14. (a) On $\mathbb{R}P^n$ and $\mathbb{C}p^n$, construct functions having only nondegenerate critical points so that at all critical points, the values of the functions are different.

(b) On $\mathbb{R}P^n$ and $\mathbb{C}P^n$, construct functions for which $f(x_\lambda) = \lambda = \mathrm{ind}\,(x_\lambda)$, where x_λ are nondegenerate critical points of index λ.

34.15. Let $F(x,y)$ be a nondegenerate bilinear form on $\mathbb{R}^n$. Consider the smooth function $f(x) = F(x,x)$, where $|x| = 1$, i.e., $f(x)$ is a function on the sphere $S^{n-1} \subset \mathbb{R}^n$. Let $\lambda_0 \leq \lambda_1 \leq \ldots \leq \lambda_{n-1}$ be all eigenvalues of the form F (recall that all λ_i, $0 \leq i \leq n-1$ are real).

Prove that λ_i are critical values of the function $f(x)$ on the sphere S^{n-1}. Find all critical points of the function $f(x)$. Prove that $\lambda_i = \inf_{S^i}\{\max_{x \in S^i} f(x)\}$, where S^i are standard i-dimensional equators in the sphere S^{n-1}.

34.16. Prove that if a point p is a nondegenerate critical point of a smooth function $f(x)$ on a smooth manifold, then there exists a local coordinate system in which the function $f(x)$ in a neighborhood of the point p is represented as a nondegenerate quadratic form.

34.17. Prove that if M_c is a non-critical level for a function $f(x)$ on a manifold M (i.e., the level hypersurfaces $f(x) = c = \mathrm{const}$ contain no critical points of $f(x)$), then a neighborhood of M_c is diffeomorphic to $M_c \times I$.

34.18. If M_{c_1} and M_{c_2} are consecutive critical levels, then the domain between them is diffeomorphic to $M_c \times I$, where $c_1 < c < c_2$.

34.19. If between M_{c_1} and M_{c_2}, there are no critical levels (i.e., level hypersurfaces $f(x) = \mathrm{const}$ with critical points), then M_{c_1} and M_{c_2} are also noncritical and they are diffeomorphic.

34.20. (a) Construct on an arbitrary compact orientable 2-dimensional smooth manifold M^2 a smooth function $f(x)$ having only one minimum point, one maximum point (nondegenerate points), and one more critical point, probably, degenerate. Find the connection between such a function and the representation of M^2 in the form of the Riemann surface of a certain multivalued analytic function. Reveal the situation for the case of a non-orientable 2-dimensional Riemannian manifold M^2 (for example, the case of the projective plane $\mathbb{R}P^2$).

(b) Construct a smooth function $f(x)$ on an arbitrary compact manifold M^2 having only nondegenerate critical points: exactly one maximum point, one minimum point, and s saddle points (find the number s). Construct the function $f(x)$ in such a way that it assumes the same value at saddle points. Study the non-orientable case. Indicate the connection with the problem of item (a); construct the confluence of all saddle points to a single degenerate critical point.

Let X be a Hausdorff topological space. We are mainly interested in the case where X is a smooth manifold. Let f be a continuous function on X. Define the Reeb graph $\Gamma = \Gamma_f$ for the function f. For this purpose, introduce the following equivalence relation on X. Two points x_1 and x_2 of the space X are equivalent if

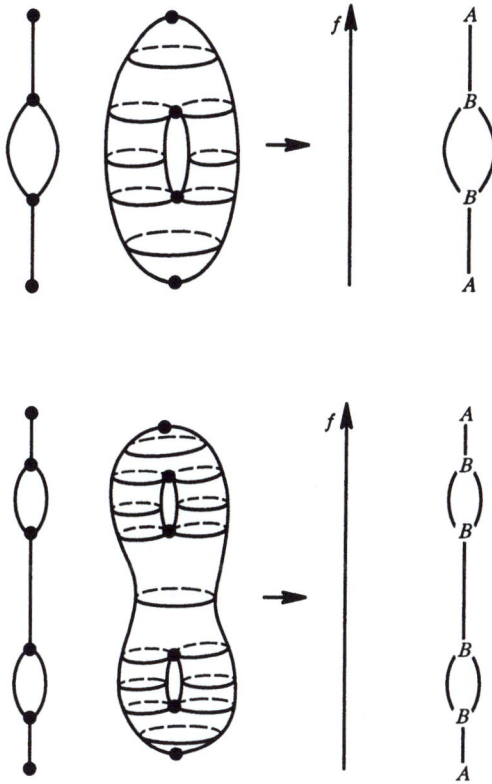

Figure 109

$f(x_1) = f(x_2)$ and x_1 and x_2 belong to the same connected component of the set $f^{-1}(c)$, where $c = f(x_1) = f(x_2)$. The corresponding quotient space X by this equivalence relation is called the Reeb graph of the function f. It is easy to see that Γ is a one-dimensional graph. Moreover, its endpoint vertices (vertices of multiplicity 1) correspond to local minima and maxima of the function f; other vertices (vertices of multiplicity greater than 1) correspond to "saddle" critical values of the function f. In Fig. 109, we show examples of Reeb graphs for height functions (in the figure, to the left from the surface) on the torus and the pretzel embedded in $\mathbb{R}^3$ in the standard way.

Now let f be a smooth function on a manifold. Then its Reeb graph can be equipped with additional information about the structure of critical levels of the function f. Let us show how to do this examining a Morse function f on a two-dimensional closed surface M. Let $c \in \mathbb{R}$ be an arbitrary critical value of the function f. In this case, $f^{-1}(c)$ can contain several critical points of the function f. Consider the inverse image $f^{-1}([c - \varepsilon, c + \varepsilon])$, where $\varepsilon > 0$ is a number such that there are no other critical values on the closed interval $[c - \varepsilon, c + \varepsilon]$, except for c.

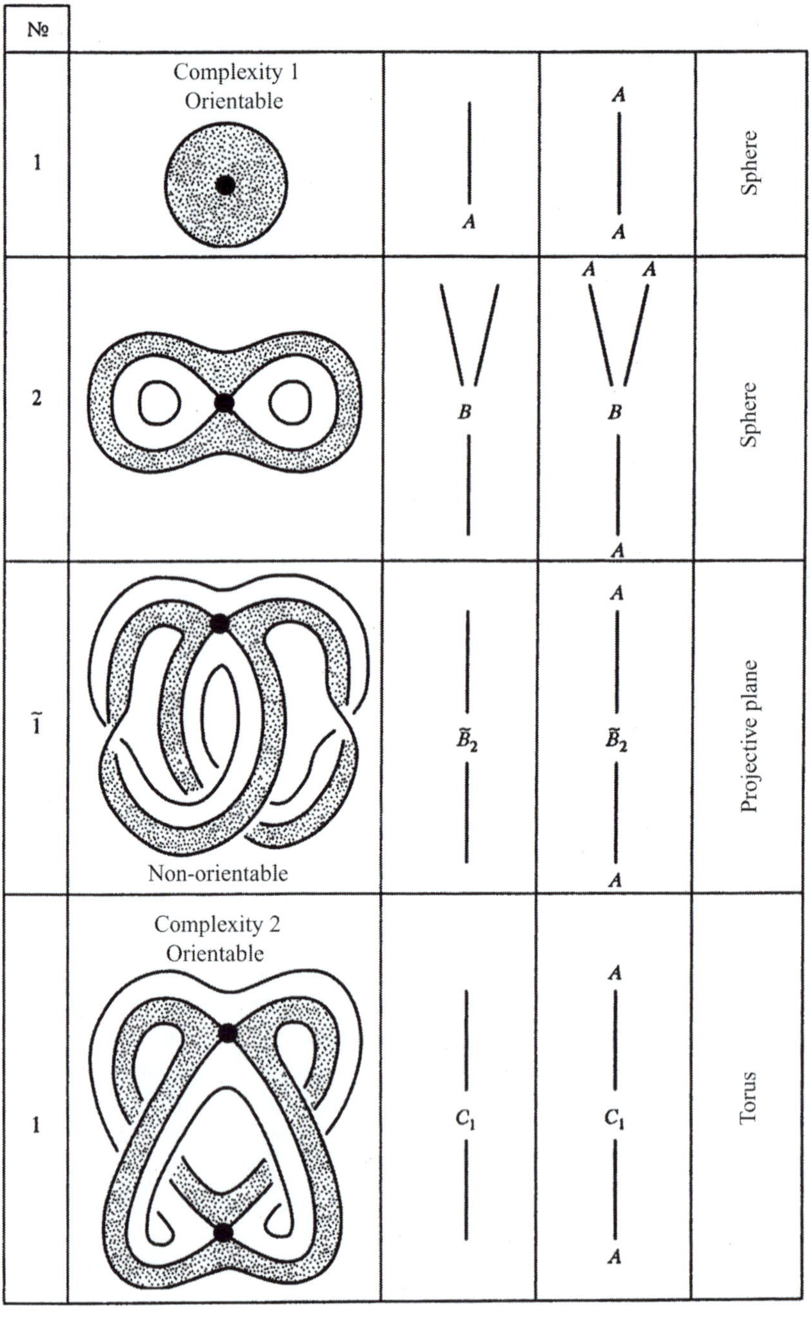
№
Complexity 1
Orientable
1
A
A
A
Sphere
2
A A
B
B
A
Sphere
1̃
$\tilde{B}_2$
A
$\tilde{B}_2$
A
Projective plane
Non-orientable
Complexity 2
Orientable
1
C_1
A
C_1
A
Torus

Figure 110

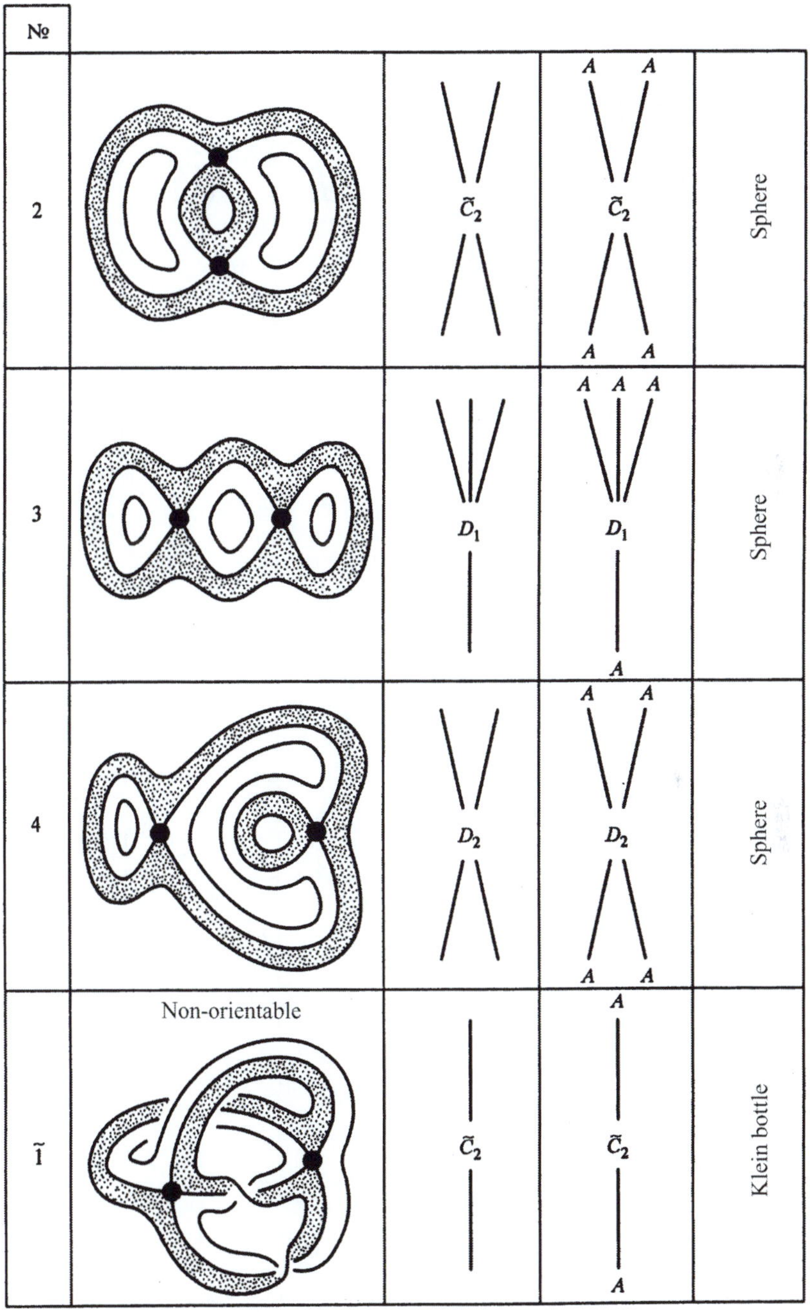
№
2
$\tilde{C}_2$
A A
A A
Sphere
3
D_1
A A A
A
Sphere
4
D_2
A A
A A
Sphere
$\tilde{1}$
Non-orientable
$\tilde{C}_2$
A
A
Klein bottle

Figure 111

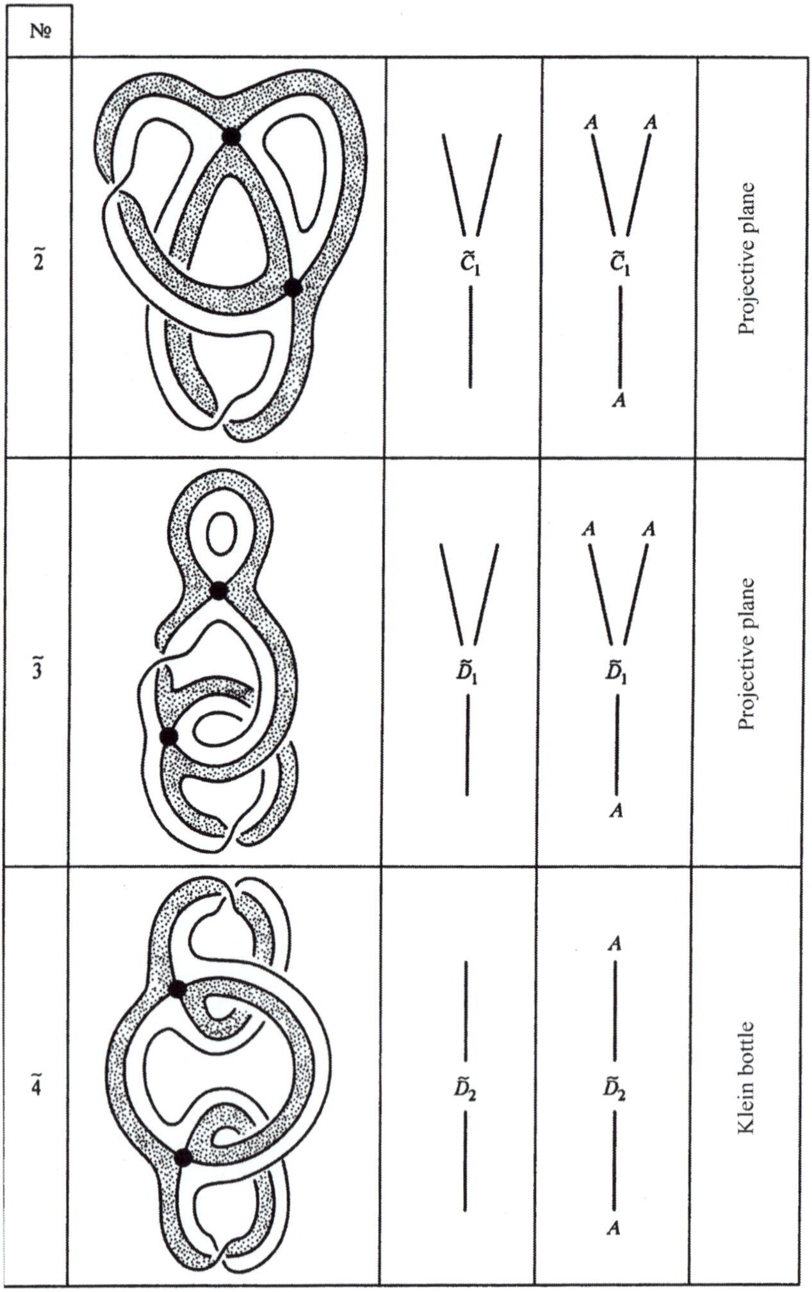

Figure 112

It is easy to see that $f^{-1}([c - \varepsilon, c + \varepsilon])$ is a two-dimensional surface with boundary. The boundary of $f^{-1}([c - \varepsilon, c + \varepsilon])$ is the union of two circles. A circle is said to be *positive* if f is equal to $c + \varepsilon$ on it and *negative* if f is equal to $c - \varepsilon$. The surface $f^{-1}([c - \varepsilon, c + \varepsilon])$ with such an additional structure on the boundary is called an *atom*.

Replacing all vertices by atoms corresponding to critical values of the function f in the Reeb graph, we obtain the graph containing the additional information about the critical points of the function. Such a graph is called a *molecule* (see the example in Fig. 109 to the right from the surface).

34.21. (a) Construct a Morse function on the torus T^2 having exactly four critical points in such a way that two saddle point lie on the same critical level. Draw level line of such a function on the square whose glued sides compose the torus. Draw the molecule for such a function.

(b) Realize such a function as the height function on the torus for its appropriate embedding in $\mathbb{R}^3$.

34.22. Construct a Morse function on the Klein bottle having four critical points with four different critical values. Draw its level lines on the square whose glued sides compose the Klein bottle. Draw the corresponding atoms and molecule.

34.23. Construct a Morse function on the Klein bottle having four critical points and three critical values. Draw its level lines on the square whose glued sides compose the Klein bottle. Draw the corresponding atoms and molecule.

34.24. Construct a Morse function on the projective plane $\mathbb{R}P^2$ having three critical points and three different critical values. Draw its level lines on the square whose glued sides yield $\mathbb{R}P^2$. Draw the corresponding atoms and molecule.

34.25. Construct Morse functions on the sphere, on the torus, on the projective plane, and on the Klein bottle having atoms (neighborhoods of saddle critical levels) and molecules depicted in Figs. 110–112. Draw their level lines in the square whose glued sides compose the surface considered.

§ 35. Simplest Variational Problems

35.1. Prove that the extremal trajectories of the action functional $E[\gamma] = \int_0^1 |\dot{\gamma}|^2 \, dt$ on a smooth Riemannian manifold M^n (where $\gamma(t)$ are the smooth trajectories on M^n, $0 \le t \le 1$; $\dot{\gamma}(t)$ is the velocity vector of $\gamma(t)$) are geodesics.

35.2. Find the connection between the extremal trajectories of the length functional $L[\gamma] = \int_0^1 |\dot{\gamma}| \, dt$ and those of the action functional $E[\gamma] = \int_0^1 |\dot{\gamma}|^2 \, dt$. Prove that any extremal $\gamma_0(t)$ of $E[\gamma]$ is an extremal of $L[\gamma]$. Prove that if $s_0(t)$ is an extremal of $L[\gamma]$, then, by the change of the parameter $t = t(\tau)$

by $s_0(t)$, one can transform this trajectory into an extremal of $E[\gamma]$.

35.3. Let

$$S(f) = \int_D \sqrt{EG - F^2}\, du\, dv$$

be the functional such that to each smooth function $z = f(x, y)$ defined on a bounded domain $D = D(x, y) \subset \mathbb{R}^2(x, y)$ (here, x, y, z are Cartesian coordinates in $\mathbb{R}^3$), it puts in correspondence the area of the graph of the function $z = f(x, y)$. Prove that the extremality of a function f_0 with respect to the functional S is equivalent to the condition $H = 0$, where H denotes the mean curvature of the graph of $z = f_0(x, y)$ considered as a two-dimensional smooth submanifold in $\mathbb{R}^3$.

35.4. Prove the assertion formulated in the previous problem for the case of $(n-1)$-dimensional graphs $x^n = f(x^1, \dots, x^{n-1})$ in $\mathbb{R}^n$.

35.5. Prove that the action functional $E[\gamma]$ and the length functional $L[\gamma]$ are connected by the relation $(L[\gamma])^2 \leq E[\gamma]$, and, moreover, the equality holds if and only if $\gamma(t)$ is a geodesic.

35.6. Prove that the area functional

$$S[\mathbf{r}] = \int_D \sqrt{EG - F^2}\, du\, dv$$

(where $\mathbf{r} = \mathbf{r}(u, v)$ is the radius-vector in $\mathbb{R}^3$ smoothly depending on (u, v)) and the Dirichlet functional

$$D[\mathbf{r}] = \int_D \frac{E + G}{2}\, du\, dv$$

are connected by the relation $S[\mathbf{r}] \leq D[\mathbf{r}]$.

35.7. The radius-vector $\mathbf{r}(u, v)$ defining a two-dimensional surface M^2 in the three-dimensional Euclidean space is said to be *harmonic* if $\mathbf{r}(u, v)$ is an extremal of the Dirichlet functional $D[\mathbf{r}] = \frac{1}{2} \int_D (E + G)\, du\, dv$. Prove that if the mean curvature H of a surface M^2 given by a radius vector $\mathbf{r}(u, v)$ is equal to 0, then in a neighborhood of each point on the surface, it is possible to introduce local coordinates (p, q) in which the radius-vector $\mathbf{r}(p, q)$ becomes harmonic.

35.8. Construct an example of a harmonic radius-vector $\mathbf{r}(u, v)$ such that the surface $M^2 \subset \mathbb{R}^3$ described by it be not minimal (i.e., $H \not\equiv 0$).

35.9. W i r t i n g e r i n e q u a l i t y. Let H be a Hermitian symmetric positive definite form in $\mathbb{C}^n$, and let $\alpha\colon \mathbb{C}^n \to \mathbb{R}^{2n}$ be the realification of $\mathbb{C}^n$; then

$$H \to H^R = \begin{pmatrix} S & A \\ -A & S \end{pmatrix},$$

where $H = S + iA$; S, A are real matrices; $S^\top = S$; $A^\top = -A$; $\overline{H}^\top = H$.

The form S defines the Euclidean inner product on $\mathbb{R}^{2n}$; the form A defines the exterior 2-form $\omega^{(2)}$ on $\mathbb{R}^{2n}$. For simplicity, we can assume that $\omega^{(2)} = \sum_{k=1}^{n} dz^k \wedge d\bar{z}^k$. Consider the form

$$\Omega^{(2r)} = \frac{1}{r!} \underbrace{\omega \wedge \ldots \wedge \omega}_{r}, \quad r \le n.$$

(a) If $\omega_1, \ldots, \omega_{2n}$ is an arbitrary orthonormal basis in $\mathbb{R}^{2n} \cong \mathbb{C}^n$ with respect to the inner product $S = \operatorname{Re} H$, then

$$\left| \Omega^{(2r)}(\omega_1, \ldots, \omega_{2r}) \right| \le 1 \text{ and } \left| \Omega^{(2r)}(\omega_1, \ldots, \omega_{2r}) \right| = 1$$

if and only if the plane $L(\omega_1, \ldots, \omega_{2r})$ generated by the vectors $\omega_1, \ldots, \omega_{2r}$ is a complex subspace in $\mathbb{R}^{2n} \cong \mathbb{C}^n$.

(b) Let $W^r \subset \mathbb{C}^n$, $r < n$ (r is the complex dimension), be a complex submanifold in $\mathbb{C}^n$ (if W^r is an algebraic variety, then singular points are admitted on W^r). Let V^{2r} be a real submanifold in $\mathbb{C}^n$ such that $V \cup W = \partial Z^{2r+1}$, where Z^{2r+1} is a real $(2r+1)$-dimensional submanifold in $\mathbb{C}^n$ whose boundary is $V \cup W$. Let $K = V \cap W$; then $\operatorname{vol}_{2r}(V \setminus K) \ge \operatorname{vol}_{2r}(W \setminus K)$.

Remark. This assertion means that complex submanifolds W in the complex space $\mathbb{C}^n$ are minimal submanifolds, i.e., for any "perturbation" of V, the $2r$-dimensional volume (vol_{2r}) does not decrease.

(c) Prove that the assertion of item (b) remains valid if we replace $\mathbb{C}^n$ by any Kähler manifold equipped with an exterior 2-form $\omega^{(2)}$ being nondegenerate and closed.

35.10. On $\mathbb{R}^n(x^1, \ldots, x^n)$, consider functions of the form $F(x^1, \ldots, x^n)$ and the functional $J[F] = \int_D |\operatorname{grad} F| \, d\sigma^n$, where D is the domain of the functions F. Let F_0 be an extremal of the functional J. Prove that the level surfaces $F_0(x^1, \ldots, x^n) = \operatorname{const}$ considered as hypersurfaces in $\mathbb{R}^n(x^1, \ldots, x^n)$ are locally minimal surfaces.

35.11. Let $n(x, y)$ be the refraction exponent of a transparent plane isotropic but inhomogeneous substance filling in the two-dimensional plane. The trajectory of light rays are integral trajectories of the vector field $\operatorname{grad} n(x, y)$. Prove that they are geodesic curves of the metric

$$ds^2 = n(x, y)\left(dx^2 + dy^2\right).$$

§ 36. General Topology

36.1. Let $M = X \times Y$, where X and Y are topological spaces. Let a set from M be open if it is a product of open sets from X and Y or a finite union of such sets. Prove that such a system satisfies all the axioms defining a topology on the set M.

36.2. Prove that if spaces X and Y are Hausdorff, and, besides, X is locally compact, then for any space T the spaces $H(X \times Y, T)$ and $H(Y, H(X, T))$ are homeomorphic; here, $H(X, Y) = Y^X$.

36.3. Prove that there exists a homeomorphism of the Cantor discontinuum onto itself that interchanges two given points.

36.4. Let X be a locally arcwise connected metric space. Prove that if X is connected, then X is arcwise connected.

36.5. Let $A \cap B \neq \varnothing$, and let $X = A \cup B$. Prove that if A and B are arcwise connected spaces, then X is a connected space.

36.6. Prove that the Hamming metric in the n-dimensional cube is not embedded in any $\mathbb{R}^n$, i.e., there is no embedding such that the Hamming metric is induced by the standard Euclidean metric (the cube is considered only as the set of its vertices, i.e., as a discrete set, and then the distance $p(a, b)$, where a and b are vertices of the cube is equal to the number of coordinates different for the vertices).

36.7. Prove that the group of unitary matrices $U(n)$ considered as a topological space is homeomorphic to the direct product $S^1 \times SU(n)$ (as topological spaces).

36.8. Prove that the group $GL(n, \mathbb{R})$ of real nonsingular matrices of size $n \times n$ is a topological space consisting of two connected components.

36.9. Prove that any finite CW-complex can be embedded in a finite-dimensional Euclidean space $\mathbb{R}^N$ (of a sufficiently large dimension).

36.10. If as CW-complex, we take a compact smooth closed manifold M^n, then the result formulated in the previous problem can be refined,

(a) Prove that M^n can be embedded in the Euclidean space $\mathbb{R}^{2nk}$, where k is the number of open ball D^n composing a covering of M^n.

(b) Prove that M^n can be embedded in $\mathbb{R}^{nk}$, where the number k is defined in item (a).

36.11. Prove that any finite simplicial complex is a subcomplex of a simplex of a sufficiently large dimension. In particular, it can be embedded in the Euclidean space so that the embedding is linear on each simplex.

36.12. Let $f: M^2 \to S^2$ be a mapping of class C^2 of a closed smooth compact manifold M^2 on to S^2, and, moreover, let f be open (the image of any open set is open) and of finite multiplicity (the image of each point $x \in S^2$ is a finite number of points). Prove that M^2 is diffeomorphic to the sphere S^2. What can be said about an analogous mapping $f: M^n \to S^n$?

Let X be a topological space. Denote by $\exp X$ the set of all nonempty closed subsets of the space X. For open sets $U_1, \ldots, U_k \subset X$, we set

$$O \langle U_1, \ldots, U_k \rangle = \left\{ F \in \exp X \mid F \subset \bigcup_{i=1}^{k} U_i, F \cap U_i \neq \varnothing \forall i \right\}.$$

36.13. Prove that sets of the form $O \langle U_1, \ldots, U_k \rangle$ compose a base of a certain topology. This topology is called the *Vietoris topology*, and the set $\exp X$ equipped with the Vietoris topology is called the *hyperspace* (of closed sets) of the space X.

36.14. Let X be a T_1-space. Prove that the natural embedding $X \subset \exp X$ transforming a point x to the set $\{x\}$ is topological.

36.15. Let X be a T_1-space. Then X is regular if and only if $\exp X$ is Hausdorff.

36.16. Prove that for a T_1-space X, the following conditions are equivalent:

(a) X is normal; (b) $\exp X$ is completely regular: (c) $\exp X$ is regular.

Let n be a natural number. Denote by $\exp_n X$ the set of all nonempty closed subsets of X consisting of no more than n points. Assume that the identical embedding $\exp_n X \subset \exp X$ is topological, i.e., equip $\exp_n X$ with the Vietoris topology.

36.17. Prove that if X is Hausdorff, then $\exp_n X$ is closed in $\exp X$.

36.18. Prove that if X is a T_1-space and $\exp_1 X$ is closed in $\exp_2 X$, then X is Hausdorff.

Denote by $\pi_n \colon X^n \to \exp_n X$ the mapping transforming a point $(x_1, \ldots, x_n) \in X^n$ into the set $\{x_1, \ldots, x_n\}$ of its coordinates (it is assumed that X is a T_1-space.

36.19. Prove that the mapping π_n is continuous. Moreover, it is factor, i.e., a set $V \subset \exp_n X$ is open if and only if $\pi_n^{-1}(V)$ is open.

Therefore, the hyperspace $\exp_n X$ is obtained from the n-th power of the space X by factorization defined by the following equivalence relation: a point $(x_1, \ldots, x_n)$ is equivalent to a point $(x_1', \ldots, x_n')$ if and only if the sets $\{x_1, \ldots, x_n\}$ and $\{x_1', \ldots, x_n'\}$ coincide.

This justifies that the hyperspace $\exp_n X$ is called the *hypersymmetric n-th power* of the space X.

Along with the equivalence relation just described, on the product X^n, there naturally arise other symmetries. A free group S_n acts on X^n as the group of permutations of coordinates. More generally, let G be a subgroup of the group S_n Define the equivalence relation $\sim_G$ on X^n as follows: $(x_1, \ldots, x_n) \sim_G (x_1', \ldots, x_n')$ if and only if there exists a permutation $\sigma \in G$ for which $x_i = x_{\sigma(i)}'$.

The quotient space X^n/G is denoted by $SP_G^n X$ and is called the *G-symmetric power* of the space X.

If $G = \{e\}$, then $SP_G^n X = X^n$.

If $G = S_n$, then $SP_G^n X$ is denoted by $SP^n X$ and is called the *symmetric n-th power* of the space X.

The natural projection $X^n \to X^n/G$ will be denoted by π_G^n. If $G = S_n$, then $\pi_G^n = \pi^n$.

36.20. Prove that if $G' \subset G$, then there exists a unique mapping $\pi_{G'G}^n$: $SP_{G'}^n X \to SP_G^n X$ satisfying the condition

$$\pi_{G'G}^n \circ \pi_{G'}^n = \pi_G^n.$$

36.21. Prove that the mapping π_G^n is open and closed.

36.22. Prove that the mapping $\pi_{G'G}^n$ is open and closed.

Since the hypersymmetry relation is stronger that symmetry relation, there exists a unique mapping

$$q_n : SP^n X \to \exp_n X$$

such that $\pi_n = q_n \circ \pi^n$.

36.23. Prove that the mapping q_n is factor.

Let us return to the hyperspaces. A bicompact Hausdorff space is called the *compactum*.

36.24. Prove that if X is a compactum, then $\exp X$ is also a compactum.

36.25. Prove that if X is a compactum, then $\exp_n X$ is also a compactum.

36.26. Prove that if X is a compactum, then $SP_G^n X$ is also a compactum.

36.27. Prove that in the definition of the Vietoris topology (in the case where X is compact), as base sets, we can take sets of the form $O\langle U_1, \ldots, U_k \rangle$, where $U_1, \ldots, U_k$ runs over a certain base of the space X.

Denote by wX the *weight* of the space X, i.e., the minimum of the cardinality of all bases of the space X.

36.28. Prove that if X is an infinite compactum, then

$$wX = w \exp X = w \exp_n X = wSP_G^n X.$$

Let (X, ρ) be a bounded metric space. Denote by ρ_H the following distance function on $\exp X$:

$$\rho_H(F_1, F_2) = \inf \{\varepsilon > 0 \,|\, F_1 \subset O_\varepsilon F_2, \ F_2 \subset O_\varepsilon F_1\},$$

where $O_\varepsilon A$ denotes the open ε-ball around the set A.

36.29. Prove that the function ρ_H is a metric on the set $\exp X$ coinciding with the metric ρ on $X = \exp_1 X$.

The function ρ_H is called the *Hausdorff metric* generated by the metric ρ.

36.30. Prove that the Hausdorff metric on $\exp_n X$ generates the Vietoris topology.

36.31. Prove that if a space X is metrizable, then the space $SP_G^n X$ is also metrizable.

36.32. Prove that if (X, ρ) is a metric compactum, then the Hausdorff metric on $\exp X$ generates the Vietoris topology.

36.33. Prove that if X is a metrizable compactum, then $\exp X$, $\exp_n X$, and $SP^n_G X$ are also metrizable compacta.

Denote by H^n the n-dimensional closed half-space of $\mathbb{R}^n$, i.e., $H^n = \{x \mid x_1 > 0\}$. Denote by $X(n)$ the *cyclic n-th power of the space* X, i.e.,

$$X(n) = SP^n_{\mathbb{Z}_n} X,$$

where $\mathbb{Z}_n$ is a free cyclic subgroup of S_n with the generator

$$\begin{pmatrix} 1 & 2 & \dots & n \\ n & 1 & \dots & n-1 \end{pmatrix}.$$

In Problems 36.34–36.46, prove that the topological spaces are homeomorphic:

36.34. $\exp_2 I \approx I^2$.

36.35. $\exp_2 H^1 \approx H^2$.

36.36. $\exp_2 \mathbb{R} \approx H^2$.

36.37. $\exp_2 S^1$ is homeomorphic to the Möbius band.

36.38. $I(3) \approx I^3$.

36.39. $SP^3(I) \approx I^3$.

36.40. $\exp_3 I \approx I^3$.

36.41. $H^1(3) \approx H^3$.

36.42. $SP^3(H^1) \approx H^3$.

36.43. $\exp_3(H^1) \approx H^3$.

36.44. $\mathbb{R}(3) \approx \mathbb{R}^3$.

36.45. $SP^3(\mathbb{R}) \approx H^3$.

36.46. $\exp_3(\mathbb{R}) \approx \mathbb{R}^3$.

36.47. Prove that the space $\exp_3(S^1)$ is simply connected.

36.48. Prove that the space $SP^3(S^1)$ is not simply connected.

36.49. Prove that the space $S^1(3)$ is not simply connected.

In Problems 36.50–36.58, prove that topological spaces are homeomorphic.

36.50. $S^1(3) \approx S^1 \times S^2$.

36.51. $SP^3(S^1) \approx S^1 \times I^2$.

36.52. $\exp_3(S^1) \approx S^3$.

36.53. $\exp_4 I \not\subset \mathbb{R}^4$.

36.54. $\exp_4 I$ is not a topological manifold.

36.55. $SP^n I \approx I^n$, $n \geq 4$.

36.56. $SP^n H^1 \approx H^n$, $n \geq 4$.

36.57. $SP^n \mathbb{R} \approx H^n$, $n \geq 4$.

36.58. $SP^n S^1 \approx \begin{cases} S^1 \bowtie I^{n-1} & \text{if } n \text{ is even} \\ S^1 \times I^{n-1} & \text{if } n \text{ is odd.} \end{cases}$

Here, $S^1 \bowtie I^{n-1}$ denotes a unique nontrivial bundle over S^1 with fiber I^{n-1}.

36.59. Prove that the space $M(4)$ is not a topological manifold for any topological manifold M.

36.60. Show that the space $\exp_3 M^2$ is not a topological manifold.

In Problems 36.61–36.66, prove that the topological spaces are homeomorphic:

36.61. $SP^n I^2 \approx I^{2n}$.

36.62. $SP^n H^2 \approx H^{2n}$.

36.63. $SP^n \mathbb{R}^2 \approx \mathbb{R}^{2n}$.

36.64. $SP^n S^2 \approx \mathbb{C}P^n$.

36.65. $SP^n \left(\mathbb{R}^2 \setminus \{0\}\right) \approx \left(\mathbb{R}^2 \setminus \{0\}\right) \times \mathbb{R}^{2n-2}$.

36.66. $SP^n \left(S^1 \times I\right) \approx S^1 \times I^{2n-1}$.

Answers and Solutions[*]

Part 1

1.1. (a) $x_1 = a_1 u_1 \cos u_2$, $x_2 = a_2 u_1 \sin u_2$; (b) see Fig. 113; (c) $a_1 a_2 u_1$, $\dfrac{1}{a_1 a_2 u_1}$, the one-to-oneness is violated at the origin of the coordinates $x_1 = x_2 = 0$.

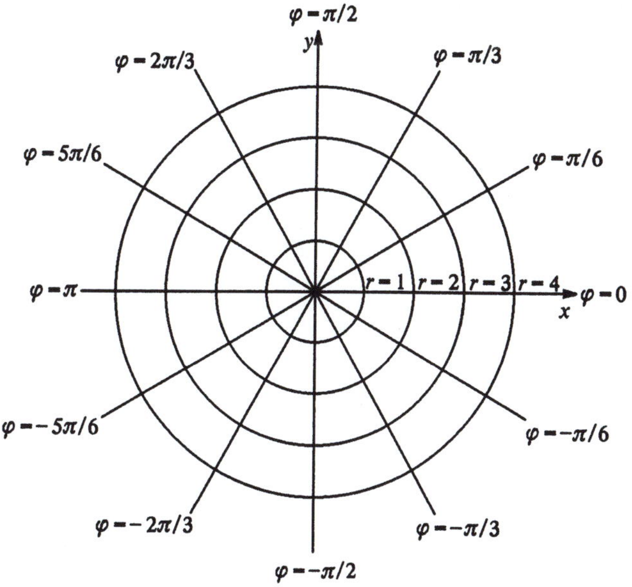

Figure 113 A polar coordinate system, $a_1 = a_2 = 1$, $x_1 = x$, $x_2 = y$, $u_1 = r$, $u_2 = \varphi$

1.2. (a) $x_1 = \cos u_2 \cosh u_1$, $x_2 = \sin u_2 \sinh u_1$; (b) see Fig. 114; (c) $\sinh^2 u_1 + \sin^2 u_2$, $\dfrac{1}{\sinh^2 u_1 + \sin^2 u_2}$.

1.3. (a) $x_1 = u_1^2 - u_2^2$, $x_2 = 2u_1 u_2$; (b) see Fig. 115; (c) $4(u_1^2 + u_2^2)$, $\dfrac{1}{4(u_1^2 + u_2^2)}$.

_____

[*]Answers and solutions are not given for all of the problems presented in Part 1 and Part 2, since many of the problems are similar.

217

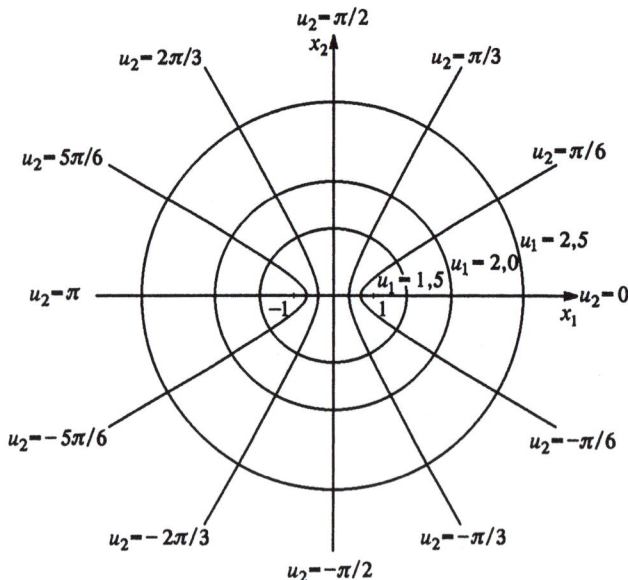

Figure 114 An elliptic coordinate system

1.4. (a) $x_1 = \dfrac{\sinh u_1}{\cos u_2 + \cosh u_1}$, $x_2 = \dfrac{\sin u_2}{\cos u_2 + \cosh u_1}$; (b) see Fig. 116;

(c) $\dfrac{1}{(\cos u_2 + \cosh u_1)^2}$, $(\cos u_2 + \cosh u_1)^2$.

1.5. (a) $x_1 = u_1^3 - 3u_1 u_2^2$, $x_2 = 3u_1^2 u_2 - u_2^3$; (c) $9(u_1^2 + u_2^2)^2$, $\dfrac{1}{9(u_1^2 + u_2^2)^2}$.

1.6. (b) $a_1 a_2 u_1$, $\dfrac{1}{a_1 a_2 u_1}$, the one-to-oneness is violated at $x_1 = x_2 = 0$; (c) no, if $a_1 \neq a_2$; yes, if $a_1 = a_2$.

1.7. (b) $a_1 a_2 a_3 u_1^2 \sin u_2$, $\dfrac{1}{a_1 a_2 a_3 u_1^2 \sin u_2}$, the one-to-oneness is violated at $x_1 = x_2 = 0$; (c) no, if $a_1 \neq a_2$; yes, if $a_1 = a_2$.

1.8. (b) $\dfrac{(u_1 - u_2)(u_3 - u_1)(u_2 - u_3)}{8\sqrt{-\prod\limits_{i,j=1}^{3}(a_i - u_j)}}$, $\dfrac{8\sqrt{-\prod\limits_{i,j=1}^{3}(a_i - u_j)}}{(u_1 - u_2)(u_3 - u_1)(u_2 - u_3)}$; (c) yes.

1.9. (b) $u_1^3 u_2 + u_2^3 u_1$, $\dfrac{1}{u_1^3 u_2 + u_2^3 u_1}$, the one-to-oneness is violated at $x_1 = x_2 = 0$; (c) yes.

1.10. (b) $\sinh u_1 \sin u_2 (\sinh^2 u_1 + \sin^2 u_2)$, $\dfrac{1}{\sinh u_1 \sin u_2 (\sinh^2 u_1 + \sin^2 u_2)}$; (c) yes.

1.11. (b) $\cosh u_1 \sin u_2 (\sinh^2 u_1 - \sin^2 u_2)$, $\dfrac{1}{\cosh u_1 \sin u_2 (\sinh^2 u_1 - \sin^2 u_2)}$;

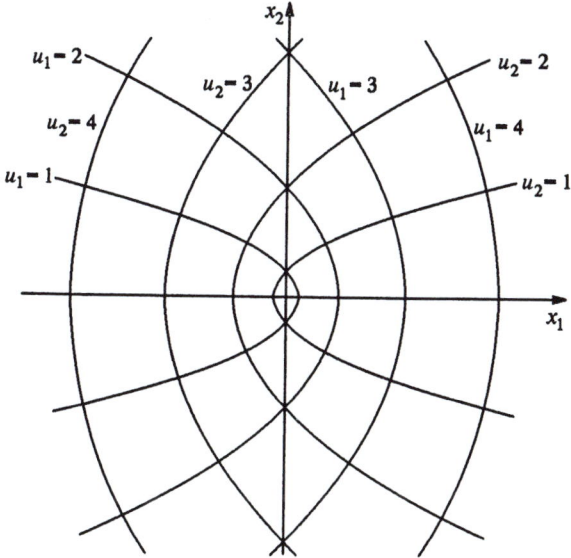

Figure 115 A parabolic coordinate system

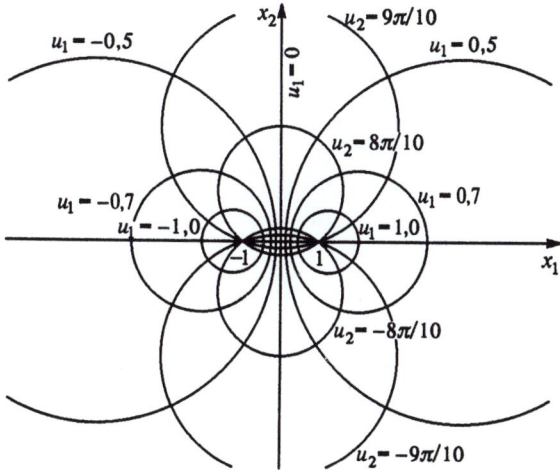

Figure 116 A bipolar coordinate system

(c) yes.

1.12. (b) $\dfrac{\sinh u_1}{(\cosh u_1 - \cos u_2)^3}$, $\dfrac{(\cosh u_1 - \cos u_2)^3}{\sinh u_1}$; (c) yes.

1.13. $\sqrt{v - u^2}\,\dfrac{\partial z}{\partial u}$. Range of definition: $y > 0$; range of values: $v > u^2$.

1.14. (a) $\dfrac{\partial u}{\partial \varphi}$; (b) $r\cos 2\varphi\,\dfrac{\partial u}{\partial r} - \sin 2\varphi\,\dfrac{\partial u}{\partial \varphi}$; (c) $\left(\dfrac{\partial u}{\partial r}\right)^2 + \dfrac{1}{r^2}\left(\dfrac{\partial u}{\partial \varphi}\right)^2$;

(d) $\dfrac{\partial^2 u}{\partial r^2} + \dfrac{1}{r^2}\dfrac{\partial^2 u}{\partial \varphi^2} + \dfrac{1}{r}\dfrac{\partial u}{\partial r}$.

1.15. $-4a^2\dfrac{\partial^2 z}{\partial u\partial v}$, $a \neq 0$.

1.16. $\dfrac{\partial^2 z}{\partial u^2} + \dfrac{\partial^2 z}{\partial v^2} + (u^2 + v^2)k^2 z = 0$.

1.17. (a) $\left(\dfrac{\partial V}{\partial r}\right)^2 + \dfrac{1}{r^2}\left(\dfrac{\partial V}{\partial \theta}\right)^2 + \dfrac{1}{r^2\sin^2\theta}\left(\dfrac{\partial V}{\partial \varphi}\right)^2$;

(b) $\dfrac{\partial^2 V}{\partial r^2} + \dfrac{1}{r^2}\dfrac{\partial^2 V}{\partial \theta^2} + \dfrac{1}{r^2\sin^2\theta}\dfrac{\partial^2 V}{\partial \varphi^2} + \dfrac{\cot\theta}{r^2}\dfrac{\partial V}{\partial \theta} + \dfrac{2}{r}\dfrac{\partial V}{\partial r}$.

1.18.
$$\dfrac{\partial^2 V}{\partial u^2} = \dfrac{\partial^2 V}{\partial x^2}\left(\dfrac{\partial f}{\partial u}\right)^2 + \dfrac{\partial^2 V}{\partial y^2}\left(\dfrac{\partial f}{\partial v}\right)^2 - 2\dfrac{\partial^2 V}{\partial x\,\partial y}\dfrac{\partial f}{\partial u}\dfrac{\partial f}{\partial v} + \dfrac{\partial V}{\partial x}\dfrac{\partial^2 f}{\partial u^2} - \dfrac{\partial V}{\partial y}\dfrac{\partial^2 f}{\partial v\,\partial u},$$

$$\dfrac{\partial^2 V}{\partial v^2} = \dfrac{\partial^2 V}{\partial x^2}\left(\dfrac{\partial f}{\partial v}\right)^2 + \dfrac{\partial^2 V}{\partial y^2}\left(\dfrac{\partial f}{\partial u}\right)^2 + 2\dfrac{\partial^2 V}{\partial x\,\partial y}\dfrac{\partial f}{\partial u}\dfrac{\partial f}{\partial v} + \dfrac{\partial V}{\partial x}\dfrac{\partial^2 f}{\partial v^2} + \dfrac{\partial V}{\partial y}\dfrac{\partial^2 f}{\partial u\,\partial v},$$

$$\dfrac{\partial^2 f}{\partial u^2} + \dfrac{\partial^2 f}{\partial v^2} = 0.$$

1.19. $\dfrac{\partial^2 V}{\partial r^2} + \dfrac{1}{r^2}\dfrac{\partial^2 V}{\partial \varphi^2} + \dfrac{\partial^2 V}{\partial z^2} + \dfrac{1}{r}\dfrac{\partial V}{\partial r}$.

1.20. *Hint:* use the formulas
$$\dfrac{\partial u}{\partial x} = \dfrac{\partial u}{\partial \rho}\cos\varphi - \dfrac{\sin\varphi}{\rho}\dfrac{\partial u}{\partial \varphi}, \quad \dfrac{\partial u}{\partial y} = \dfrac{\partial u}{\partial \rho}\sin\varphi + \dfrac{\cos\varphi}{\rho}\dfrac{\partial u}{\partial \varphi},$$
$$\dfrac{\partial v}{\partial x} = \dfrac{\partial v}{\partial \rho}\cos\varphi - \dfrac{\sin\varphi}{\rho}\dfrac{\partial v}{\partial \varphi}, \quad \dfrac{\partial v}{\partial y} = \dfrac{\partial v}{\partial \rho}\sin\varphi + \dfrac{\cos\varphi}{\rho}\dfrac{\partial v}{\partial \varphi}.$$

1.21. $\dfrac{1}{4(u^2 + v^2)}\left(\dfrac{\partial^2 V}{\partial u^2} + \dfrac{\partial^2 V}{\partial v^2}\right)$.

1.22. $\dfrac{1}{a^2(\sinh^2 u + \sin^2 v)}\left(\dfrac{\partial^2 V}{\partial u^2} + \dfrac{\partial^2 V}{\partial v^2}\right)$.

1.23. $e^{-2u}\left(\dfrac{\partial^2 V}{\partial u^2} + \dfrac{\partial^2 V}{\partial v^2}\right)$.

2.1. $r = a\varphi$.

2.2. $r = r_0 e^{k\varphi}$, where $\varphi = \omega t$.

2.3. $x = at - d\sin t$, $y = a - d\cos t$.

2.4. $x = (R + r)\cos\dfrac{rt}{R} - r\cos\dfrac{(R + r)t}{R}$, $y = (R + r)\sin\dfrac{rt}{R} - r\sin\dfrac{(R + r)t}{R}$.

2.5. $x = (R - mR)\cos mt - mR\cos(t - mt)$, $y = (R - mR)\sin mt - mR\sin(t - mt)$, $m = r/R$. Cf. Problem 2.4.

2.6. The equation of the sought-for curve is

$$\mathbf{r}(t) = \mu(t)\mathbf{a} + \mathbf{b},$$

where $\mathbf{b}$ is a constant vector, $\mu(t)$ is the primitive for the function $\lambda(t)$, $c < t < d$.

Geometrically, the following cases are possible:

- a line collinear to $\mathbf{a}$, if $\int_c^d \lambda(t)\,dt$ diverges at $t = c$ and at $t = d$;

- a ray having the direction of the vector $\mathbf{a}$, if $\int_c^d \lambda(t)\,dt$ converges at $t = c$ but diverges at $t = d$;

- a ray having the direction of the vector $-\mathbf{a}$, if $\int_c^d \lambda(t)\,dt$ diverges at $t = c$ but converges at $t = d$;

- an open segment collinear to $\mathbf{a}$, if $\int_c^d \lambda(t)\,dt$ converges.

2.7. The equation of the sought-for curve is

$$\mathbf{r}(t) = \frac{t^2}{2}\mathbf{a} + t\mathbf{b} + \mathbf{c},$$

where $\mathbf{b}$ and $\mathbf{c}$ are arbitrary constant vectors. If $\mathbf{a}$ and $\mathbf{b}$ are linearly independent, then this equation (at fixed $\mathbf{b}$, $\mathbf{c}$) sets a parabola with the axis having the direction of the vector $\mathbf{a}$. If $\mathbf{a}$ and $\mathbf{b}$ are linearly dependent, then we get a twice taken ray parallel to $\mathbf{a}$.

2.8. (a) $|\mathbf{r}'|^2\,|\mathbf{r}' \times \mathbf{a}|^2$; b) $-\langle \mathbf{r}', \mathbf{a}\rangle\,|\mathbf{r}' \times \mathbf{a}|^2$.

2.9. $\mathbf{r}' = (\varphi', \varphi + t\varphi')$, $\mathbf{r}'' = (\varphi'', 2\varphi' + t\varphi'')$, $|\mathbf{r}' \times \mathbf{r}''| = 2\varphi'^2 - \varphi\varphi''$. This equation defines a line then and only then, when $2\varphi'^2 - \varphi\varphi'' = 0$. By solving this equation, we find $\varphi = \dfrac{a}{1 + bt}$, where a and b are constants.

2.10. We shall designate the length of the vector $\mathbf{r}$ as r. Let the vector $\mathbf{e}$ be defined by the formula $\mathbf{r} = r\mathbf{e}$. Then $\dfrac{d\mathbf{r}}{d\varphi} = r'\mathbf{e} + r\dfrac{d\mathbf{e}}{d\varphi}$. As $\mathbf{e} = (\cos\varphi, \sin\varphi)$, then $\dfrac{d\mathbf{e}}{d\varphi} = (-\sin\varphi, \cos\varphi)$, i.e., $\dfrac{d\mathbf{e}}{d\varphi}$ is obtained from $\mathbf{e}$ by the turn to an angle $\pi/2$. We designate the vector obtained from $\mathbf{e}$ by the turn to an angle $\pi/2$ as $\mathbf{f}$. Therefore, $\dfrac{d\mathbf{r}}{d\varphi} = r'\mathbf{e} + r\mathbf{f}$. Further,

$$\frac{d^2\mathbf{r}}{d\varphi^2} = r''\mathbf{e} + 2r'\mathbf{f} - r\mathbf{e} = (r'' - r)\mathbf{e} + 2r'\mathbf{f},$$

$$\left| \frac{d\mathbf{r}}{d\varphi} \times \frac{d^2\mathbf{r}}{d\varphi^2} \right| = \left| \begin{matrix} r' & r \\ r'' - r & 2r' \end{matrix} \right| = 2r'^2 - rr'' + r^2 = 0.$$

Assuming $r' = \omega$, we find

$$r'' = \frac{d\omega}{d\varphi} = \frac{d\omega}{dr}r' = \omega\frac{d\omega}{dr},$$

$$2\omega^2 - \omega r \frac{d\omega}{dr} + r^2 = 0, \qquad \frac{2\omega^2}{r^2} - \frac{\omega}{r}\frac{d\omega}{dr} + 1 = 0.$$

Let $\omega^2 = p$, $r^2 = q$, then $dp/dq = 2p/q+1$. Solving this equation, we find $p = aq^2 - q$ or $\omega^2 = ar^4 - r^2$, $r' = r\sqrt{ar^2 - 1}$. Performing in this equation the substitution $1/r = \xi$, we readily obtain

$$\frac{1}{r} = C_1 \sin(\varphi + C_2),$$

where C_1, C_2 are arbitrary numbers.

The case of a line passing through the origin of the coordinates is left to be considered by the reader.

2.11. (a) Consider the vector $\mathbf{r}(t) \times \mathbf{r}'(t)$. Its derivative is $\mathbf{r}(t) \times \mathbf{r}''(t)$. According to Newton's second law, $\mathbf{F} = m\mathbf{r}''(t)$, where m is the mass of a material point M. Hence,

$$\frac{d}{dt}\left(\mathbf{r}(t) \times \mathbf{r}'(t)\right) = \mathbf{r}(t) \times \frac{\mathbf{F}}{m}.$$

By the condition $\mathbf{F} = F\mathbf{r}$ and, therefore, $\frac{d}{dt}(\mathbf{r}(t) \times \mathbf{r}'(t)) = 0$. Thus, the vector $\mathbf{r}(t) \times \mathbf{r}'(t)$ is constant. Thereby, the material point M moves in a fixed plane orthogonal to the vector $\mathbf{r}(t) \times \mathbf{r}'(t)$;

(b) $u^2 \left(\frac{d^2 u}{d\varphi^2} + u\right) = -\frac{F}{mc^2}$, $u = \frac{1}{r}$, $c = \mathrm{const}$;

(c) denote the constant vector $\mathbf{r}(t) \times \mathbf{r}'(t)$ as $\mathbf{a}$. Here r designates the length of the vector $\mathbf{r}$. We have

$$\mathbf{a} \times \mathbf{r}'' = \mathbf{a} \times \frac{\mathbf{F}}{m} = -\frac{k}{r^3}\mathbf{a} \times \mathbf{r} = -\frac{k}{r^3}\left(\mathbf{r} \times \mathbf{r}'\right) \times \mathbf{r} = -k\frac{r\mathbf{r}' - r'\mathbf{r}}{r^2} = -k\frac{d}{dt}\left(\frac{\mathbf{r}}{r}\right).$$

Just in case, note that r' is $\dfrac{dr}{dt}$, not $\left|\dfrac{d\mathbf{r}}{dt}\right|$.

Thus,

$$\frac{d}{dt}\left(\mathbf{a} \times \mathbf{r}' + k\frac{\mathbf{r}}{r}\right) = 0.$$

Integrating, we get

$$\mathbf{a} \times \mathbf{r}' + k\frac{\mathbf{r}}{r} = \mathbf{b} = \mathrm{const}.$$

Multiplying scalarwise both parts of this equality by $\mathbf{r}$ and noting that $\langle \mathbf{a} \times \mathbf{r}', \mathbf{r}\rangle = \langle \mathbf{a}, \mathbf{r}' \times \mathbf{r}\rangle = -|\mathbf{a}|^2$, we have $-|\mathbf{a}|^2 + k|\mathbf{r}| = \langle \mathbf{b}, \mathbf{r}\rangle$. Recall that the motion occurs in a fixed plane perpendicular to the vector $\mathbf{a}$ (as from the ratio $\mathbf{r} \times \mathbf{r}' = \mathbf{a}$ it follows that $\langle \mathbf{a}, \mathbf{r}\rangle = 0$). We introduce in this plane a polar coordinate system, aligning its origin of coordinates with the origin of the position vectors and directing the polar axis along the vector $\mathbf{b}$. We get $-|\mathbf{a}|^2 + k|\mathbf{r}| = |\mathbf{b}| \cdot |\mathbf{r}| \cos \varphi$, whence $|\mathbf{r}| = |\mathbf{a}|^2 / (k - |\mathbf{b}| \cos \varphi)$ is a curve of the second order. Hence, it follows, in particular, that the vector $\mathbf{b}$ is directed along the axis of the found second-order curve.

2.12. We introduce a Cartesian rectangular coordinate system, positioning the axis Oz collinearly to the vector $\mathbf{H}$. Then $\mathbf{r} = x\mathbf{i} + y\mathbf{j} + z\mathbf{k}$, and this differential equation takes the following form:

$$x'' = ay', \quad y'' = -ax', \quad z'' = 0, \quad a = |\mathbf{H}|.$$

From the relation $z'' = 0$, we find $z = C_1 t + C_2$; from the relations $x'' = ay'$, $y'' = -ax'$, we find

$$x = C_3 \cos at + C_4 \sin at + C_5, \quad y = -C_3 \sin at + C_4 \cos at + C_6.$$

At $C_1 \neq 0$, this is a family of helical lines whose axes are collinear to the vector $\mathbf{H}$; at $C_1 = 0$, we obtain a family of circles lying in the planes orthogonal to the vector $\mathbf{H}$.

2.13. The circles whose centres are on the curve passing through the origin of position vectors collinearly to the vector ω, and the planes of these circles are perpendicular to the given line.

2.14. The lines by which the planes, perpendicular to the vector $\mathbf{e}$, intersect with the planes passing through the line drawn through the origin of the coordinates collinearly to the vector $\mathbf{e}$.

2.15. Introduce the Cartesian rectangular coordinate system, placing the axis Oz collinearly to the vector $\mathbf{e}$. Then $a\mathbf{e} + \mathbf{e} \times \mathbf{r} = -y\mathbf{i} + x\mathbf{j} + a\mathbf{e}$, and this differential equation takes the following form: $x' = -y$, $y' = x$, $z' = a$. From the relations $x' = -y$, $y' = x$ we find $x^2 + y^2 = C_1$ – a family of circular cylinders, whose axes coincide with the line passing through the origin of the position vectors collinearly to the vector $\mathbf{e}$. Now,

$$\frac{dx}{dz} = -\frac{y}{a}, \quad \frac{dy}{dz} = \frac{x}{a},$$

whence

$$\frac{x\,dy - y\,dx}{dz} = \frac{x^2 + y^2}{a}, \quad a\frac{x\,dy - y\,dx}{x^2} = \left(1 + \frac{y^2}{x^2}\right)dz,$$

$$\frac{ad\,(y/x)}{1 + y^2/x^2} = dz, \quad z + C_2 = a\arctan\frac{y}{x}$$

is a family of right helicoids, for which the above-mentioned axis of cylinders is the axis. The integral lines are helical. Finally, $z = at + C_3$. From the relations obtained it is easy to express x, y, z through t.

2.16. The circles tangent to the axis Oz (collinear to the vector $\mathbf{e}$) at the origin of the coordinates.

2.17. $\arctan 3$.

2.18. $\pi/4$ and $\pi/2$.

2.19. $\arctan 3$.

2.20. (a) $s = \displaystyle\int_0^x \sqrt{1 + y'^2}\,dx = \frac{a}{2}\left(e^{x/a} - e^{-x/a}\right);$

(b) $s = \displaystyle\int_0^x \sqrt{1 + y'^2}\,dx = \frac{1}{27}\left((4 + 9x)^{3/2} - 8\right);$

(c) $s = \displaystyle\int_0^x \sqrt{1 + y'^2}\,dx = \frac{x}{2}\sqrt{1 + 4x^2} + \frac{1}{4}\ln\left(2 - x + \sqrt{1 + 4x^2}\right);$

(d) $s = \displaystyle\int_1^x \sqrt{1 + y'^2}\,dx = \sqrt{1 + x^2} + \ln\frac{\sqrt{1 + x^2}}{x} - \sqrt{2} - \ln\left(\sqrt{2} - 1\right);$

(e) $s = \displaystyle\int_0^{\varphi} \sqrt{r^2 + (dr/d\varphi)^2}\, d\varphi = 4a \sin \dfrac{\varphi}{2}$;

(f) $s = \displaystyle\int_0^{t} \sqrt{x'^2 + y'^2}\, dt = 4a \left(1 - \cos \dfrac{t}{2}\right)$;

(g) $s = \displaystyle\int_0^{t} \sqrt{x'^2 + y'^2}\, dt = \dfrac{at^2}{2}$;

(h) $s = \displaystyle\int_0^{t} \sqrt{x'^2 + y'^2}\, dt = \dfrac{8a}{3} \sin \dfrac{t}{2}$;

(i) $s = \displaystyle\int_0^{t} \sqrt{x'^2 + y'^2}\, dt = \dfrac{3a}{2} \sin^2 t$;

(j) $s = \displaystyle\int_0^{x} \sqrt{1 + y'^2}\, dx = \sqrt{1 + e^{2x}} + \dfrac{1}{2}\ln \dfrac{\sqrt{1 + e^{2x}} - 1}{\sqrt{1 + e^{2x}} + 1} - \sqrt{2} - \ln\left(\sqrt{2} - 1\right)$;

(k) $s = \displaystyle\int_{\pi/2}^{t} \sqrt{x'^2 + y'^2}\, dt = a \ln\left(\sin t\right)$.

2.21. $f(\alpha) + f''(\alpha),\ f'(\alpha) + f'''(\alpha) > 0$.

2.22. $\mathbf{r}(u, v) = ((a + b\cos v)\cos u, (a + b\cos v)\sin u, b\sin v)$.

2.23. Let at the initial moment of time the moving line coincide with the axis Ox; the second line mentioned in the condition, with the axis Oz. Then the equation of the right helicoid has the form

$$\mathbf{r} = (v \cos u, v \sin u, ku).$$

Here v is the distance from the point of the helicoid to its axis (the axis Oz); u is the longitude of the point.

2.24. $\mathbf{r} = (v, a\cos u\, \cosh\dfrac{v}{a}, a\sin u\, \cosh\dfrac{v}{a})$, where u is the longitude; v, the oriented distance from the point of the surface to the throat cross section of the catenoid.

2.25. $\mathbf{r} = \left(a \ln\tan\left(\dfrac{\pi}{4} + \dfrac{t}{2}\right) - a\sin t, a\cos t\cos u, a\cos t\sin u\right)$.

2.26. Let

$$x = (a + R\cos\theta)\cos\varphi, \quad y = (a + R\cos\theta)\sin\varphi, \quad z = R\sin\theta$$

be the parametric representation of the torus. Here R is the radius of the meridian, i.e., of the rotated circle; $a > R$ is the distance from the centre of the meridian to the axis of revolution passing along Oz. The pairs of points at which the curvature of the torus is negative (at the points $K > 0$ the tangent plane does not intersect with the torus) and the tangent planes coincide, have the intrinsic coordinates (φ_0, θ_0) and $(\varphi_0 + \pi, \pi + \theta_0)$ with $\cos\theta_0 = -R/a$. To obtain these points, we need to make use of the fact that this tangent plane intersects the plane xOz along the line tangent to two diametrically opposite meridians of the torus. Further on, by direct calculations we obtain that the tangent plane has the equation $-R(x\cos\varphi_0 + y\sin\varphi_0) + \sqrt{a^2 - R^2}\,z =$

0, and the intersection of this plane with the torus gives a union of two intersecting curves with the equations

$$x = \sqrt{a^2 - R^2}\sin\theta\cos\varphi_0 \mp (a\cos\theta + R)\sin\varphi_0,$$

$$y = \sqrt{a^2 - R^2}\sin\theta\sin\varphi_0 \pm (a\cos\theta + R)\cos\varphi_0,$$

$$z = R\sin\theta.$$

From here, it is easy to obtain the Villarceau radius (it is equal to a) and the angle between this circle and the parallel of the torus: $\arcsin(R/a)$.

————

3.1. (a) The equation of a sphere of unit radius $x^2 + y^2 + z^2 = 1$:

$$z = \pm\sqrt{1 - x^2 - y^2}, \quad \mathbf{r}(x,y) = \left(x, y \pm \sqrt{1 - x^2 - y^2}\right),$$

$$\mathbf{r}'_x = \left(1, 0, \frac{\mp x}{\sqrt{1 - x^2 - y^2}}\right), \quad \mathbf{r}'_y = \left(0, 1, \frac{\mp y}{\sqrt{1 - x^2 - y^2}}\right).$$

Hence (at any choice of the sign of the coordinate z)

$$g_{11} = \langle \mathbf{r}_x, \mathbf{r}_x \rangle = 1 + \frac{x^2}{1 - x^2 - y^2} = \frac{1 - y^2}{1 - x^2 - y^2},$$

$$g_{12} = g_{21} = \langle \mathbf{r}_x, \mathbf{r}_y \rangle = \frac{xy}{1 - x^2 - y^2},$$

$$g_{22} = \langle \mathbf{r}_y, \mathbf{r}_y \rangle = \frac{1 - x^2}{1 - x^2 - y^2},$$

$$ds^2 = \frac{\left(1 - y^2\right)dx^2 + 2xy\,dx\,dy + \left(1 - x^2\right)dy^2}{1 - x^2 - y^2}.$$

(b) Consider the following parameterization of the sphere:

$$x = \sin\theta\cos\varphi, \quad y = \sin\theta\sin\varphi, \quad z = \cos\theta.$$

Hence

$\mathbf{r} = (x(\theta,\varphi), y(\theta,\varphi), z(\theta,\varphi)),$

$\mathbf{r}'_\theta = (\cos\theta\cos\varphi, \cos\theta\sin\varphi, -\sin\theta),$

$\mathbf{r}'_\varphi = (-\sin\theta\sin\varphi, \sin\theta\cos\varphi, 0),$

$g_{11} = \langle \mathbf{r}'_\theta, \mathbf{r}'_\theta \rangle = 1, \quad g_{12} = g_{21} = \langle \mathbf{r}'_\theta, \mathbf{r}'_\varphi \rangle = 0, \quad g_{22} = \langle \mathbf{r}'_\varphi, \mathbf{r}'_\varphi \rangle = \sin^2\theta,$

$ds^2 = d\theta^2 + \sin^2\theta d\varphi^2;$

(c) $ds^2 = \dfrac{4\left(du^2 + dv^2\right)}{(1 + u^2 + v^2)^2};$ (d) $ds^2 = \dfrac{4\left(dr^2 + r^2d\varphi^2\right)}{(1 + r^2)^2};$

(e) $ds^2 = \dfrac{4dw\,d\bar{w}}{(1 + w\bar{w})^2}.$

3.2. (a) $t^2 - x^2 - y^2 = 1$, $ds^2 = dx^2 + dy^2 - dt^2$,

$$ds^2 = \frac{(y^2 + 1)dx^2 - 2xy\,dx\,dy + (x^2 + 1)dy^2}{1 + x^2 + y^2};$$

(b) $\mathbf{r} = (x,y,t)$, $t = \cosh\chi$, $x = \sinh\chi\cos\varphi$, $y = \sinh\chi\sin\varphi$, $ds^2 = d\chi^2 + \sinh^2\chi d\varphi^2;$

(c) $ds^2 = \dfrac{4\left(du^2 + dv^2\right)}{\left(1 - u^2 - v^2\right)^2}$; (d) $ds^2 = \dfrac{4\left(dr^2 + r^2 d\varphi^2\right)}{\left(1 - r^2\right)^2}$;

(e) $ds^2 = \dfrac{4 dz d\bar{z}}{\left(1 - z\bar{z}\right)^2}$; (f) $ds^2 = \dfrac{dz d\bar{z}}{(\operatorname{Im} z)^2}$.

3.3. *Hint:* prove a more general proposition, on the coincidence of the angles in the metrics $ds^2 = dx^2 + dy^2$ and $dl^2 = G(x,y)(dx^2 + dy^2)$.*

3.18. (a) We shall solve the problem in the Poincaré model. Let us use the results of Problem 3.2d:

$$R = \int_0^r \sqrt{\frac{4}{(1 - \tau^2)^2}}\, d\tau = \int_0^r \frac{2}{1 - \tau^2}\, d\tau = \ln\left|\frac{1 + \tau}{1 - \tau}\right|\Big|_0^r = \ln\frac{1 + r}{1 - r},$$

hence $r = \dfrac{e^R - 1}{e^R + 1}$,

$$l = \int_0^{2\pi} \sqrt{\frac{4 r^2}{(1 - r^2)^2}}\, d\varphi = \int_0^{2\pi} \frac{2r}{1 - r^2}\, d\varphi =$$

$$4\pi \frac{\left(e^R - 1\right)/\left(e^R + 1\right)}{1 - \left(\left(e^R - 1\right)/\left(e^R + 1\right)\right)^2} = 2\pi \frac{e^{2R} - 1}{2 e^R} = 2\pi \sinh R,$$

$$S = \int_0^r \int_0^{2\pi} \frac{4\tau}{(1 - \tau^2)^2}\, d\varphi\, d\tau = 4\pi \int_0^{r^2} \frac{dt}{(1 - t)^2} =$$

$$4\pi \left(\frac{1}{1 - t}\right)\Big|_0^{r^2} = 4\pi \frac{r^2}{1 - r^2} = 4\pi \frac{\left(e^R - 1\right)^2}{4 e^R} = 4\pi \sinh^2 \frac{R}{2};$$

(b) $l = 2\pi \sin R$, $S = 4\pi \sin^2 \dfrac{R}{2}$.

3.19. (a) Deduce the formula of the distance for the points $(0, y_1), (0, y_2)$:

$$l = \int_{y_1}^{y_2} \frac{d\tau}{\tau} = \ln |\tau|\Big|_{y_1}^{y_2} = \ln \frac{y_2}{y_1},$$

$$\ln \frac{3}{y} = \ln \frac{y}{1}, \quad \frac{3}{y} = \frac{y}{1}, \quad y^2 = 3, \quad y = \sqrt{3}.$$

Answer: $(0, \sqrt{3})$.

(c) Using the symmetry considerations, we conclude that the line $x = 0$ intersects our circle by the diameter. It remains to find the midpoint of this diameter:

$$\int_{r_1}^{r_2} \sqrt{\frac{4}{(1 + r^2)^2}}\, dr = 2 \arctan r\Big|_{r_1}^{r_2} = 2 \arctan \frac{r_2 - r_1}{1 + r_1 r_2}.$$

---

*Answers and solutions are not given for Problems 3.4 to 3.17 since they are similar to others in this section.

$$2\arctan\frac{3-r}{1+3r} = 2\arctan\frac{r-1}{1+r},$$

$$-r^2 + 3r - r + 3 = 3r^2 - 3r + r - 1,$$

$$4r^2 - 4r - 4 = 0, \quad r_{1,2} = \frac{1 \pm \sqrt{5}}{2}.$$

The centre of the circle has the coordinates $\left(0, \dfrac{1+\sqrt{5}}{2}\right)$.

3.20. Let $A = 0.5 + 0.5i$, $B = 0.9 + 0.3i$. Let us find the linear-fractional transformation of the upper half-plane into a unit circle, which transforms A into 0, and B into a point of a real line. It is easy to show that, as such a transformation, we can take ρ: $z \longmapsto \dfrac{(1+3i)z + (1-2i)}{(3-i)z - (1-2i)}$. Then ρ: $A \longmapsto 0$, ρ: $B \longmapsto \dfrac{1}{2}$.

The distance in the Poincaré metric between 0 and $\dfrac{1}{2}$ is equal to $\ln 3$ (see Problem 3.18). Then the midpoint of the segment AB under the action of the linear-fractional transformation ρ passes into a point of the form $c + 0i$, where $c > 0$. The distance from this point to 0 should be equal to $\dfrac{1}{2}\ln 3$. Hence (see Problem 3.18) we obtain the condition for c: $\ln\dfrac{1+c}{1-c} = \dfrac{1}{2}\ln 3$. Therefore, $c = \dfrac{\sqrt{3}-1}{\sqrt{3}+1}$. Making the inverse linear-fractional transformation, we get that the midpoint of the segment AB is the point $\rho^{-1}\left(\dfrac{\sqrt{3}-1}{\sqrt{3}+1} + 0i\right) = \dfrac{3+\sqrt{3}i}{4}$.

3.21. Let the angles of the triangle be equal to α, β and γ:

(a) $S_\Delta = -\pi + \alpha + \beta + \gamma$; (b) $S_\Delta = \pi - \alpha - \beta - \gamma$.

3.27. (a) At $n \geq 2$ (there exists a right digon, which at $n = 2$ is a circle); (b) at $n \geq 2$ (there exists a regular k-gon at any $k > \dfrac{2n}{n-2}$; at $n = 2$, this is a circle).

3.28. *Hint:* make use of the geometric criterion of the existence of a circumscribed circle for the tetragon and the expression for the argument of the quotient of two complex numbers.

3.29. *Hint:* make use of the formula of sines and the expressions for the length of circle from Problem 3.18.

3.30. *Hint:* make use of the result of Problem 3.21.

3.31. (a) Let $y_1 = OA$, $y_2 = OB$. Then

$$\rho(A, B) = \int\limits_{y_1}^{y_2} \sqrt{\frac{1}{y^2}}\, dy = \ln|y|\Big|_{y_1}^{y_2} = \left|\ln\frac{OB}{OA}\right|;$$

(b) let O and O_1 be points of intersection of the real line with the semi-circle passing through A and B and lying on the real line as on the diameter. What is more, O_1 lies more to the left than O. Performing the inversion I with the centre at O and the radius OO_1, we will map our circle onto the vertical line; the ordinates of the images of A and B in this case will be equal to $OO_1 \cdot |\tan\alpha|$ and $|\tan\beta|$. Further, we will make use of the result of (a) and of the fact that I is the motion of the metric of the Lobachevskii plane;

(c) transfer the line connecting the points A and B into a vertical line using an appropriate linear-fractional transformation from Problem 3.5.

3.32. (a) Yes, true; (b) no, untrue. Consider in the model on the upper half-plane the points $1+i$, $-1+i$, $2i$. Then, on the one hand, we know that in this model the circles in the Lobachevskii metric coincide with the "usual" ones; on the other hand, the circle passing through the three points considered has an "osculation point" on the absolute and, therefore, is not a circle in the sense of the Lobachevskii-plane metric.

3.33. For example, we can consider in the model on the upper half-plane the lines $\mathrm{Re}\, z = 0$ and $z\bar{z} = 1$ and the point $5+i$ together with the cluster of lines passing through it.

3.34. (a) Let a be a side of a regular hexagon, and R the radius of a circle circumscribed around it. Then by Problem 3.24

$$\cosh a = \cosh^2 R - \cos\frac{\pi}{3}\sinh^2 R,$$

$$2\cosh a = 2\cosh^2 R - \sinh^2 R = 1 + \cosh^2 R,$$

$$\cosh a = \frac{1+\cosh^2 R}{2} > \cosh R \quad (R \neq 0), \quad a > R;$$

(b) the solution is analogous to (a). Answer: $a < R$.

3.35. (a) By 3.24b, the cosine of the angle of a triangle is equal to $\dfrac{\cosh a}{\cosh a + 1}$. By

3.21b, $S_\Delta = \pi - 3\arccos\dfrac{\cosh a}{1+\cosh a}$;

(b) the solution is analogous. Answer: $S_\Delta = -\pi + 3\arccos\dfrac{\cos a}{\cos a + 1}$.

3.36. (a) It is clear that the centre of the circle lies on the line $x = -1$. Let it have the coordinates $(-1, 4+a)$. As the distances from the centre of the circle to the points $(-1,4)$ and $(-1,6)$ should be equal, we obtain the equation

$$\ln\frac{6}{4+a} = \ln\frac{4+a}{4}.$$

We solve it:

$$(4+a)^2 = 24, \quad a = 2\sqrt{6} - 4.$$

Thus, the centre of the circle has the coordinates $(-1, 2\sqrt{6})$. Hence, it is easy to find the radius of the circle in the sense of the Lobachevskii metric. It remains to make use of Problem 3.18a.

Answer: $S = 4\pi\sinh^2\left(\dfrac{1}{4}\ln\dfrac{3}{2}\right) = \left(\dfrac{5}{\sqrt{6}} - 2\right)\pi$, $l = 2\pi\sinh\left(\dfrac{1}{2}\ln\dfrac{3}{2}\right) = \dfrac{\pi}{\sqrt{6}}$;

(b) using symmetry considerations, we conclude that the line $5x + y = 0$ intersects our circle by the diameter whose length is equal to the distance between the points $\left(-1+\dfrac{1}{\sqrt{26}}, 5 - \dfrac{5}{\sqrt{26}}\right)$ and $\left(-1 - \dfrac{1}{\sqrt{26}}, 5 + \dfrac{5}{\sqrt{26}}\right)$ in the sense of the metric from Problem 3.1c, i.e., is equal to

$$\int_{-1-\frac{1}{\sqrt{26}}}^{-1+\frac{1}{\sqrt{26}}}\sqrt{\frac{4\left(1^2 + 5^2\right)}{\left(1 + x^2 + 25x^2\right)^2}}\,dx = 2\arctan\left(x\sqrt{26}\right)\Big|_{-1-\frac{1}{\sqrt{26}}}^{-1+\frac{1}{\sqrt{26}}} =$$

$$2\arctan\frac{1 - \sqrt{26} + 1 + \sqrt{26}}{1 - \left(1 + \sqrt{26}\right)\left(1 - \sqrt{26}\right)} = 2\arctan\frac{1}{13}.$$

Thus, the circle has the radius $R = \arctan \dfrac{1}{13}$. By Problem 3.18b, we have

$$S = 4\pi \sin^2 \frac{R}{2} = 2\pi \left(1 - \frac{13}{\sqrt{170}}\right), \quad l = 2\pi \sin R = \frac{2\pi}{\sqrt{170}}.$$

4.1. In Problems (a), (b), (f) and (g), prove and use the formula for the curvature of a curve $\mathbf{r} = \mathbf{r}(t)$:

$$k = \frac{\left|\,|\dot{\mathbf{r}}|^2\, \ddot{\mathbf{r}} - (\dot{\mathbf{r}}, \ddot{\mathbf{r}})\, \dot{\mathbf{r}}\,\right|}{|\dot{\mathbf{r}}|^4} = \frac{|[\dot{\mathbf{r}}, \ddot{\mathbf{r}}]|}{|\dot{\mathbf{r}}|^3}$$

In Problems (c), (d) and (e), prove and use the general formula for the curvature of a curve given in polar coordinates by the equation $r = r(\varphi)$:

$$k(\varphi) = \frac{\left| r^2 + 2\,(r')^2 - rr'' \right|}{(r^2 + r'^2)^{3/2}}.$$

(a) 1; (b) $\dfrac{a}{y^2}$; (c) $\dfrac{3r}{a^2}$; (d) $\dfrac{3}{4a\left|\cos(\varphi/2)\right|}$; (e) $\dfrac{2 + \varphi^2}{a\,(1 + \varphi^2)^{3/2}}$; (f) $\dfrac{1}{3a\left|\sin t \cos t\right|}$;

(g) $\dfrac{1}{4a\left|\sin(t/2)\right|}$.

4.3. $\dfrac{a}{b^2}$ and $\dfrac{b}{a^2}$.

4.4. $k = \dfrac{\left| F_x^2 F_{yy} - 2F_x F_y F_{xy} + F_y^2 F_{xx} \right|}{\left(F_x^2 + F_y^2\right)^{3/2}}$.

4.5. $k = \dfrac{\left| P\,(Q\,(\partial Q/\partial x) - P\,(\partial Q/\partial y)) + Q\,(P\,(\partial P/\partial y) - Q\,(\partial P/\partial x)) \right|}{\left(P^2 + Q^2\right)^{3/2}}$.

4.6. $k(\varphi) = \dfrac{\left| 2\,(r')^2 - r''r + r^2 \right|}{\left((r')^2 + r^2\right)^{3/2}}$.

4.7. $\mathbf{r}(s) = \left(a\cos \dfrac{s}{\sqrt{a^2 + b^2}},\, a\sin \dfrac{s}{\sqrt{a^2 + b^2}},\, \dfrac{bs}{\sqrt{a^2 + b^2}} \right)$.

4.8. $\mathbf{r}(s) = \dfrac{s + \sqrt{3}}{\sqrt{3}} \left(\cos\left(\ln \dfrac{s + \sqrt{3}}{\sqrt{3}}\right),\, \sin\left(\ln \dfrac{s + \sqrt{3}}{\sqrt{3}}\right),\, 1 \right)$.

4.9. $\mathbf{r}(s) = \left(\sqrt{\dfrac{2 + s^2}{2}},\, \dfrac{s}{\sqrt{2}},\, \ln\left(\dfrac{s}{\sqrt{2}} + \sqrt{\dfrac{2 + s^2}{2}}\right) \right)$.

4.10. (a) $k = \dfrac{\sqrt{2}}{(e^t + e^{-t})^2}$, $\varkappa = -\dfrac{\sqrt{2}}{(e^t + e^{-t})^2}$; (b) $k = \dfrac{2t}{(1 + 2t^2)^2}$, $\varkappa = -\dfrac{2t}{(1 + 2t^2)^2}$;

(c) $k = \dfrac{\sqrt{2}}{3e^t}$, $\varkappa = -\dfrac{1}{3e^t}$; (d) $k = \varkappa = \dfrac{1}{3\,(t^2 + 1)^2}$; (e) $k = \dfrac{3}{25\left|\sin t \cos t\right|}$, $\varkappa = \dfrac{3}{25\sin t \cos t}$.

4.11. $k = \dfrac{1}{\sqrt{6}}$, $\varkappa = 1$.

4.12. $k = \dfrac{\sqrt{6}}{9}$, $\varkappa = -\dfrac{1}{2}$.

4.13. $k = \dfrac{\sqrt{y''^2 + z''^2 + (y'z'' - y''z')^2}}{(1 + y'^2 + z'^2)^{3/2}}$, $\varkappa = \dfrac{y''z''' - y'''z''}{y''^2 + z''^2 + (y'z'' - y''z')^2}$;

$\mathbf{v} = \dfrac{(1, y', z')}{\sqrt{1 + y'^2 + z'^2}}$, $\mathbf{b} = \dfrac{(y'z'' - y''z', -z'', y'')}{\sqrt{y''^2 + z''^2 + (y'z'' - y''z')^2}}$,

$\mathbf{n} = \dfrac{(-z'z'' - y'y'', y'' - z'(y'z'' - y''z'), z'' + y'(y'z'' - y''z'))}{\sqrt{(z'z'' + y'y'')^2 + (y'' - z'(y'z'' - y''z'))^2 + (z'' + y'(y'z'' - z'y''))^2}}$.

4.14. $\mathbf{v} = \dfrac{(2t, -1, 3t^2)}{\sqrt{1 + 4t^2 + 9t^4}}$, $\mathbf{n} = \dfrac{(1 - 9t^4, 2t + 9t^3, 3t + 6t^3)}{\sqrt{(1 - 9t^4)^2 + (2t + 9t^3)^2 + (3t + 6t^3)^2}}$,

$\mathbf{b} = \dfrac{(-3t, -3t^2, 1)}{\sqrt{1 + 9t^2 + 9t^4}}$, $k = \dfrac{2(1 + 9t^2 + 9t^4)^{1/2}}{(1 + 4t^2 + 9t^4)^{3/2}}$, $\varkappa = \dfrac{3}{1 + 9t^2 + 9t^4}$.

4.15. (a) $k = \dfrac{a}{a^2 + h^2}$, $\varkappa = \dfrac{h}{a^2 + h^2}$; (b) at $h = a$;

(c) $\mathbf{v} = \dfrac{1}{\sqrt{a^2 + h^2}}(-a\sin t, a\cos t, h)$, $\mathbf{n} = (-\cos t, -\sin t, 0)$,

$\mathbf{b} = \dfrac{1}{\sqrt{a^2 + h^2}}(h\sin t, -h\cos t, a)$.

4.16. If the circumference of the cylinder is equal to the pitch of the helical line.

4.18. *Hint:*

$$Y : \mathbf{x} = \begin{pmatrix} x^1 \\ x^2 \\ x^3 \end{pmatrix} \longmapsto \mathbf{y} \times \mathbf{x} = \begin{pmatrix} y^2 x^3 - y^3 x^2 \\ y^3 x^1 - y^1 x^3 \\ y^1 x^2 - y^2 x^1 \end{pmatrix} =$$

$$\begin{pmatrix} 0 & -y^3 & y^2 \\ y^3 & 0 & -y^1 \\ -y^2 & y^1 & 0 \end{pmatrix} \begin{pmatrix} x^1 \\ x^2 \\ x^3 \end{pmatrix} = Y \cdot \mathbf{x},$$

so the operator Y is represented by the skew-symmetric matrix

$$\begin{pmatrix} 0 & -y^3 & y^2 \\ y^3 & 0 & -y^1 \\ -y^2 & y^1 & 0 \end{pmatrix}.$$

4.19. *Hint:* check the Jacobi identity:

$$\mathbf{x} \times (\mathbf{y} \times \mathbf{z}) + \mathbf{y} \times (\mathbf{z} \times \mathbf{x}) + \mathbf{z} \times (\mathbf{x} \times \mathbf{y}) = 0.$$

Deduce from here that $(\mathbf{y} \times \mathbf{z}) \times \mathbf{x} = \mathbf{y} \times (\mathbf{z} \times \mathbf{x}) - \mathbf{z} \times (\mathbf{y} \times \mathbf{x})$.

4.21. (a) $R^2 + 4s^2 - 6as = 0$ at $0 \le t \le \dfrac{\pi}{2}$;

(b) $(27s + 8)^2 = \left(4 + 9 \cdot \dfrac{36R^2}{(27s + 8)^2}\right)^3$;

(c) $s = \dfrac{1}{4}\sqrt{\sqrt[3]{4R^2} - 1} + \sqrt[3]{2R} + \dfrac{1}{4}\ln\left(\sqrt{\sqrt[3]{4R^2} - 1} + \sqrt[3]{2R}\right)$;

(d) the parametric natural equations:

$$s = \sqrt{1 + x^2} + \ln\dfrac{\sqrt{1 + x^2} - 1}{x}, \quad k = \dfrac{x}{(1 + x^2)^{3/2}};$$

(e) $R = a + s^2/a$;

(f) the parametric natural equations:

$$s = \sqrt{1 + e^{2x}} + \ln \frac{\sqrt{1 + e^{2x}} - 1}{e^x}, \quad k = \frac{e^x}{(1 + e^{2x})^{3/2}};$$

(g) $R^2 + a^2 = a^2 e^{-2s/a}$; (h) $s^2 + 9R^2 = 16a^2$ at $0 \le \varphi \le \frac{\pi}{4}$;

(i) $R^2 = 2as$; (j) $R^2 + s^2 - 8as = 0$ at $0 \le t \le 2\pi$. Here $R = \frac{1}{k}$;

4.22. (a) $r = Ce^{a\varphi}$ is a logarithmic spiral, where r, φ are the polar coordinates,

$$C = \frac{ae^{a\arctan a}}{\sqrt{a^2 + 1}};$$

(b) $x(t) = \frac{a}{2} \left(\frac{b}{a+b} \sin \frac{(a+b)t}{b} + \frac{b}{a-b} \sin \frac{(a-b)t}{b} \right),$

$y(t) = \frac{a}{2} \left(-\frac{b}{a+b} \cos \frac{(a+b)t}{b} - \frac{b}{a-b} \cos \frac{(a-b)t}{b} \right)$ at $0 \le t \le \frac{\pi}{2}$;

(c) $\mathbf{r}(s) = \left(\int_0^s \cos(s^2/2a^2)\,ds, \int_0^s \sin(s^2/2a^2)\,ds \right)$ – a clothoid;

(d) $x(t) = a \ln \left(\tan \left| \frac{\pi}{4} + \frac{t}{2} \right| \right)$, $y(t) = \frac{a}{\cos t}$ – a catenary, $0 \le t \le \frac{\pi}{2}$;

(e) $\mathbf{r}(t) = (a(\cos t + t \sin t), \ a(\sin t - t \cos t))$ – an evolvent of a circle

4.23. $y(s) = \text{const}$, $z(s) = \text{const}$. *Hint:* $\sqrt{x'(s)^2 + y'(s)^2 + z'(s)^2} = 1$.

4.25. $r = Ce^{\pm k\varphi}$, $k = \cot \alpha$, $C = \text{const}$, r and φ are the polar coordinates.

4.26. $p = |\langle \mathbf{r}, \mathbf{n} \rangle|$. Assume that $\langle \mathbf{r}, \mathbf{n} \rangle > 0$; then $p = \langle \mathbf{r}, \mathbf{n} \rangle$. Hence

$$\frac{dp}{ds} = \langle \dot{\mathbf{r}}, \mathbf{n} \rangle + \langle \mathbf{r}, \dot{\mathbf{n}} \rangle = -k \langle \mathbf{r}, \mathbf{v} \rangle = -k \langle \mathbf{r}, \dot{\mathbf{r}} \rangle.$$

Note that $\langle \mathbf{r}, \mathbf{r} \rangle = l^2$. Hence, $\langle \mathbf{r}, \dot{\mathbf{r}} \rangle = l\dot{l}$. Thus,

$$\frac{dp}{ds} = -kl \frac{dl}{ds},$$

hence, we obtain the required relation.

4.29. We will write the equation of the osculating circle as $(\rho - \mathbf{r}_0 - R_0 \mathbf{n}_0)^2 = R_0^2$. Hence, $\langle \rho - \mathbf{r}_0, \rho - \mathbf{r}_0 \rangle - 2R_0 \langle \mathbf{n}_0, \rho - \mathbf{r}_0 \rangle = 0$.

Consider the function

$$\varphi(s) = \langle \mathbf{r} - \mathbf{r}_0, \mathbf{r} - \mathbf{r}_0 \rangle - 2R_0 \langle \mathbf{n}_0, \mathbf{r} - \mathbf{r}_0 \rangle.$$

We have:
$\varphi'(s) = 2 \langle \mathbf{r} - \mathbf{r}_0, \mathbf{v} \rangle - 2R_0 \langle \mathbf{n}_0, \mathbf{v} \rangle$,
$\varphi''(s) = 2 + 2k \langle \mathbf{n}, \mathbf{r} - \mathbf{r}_0 \rangle - 2R_0 k \langle \mathbf{n}_0, \mathbf{n} \rangle$,
$\varphi'''(s) = 2\dot{k} \langle \mathbf{n}, \mathbf{r} - \mathbf{r}_0 \rangle - 2k^2 \langle \mathbf{v}, \mathbf{r} - \mathbf{r}_0 \rangle - 2R_0 \dot{k} \langle \mathbf{n}_0, \mathbf{n} \rangle + 2R_0 k^2 \langle \mathbf{n}_0, \mathbf{v} \rangle$.
Thus,

$$\varphi'(s_0) = 0, \quad \varphi''(s_0) = 0, \quad \varphi'''(s_0) = -2R_0 \dot{k}(s) \ne 0$$

and, therefore, $\varphi(s)$ changes the sign in transition of s through s_0, which proves the proposition.

4.30. See the solution of Problem 4.29. We have:

$$\varphi'(s_0) = \varphi''(s_0) = \varphi'''(s_0) = 0,$$

$$\varphi^{(4)}(s) = 2\ddot{k}\langle \mathbf{n}, \mathbf{r} - \mathbf{r}_0 \rangle - 2k\dot{k}\langle \mathbf{v}, \mathbf{r} - \mathbf{r}_0 \rangle - 4k\dot{k}\langle \mathbf{v}, \mathbf{r} - \mathbf{r}_0 \rangle - 2k^3\langle \mathbf{n}, \mathbf{r} - \mathbf{r}_0 \rangle -$$
$$2k^2 - 2R_0\ddot{k}\langle \mathbf{n}_0, \mathbf{n} \rangle + 2R_0 k\dot{k}\langle \mathbf{n}_0, \mathbf{n} \rangle + 4R_0 k\dot{k}\langle \mathbf{n}_0, \mathbf{v} \rangle + 2R_0 k^3\langle \mathbf{n}_0, \mathbf{n} \rangle,$$

$$\varphi^{(4)}(s) = -2k_0^2 - 2R_0\ddot{k}_0 + 2k_0^2 = -2R_0\ddot{k}_0 \neq 0,$$

which means that the *power of the point of the curve relative to the osculating circle* does not change its sign in transition of s through s_0.

4.31. We will consider that the vector $\mathbf{a}$ has a unit length and $\mathbf{b}$ is another constant vector, so that $\mathbf{a}$ and $\mathbf{b}$ form a positively oriented orthonormal basis in the plane. Let x, y be the coordinates of the point relative the basis $\mathbf{a}, \mathbf{b}$. In (a) and (b), it is assumed that $\dfrac{d\alpha}{ds} > 0$.

(a) $\dfrac{d\alpha}{ds} = k = \dfrac{1}{f(\alpha)}$, $ds = f(\alpha)\, d\alpha$,

$x = \displaystyle\int f(\alpha)\cos\alpha \, d\alpha$, $y = \displaystyle\int f(\alpha)\sin\alpha \, d\alpha$;

(b) $\dfrac{d\alpha}{ds} = \dfrac{1}{R}$, $f'(R)\dfrac{dR}{ds} = \dfrac{1}{R}$, $ds = Rf'(R)\, dR$,

$x = \displaystyle\int Rf'(R)\cos(f(R))\, dR$, $y = \displaystyle\int Rf'(R)\sin(f(R))\, dR$;

(c) $x = \displaystyle\int f'(\alpha)\cos\alpha \, d\alpha$, $y = \displaystyle\int f'(\alpha)\sin\alpha \, d\alpha$;

(d) $x = \displaystyle\int \cos(f(s))\, ds$, $y = \displaystyle\int \sin(f(s))\, ds$.

4.32. Choosing a coordinate system as appropriate, we write down the equations of the Viviani curve as

$$x^2 + y^2 + z^2 = a^2, \quad \left(x - \frac{a}{2}\right)^2 + y^2 = \frac{a^2}{4}.$$

For comparison of the parametric equations, let

$$x - \frac{a}{2} = \frac{a}{2}\cos t, \quad y = \frac{a}{2}\sin t.$$

Then $\dfrac{a^2}{4}(1 + \cos t)^2 + \dfrac{a^2}{4}\sin^2 t + z^2 = a^2$. Hence, $z = a\sin\dfrac{t}{2}$ (the sign can be dropped, because if 2π is added to t, then x and y would not change but z would change its sign). Thus:

$$\mathbf{r} = \left(\frac{a}{2}(1 + \cos t), \frac{a}{2}\sin t, a\sin\frac{t}{2}\right).$$

The tangent:

$$\mathbf{r} = \left(\frac{a}{2}(1 + \cos t) - \lambda\sin t, \frac{a}{2}\sin t + \lambda\cos t, a\sin\frac{t}{2} + \lambda\cos\frac{t}{2}\right).$$

The normal plane: $x\sin t - y\cos t - z\cos\dfrac{t}{2} = 0$.

The binormal:

$$\mathbf{r} = \left(\frac{a}{2} \left(1 + \cos t\right) + \lambda \left(2 + \cos t\right) \sin \frac{t}{2} , \right.$$

$$\left. \frac{a}{2} \sin t - \lambda \left(1 + \cos t\right) \cos \frac{t}{2}, \ a \sin \frac{t}{2} + 2\lambda \right).$$

The principal normal:

$$\mathbf{r} = \left(\frac{a}{2} \left(1 + \cos t\right) - \lambda \left(2 \cos t + \left(1 + \cos t\right) \cos^2 \frac{t}{2} \right), \right.$$

$$\left. \frac{a}{2} \sin t - \frac{\lambda}{2} \left(6 + \cos t\right) \sin t, \ a \sin \frac{t}{2} - \lambda \sin \frac{t}{2} \right).$$

The osculating plane:

$$x \left(2 + \cos t\right) \sin \frac{t}{2} - y \left(1 + \cos t\right) \cos \frac{t}{2} + 2z - \frac{a}{2} \left(5 + \cos t\right) \sin \frac{t}{2} = 0.$$

$$\mathbf{v} = \frac{\left(- \sin t, \cos t, \cos \left(t/2\right)\right)}{\sqrt{1 + \cos^2 \left(t/2\right)}},$$

$$\mathbf{n} = \frac{\left(-\cos^2 t - 6 \cos t - 1, - \sin t \left(6 + \cos t\right), -2 \sin \left(t/2\right)\right)}{\sqrt{\left(\cos^2 t + 6 \cos t + 1\right)^2 + \sin^2 t \left(6 + \cos t\right)^2 + 4 \sin^2 \left(t/2\right)}},$$

$$\mathbf{b} = \sqrt{\frac{2}{13 + 3 \cos t}} \left(\left(2 + \cos t\right) \sin \frac{t}{2} - \left(1 + \cos t\right) \cos \frac{t}{2}, 2 \right);$$

$$k = \frac{1}{a} \sqrt{\frac{13 + 3 \cos t}{2 \left(1 + \cos^2 \left(t/2\right)\right)^3}}, \ \varkappa = \frac{12 \cos \left(t/2\right)}{a \left(13 + 3 \cos t\right)}.$$

4.33. Parameterization of the equation of the curve: $\mathbf{r}\left(t\right) = \left(\sqrt{2a}t, \sqrt{2b}t, t^2 \right)$. This means that the curve lies in the plane $x\sqrt{b} - y\sqrt{a} = 0$.

4.34. Consider the function

$$\varphi\left(s\right) = \langle \mathbf{b}\left(s_0\right), \mathbf{r}\left(s\right) - \mathbf{r}\left(s_0\right) \rangle,$$

where $\mathbf{r}\left(s_0\right) = M$. We have:

$$\varphi'\left(s\right) = \langle \mathbf{b}\left(s_0\right), \mathbf{v}\left(s\right) \rangle, \quad \varphi''\left(s\right) = \langle \mathbf{b}\left(s_0\right), k\left(s\right) \mathbf{n}\left(s\right) \rangle,$$

$$\varphi'''\left(s\right) = \langle \mathbf{b}\left(s_0\right), k'\left(s\right) \mathbf{n}\left(s\right) - k\left(s\right) \left(-k\left(s\right) \mathbf{v}\left(s\right) + \varkappa\left(s\right) b\left(s\right)\right) \rangle.$$

Thus,

$$\varphi\left(s_0\right) = \varphi'\left(s_0\right) = \varphi''\left(s_0\right) = 0, \quad \varphi'''\left(s_0\right) = -k\left(s_0\right) \varkappa\left(s_0\right) \neq 0$$

and, therefore, $\varphi\left(s\right)$ changes its sign in transition of s through s_0, which proves the proposition.

4.35. Let $k \neq 0$, i.e., the curve be biregular. Let all osculating planes pass through the point $\mathbf{r}_0$. We have:

$$\langle \mathbf{b}, \mathbf{r}_0 - \mathbf{r} \rangle = 0, \quad \langle -\varkappa \mathbf{n}, \mathbf{r}_0 - \mathbf{r} \rangle - \langle \mathbf{b}, \mathbf{v} \rangle = 0, \quad \varkappa \langle \mathbf{n}, \mathbf{r}_0 - \mathbf{r} \rangle = 0.$$

If $\varkappa \neq 0$, then

$$\langle \mathbf{n}, \mathbf{r}_0 - \mathbf{r} \rangle = 0, \quad \langle -k\mathbf{v} + \varkappa \mathbf{b}, \mathbf{r}_0 - \mathbf{r} \rangle - \langle \mathbf{n}, \mathbf{v} \rangle = 0,$$

$$\langle \mathbf{v}, \mathbf{r}_0 - \mathbf{r} \rangle = 0, \quad \langle k\mathbf{n}, \mathbf{r}_0 - \mathbf{r} \rangle - \langle \mathbf{v}, \mathbf{v} \rangle = 0, \quad \langle k\mathbf{n}, \mathbf{r}_0 - \mathbf{r} \rangle = 1.$$

Contradiction: This means that $\varkappa = 0$, the curve is plane, see Problem 4.43.

4.36. $\dfrac{d}{ds}\mathbf{r} = \mathbf{v}, \quad \dfrac{d^2}{ds^2}\mathbf{r} = k\mathbf{n}, \quad \dfrac{d^3}{ds^3}\mathbf{r} = -k^2\mathbf{v} + \dot{k}\mathbf{n} + k\varkappa\mathbf{b}.$

4.37. $\dot{\mathbf{b}} = -\varkappa\mathbf{n}$, wherefrom follows the required equality.

4.38. $\varkappa^5\dfrac{d}{ds}\left(\dfrac{k}{\varkappa}\right).$

4.39. See the answer to Problem 4.36. We have:

$$\mathbf{v}' = \mathbf{r}'' = k\mathbf{n}, \quad \mathbf{v}'' = \mathbf{r}''' = -k^2\mathbf{v} + k'\mathbf{n} + k\varkappa\mathbf{b},$$

$$\mathbf{v}''' = \mathbf{r}^{(4)} = -2kk'\mathbf{v} - k^3\mathbf{n} + k''\mathbf{n} + k'\left(-k\mathbf{v} + \varkappa\mathbf{b}\right) + k'\varkappa\mathbf{b} + k\varkappa'\mathbf{b} - k\varkappa^2\mathbf{n},$$

wherefrom we obtain the required answer.

4.40. $\langle \mathbf{e}, \mathbf{n} \rangle = C, \quad \langle \mathbf{e}, -k\mathbf{v} + \varkappa\mathbf{b} \rangle = 0, \quad \langle \mathbf{e}, \mathbf{v} \rangle - \dfrac{\varkappa}{k}\langle \mathbf{e}, \mathbf{b} \rangle = 0, \quad \langle \mathbf{e}, \mathbf{n} \rangle\,k =$

$\left(\dfrac{\varkappa}{k}\right)'\langle \mathbf{e}, \mathbf{b} \rangle - \dfrac{\varkappa^2}{k}\langle \mathbf{e}, \mathbf{n} \rangle = 0, \quad \langle \mathbf{e}, \mathbf{b} \rangle = C\dfrac{k^2 + \varkappa^2}{k\,(\varkappa/k)'}.$ Differentiating once more, we

obtain the required relation. Note that, in consequence of the above relations, we can consider that

$$\mathbf{e} = \dfrac{\varkappa}{k}\dfrac{k^2 + \varkappa^2}{k\,(\varkappa/k)'}\mathbf{v} + \mathbf{n} + \dfrac{k^2 + \varkappa^2}{k\,(\varkappa/k)'}\mathbf{b}.$$

If the relation $\left(\dfrac{k^2 + \varkappa^2}{k\,(\varkappa/k)'}\right)' + \varkappa = 0$ is satisfied, then this vector is constant. This

constant vector $\mathbf{e}$ forms with the vector $\mathbf{n}$ and angle, the cosine of which is equal to $1/|\mathbf{e}| = $ const.

4.41. Let the curve be biregular. Then $\langle \mathbf{e}, \mathbf{v} \rangle = 0, \; k\langle \mathbf{e}, \mathbf{n} \rangle = 0$, hence $\langle \mathbf{e}, \mathbf{n} \rangle = 0$. If $\langle \mathbf{e}, \mathbf{n} \rangle = 0$, then $\langle \mathbf{e}, -k\mathbf{v} + \varkappa\mathbf{b} \rangle = 0$, wherefrom $\varkappa = 0$ (the curve is plane).

4.42. Let the curve be biregular. Then $\langle \mathbf{e}, \mathbf{b} \rangle = 0, \; \varkappa\langle \mathbf{e}, \mathbf{n} \rangle = 0$; hence, $\varkappa = 0$. As, if it would have been $\varkappa \neq 0$, then $\langle \mathbf{e}, \mathbf{n} \rangle = 0, \; \langle \mathbf{e}, -k\mathbf{v} + \varkappa\mathbf{b} \rangle = 0, \; k\langle \mathbf{e}, \mathbf{v} \rangle = 0$, but $\langle \mathbf{e}, \mathbf{v} \rangle \neq 0$, which means that $k = 0$ is a straight line.

4.43. (a) $\dot{\mathbf{v}} = k\mathbf{n} = 0, \; \mathbf{v} = $ const, $\mathbf{r} = \mathbf{r}_0 + \mathbf{v}s$ is a straight line.

(b) $\langle \mathbf{r}, \mathbf{b} \rangle' = \langle \mathbf{v}, \mathbf{b} \rangle = 0, \; \langle \mathbf{r}, \mathbf{b} \rangle = \langle \mathbf{r}_0, \mathbf{b} \rangle$ is a plane line;

(c) $\dot{\mathbf{b}} = -\varkappa\mathbf{n} = 0, \; \mathbf{b} = $ const.

4.44. $\langle \mathbf{b}, \mathbf{e} \rangle = $ const, $\varkappa\langle \mathbf{e}, \mathbf{n} \rangle = 0$. Further, see the solution of Problem 4.42.

4.45. (a) $f(t) = C_1 e^{6t} + C_2 e^{-t} + C_3;$

(b) $f(t) = C_1 \sin t + C_2 \cos t + C_3.$

4.47. (a) $\dfrac{d\bar{s}}{ds} = \left|\dfrac{d\mathbf{v}}{ds}\right| = |k\mathbf{n}| = k;$

(b) $k \neq 0$ is a biregular line;

(c) let $\mathbf{r}' = \mathbf{r}'(s) = \mathbf{v}(s)$ be a tangent spherical image of a closed curve $\mathbf{r} = \mathbf{r}(s)$. Then $\mathbf{r}' = \mathbf{r}'(s)$ is a closed curve of length $\int |\mathbf{v}'(s)|\, ds = \int k\, ds$, lying on a sphere of unit radius.

M e t h o d I. Assume that $\int k\, ds < 2\pi$, i.e., the curve $\mathbf{v}$ has the length smaller than the length of a large circle. It could be shown that this curve $\mathbf{v}$ totally lies in some half-sphere. This means that for a suitable non-zero vector $\mathbf{e}$ we have $\langle \mathbf{v}(s), \mathbf{e} \rangle > 0$. As the curve $\mathbf{r}(s)$ is closed, then $\int \mathbf{v}(s)\, ds = \int \mathbf{r}'(s)\, ds = 0$, whence, $\int \langle \mathbf{v}(s), \mathbf{e} \rangle\, ds = 0$ is a contradiction.

M e t h o d I I. As $\int \mathbf{v}(s)\, ds = 0$ and the curve $\mathbf{v}(s)$ is compact, the point O belongs to the convex envelope conv($\mathbf{v}$) of the curve $\mathbf{v}(s)$. We show that $\int |\mathbf{v}'(s)|\, ds \geq 2\pi$ for any closed piecewise regular curve $\mathbf{v}(s)$ lying on a sphere of radius 1 and such that $O \in \text{conv}(\mathbf{v})$. If O is an inner point of this convex envelope, it is easy to construct a new piecewise regular closed curve $\tilde{\mathbf{v}}(s)$, whose length is smaller than the length of $\mathbf{v}(s)$, such that $O \in \text{conv}(\tilde{\mathbf{v}})$, but O is not an inner point of conv($\tilde{\mathbf{v}}$). If O is not an inner point of conv($\tilde{\mathbf{v}}$), we put $\tilde{\mathbf{v}} = \mathbf{v}$. We show that the length of the curve $\tilde{\mathbf{v}}$ is not less than 2π. As $O \in \text{conv}(\tilde{\mathbf{v}})$, we will find $0 \leq s_0 < s_1 < s_2 < s_3 \leq l$, such that $O \in \Delta = \text{conv}(\tilde{\mathbf{v}}(s_0), \tilde{\mathbf{v}}(s_1), \tilde{\mathbf{v}}(s_2), \tilde{\mathbf{v}}(s_3))$. As the point O is not an inner point of the tetrahedron Δ, it belongs to one of the faces of Δ, say, the face $\Delta_0 = \text{conv}(\tilde{\mathbf{v}}(s_1), \tilde{\mathbf{v}}(s_2), \tilde{\mathbf{v}}(s_3))$. Consider the plane passing through the points $\tilde{\mathbf{v}}(s_1), \tilde{\mathbf{v}}(s_2), \tilde{\mathbf{v}}(s_3)$ and O. It intersects the sphere by the large circumference and, therefore, the length of its arc enclosed between the points $\tilde{\mathbf{v}}(s_i)$ and $\tilde{\mathbf{v}}(s_j)$ is not larger that the length of any curve connecting these points. In particular, the sum of all three arcs is not greater that the length of the curve $\tilde{\mathbf{v}}(s)$. This proves the inequality $\int |\tilde{\mathbf{v}}'(s)|\, ds \geq 2\pi$.

4.48. $\dfrac{ds^*}{ds} = \left| \dfrac{d\mathbf{n}^*}{ds} \right| = |-k\mathbf{v} + \varkappa \mathbf{b}| = \sqrt{k^2 + \varkappa^2}, \quad \left| \dfrac{d\mathbf{b}}{ds} \right| = |-\varkappa \mathbf{n}| = |\varkappa^2|.$

4.49. Let $\mathbf{r}(s)$ be a spherical curve. We have:
$$\langle \mathbf{r} - \mathbf{m}, \mathbf{r} - \mathbf{m} \rangle = R^2, \quad \langle \mathbf{v}, \mathbf{r} - \mathbf{m} \rangle = 0, \quad \langle k\mathbf{n}, \mathbf{r} - \mathbf{m} \rangle + \langle \mathbf{v}, \mathbf{v} \rangle = 0,$$
so
$$\langle \mathbf{n}, \mathbf{r} - \mathbf{m} \rangle = -\frac{1}{k}, \quad \left\langle \dot{k}\mathbf{n} - k^2 \mathbf{v} + k\varkappa \mathbf{b}, \mathbf{r} - \mathbf{m} \right\rangle + \langle k\mathbf{n}, \mathbf{v} \rangle = 0,$$

$$\langle \mathbf{b}, \mathbf{r} - \mathbf{m} \rangle = -\frac{\dot{k}}{k^2 \varkappa}, \quad \langle -\varkappa \mathbf{n}, \mathbf{r} - \mathbf{m} \rangle + \langle \mathbf{b}, \mathbf{v} \rangle = \frac{d}{ds}\left(\frac{\dot{k}}{k^2 \varkappa} \right).$$

Hence, we obtain the required relation $\dfrac{\varkappa}{k} = \dfrac{d}{ds}\left(\dfrac{\dot{k}}{k^2 \varkappa} \right)$. If this relation is satisfied, then the vector $\mathbf{m} = \mathbf{r} + \dfrac{1}{k}\mathbf{n} - \dfrac{\dot{k}}{k^2 \varkappa}\mathbf{b}$ is constant. As the vector $\mathbf{r} - \mathbf{m}$ is orthogonal to the vector $\mathbf{v}$, its length is constant, i.e., $\mathbf{r}(s)$ is a spherical curve.

4.50. Let $\mathbf{r}(s)$ be a curve γ parameterized by a natural parameter. Parameterization of the curve γ^* has the form $\mathbf{r}^*(s) = \mathbf{r}(s) + \dfrac{1}{k}\mathbf{n}(s)$. We have:

$$\frac{d}{ds}\mathbf{r}^* = \mathbf{v} + \frac{1}{k}(-k\mathbf{v} + \varkappa \mathbf{b}) = \frac{\varkappa}{k}\mathbf{b}, \quad \frac{d^2}{ds^2}\mathbf{r}^* = \frac{\dot{\varkappa}}{k}\mathbf{b} - \frac{\varkappa^2}{k}\mathbf{n},$$

this means that $\left[\dfrac{d}{ds}\mathbf{r}^*, \dfrac{d^2}{ds^2}\mathbf{r}^* \right] = \dfrac{\varkappa^3}{k^2}\mathbf{v}\,;$

$$\frac{d^3}{ds^3}\mathbf{r}^* = \left(\frac{\dot{\varkappa}}{k} \right)^{\!\cdot} \mathbf{b} - \frac{\dot{\varkappa}\varkappa}{k}\mathbf{n} - \left(\frac{\varkappa^2}{k} \right)^{\!\cdot} \mathbf{n} - \frac{\varkappa^2}{k}(-k\mathbf{v} + \varkappa \mathbf{b}).$$

We obtain $\left(\dfrac{d}{ds}\mathbf{r}^*, \dfrac{d^2}{ds^2}\mathbf{r}^*, \dfrac{d^3}{ds^3}\mathbf{r}^* \right) = \dfrac{\varkappa^3}{k^2}\mathbf{v}$. The curvature and torsion:

$$k^* = \frac{\left| \left[\dfrac{d}{ds}\mathbf{r}^*, \dfrac{d^2}{ds^2}\mathbf{r}^* \right] \right|}{\left| \dfrac{d}{ds}\mathbf{r}^* \right|^3} = k, \quad \varkappa^* = \frac{\left(\dfrac{d}{ds}\mathbf{r}^*, \dfrac{d^2}{ds^2}\mathbf{r}^*, \dfrac{d^3}{ds^3}\mathbf{r}^* \right)}{\left| \left[\dfrac{d}{ds}\mathbf{r}^*, \dfrac{d^2}{ds^2}\mathbf{r}^* \right] \right|^2} = \frac{k^2}{\varkappa}.$$

4.51. $c(s) = \langle \mathbf{r}(s) - \mathbf{m}, \mathbf{r}(s) - \mathbf{m} \rangle$,

$c'(s) = \langle \mathbf{v}(s), \mathbf{r}(s) - \mathbf{m} \rangle$,

$c''(s) = k(s)\langle \mathbf{n}(s), \mathbf{r}(s) - \mathbf{m} \rangle + 1$,

$c'''(s) = \langle k'(s)\mathbf{n}(s) - k(s)^2 \mathbf{v}(s) + k(s)\varkappa(s)\mathbf{b}(s), \mathbf{r}(s) - \mathbf{m} \rangle$.

4.53. See the solution of Problem 4.51. The condition $c'(s_0) = c''(s_0) = 0$ is equivalent to the following:

$$\mathbf{r}(s_0) - \mathbf{m} = -\frac{1}{k(s_0)}\mathbf{n}(s_0) - \lambda \mathbf{b}(s_0),$$

where λ is an arbitrary number.

4.54. See the solution of Problem 4.53. The condition $c'(s_0) = c''(s_0) = c'''(s_0) = 0$ is equivalent to the following:

$$\mathbf{r}(s_0) - \mathbf{m} = -\frac{1}{k(s_0)}\mathbf{n}(s_0) + \frac{\dot{k}(s_0)}{k^2(s_0)\varkappa(s_0)}\mathbf{b}(s_0).$$

The relation for the radius R^* follows from the condition $\sqrt{c(s_0)} = R^*$.

4.55. Let $\mathbf{r} = \mathbf{r}(s)$ be a closed curve on a sphere. In consequence of Problem 4.53, the curvature of the curve is everywhere positive. If $\varkappa \neq 0$, then the function $\dfrac{\varkappa(s)}{k(s)}$ does not change its sign, and then, by Problem 4.49, the function $\dfrac{1}{\varkappa k^2}\dfrac{dk}{ds}$ is strictly monotonous, which contradicts the closedness of the curve $\mathbf{r} = \mathbf{r}(s)$. Thus, there exists a point of the curve, in which the torsion is equal to 0.

———

5.1. (a) $a^2\left(\cos^2 v\, du^2 + dv^2\right)$;

(b) $\left(a^2\sin^2 u + b^2\cos^2 u\right)\cos^2 v\, du^2 + 2\left(a^2 - b^2\right)\sin u \cos u \sin v \cos v\, du\, dv +$
$$\left(\left(a^2\cos^2 u + b^2\sin^2 u\right)\sin^2 v + c^2\cos^2 v\right)dv^2;$$

(c) $v^2\left(a^2\sin^2 u + b^2\cos^2 u\right)du^2 + 2\left(b^2 - a^2\right)\sin u \cos u\, du\, dv +$
$$\left(a^2\cos^2 u + b^2\sin^2 u + c^2\right)dv^2;$$

(d) $\left(a^2\sin^2 u + b^2\cos^2 u\right)du^2 + c^2 dv^2$.

5.2. (a) $ds^2 + 2\langle \mathbf{v}, \mathbf{e}\rangle ds\, d\lambda + \langle \mathbf{e}, \mathbf{e}\rangle d\lambda^2$;

(b) $v^2 ds^2 + 2v\langle \mathbf{v}, \boldsymbol{\rho}\rangle ds\, dv + \langle \boldsymbol{\rho}, \boldsymbol{\rho}\rangle dv^2$;

(c) $\left|\mathbf{v} + \lambda\dfrac{d\mathbf{e}}{ds}\right|^2 ds^2 + 2\langle \mathbf{e}, \mathbf{v}\rangle ds\, d\lambda - d\lambda^2$;

(d) $\left((1 - k\cos\varphi)^2 + \varkappa^2\right)ds^2 + 2\varkappa\, ds\, d\varphi + d\varphi^2$;

(e) $\varphi^2 du^2 + \left(\varphi'^2 + \psi'^2\right)dv^2$; (f) $(a + b\cos v)^2 du^2 + b^2 dv^2$;

(g) $du^2 + \left(u^2 + a^2\right)dv^2$; (h) $\left((1 - \lambda k)^2 + \varkappa^2\lambda^2\right)ds^2 + d\lambda^2$;

(i) $\left(1 + \lambda^2\varkappa^2\right)ds^2 + d\lambda^2$; (j) $a^2\cosh^2\dfrac{z}{a}d\varphi^2 + \sinh^2\dfrac{z}{a}dz^2$.

5.3. $a^2\cot^2 u\, du^2 + a^2\sin^2 u\, dv^2$.

5.4. $\cos\theta = \dfrac{2}{3}$.

5.5. $\cos\theta = \dfrac{7}{9}$.

5.6. $\cos\theta = \dfrac{1-a^2}{1+a^2}$.

5.7. $\cos\theta = -\dfrac{\sqrt{7}}{\sqrt{11}}$.

5.8. $l = \pi R$.

5.9. (a) $l = \pi R$; (b) $l_\alpha = 2\pi R\sqrt{\alpha}$; $\theta = \dfrac{\pi}{2}$.

5.10. $l = \pi R$.

5.11. $l = 2\pi a$; $\theta = \dfrac{\pi}{2}$.

5.12. $\cos\theta = \dfrac{2-4a^2}{\sqrt{(4+a^2)(1+16a^2)}}$.

5.13. $\pi/2$.

5.14. $v = \pm\left(\sqrt{u^2+1} + \ln\dfrac{\sqrt{u^2+1}-1}{u}\right) + \text{const}$.

5.15. $v = \dfrac{\tan\theta}{a}\cdot\ln\left(u+\sqrt{u^2-a^2}\right) + \text{const}$.

5.16. Two families of curves:

(a) $y^2 + z^2 = \text{const}$, $xy = az$;

(b) $x^2 + z^2 = \text{const}$, $xy = az$.

5.17. Let the equation of the sphere have the form:

$$\mathbf{r} = (a\cos v\cos u,\, a\cos v\sin u,\, a\sin v).$$

The equation of the loxodrome:

$$u = \tan\theta\cdot\ln\left(\tan\left(\frac{\pi}{4}+\frac{v}{2}\right)\right),$$

where θ is the given angle,

$$\mathbf{v} = \cos\theta\,(-\sin v\cos u - \sin u\tan\theta,\ -\sin v\sin u + \cos u\tan\theta,\ \cos v),$$

$$\mathbf{n} = \frac{\cos\theta}{\sqrt{1+\tan^2\theta/\cos^2 v}}\left(-\cos u\cos v + \tan v\sin u\tan\theta - \frac{\cos u}{\cos v}\tan^2\theta,\right.$$

$$\left.-\sin u\cos v - \tan v\cos u\tan\theta - \frac{\sin u}{\cos v}\tan^2\theta, -\sin v\right),$$

$$\mathbf{b} = \frac{(\sin u,\, -\cos u,\, \tan\theta/\cos v)}{\sqrt{1+\tan^2\theta/\cos^2\theta}};$$

$$k = \frac{\cos\theta}{a}\sqrt{1+\frac{\tan^2\theta}{\cos^2 v}}, \quad \varkappa = \frac{\tan\theta}{a(\cos^2 v + \tan^2\theta)}.$$

5.19. (a) The curves $u = av^2/2$, $u = -av^2/2$ and $v = 1$ intersect at points $A\,(u = 0, v = 0)$, $B\,(u = a/2, v = 1)$, $C\,(u = -a/2, v = 1)$. Herewith, the differentials of the curvilinear coordinates at these curves are related as: $du = av\,dv$, $du = -av\,dv$, $dv = 0$

on the curve AB $\left(\text{the equation } u = \dfrac{av^2}{2}\right)$;

on the curve AC $\left(\text{the equation } u = -\dfrac{av^2}{2}\right)$;

on the curve BC (the equation $v = 1$).

Substituting these relations into the first quadratic form, we obtain:

on the curve AB $\quad ds^2 = a^2\left(\dfrac{v^4}{4} + v^2 + 1\right)dv^2$, $\quad ds = \left(\dfrac{v^2}{2} + 1\right)dv$;

on the curve AC $\quad ds^2 = a^2\left(\dfrac{v^4}{4} + v^2 + 1\right)dv^2$, $\quad ds = \left(\dfrac{v^2}{2} + 1\right)dv$;

on the curve BC $\quad ds^2 = dv^2$, $\qquad\qquad\qquad\quad ds = du$.

It remains to take the integral within the limits defined by the coordinates of the points A, B, C:

$$AB = AC = a\int\limits_{v=0}^{v=1}\left(\dfrac{v^2}{2}+1\right)dv = \dfrac{7a}{6}, \quad CB = \int\limits_{u=-a/2}^{u=a/2} du = a.$$

Thus, the perimeter of the triangle is equal to $\dfrac{10a}{3}$;

(b) $\cos A = 1$, $\cos B = \cos C = \dfrac{2}{3}$, i.e., $A = 0$, $B = C = \arccos\dfrac{2}{3}$;

(c) $S = a^2\left(\dfrac{2}{3} - \dfrac{\sqrt{2}}{3} + \ln\left(1 + \sqrt{2}\right)\right)$.

5.20. (a) $(v_0^2 + \sinh^2 v_0)/4$; (b) v_0, $\sinh v_0$, $\sqrt{2}\sinh v_0$; (c) $\pi/2$, $\pi/4$, $\pi/4$.

5.21. $S = 2\alpha R^2$, where R is the radius of the sphere.

5.22. $4\pi^2 rR$.

5.23. Let the image $f : X \to Y$ be given in the local coordinates by the equations $y^1 = y^1\left(x^1, x^2\right)$, $y^2 = y^2\left(x^1, x^2\right)$. Let the quadratic forms $g = g_{ij}dx^i dx^j$ and $h = h_{ij}dx^i dx^j$ define the Riemannian metrics on X and Y, respectively. The mapping f induces on X the Riemannian metric

$$h' = h'_{ij}\left(x\right)dx^i dx^j = h_{kl}\left(g\left(x\right)\right)\dfrac{\partial y^k\left(x\right)}{\partial x^i}\dfrac{\partial y^l\left(x\right)}{\partial x^j}dx^i dx^j.$$

The conformity of the mapping f is equivalent to that the metrics g and h' on X are conformally equivalent, i.e., the angle between the intersecting curves with respect to the metric g is equal to the angle between them with respect to the metric h'. Hence follows the existence of the conformal multiplier $\lambda(x) > 0$, smoothly depending on the point and relating the coefficients of the metrics by the equality $h'_{ij}\left(x\right) = \lambda\left(x\right)g_{ij}\left(x\right)$. The shape of the area on X, induced by the mapping f, is equal to $d\sigma' = \sqrt{\det h'\left(x\right)}dx^1 \wedge dx^2 = \lambda\left(x\right)d\sigma$. As f preserves the area, $d\sigma = d\sigma'$, hence $\lambda = 1$. Therefore, $h' = g$, i.e., f is a local isometry.

5.25. Let $S \ni (x, y, z) \mapsto (x_1, y_1, z_1) \in S$. At a stereographic projection from the north pole to the equatorial plane P with the coordinates ξ, η, we have:

$$\xi = \dfrac{x}{1-z}, \quad \eta = \dfrac{y}{1-z}.$$

and

$$x_1 = \frac{2\xi_1}{1+\xi_1^2+\eta_1^2}, \quad y_1 = \frac{2\eta_1}{1+\xi_1^2+\eta_1^2}, \quad z_1 = \frac{\xi_1^2+\eta_1^2-1}{1+\xi_1^2+\eta_1^2}.$$

As the conformal mapping of the sphere onto itself generates a conformal mapping of the plane P onto itself with preservation of zero and infinity, then $\zeta_1 = \xi_1 + i\eta_1 = a\zeta = a(\xi + i\eta)$, where a is a nonzero complex number. It remains to substitute into the above formula instead of ξ_1 and η_1 their values through $\xi(x,y)$ and $\eta(x,y)$.

5.26. The metric of a hyperbolic paraboloid is given by the matrix-valued function

$$G = \begin{pmatrix} 1+u^2 & -uv \\ -uv & 1+v^2 \end{pmatrix},$$

so the shape of the area is equal to

$$d\sigma = \sqrt{\det G}\,du \wedge dv = \sqrt{1+u^2+v^2}\,du \wedge dv.$$

The deformation induces the metric given by the matrix

$$G' = \begin{pmatrix} 1+(u\sin t + v\cos t)^2 & (u\sin t + v\cos t)(u\cos t - v\sin t) \\ (u\sin t + v\cos t)(u\cos t - v\sin t) & 1+(u\cos t - v\sin t)^2 \end{pmatrix};$$

the induced shape of the area is equal to

$$d\sigma' = \sqrt{\det G'}\,du \wedge dv = \sqrt{1+u^2+v^2}\,du \wedge dv.$$

Thus, $d\sigma = d\sigma'$, i.e., this deformation preserves the area.

5.27. $\mathbf{R}(u,v) = \left(\sqrt{a^2+u^2}\cos v, \sqrt{a^2+u^2}\sin v, a\ln\frac{u+\sqrt{a^2+u^2}}{a} \right)$, or $\mathbf{R}(z,\varphi) = \left(a\cosh\frac{z}{a}\cos\varphi, \cosh\frac{z}{a}\sin\varphi, z \right)$. This is the catenoid (the surface of revolution of the catenary $x = a\cosh\frac{z}{a}$).

5.28. The metric of the conoid: $2d\rho^2 + 2d\rho\,dv + (\rho^2+1)\,dv^2$; the metric of the hyperboloid of revolution: $\frac{2r^2-1}{r^2-1}dr^2 + r^2 d\varphi^2$. It remains to verify that at the given correspondence of points the second metric acquires the shape of the first metric.

5.29. *Hint:* the first of the equations defining the correspondence of the points has the form $r^2 + \rho^2 + a^2$.

5.30. The metric of the helical surface

$$\left(1+F'(u)^2\right)du^2 + 2aF'\,du\,dv + \left(u^2+a^2\right)dv^2.$$

The metric of the surface of revolution $\mathbf{r} = (r\cos\varphi, r\sin\varphi, G(r))$ is equal to

$$\left(1+G'(r)^2\right)dr^2 + r^2 d\varphi^2.$$

Let the correspondence of the points be established by the equations $r^2 = u^2 + a^2$, $\varphi = v + H(u)$. It defines the local isometry, if in the coordinates (u,v) the second metric acquires the shape of the first. Hence, we obtain the condition for the functions G and H:

$$G'\left(r\right)^2 = F'\left(u\right)^2 + \frac{a^2}{u^2}, \quad H'\left(u\right) = \frac{a}{u^2 + a^2}F'\left(u\right).$$

5.32. *Hint:* take the equation of a cylindrical surface in the form $\mathbf{r} = \gamma + v\mathbf{e}$, where $\mathbf{e}$ is a constant vector of unit length, $\gamma(s)$ is a curve parameterized by a natural parameter and lying in the plane orthogonal to vector $\mathbf{e}$. Compare the first quadratic form of this surface with the quadratic form of the plane.

5.33. *Hint:* take the equation of a conical surface in the form $\mathbf{r} = v\mathbf{e}\left(u\right)$, where $\left|\mathbf{e}\left(u\right)\right| = 1$, and compare its first quadratic form with the quadratic form of a plane in polar coordinates.

5.34. No, as segments of the curves of the same length should correspond to the edges of the cylindrical surface and the respective domain D on the conical surface on their development, which is impossible to be done for the development of a cone.

––––––––––

6.1. (a) We have:

$$\mathbf{r}_u = \left(-R\sin u \cos v, -R\sin u \sin v, R\cos u\right),$$
$$\mathbf{r}_v = \left(-R\cos u \sin v, R\cos u \cos v, 0\right),$$
$$\mathbf{r}_u \times \mathbf{r}_v = \left(-R^2 \cos^2 u \cos v, -R^2 \cos^2 u \sin v, -R^2 \cos u \sin u\right),$$
$$\left|\mathbf{r}_u \times \mathbf{r}_v\right| = R^2 \cos u,$$
$$\mathbf{n} = \frac{\mathbf{r}_u \times \mathbf{r}_v}{\left|\mathbf{r}_u \times \mathbf{r}_v\right|} = \left(-\cos u \cos v, -\cos u \sin v, -\sin u\right),$$
$$\mathbf{r}_{uu} = \left(-R\cos u \cos v, -R\cos u \sin v, -R\sin u\right),$$
$$\mathbf{r}_{uv} = \left(R\sin u \sin v, -R\sin u \cos v, 0\right),$$
$$\mathbf{r}_{vv} = \left(-R\cos u \cos v, -R\cos u \sin v, 0\right),$$

$$L = \langle\mathbf{r}_{uu}, \mathbf{n}\rangle = R\left(\cos^2 u \cos^2 v + \cos^2 u \sin^2 v + \sin^2 u\right) = R,$$

$$M = \langle\mathbf{r}_{uv}, \mathbf{n}\rangle = R\left(-\cos u \cos v \sin u \sin v + \left(-\cos u \sin v\right)\left(-\sin u \cos v\right)\right) = 0,$$

$$N = \langle\mathbf{r}_{vv}, \mathbf{n}\rangle = R\left(\left(-\cos u \cos v\right)\left(-\cos u \cos v\right) + \right.$$
$$\left.\left(-\cos u \sin v\right)\left(-\cos u \sin v\right)\right) = R\cos^2 u.$$

Answer: $R\,du^2 + R\cos^2 u\,dv^2$.

(b) We have:

$$\mathbf{r}_u = \left(-a\sin u \cos v, -a\sin u \sin v, c\cos u\right),$$
$$\mathbf{r}_v = \left(-a\cos u \sin v, a\cos u \cos v, 0\right),$$
$$\mathbf{r}_u \times \mathbf{r}_v = \left(-ac\cos^2 u \cos v, -ac\cos^2 u \sin v, -a^2 \cos u \sin u\right),$$
$$\left|\mathbf{r}_u \times \mathbf{r}_v\right| = a\cos u\sqrt{c^2 \cos^2 u + a^2 \sin^2 u},$$
$$\mathbf{n} = \frac{\mathbf{r}_u \times \mathbf{r}_v}{\left|\mathbf{r}_u \times \mathbf{r}_v\right|} = \frac{1}{\sqrt{c^2 \cos^2 u + a^2 \sin^2 u}}\left(-c\cos u \cos v, -c\cos u \sin v, -a\sin u\right),$$

$$\mathbf{r}_{uu} = (-a\cos u\cos v, -a\cos u\sin v, -c\sin u),$$
$$\mathbf{r}_{uv} = (a\sin u\sin v, -a\sin u\cos v, 0),$$
$$\mathbf{r}_{vv} = (-a\cos u\cos v, -a\cos u\sin v, 0),$$

$$L = \langle \mathbf{r}_{uu}, \mathbf{n} \rangle = \frac{1}{\sqrt{c^2\cos^2 u + a^2\sin^2 u}} ac\left(\cos^2 u\cos^2 v + \cos^2 u\sin^2 v + \sin^2 u\right) =$$

$$\frac{ac}{\sqrt{c^2\cos^2 u + a^2\sin^2 u}},$$

$$M = \langle \mathbf{r}_{uv}, \mathbf{n} \rangle = \frac{1}{\sqrt{c^2\cos^2 u + a^2\sin^2 u}}(-c\cos u\cos v\, a\sin u\sin v +$$

$$(-c\cos u\sin v)(-a\sin u\cos v)) = 0,$$

$$N = \langle \mathbf{r}_{vv}, \mathbf{n} \rangle = \frac{1}{\sqrt{c^2\cos^2 u + a^2\sin^2 u}}((-c\cos u\cos v)(-a\cos u\cos v) +$$

$$(-c\cos u\sin v)(-a\cos u\sin v)) = \frac{ac\cos^2 u}{\sqrt{c^2\cos^2 u + a^2\sin^2 u}}.$$

Answer: $\dfrac{ac}{\sqrt{c^2\cos^2 u + a^2\sin^2 u}}\left(du^2 + \cos^2 u\, dv^2\right).$

(c) We have:

$$\mathbf{r}_u = (-b\sin u\cos v, -b\sin u\sin v, b\cos u),$$
$$\mathbf{r}_v = (-(a+b\cos u)\sin v, (a+b\cos u)\cos v, 0),$$
$$\mathbf{r}_u \times \mathbf{r}_v = b(a+b\cos u)(-\cos u\cos v, -\cos u\sin v, -\sin u),$$
$$|\mathbf{r}_u \times \mathbf{r}_v| = b(a+b\cos u),$$
$$\mathbf{n} = \frac{\mathbf{r}_u \times \mathbf{r}_v}{|\mathbf{r}_u \times \mathbf{r}_v|} = (-\cos u\cos v, -\cos u\sin v, -\sin u),$$
$$\mathbf{r}_{uu} = (-b\cos u\cos v, -b\cos u\sin v, -b\sin u),$$
$$\mathbf{r}_{uv} = (b\sin u\sin v, -b\sin u\cos v, 0),$$
$$\mathbf{r}_{vv} = (-(a+b\cos u)\cos v, -(a+b\cos u)\sin v, 0),$$

$$L = \langle \mathbf{r}_{uu}, \mathbf{n} \rangle = b\left(\cos^2 u\cos^2 v + \cos^2 u\sin^2 v + \sin^2 u\right) = b,$$

$$M = \langle \mathbf{r}_{uv}, \mathbf{n} \rangle = -\cos u\cos v\, b\sin u\sin v + (-\cos u\sin v)(-b\sin u\cos v) = 0,$$

$$N = \langle \mathbf{r}_{vv}, \mathbf{n} \rangle = (-\cos u\cos v)(-(a+b\cos u)\cos v) +$$

$$(-\cos u\sin v)(-(a+b\cos u)\sin v) = (a+b\cos u)\cos u.$$

Answer: $b\, du^2 + (a+b\cos u)\cos u\, dv^2.$

(d) We have:

$$\mathbf{r}_u = \left(\sinh \frac{u}{a} \cos v, \sinh \frac{u}{a} \sin v, 1 \right),$$

$$\mathbf{r}_v = \left(-a \cosh \frac{u}{a} \sin v, a \cosh \frac{u}{a} \cos v, 0 \right),$$

$$\mathbf{r}_u \times \mathbf{r}_v = \left(-a \cosh \frac{u}{a} \cos v, -a \cosh \frac{u}{a} \sin v, a \cosh \frac{u}{a}, \sinh \frac{u}{a} \right),$$

$$|\mathbf{r}_u \times \mathbf{r}_v| = a \cosh \frac{u}{a} \sqrt{\cos^2 v + \sin^2 v + \sinh^2 \frac{u}{a}} = a \cosh^2 \frac{u}{a},$$

$$\mathbf{n} = \frac{\mathbf{r}_u \times \mathbf{r}_v}{|\mathbf{r}_u \times \mathbf{r}_v|} = \frac{1}{\cosh(u/a)} \left(-\cos v, -\sin v, \sinh \frac{u}{a} \right),$$

$$\mathbf{r}_{uu} = \left(\frac{1}{a} \cosh \frac{u}{a} \cos v, \frac{1}{a} \cosh \frac{u}{a} \sin v, 0 \right),$$

$$\mathbf{r}_{uv} = \left(-\sinh \frac{u}{a} \sin v, \sinh \frac{u}{a} \cos v, 0 \right),$$

$$\mathbf{r}_{vv} = \left(-a \cosh \frac{u}{a} \cos v, -a \cosh \frac{u}{a} \sin v, 0 \right),$$

$$L = \langle \mathbf{r}_{uu}, \mathbf{n} \rangle = \frac{1}{a \cosh(u/a)} \left(-\cosh \frac{u}{a} \cos^2 v - \cosh \frac{u}{a} \sin^2 v \right) = -\frac{1}{a},$$

$$M = \langle \mathbf{r}_{uv}, \mathbf{n} \rangle =$$
$$\frac{1}{\cosh(u/a)} \left((-\cos v)\left(-\sinh \frac{u}{a} \sin v \right) + (-\sin v)\left(\sinh \frac{u}{a} \cos v \right) \right) = 0,$$

$$N = \langle \mathbf{r}_{vv}, \mathbf{n} \rangle =$$
$$\frac{1}{\cosh(u/a)} \left((-\cos v)\left(-a \cosh \frac{u}{a} \cos v \right) + (-\sin v)\left(a \cosh \frac{u}{a} \sin v \right) \right) = a,$$

Answer: $-\frac{1}{a} du^2 + a\, dv^2$.

(e) We have:

$$\mathbf{r}_u = \left(a \cos u \cos v, a \cos u \sin v, -a \sin u + \frac{a}{\sin u} \right),$$

$$\mathbf{r}_v = (-a \sin u \sin v, a \sin u \cos v, 0),$$

$$\mathbf{r}_u \times \mathbf{r}_v = \left(a^2 \sin^2 u \cos v - a^2 \cos v, a^2 \sin^2 u \sin v - a^2 \sin v, a^2 \cos u \sin u \right) =$$
$$\left(-a^2 \cos^2 u \cos v, -a^2 \cos^2 u \sin v, a^2 \cos u \sin u \right),$$

$$|\mathbf{r}_u \times \mathbf{r}_v| = a^2 \cos u \sqrt{\cos^2 u \cos^2 u + \cos^2 u \sin^2 u + \sin^2 u} = a^2 \cos u,$$

$$\mathbf{n} = \frac{\mathbf{r}_u \times \mathbf{r}_v}{|\mathbf{r}_u \times \mathbf{r}_v|} = (-\cos u \cos v, -\cos u \sin v, \sin u),$$

$$\mathbf{r}_{uu} = \left(-a \sin u \cos v, -a \sin u \sin v, -a \cos u - \frac{a \cos u}{\sin^2 u} \right),$$

$$\mathbf{r}_{uv} = (-a \cos u \sin v, a \cos u \cos v, 0),$$

$$\mathbf{r}_{vv} = (-a \sin u \cos v, -a \sin u \sin v, 0),$$

$L = \langle \mathbf{r}_{uu}, \mathbf{n} \rangle =$

$$a \cos u \sin u \cos^2 v + a \cos u \sin u \sin^2 v - a \sin u \cos u - \frac{a \cos u \sin u}{\sin^2 u} = -a \cot u,$$

$M = \langle \mathbf{r}_{uv}, \mathbf{n} \rangle = (-\cos u \cos v)(-a \cos u \sin v) + (-\cos u \sin v) a \cos u \cos v = 0,$

$N = \langle \mathbf{r}_{vv}, \mathbf{n} \rangle =$

$$(-\cos u \cos v)(-a \sin u \cos v) + (-\cos u \sin v)(-a \sin u \sin v) = a \cos u \sin u.$$

Answer: $-a \cot u \, du^2 + a \cos u \sin u \, dv^2$.

(f) We have:

$\mathbf{r}_u = (\cos v, \sin v, 0),$
$\mathbf{r}_v = (-u \sin v, u \cos v, a),$
$\mathbf{r}_u \times \mathbf{r}_v = (a \sin v, -a \cos v, u),$
$|\mathbf{r}_u \times \mathbf{r}_v| = \sqrt{u^2 + a^2},$
$\mathbf{n} = \dfrac{\mathbf{r}_u \times \mathbf{r}_v}{|\mathbf{r}_u \times \mathbf{r}_v|} = \dfrac{(a \sin v, -a \cos v, u)}{\sqrt{u^2 + a^2}},$
$\mathbf{r}_{uu} = (0, 0, 0), \quad \mathbf{r}_{uv} = (-\sin v, \cos v, 0),$
$\mathbf{r}_{vv} = (-u \cos v, -u \sin v, 0),$

$L = \langle \mathbf{r}_{uu}, \mathbf{n} \rangle = 0,$

$$M = \langle \mathbf{r}_{uv}, \mathbf{n} \rangle = \frac{a \sin v (-\sin v) + (-a \cos v) \cos v}{\sqrt{u^2 + a^2}} = -\frac{a}{\sqrt{u^2 + a^2}},$$

$$N = \langle \mathbf{r}_{vv}, \mathbf{n} \rangle = \frac{a \sin v (-u \cos v) + (-a \cos v)(-u \sin v)}{\sqrt{u^2 + a^2}} = 0,$$

Answer: $\dfrac{-2a \, du \, dv}{\sqrt{u^2 + a^2}}$.

(g) Expressing z through the other variables, we have $z = f(x, y)$, where $f(x, y) = \dfrac{a^3}{xy}$. Then

$$f_x = -\frac{a^3}{x^2 y}, \quad f_y = -\frac{a^3}{xy^2}, \quad f_{xx} = \frac{2a^3}{x^3 y}, \quad f_{xy} = \frac{a^3}{x^2 y^2}, \quad f_{yy} = \frac{2a^3}{xy^3}.$$

By the formulas for the second quadratic form of the surface, $z = f(x, y)$, we obtain:

$$\sqrt{1 + f_x^2 + f_y^2} = \frac{\sqrt{x^4 y^4 + a^6 x^2 + a^6 y^2}}{x^2 y^2},$$

$$L = \frac{f_{xx}}{\sqrt{1+f_x^2+f_y^2}} = \frac{2a^3 y}{x\sqrt{x^4 y^4 + a^6 x^2 + a^6 y^2}},$$

$$M = \frac{f_{xy}}{\sqrt{1+f_x^2+f_y^2}} = \frac{a^3}{\sqrt{x^4 y^4 + a^6 x^2 + a^6 y^2}},$$

$$N = \frac{f_{yy}}{\sqrt{1+f_x^2+f_y^2}} = \frac{2a^3 x}{y\sqrt{x^4 y^4 + a^6 x^2 + a^6 y^2}}.$$

Answer: $\dfrac{2a^3}{\sqrt{x^4 y^4 + a^6 x^2 + a^6 y^2}} \left(\dfrac{y}{x} dx^2 + dx\,dy + \dfrac{x}{y} dy^2 \right).$

6.4. (a) $\mathbf{r}_u = (x', \rho' \cos\varphi, \rho' \sin\varphi)$, $\mathbf{r}_\varphi = (0, -\rho \sin\varphi, \rho \cos\varphi)$. Here the prime implies the differentiation by u. The matrix of the first quadratic form

$$\mathbf{I} = \begin{pmatrix} (x')^2 + (\rho')^2 & 0 \\ 0 & \rho^2 \end{pmatrix};$$

$$\mathbf{r}_u \times \mathbf{r}_v = (\rho\rho', -\rho x' \cos\varphi, -\rho x' \sin\varphi), \quad |\mathbf{r}_u \times \mathbf{r}_v| = \rho\sqrt{(x')^2 + (\rho')^2},$$

$$\mathbf{n} = \frac{\mathbf{r}_u \times \mathbf{r}_v}{|\mathbf{r}_u \times \mathbf{r}_v|} = \frac{(\rho', -x' \cos\varphi, -x' \sin\varphi)}{\sqrt{(x')^2 + (\rho')^2}},$$

$$\mathbf{r}_{uu} = (x'', \rho'' \cos\varphi, \rho'' \sin\varphi), \quad \mathbf{r}_{u\varphi} = (0, -\rho' \sin\varphi, \rho' \cos\varphi),$$

$$\mathbf{r}_{\varphi\varphi} = (0, -\rho \cos\varphi, -\rho \sin\varphi), \quad L = \langle \mathbf{r}_{uu}, \mathbf{n} \rangle = \frac{x'' \rho' - x' \rho''}{\sqrt{(x')^2 + (\rho')^2}},$$

$$M = \langle \mathbf{r}_{u\varphi}, \mathbf{n} \rangle = 0, \quad N = \langle \mathbf{r}_{\varphi\varphi}, \mathbf{n} \rangle = \frac{\rho x'}{\sqrt{(x')^2 + (\rho')^2}}.$$

The second quadratic form: $\dfrac{(x'' \rho' - x' \rho'')\,du^2 + \rho x'\,d\varphi^2}{\sqrt{(x')^2 + (\rho')^2}}.$

(b) We find the Gaussian curvature of the surface of revolution:

$$K = \frac{LN - M^2}{EG - F^2} = \frac{\rho x'(x'' \rho' - x' \rho'')}{\rho^2\left((x')^2 + (\rho')^2\right)^2} = \frac{x'(x'' \rho' - x' \rho'')}{\rho\left((x')^2 + (\rho')^2\right)^2}.$$

The direction of the convexity of the meridian is determined by the sign of the expression $\dfrac{d^2\rho}{dx^2}$. Calculating it by the formulas for the derivative of the implicit function, we obtain:

$$\frac{d^2\rho}{dx^2} = \frac{d}{dx}\left(\frac{d\rho}{dx}\right) = \frac{d}{du}\left(\frac{\rho'}{x'}\right)\frac{du}{dx} = \frac{\rho'' x' - \rho' x''}{(x')^2} \frac{1}{x'} = \frac{\rho'' x' - \rho' x''}{(x')^3}.$$

Comparing the obtained expression with the formula for K, we find that $K = -\dfrac{(x')^4}{\rho\left((x')^2 + (\rho')^2\right)^2} \dfrac{d^2\rho}{dx^2}$. Hence, we have that $K > 0$, when $\dfrac{d^2\rho}{dx^2} < 0$, i.e., when the convexity of the meridian is directed away from the axis of revolution; $K < 0$, when the convexity is directed towards the axis of revolution, $K = 0$, if the meridian has a point of inflection or is orthogonal to the axis of revolution $(x' = 0)$.

(c) Let $\rho(u) = u$,

$$x(u) = \pm \left(a \ln \frac{a + \sqrt{a^2 - u^2}}{u} - \sqrt{a^2 - u^2} \right), \quad a > 0.$$

Then $\rho' = 1$, $\rho'' = 0$,

$$x\prime = \pm \left(-\frac{a}{u} + \frac{a}{a + \sqrt{a^2 - u^2}} \frac{-u}{\sqrt{a^2 - u^2}} + \frac{u}{\sqrt{a^2 - u^2}} \right) =$$

$$\pm \left(-\frac{a}{u} + \frac{-au + u\left(a + \sqrt{a^2 - u^2}\right)}{\left(a + \sqrt{a^2 - u^2}\right)\sqrt{a^2 - u^2}} \right) = \pm \left(-\frac{a}{u} + \frac{u}{a + \sqrt{a^2 - u^2}} \right) =$$

$$\pm \left(-\frac{a}{u} + \frac{u}{a^2 - (a^2 - u^2)} \left(a - \sqrt{a^2 - u^2}\right) \right) = \mp \frac{\sqrt{a^2 - u^2}}{u},$$

$$x'' = \pm \frac{1}{\sqrt{a^2 - u^2}} + \frac{\sqrt{a^2 - u^2}}{u^2},$$

$$(x')^2 + (\rho')^2 = (x')^2 + 1 = \frac{a^2 - u^2}{u^2} + 1 = \frac{a^2}{u^2},$$

$$x'x'' = -\left(\frac{1}{u} + \frac{a^2 - u^2}{u^3} \right) = -\frac{a^2}{u^3},$$

$$K = \frac{x'x''}{u\left((x')^2 + 1\right)^2} = \frac{-a^2/u^3}{u\left(a^4/u^4\right)} = -\frac{1}{a^2}.$$

(d) We calculate the mean curvature of the surface of revolution:

$$H = \frac{EN + GL - 2FM}{EG - F^2} = \frac{\rho x'\left((x')^2 + (\rho')^2\right) + \rho^2\left(x''\rho' - x'\rho''\right)}{\rho^2\left((x')^2 + (\rho')^2\right)^{3/2}} =$$

$$\frac{x'\left((x')^2 + (\rho')^2\right) + \rho\left(x''\rho' - x'\rho''\right)}{\rho\left((x')^2 + (\rho')^2\right)^{3/2}}.$$

e) We find the solution of the equation $H = 0$, when $x = u$. Substituting the expression of x into the formula for H, we obtain $H = \frac{1 + (\rho')^2 - \rho\rho''}{\rho\left(1 + (\rho')^2\right)^{3/2}}$. From the condition $H = 0$, we obtain a differential equation $\rho''\rho - (\rho')^2 - 1 = 0$. As $\rho > 0$, then $\rho'' > 0$. After the differentiation of the equation, we obtain $\rho'''\rho + \rho''\rho' - 2\rho''\rho' = 0$. Hence, $\rho'''\rho - \rho''\rho' = 0$, $\frac{\rho'''}{\rho''} = \frac{\rho'}{\rho}$, $(\ln \rho'')' = (\ln \rho)'$. Thus, the problem was reduced to the solution of the differential equation $\rho'' = C\rho$, where $C = k^2 > 0$, $k > 0$ is a constant. The general solution of the obtained differential equation has the form:

$$\rho(u) = A \cosh ku + B \sinh ku.$$

Substituting this solution into the initial equation and collecting like terms, we obtain a condition for the coefficients:

$$1 = \rho''\rho - (\rho')^2 = k^2\left((A^2 - B^2)\cosh^2 ku + (B^2 - A^2)\sinh^2 ku\right),$$

whence $k^2(A^2 - B^2) = 1$. As $\rho(u) > 0$ for all u, $A > 0$. Then A and B can be presented as

$$A = \frac{1}{k}\cosh ku_0, \quad B = -\frac{1}{k}\sinh ku_0;$$

$$\rho(u) = \frac{1}{k}\left(\cosh ku_0 \cosh ku - \sinh ku_0 \sinh ku\right) = \frac{1}{k}\cosh\left(k(u - u_0)\right).$$

Thus, the solution of the equation $H = 0$, when $x = u$, are the functions $\rho(u) = \frac{1}{k}\cosh(k(u - u_0))$, where $k > 0$, u_0 is arbitrary.

6.7. In this problem, the normal to the surface will be denoted by $\mathbf{m}$. Let $\mathbf{v}, \mathbf{n}, \mathbf{b}$ be the Frénet frame of this curve. Then $\mathbf{r}_s = \mathbf{v} + ku\mathbf{n}$, $\mathbf{r}_u = \mathbf{v}$. The matrix of the first quadratic form

$$\mathbf{I} = \begin{pmatrix} 1 + k^2u^2 & 1 \\ 1 & 1 \end{pmatrix};$$

$$\mathbf{r}_s \times \mathbf{r}_u = ku\mathbf{n} \times \mathbf{v} = -ku\mathbf{b}, \ |\mathbf{r}_u \times \mathbf{r}_v| = ku, \ \mathbf{m} = \frac{\mathbf{r}_s \times \mathbf{r}_u}{|\mathbf{r}_s \times \mathbf{r}_u|} = -\mathbf{b},$$

$$\mathbf{r}_{ss} = k\mathbf{n} + k_su\mathbf{n} + ku(-k\mathbf{v} + \varkappa\mathbf{b}) = -k^2u\mathbf{v} + (1 + k_su)\mathbf{n} + uk\varkappa\mathbf{b},$$

$$\mathbf{r}_{su} = k\mathbf{n}, \ \mathbf{r}_{uu} = 0,$$

$L = \langle \mathbf{r}_{ss}, \mathbf{m} \rangle = -kv\varkappa$, $M = \langle \mathbf{r}_{su}, \mathbf{m} \rangle = 0$, $N = \langle \mathbf{r}_{uu}, \mathbf{m} \rangle = 0$, where $\mathbf{m}$ is the normal to the surface.

Hence,

$$K = \frac{LN - M^2}{EG - F^2} = 0,$$

$$H = \frac{EN + GL - 2FM}{EG - F^2} = \frac{-kv\varkappa}{k^2v^2 + 1 - 1} = -\frac{\varkappa}{kv}.$$

6.8. The surface defined by the equation $z = f(x,y)$ can be given parametrically in the form

$$\mathbf{r}(x,y) = (x, y, f(x,y)), \ \mathbf{r}_x = (1, 0, f_x), \ \mathbf{r}_y = (1, 0, f_y).$$

The matrix of the first quadratic form is given by

$$\mathbf{I} = \begin{pmatrix} 1 + f_x^2 & f_xf_y \\ f_xf_y & 1 + f_y^2 \end{pmatrix};$$

$$\mathbf{r}_x \times \mathbf{r}_y = (-f_x, -f_y, 1), \ |\mathbf{r}_x \times \mathbf{r}_y| = \sqrt{1 + f_x^2 + f_y^2},$$

$$\mathbf{n} = \frac{\mathbf{r}_x \times \mathbf{r}_y}{|\mathbf{r}_x \times \mathbf{r}_y|} = \frac{1}{\sqrt{1 + f_x^2 + f_y^2}}(-f_x, -f_y, 1),$$

$$\mathbf{r}_{xx} = (0, 0, f_{xx}), \ \mathbf{r}_{xy} = (0, 0, f_{xy}), \ \mathbf{r}_{yy} = (0, 0, f_{yy}),$$

$$L = \langle \mathbf{r}_{xx}, \mathbf{n} \rangle = \frac{f_{xx}}{\sqrt{1 + f_x^2 + f_y^2}},$$

$$M = \langle \mathbf{r}_{xy}, \mathbf{n} \rangle = \frac{f_{xy}}{\sqrt{1 + f_x^2 + f_y^2}},$$

$$N = \langle \mathbf{r}_{yy}, \mathbf{n} \rangle = \frac{f_{yy}}{\sqrt{1 + f_x^2 + f_y^2}},$$

The matrix of the second quadratic form is given by

$$\mathbf{II} = \frac{1}{\sqrt{1 + f_x^2 + f_y^2}}\begin{pmatrix} f_{xx} & f_{xy} \\ f_{xy} & f_{yy} \end{pmatrix};$$

$$K = \frac{\det \mathbf{II}}{\det \mathbf{I}} = \frac{LN - M^2}{EG - F^2} = \frac{f_{xx}f_{yy} - f_{xy}^2}{(1 + f_x^2 + f_y^2)^2},$$

$$H = \mathrm{tr}\left(\mathrm{II} \cdot \mathrm{I}^{-1}\right) = \frac{EN + GL - 2FM}{EG - F^2} =$$

$$\frac{\left(1 + f_x^2\right) f_{yy} + \left(1 + f_y^2\right) f_{xx} - 2f_x\, f_y\, f_{xy}}{\left(1 + f_x^2 + f_y^2\right)^{3/2}}.$$

6.9. Let $F_z \neq 0$. Then by the implicit function theorem in some neighbourhood the surface can be given by the equation $z = f(x, y)$, where $z_x = f_x = -\dfrac{F_x}{F_z}$, $z_y = f_y = -\dfrac{F_y}{F_z}$. We find the second derivatives of the function f:

$$f_{xx} = -\frac{\partial}{\partial x}\left(\frac{F_x}{F_z}\right) = -\frac{F_{xx} + F_{xz} z_x}{F_z} + \frac{F_x\left(F_{xz} + F_{zz} z_x\right)}{F_z^2} =$$

$$-\frac{F_{xx} - F_{xz}\left(F_x/F_z\right)}{F_z} + \frac{F_x\left(F_{xz} - F_{zz}\left(F_x/F_z\right)\right)}{F_z^2} =$$

$$-\frac{F_{xx} F_z^2 - 2F_{xz} F_x F_z + F_{zz} F_x^2}{F_z^3},$$

$$f_{xy} = -\frac{F_{xy} F_z^2 - F_{xz} F_y F_z - F_{yz} F_x F_z + F_{zz} F_x F_y}{F_z^3},$$

$$f_{yy} = -\frac{F_{yy} F_z^2 - 2F_{yz} F_y F_z + F_{zz} F_y^2}{F_z^3}.$$

We make use of the results of Problem 6.8 to find the Gaussian and mean curvatures:

$$K = \frac{f_{xx} f_{yy} - f_{xy}^2}{\left(1 + f_x^2 + f_y^2\right)^2}, \qquad H = \frac{\left(1 + f_x^2\right) f_{yy} + \left(1 + f_y^2\right) f_{xx} - f_x f_y f_{xy}}{\left(1 + f_x^2 + f_y^2\right)^{3/2}},$$

$$1 + f_x^2 + f_y^2 = \frac{F_x^2 + F_y^2 + F_z^2}{F_z^2},$$

$$f_{xx} f_{yy} - f_{xy}^2 =$$

$$\frac{1}{F_z^6}\left(\left(F_{xx} F_z^2 - 2F_{xz} F_x F_z + F_{zz} F_x^2\right)\left(F_{yy} F_z^2 - 2F_{yz} F_y F_z + F_{zz} F_y^2\right) - \right.$$

$$\left(F_{xy} F_z^2 - F_{xz} F_y F_z - F_{yz} F_x F_z + F_{zz} F_x F_y\right)^2\Big) =$$

$$\frac{1}{F_z^6}\left(\left(F_{xx} F_{yy} F_z^4 + 4F_{xz} F_{yz} F_x F_y F_z^2 + F_{zz}^2 F_x^2 F_y^2 - 2F_{xx} F_{yz} F_y F_z^3 - \right.\right.$$

$$2F_{xz} F_{yy} F_x F_z^3 + F_{xx} F_{zz} F_y^2 F_z^2 + F_{yy} F_{zz} F_x^2 F_z^2 - 2F_{xz} F_{zz} F_x F_y^2 F_z -$$

$$2F_{yz} F_{zz} F_x^2 F_y F_z\big) - \big(F_{xy}^2 F_z^4 + F_{xz}^2 F_y^2 F_z^2 + F_{yz}^2 F_x^2 F_z^2 + F_{zz}^2 F_x^2 F_y^2 -$$

$$2F_{xy} F_{xz} F_y F_z^3 - 2F_{xy} F_{yz} F_x F_z^3 + 2F_{xy} F_{zz} F_x^2 F_y^2 F_z^2 -$$

$$2F_{xz} F_{zz} F_x F_y^2 F_z - 2F_{yz} F_{zz} F_x^2 F_y F_z + 2F_{xz} F_{yz} F_x F_y F_z^2\big)\Big) =$$

$$\frac{1}{F_z^4}\left(F_{xx} F_{yy} F_z^2 + F_{xx} F_{zz} F_y^2 + F_{yy} F_{zz} F_x^2 - F_{xy}^2 F_z^2 - F_{xz}^2 F_y^2 - F_{yz}^2 F_x^2 + \right.$$

$$2F_{xz} F_{yz} F_x F_y + 2F_{xy} F_{xz} F_y F_z + 2F_{xy} F_{xz} F_x F_z -$$

$$2F_{xx} F_{yz} F_y F_z - 2F_{xz} F_{yy} F_x F_z - 2F_{xy} F_{zz} F_x F_y\Big) =$$

$$-\frac{1}{F_z^4}\begin{vmatrix} F_{xx} & F_{xy} & F_{xz} & F_x \\ F_{xy} & F_{yy} & F_{yz} & F_y \\ F_{xz} & F_{yz} & F_{zz} & F_z \\ F_x & F_y & F_z & 0 \end{vmatrix}.$$

Hence,

$$K = -\frac{\begin{vmatrix} F_{xx} & F_{xy} & F_{xz} & F_x \\ F_{xy} & F_{yy} & F_{yz} & F_y \\ F_{xz} & F_{yz} & F_{zz} & F_z \\ F_x & F_y & F_z & 0 \end{vmatrix}}{\left(F_x^2 + F_y^2 + F_z^2\right)^2}$$

Let $F_z > 0$:

$$\left(1 + f_x^2\right) f_{yy} + \left(1 + f_y^2\right) f_{xx} - f_x f_y f_{xy} =$$

$$-\frac{1}{F_z^5}\left(\left(F_x^2 + F_z^2\right)\left(F_{yy}F_z^2 - 2F_{yz}F_yF_z + F_{zz}F_y^2\right) + \right.$$

$$\left(F_y^2 + F_z^2\right)\left(F_{xx}F_z^2 - 2F_{xz}F_xF_z + F_{zz}F_x^2\right) -$$

$$\left. 2F_xF_y\left(F_{xy}F_z^2 - F_{xz}F_yF_z - F_{yz}F_xF_z + F_{zz}F_xF_y\right)\right) =$$

$$-\frac{1}{F_z^5}\left(F_{yy}F_x^2F_z^2 - 2F_{yz}F_x^2F_yF_z + F_{zz}F_x^2F_y^2 + \right.$$

$$F_{xx}F_y^2F_z^2 - 2F_{xz}F_xF_y^2F_z + F_{zz}F_x^2F_y^2 +$$

$$\left(F_{xx} + F_{yy}\right)F_z^4 - 2\left(F_{xz}F_x + F_{yz}F_y\right)F_z^3 + F_{zz}\left(F_x^2 + F_y^2\right)F_z^2 -$$

$$\left. 2\left(F_{xy}F_xF_yF_z^2 - F_{xz}F_xF_y^2F_z - F_{yz}F_x^2F_yF_z + F_{zz}F_x^2F_y^2\right)\right) =$$

$$-\frac{1}{F_z^3}\left(\left(F_{yy} + F_{zz}\right)F_x^2 + \left(F_{xx} + F_{zz}\right)F_y^2 + \left(F_{xx} + F_{yy}\right)F_z^2 - \right.$$

$$\left. 2F_{xy}F_xF_y - 2F_{xz}F_xF_z - 2F_{yz}F_yF_z\right) =$$

$$-\frac{1}{F_z^3}\left(\left(F_{xx} + F_{yy} + F_{zz}\right)\left(F_x^2 + F_y^2 + F_z^2\right) - \left(F_{xx}F_x^2 + F_{yy}F_y^2 + \right.\right.$$

$$\left.\left. F_{zz}F_z^2 + 2F_{xy}F_xF_y + 2F_{xz}F_xF_z + 2F_{yz}F_yF_z\right)\right),$$

$$H = -\frac{F_{xx} + F_{yy} + F_{zz}}{\sqrt{F_x^2 + F_y^2 + F_z^2}} +$$

$$\frac{F_{xx}F_x^2 + F_{yy}F_y^2 + F_{zz}F_z^2 + 2F_{xy}F_xF_y + 2F_{xz}F_xF_z + 2F_{yz}F_yF_z}{\left(F_x^2 + F_y^2 + F_z^2\right)^{3/2}}.$$

The mean curvature can also be written down as

$$H = \frac{\mathbf{H}\left(\operatorname{grad} F, \operatorname{grad} F\right)}{|\operatorname{grad} F|^3} - \frac{\operatorname{tr}\mathbf{H}}{|\operatorname{grad} F|},$$

where $\operatorname{grad} F = \left(F_x, F_y, F_z\right)$ is the gradient of the function F; by letter $\mathbf{H}$, we denote the matrix of the second derivatives (Hessian), as well as the bilinear form corresponding to it. If $F_z < 0$, the expression for H simply changes the sign.

6.10. We have

$$\mathbf{r}_u = \left(\frac{u \cos v}{\sqrt{u^2 + a^2}}, \frac{u \sin v}{\sqrt{u^2 + a^2}}, \frac{a}{\sqrt{u^2 + a^2}} \right),$$

$$\mathbf{r}_v = \left(-\sqrt{u^2 + a^2} \sin v, \sqrt{u^2 + a^2} \cos v, 0 \right),$$

$$\mathbf{r}_u \times \mathbf{r}_v = (-a \cos v, -a \sin v, u),$$

$$|\mathbf{r}_u \times \mathbf{r}_v| = \sqrt{u^2 + a^2},$$

$$\mathbf{n} = \frac{\mathbf{r}_u \times \mathbf{r}_v}{|\mathbf{r}_u \times \mathbf{r}_v|} = \frac{(-a \cos v, -a \sin v, u)}{\sqrt{u^2 + a^2}},$$

$$\mathbf{r}_{uu} = \left(\frac{a^2 \cos v}{(u^2 + a^2)^{3/2}}, \frac{a^2 \sin v}{(u^2 + a^2)^{3/2}}, -\frac{au}{(u^2 + a^2)^{3/2}} \right),$$

$$\mathbf{r}_{uv} = \left(-\frac{u \sin v}{\sqrt{u^2 + a^2}}, \frac{u \cos v}{\sqrt{u^2 + a^2}}, 0 \right),$$

$$\mathbf{r}_{vv} = \left(-\cos v \sqrt{u^2 + a^2}, -\sin v \sqrt{u^2 + a^2}, 0 \right),$$

$$L = \langle \mathbf{r}_{uu}, \mathbf{n} \rangle = \frac{-a \cos v a^2 \cos v + (-a \sin v) a^2 \sin v + u(-au)}{(u^2 + a^2)^2} = -\frac{a}{u^2 + a^2},$$

$$M = \langle \mathbf{r}_{uv}, \mathbf{n} \rangle = \frac{-a \cos v (-u \sin v) + (-a \sin v) u \cos v}{u^2 + a^2} = 0,$$

$$N = \langle \mathbf{r}_{vv}, \mathbf{n} \rangle = -a \cos v (-\cos v) + (-a \sin v)(-\sin v) = a.$$

Answer: $-\dfrac{a}{u^2 + a^2} du^2 + a dv^2$.

6.11. We proceed to the cylindrical coordinates $x = u \cos v$, $y = u \sin v$, $z = z$. Then the equation of the surface can be written down in the following parametric form: $x = u \cos v$, $y = u \sin v$, $z = a \arctan \frac{y}{x} = av$. Thus, the surface represents a right helicoid. Making use of the results of Problem 6.1, we get:

$$\mathbf{r} = (u \cos v, u \sin v, av), \ \mathbf{r}_u = (\cos v, \sin v, 0), \ \mathbf{r}_v = (-u \sin v, u \cos v, a),$$

$$E = 1, \ F = 0, \ G = a^2 + u^2, \ L = 0, \ M = -\frac{a}{\sqrt{u^2 + a^2}}, \ N = 0.$$

Then

$$K = \frac{LN - M^2}{EG - F^2} = -\frac{a^2}{(u^2 + a^2)^2}, \quad H = \frac{EN + GL - 2FM}{EG - F^2} = 0.$$

The main curvatures $\lambda_{1,2}$ are solutions of the equations $\lambda^2 - H\lambda + K = 0$. As $H = 0$,

$$\lambda_{1,2} = \pm \sqrt{-K} = \pm \frac{a}{u^2 + a^2},$$

$$R_{1,2} = \frac{1}{\lambda_{1,2}} = \pm \frac{u^2 + a^2}{a} = \pm \frac{x^2 + y^2 + a^2}{a}.$$

Answer: $R_1 = \dfrac{x^2 + y^2 + a^2}{a}$, $R_2 = -\dfrac{x^2 + y^2 + a^2}{a}$.

6.12. We have:

$$\mathbf{r}(u, v) = (\cos v - u \sin v, \sin v + u \cos v, u + v),$$

$\mathbf{r}_u = (-\sin v, \cos v, 1)\,,\ \mathbf{r}_v = (-u\cos v - \sin v, -u\sin v + \cos v, 1),$

$E = \langle \mathbf{r}_u, \mathbf{r}_u \rangle = 2,\ F = \langle \mathbf{r}_u, \mathbf{r}_v \rangle = 2,\ G = \langle \mathbf{r}_v, \mathbf{r}_v \rangle = 2 + u^2,$

$\mathbf{r}_u \times \mathbf{r}_v = (u\sin v, -u\cos v, u)\,,\ |\mathbf{r}_u \times \mathbf{r}_v| = u\sqrt{2},$

$\mathbf{n} = \dfrac{\mathbf{r}_u \times \mathbf{r}_v}{|\mathbf{r}_u \times \mathbf{r}_v|} = \dfrac{1}{\sqrt{2}}\,(\sin v, -\cos v, 1),$

$\mathbf{r}_{uu} = (0,0,0)\,, \mathbf{r}_{uv} = (-\cos v, -\sin v, 0),$

$\mathbf{r}_{vv} = (u\sin v - \cos v, -u\cos v - \sin v, 0),$

$L = \langle \mathbf{r}_{uu}, \mathbf{n} \rangle = 0,\ M = \langle \mathbf{r}_{uv}, \mathbf{n} \rangle = 0,\ N = \langle \mathbf{r}_{vv}, \mathbf{n} \rangle = \dfrac{1}{\sqrt{2}}u,$

$K = \dfrac{LN - M^2}{EG - F^2} = 0,$

$H = \dfrac{EN + GL - 2FM}{EG - F^2} = \dfrac{u\sqrt{2}}{2\left(2 + u^2\right) - 4} = \dfrac{1}{\sqrt{2}u}.$

The main curvatures $\lambda_{1,2}$ are solutions of the equation $\lambda^2 - H\lambda + K = 0.$ Then

$$\lambda^2 - \dfrac{1}{\sqrt{2}u}\lambda = 0,\quad \dfrac{1}{R_1} = \lambda_1 = 0,\quad \dfrac{1}{R_2} = \lambda_2 = \dfrac{1}{\sqrt{2}u}.$$

Answer: $\dfrac{1}{R_1} = 0,\ \dfrac{1}{R_2} = \dfrac{1}{\sqrt{2}u}.$

6.13. We have:

$\mathbf{r} = (u\cos v, u\sin v, u + v)\,,\ \mathbf{r}_u = (\cos v, \sin v, 1),$

$\mathbf{r}_v = (-u\sin v, u\cos v, 1),$

$E = \langle \mathbf{r}_u, \mathbf{r}_u \rangle = 2,\ F = \langle \mathbf{r}_u, \mathbf{r}_v \rangle = 1,\ G = \langle \mathbf{r}_v, \mathbf{r}_v \rangle = 1 + u^2,$

$\mathbf{r}_u \times \mathbf{r}_v = (\sin v - u\cos v, -\cos v - u\sin v, u),$

$|\mathbf{r}_u \times \mathbf{r}_v| = \sqrt{1 + 2u^2},$

$\mathbf{n} = \frac{\mathbf{r}_u \times \mathbf{r}_v}{|\mathbf{r}_u \times \mathbf{r}_v|} = \frac{1}{\sqrt{1+2u^2}}\,(\sin v - u\cos v, -\cos v - u\sin v, u),$

$\mathbf{r}_{uu} = (0,0,0)\,,\ \mathbf{r}_{uv} = (-\sin v, \cos v, 0)\,,\ \mathbf{r}_{vv} = (-u\cos v, -u\sin v, 0),$

$L = \langle \mathbf{r}_{uu}, \mathbf{n} \rangle = 0,\ \ M = \langle \mathbf{r}_{uv}, \mathbf{n} \rangle = -\dfrac{1}{\sqrt{1 + 2u^2}},$

$N = \langle \mathbf{r}_{vv}, \mathbf{n} \rangle = \dfrac{u^2}{\sqrt{1 + 2u^2}},\ \ K = \dfrac{LN - M^2}{EG - F^2} = -\dfrac{1}{\left(1 + 2u^2\right)^2},$

$H = \dfrac{EN + GL - 2FM}{EG - F^2} = \dfrac{2u^2 + 2}{\left(1 + 2u^2\right)^{3/2}} = \dfrac{2\left(1 + u^2\right)}{\left(1 + 2u^2\right)^{3/2}}.$

Answer: $K = -\dfrac{1}{\left(1 + 2u^2\right)^2},\ H = \dfrac{2\left(1 + u^2\right)}{\left(1 + 2u^2\right)^{3/2}}.$

6.14. We have:

$\mathbf{r} = \left(3u + 3uv^2 - u^3, v^3 - 3v - 3u^2 v, 3\left(u^2 - v^2\right)\right),$

$\mathbf{r}_u = \left(3 + 3v^2 - 3u^2, -6uv, 6u\right)\,,\quad \mathbf{r}_v = \left(6uv, 3v^3 - 3u^2 - 3, -6v\right),$

$E = \langle \mathbf{r}_u, \mathbf{r}_u \rangle = 9\left(\left(1 + v^2 - u^2\right)^2 + 4u^2 v^2 + 4u^2\right) = 9\left(u^2 + v^2 + 1\right)^2,$

$F = \langle \mathbf{r}_u, \mathbf{r}_v \rangle = 9\left(\left(1 - u^2 + v^2\right)2uv + (-2uv)\left(v^2 - u^2 - 1\right) - 4uv\right) = 0,$

$G = \langle \mathbf{r}_v, \mathbf{r}_v \rangle = 9\left(4u^2 v^2 + \left(u^2 - v^2 + 1\right)^2 + 4v^2\right) = 9\left(u^2 + v^2 + 1\right)^2,$

$\mathbf{r}_u \times \mathbf{r}_v = 9\left(2u\left(u^2 + v^2 + 1\right), 2v\left(u^2 + v^2 + 1\right), \left(u^2 + v^2\right)^1 - 1\right),$

$|\mathbf{r}_u \times \mathbf{r}_v| = 9\left(u^2 + v^2 + 1\right)^2,$

$$\mathbf{n} = \frac{\mathbf{r}_u \times \mathbf{r}_v}{|\mathbf{r}_u \times \mathbf{r}_v|} = \frac{1}{u^2 + v^2 + 1} \left(2u, 2v, u^2 + v^2 - 1\right),$$

$$\mathbf{r}_{uu} = (-6u, -6v, 6), \ \mathbf{r}_{uv} = (6v, -6u, 0), \ \mathbf{r}_{vv} = (6u, 6v, -6),$$

$$L = \langle \mathbf{r}_{uu}, \mathbf{n} \rangle = 6 \frac{2u(-u) + 2v(-v) + (u^2 + v^2 - 1) 1}{u^2 + v^2 + 1} = -6,$$

$$M = \langle \mathbf{r}_{uv}, \mathbf{n} \rangle = 6 \frac{2uv + 2v(-u)}{u^2 + v^2 + 1} = 0,$$

$$N = \langle \mathbf{r}_{vv}, \mathbf{n} \rangle = 6 \frac{2uu + 2vv + (u^2 + v^2 - 1)(-1)}{u^2 + v^2 + 1} = 6,$$

$$K = \frac{LN - M^2}{EG - F^2} = -\frac{36}{81 \left(u^2 + v^2 + 1\right)^4} = -\frac{4}{9 \left(u^2 + v^2 + 1\right)^4},$$

$$H = \frac{EN + GL - 2FM}{EG - F^2} = \frac{\left(u^2 + v^2 + 1\right)^2 (-6) + \left(u^2 + v^2 + 1\right)^2 6}{81 \left(u^2 + v^2 + 1\right)^4} = 0.$$

Answer: $K = -\dfrac{4}{9 \left(u^2 + v^2 + 1\right)^4}$, $H = 0$.

6.17. $H\mathbf{II} - K\mathbf{I}$. See the solution of Problem 6.22.

6.20. To make use of the explicit formula for the Gaussian curvature of the surface in coordinates (u, s), where u is the coordinate along the rectilinear generatrix and s is a natural parameter on the curve along which the straight line slides.

6.21. The parametric equation of the helicoid has the form $x = u \cos v$, $y = u \sin v$, $z = av$. Then

$$\mathbf{r}(u, v) = (u \cos v, u \sin v, av),$$

$$\mathbf{r}_u = (\cos v, \sin v, 0), \ \mathbf{r}_v = (-u \sin v, u \cos v, a),$$

$$E = 1, \ F = 0, \ G = a^2 + u^2,$$

$$\mathbf{r}_u \times \mathbf{r}_v = (a \sin v, -a \cos v, u), \ |\mathbf{r}_u \times \mathbf{r}_v| = \sqrt{u^2 + a^2},$$

$$\mathbf{n} = \frac{\mathbf{r}_u \times \mathbf{r}_v}{|\mathbf{r}_u \times \mathbf{r}_v|} = \frac{(a \sin v, -a \cos v, u)}{\sqrt{u^2 + a^2}},$$

$$\mathbf{r}_{uu} = (0, 0, 0), \mathbf{r}_{uv} = (-\sin v, \cos v, 0), \ \mathbf{r}_{vv} = (-u \cos v, -u \sin v, 0),$$

$$L = \langle \mathbf{r}_{uu}, \mathbf{n} \rangle = 0,$$

$$M = \langle \mathbf{r}_{uv}, \mathbf{n} \rangle = \frac{a \sin v (-\sin v) + (-a \cos v) \cos v}{\sqrt{u^2 + a^2}} = -\frac{a}{\sqrt{u^2 + a^2}},$$

$$N = \langle \mathbf{r}_{vv}, \mathbf{n} \rangle = \frac{a \sin v (-u \cos v) + (-a \cos v)(-u \sin v)}{\sqrt{u^2 + a^2}} = 0.$$

Hence, we get $H = \dfrac{EN + GL - 2FM}{EG - F^2} = 0.$

6.22. We have:

$$\rho_u = \mathbf{r}_u + a\mathbf{n}_u, \quad \rho_v = \mathbf{r}_v + a\mathbf{n}_v,$$

$$E^* = \langle \rho_u, \rho_u \rangle = \langle \mathbf{r}_u, \mathbf{r}_u \rangle + 2a \langle \mathbf{r}_u, \mathbf{n}_u \rangle + a^2 \langle \mathbf{n}_u, \mathbf{n}_u \rangle = E - 2aL + a^2 \langle \mathbf{n}_u, \mathbf{n}_u \rangle.$$

Similarly,

$$F^* = \langle \rho_u, \rho_v \rangle = F - 2aM + a^2 \langle \mathbf{n}_u, \mathbf{n}_v \rangle,$$

$$G^* = \langle \rho_v, \rho_v \rangle = G - 2aN + a^2 \langle \mathbf{n}_v, \mathbf{n}_v \rangle.$$

For further calculations, we need an explicit expression for $\mathbf{n}_u$ and $\mathbf{n}_v$. We find it. As $|\mathbf{n}| \equiv 1$, then $\mathbf{n}_u \perp \mathbf{n}$, $\mathbf{n}_v \perp \mathbf{n}$. Then

$$\mathbf{n}_u = b\mathbf{r}_u + c\mathbf{r}_v, \mathbf{n}_v = d\mathbf{r}_u + e\mathbf{r}_v,$$

$$-L = -\langle \mathbf{n}, \mathbf{r}_{uu} \rangle = \langle \mathbf{n}_u, \mathbf{r}_u \rangle = b\langle \mathbf{r}_u, \mathbf{r}_u \rangle + c\langle \mathbf{r}_v, \mathbf{r}_u \rangle = bE + cF,$$

$$-M = -\langle \mathbf{n}, \mathbf{r}_{uv} \rangle = \langle \mathbf{n}_u, \mathbf{r}_v \rangle = b\langle \mathbf{r}_u, \mathbf{r}_v \rangle + c\langle \mathbf{r}_v, \mathbf{r}_v \rangle = bF + cG,$$

$$\begin{pmatrix} -L \\ -M \end{pmatrix} = \mathbf{I} \begin{pmatrix} b \\ c \end{pmatrix},$$

where $\mathbf{I} = \begin{pmatrix} E & F \\ F & G \end{pmatrix}$ is the matrix of the first quadratic form;

$$\begin{pmatrix} b \\ c \end{pmatrix} = -\mathbf{I}^{-1} \begin{pmatrix} L \\ M \end{pmatrix},$$

where $\mathbf{I}^{-1} = \dfrac{1}{EG - F^2} \begin{pmatrix} G & -F \\ -F & E \end{pmatrix}$.

Similarly,

$$\begin{pmatrix} d \\ e \end{pmatrix} = -\mathbf{I}^{-1} \begin{pmatrix} M \\ N \end{pmatrix}.$$

Then

$$\langle \mathbf{n}_u, \mathbf{n}_u \rangle = (b, c)\,\mathbf{I} \begin{pmatrix} b \\ c \end{pmatrix} = (L, M)\,\mathbf{I}^{-1} \mathbf{II}^{-1} \begin{pmatrix} L \\ M \end{pmatrix} =$$

$$(L, M)\,\mathbf{I}^{-1} \begin{pmatrix} L \\ M \end{pmatrix} = \frac{EM^2 - 2FLM + GL^2}{EG - F^2},$$

$$\langle \mathbf{n}_u, \mathbf{n}_v \rangle = (L, M)\,\mathbf{I}^{-1} \begin{pmatrix} M \\ N \end{pmatrix} = \frac{EMN - FLN - FM^2 + GLM}{EG - F^2},$$

$$\langle \mathbf{n}_v, \mathbf{n}_v \rangle = (M, N)\,\mathbf{I}^{-1} \begin{pmatrix} M \\ N \end{pmatrix} = \frac{EN^2 - 2FMN + GM^2}{EG - F^2}.$$

On the other hand,

$$\langle \mathbf{n}_u, \mathbf{n}_u \rangle = \frac{EM^2 - 2FLM + GL^2}{EG - F^2} =$$

$$\frac{GL^2 + ELN - 2FLM + EM^2 - ELN}{EG - F^2} =$$

$$\frac{L\,(GL + EN - 2FM) - E\,(LN - M^2)}{EG - F^2} = \frac{LH - EK}{EG - F^2},$$

$$\langle \mathbf{n}_u, \mathbf{n}_v \rangle = \frac{EMN - FLN - FM^2 + GLM}{EG - F^2} =$$

$$\frac{M\,(GL + EN - 2FM) - F\,(LN - M^2)}{EG - F^2} = \frac{MH - FK}{EG - F^2},$$

$$\langle \mathbf{n}_v, \mathbf{n}_v \rangle = \frac{EN^2 - 2FMN + GM^2}{EG - F^2} =$$

$$\frac{N\,(GL + EN - 2FM) - G\,(LN - M^2)}{EG - F^2} = \frac{NH - GK}{EG - F^2},$$

which corresponds to Problem 6.17.

Therefore,

$$E^* = \langle \rho_u, \rho_u \rangle = E - 2aL + a^2 (LH - EK) = \left(1 - a^2 K\right) E + a (aH - 2) L,$$

$$F^* = \langle \rho_u, \rho_v \rangle = F - 2aM + a^2 (MH - FK) = \left(1 - a^2 K\right) F + a (aH - 2) M,$$

$$G^* = \langle \rho_v, \rho_v \rangle = G - 2aN + a^2 (NH - GK) = \left(1 - a^2 K\right) G + a (aH - 2) N.$$

As vectors $\mathbf{r}_u$, $\mathbf{r}_v$, $\mathbf{n}_u$, $\mathbf{n}_v$ are orthogonal to vector $\mathbf{n}$, ρ_u, ρ_v are orthogonal to $\mathbf{n}$ and $\mathbf{n}^* = \mathbf{n}$.

$$L^* = \langle \rho_{uu}, \mathbf{n} \rangle = \langle \mathbf{r}_{uu}, \mathbf{n} \rangle + a \langle \mathbf{n}_{uu}, \mathbf{n} \rangle = L - a \langle \mathbf{n}_u, \mathbf{n}_u \rangle =$$
$$L - a (LH - EK) = aKE + (1 - aH) L,$$

$$M^* = \langle \rho_{uv}, \mathbf{n} \rangle = \langle \mathbf{r}_{uv}, \mathbf{n} \rangle - a \langle \mathbf{n}_u, \mathbf{n}_v \rangle = M - a (MH - FK) =$$
$$aKF + (1 - aH) M,$$

$$N^* = \langle \rho_{vv}, \mathbf{n} \rangle = \langle \mathbf{r}_{vv}, \mathbf{n} \rangle - a \langle \mathbf{n}_v, \mathbf{n}_v \rangle = N - a (NH - GK) =$$
$$aKG + (1 - aH) N.$$

Answer: $E^* = \left(1 - a^2 K\right) E + a (aH - 2) L,$
$\quad F^* = \left(1 - a^2 K\right) F + a (aH - 2) M, \quad G^* = \left(1 - a^2 K\right) G + a (aH - 2) N,$
$\quad L^* = aKE + (1 - aH) L, \; M^* = aKF + (1 - aH) M,$
$\quad N^* = aKG + (1 - aH) N.$

6.24. As is known, $K^* = \dfrac{L^* N^* - M^{*2}}{E^* G^* - F^{*2}}$. Substituting into this formula explicit expressions for E^*, F^*, G^*, L^*, M^*, N^* from Problem 6.22 and performing elementary computations, we get the answer: $K^* = \dfrac{K}{1 - aH + a^2 K}$.

6.25. As is known, $H^* = \dfrac{E^* N^* + G^* L^* - 2F^* M^*}{E^* G^* - F^{*2}}$. Substituting into this formula explicit expressions for E^*, F^*, G^*, L^*, M^*, N^* from Problem 6.22 and performing elementary computations, we get the answer:

$$H^* = \frac{H - 2aK}{1 - aH + a^2 K}.$$

6.26. (a) Let these parallel surfaces S and S^* be given by the equations: $\mathbf{r} = \mathbf{r}(u, v)$, $\mathbf{r}^* = \mathbf{r}(u, v) + a\mathbf{n}(u, v)$. Then according to Problems 6.24 and 6.25, we have $K^* = \dfrac{K}{1 - aH + a^2 K}$, $H^* = \dfrac{H - 2aK}{1 - aH + a^2 K}$. From here, we get

$$\frac{H^{*2} - 4K^*}{K^{*2}} = \frac{(H - 2aK)^2 - 4K \left(1 - aH + a^2 K\right)}{K^2} =$$
$$\frac{H^2 - 4aKH + 4a^2 K^2 - 4 \left(K - aKH + a^2 K^2\right)}{K^2} = \frac{H^2 - 4K}{K^2}.$$

(b) The minimal surface is given by the equation $H^* = 0$. According to Problem 6.25, we have $H^* = \dfrac{H - 2aK}{1 - aH + a^2 K}$. From here it is seen that the equation $H^* = 0$

is tantamount to the fulfillment of the equality $2a = H/K$. As the ratio of the curvatures is constant, the surface S^* given by the equation

$$\rho\left(u, v\right) = \mathbf{r}\left(u, v\right) + \frac{2H}{K}\mathbf{n}\left(u, v\right),$$

shall be the minimal surface.

(c) According to Problem 6.25, we have

$$K^* = \frac{K}{1 - aH + a^2 K} = \frac{K}{1 - H/H + K/H^2} = \frac{K}{K/H^2} = H^2 = \text{const.}$$

(d) According to Problem 6.25, we have

$$H^* = \frac{H - 2aK}{1 - aH + a^2 K} = \frac{H - \sqrt{K}}{1 - H/\sqrt{K} + K/\left(\sqrt{K}\right)^2} = -\sqrt{K} = \text{const.}$$

Answer: $H^* = -\sqrt{K}$.

6.27. *Hint:* according to Problem 6.26, for parallel surfaces the following relation is fulfilled: $\dfrac{H^2 - 4K}{K^2} = \dfrac{H^{*2} - 4K^*}{K^{*2}}$. Note that a point is umbilical then and only then, when at that point $H^2 - 4K = 0$. The solution readily follows from here.

6.28. We make use of Problem 6.26. It is known as the Bonnet theorem. For simplicity, we write down the metric of the given surface in isothermic coordinates. We lay off segments of length $a = \dfrac{1}{H}$ along the normals to the surface. Then for the parallel surface we have an inequality

$$EG - F^2 = \frac{K^2\Lambda^4}{H^4} > 0, \qquad E + G = \left(1 - \frac{2K}{H^2}\right)\Lambda^2 > 0,$$

where Λ is a conformal multiplier of the metric. A regular surface of constant Gaussian curvature is obtained. It is known that such a surface is isometric to a standard sphere. Note that if the surface would have been open, it could have been isometric to part of a sphere but not superposed with it by the motion of an enveloping space. The matter is that, e.g., part of a sphere admits nontrivial deformations.

6.29. We recall that if λ_1 and λ_2 are the principal curvatures of the surface and $H = \lambda_1 + \lambda_2$, $K = \lambda_1\lambda_2$, then λ_i are the roots of the equation $\lambda^2 - H\lambda + K = 0$. As λ_i are real numbers, the discriminant of the equation is non-negative: $D = H^2 - 4K \geqslant 0$. The equality $H^2 = 4K$ (i.e., $D = 0$) is tantamount to the equality $\lambda_1 = \lambda_2$. Thus, the squared mean curvature is equal to the quadrupled Gaussian curvature in umbilical points, i.e., in the points, where $\lambda_1 = \lambda_2$ (see Fig. 117).

6.30. According to the known formula, $H = \text{tr}\,\mathbf{II} \cdot \mathbf{I}^{-1}$, where $\mathbf{I}$ is the first quadratic form, and the trace is taken from the product of the corresponding matrices. Then the mean curvature is equal to the trace of operator $\mathbf{A}$ canonically adjoint to form $\mathbf{II}$ relative to the metric given by form $\mathbf{I}$. Operator $\mathbf{A}$ and form $\mathbf{II}$ are linked by the relation $\mathbf{II}\left(\mathbf{e}_1, \mathbf{e}_2\right) = \langle\mathbf{A}\mathbf{e}_1, \mathbf{e}_2\rangle$, where the inner product $\langle\cdot, \cdot\rangle$ is given by the symmetric matrix $\mathbf{I}$. The trace of the operator in an arbitrary orthonormal basis $\mathbf{e}_1, \mathbf{e}_2$ is computed by the formula $\text{tr}\,\mathbf{A} = \langle\mathbf{A}\mathbf{e}_1, \mathbf{e}_1\rangle + \langle\mathbf{A}\mathbf{e}_2, \mathbf{e}_2\rangle$. Then

$$H = \left(\langle\mathbf{A}\mathbf{e}_1, \mathbf{e}_1\rangle + \langle\mathbf{A}\mathbf{e}_2, \mathbf{e}_2\rangle\right) = \mathbf{II}\,\langle\mathbf{e}_1, \mathbf{e}_1\rangle + \mathbf{II}\,\langle\mathbf{e}_2, \mathbf{e}_2\rangle,$$

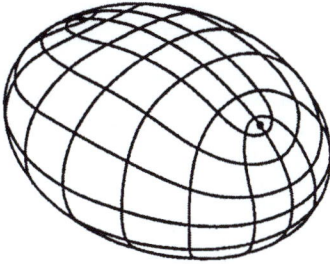

Figure 117 Umbilical points on an ellipsoid

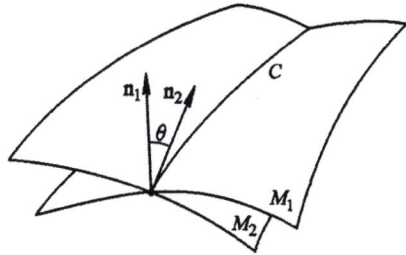

Figure 118

which is what was required to be proved.

6.33. Let $\mathbf{v}(s)$ be the vector of a tangent to curve γ, and $\mathbf{n}(s)$ be the unit vector of the normal to curve γ (see Fig. 118). Then $\dfrac{d\mathbf{v}}{ds} = k\mathbf{n}$, $\lambda_i = k\langle \mathbf{n}, \mathbf{n}_i \rangle$, where $\mathbf{n}_i$ is the normal to the surface M_i, and $\mathbf{v} \perp \mathbf{n}$. As $\mathbf{v}$ is a vector tangent to surfaces M_1 and M_2, then $\mathbf{v} \perp \mathbf{n}_1$, $\mathbf{v} \perp \mathbf{n}_2$. Therefore, $\mathbf{n}_1$, $\mathbf{n}_2$, $\mathbf{n}$ lie in one plane.

1) $\theta = 0$. Then $\mathbf{n}_1 = \mathbf{n}_2$, $\lambda_1 = \lambda_2$ and

$$\lambda_1^2 + \lambda_2^2 - 2\lambda_1\lambda_2 \cos\theta = (\lambda_1 - \lambda_2)^2 = 0 = k^2 \sin^2\theta;$$

2) $\theta = \pi$. Then $\mathbf{n}_1 = -\mathbf{n}_2$, $\lambda_1 = -\lambda_2$ and we arrive at the identity $0 = 0$;

3) $0 < \theta < \pi$. Then $\mathbf{n}_1$ and $\mathbf{n}_2$ form the basis of space orthogonal to $\mathbf{v}$ and $\mathbf{n} = a\mathbf{n}_1 + b\mathbf{n}_2$. Therefore,

$$\lambda_1 = ka\langle \mathbf{n}_1, \mathbf{n}_1 \rangle + kb\langle \mathbf{n}_2, \mathbf{n}_1 \rangle = ka + kb\cos\theta,$$
$$\lambda_2 = ka\langle \mathbf{n}_1, \mathbf{n}_2 \rangle + kb\langle \mathbf{n}_2, \mathbf{n}_2 \rangle = ka\cos\theta + kb,$$
$$\lambda_1^2 + \lambda_2^2 - 2\lambda_1\lambda_2\cos\theta = (ka + kb\cos\theta)^2 + (ka\cos\theta + kb)^2 -$$
$$2(ka + kb\cos\theta)(ka\cos\theta + kb)\cos\theta = k^2\left(a^2\left(1 + \cos^2\theta - 2\cos^2\theta\right) +\right.$$
$$2ab\left(2\cos\theta - (1 + \cos^2\theta)\cos\theta\right) + b^2\left(1 + \cos^2\theta - 2\cos^2\theta\right)\right) =$$
$$= k^2\sin^2\theta\left(a^2 + 2ab\cos\theta + b^2\right) = k^2\sin^2\theta,$$

as $1 = \langle \mathbf{n}, \mathbf{n} \rangle = \langle a\mathbf{n}_1 + b\mathbf{n}_2, a\mathbf{n}_1 + b\mathbf{n}_2 \rangle = a^2 + 2ab\cos\theta + b^2$.

6.41. *Hint:* considering that the metric of the surface has the form $du^2 + G dv^2$, express G via the Gaussian curvature K.

6.42. Consider the following surfaces of revolution:

$$x = r\cos\varphi, \quad y = r\sin\varphi,$$

$$z = \frac{1}{2}\left(\arctan\sqrt{-\frac{1 + 2C + 2r}{C + 2r}} - \sqrt{-(C + 2r)(1 + 2C + 2r)}\right).$$

In these surfaces, $K = \dfrac{1}{r}$, $ds^2 = -\dfrac{1}{C + 2r}dr^2 + r^2 d\varphi^2$. Hence, it is seen that, by changing number C, we get a one-parameter family of pairwise non-isometric surfaces with the same Gaussian curvature in corresponding points.

6.49. Show that each point of the surface is a point of flattening. In other words, in each point both principal curvatures are equal to zero.

7.1. As charts, we can take manifolds $U_k^{\pm}$, $k = 0, \ldots, n$ given by the following inequalities: $U_k^+ = \{x_k > 0\}$, $U_k^- = \{x_k < 0\}$. As coordinate functions in the chart $U_k^{\pm}$, all Cartesian coordinates except x_k should be taken. A minimal atlas contains two charts.

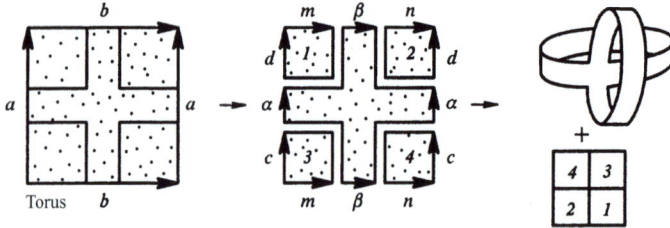

Figure 119

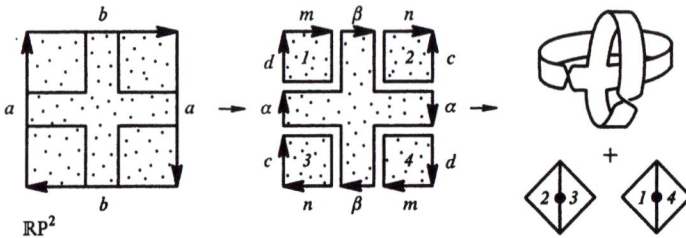

Figure 120

7.2. Make use of the fact that the torus T^2 is homeomorphic to the product $S^1 \times S^1$ and reduce the problem to the previous problem at $n = 1$.

7.3. The projective space $\mathbb{R}P^n$ is a manifold of equivalence classes of sets $(x_0 : x_1 : \ldots : x_n)$, where $x_i \in \mathbb{R}$, $\sum = x_i^2 \neq 0$, and the relation of equivalence is given as follows:

$$(x_0 : x_1 : \ldots : x_n) \backsim (\lambda x_0 : \lambda x_1 : \ldots : \lambda x_n),$$

where $\lambda \in \mathbb{R}, \lambda \neq 0$. We introduce on $\mathbb{R}P^n$ a real-analytic structure. For this, we cover $\mathbb{R}P^n$ with a set from an $(n+1)$ chart. We consider sets $(x_0 : x_1 : \ldots : x_n)$, for which $x_i \neq 0$. A manifold of such sets is naturally identified with $\mathbb{R}^n$, namely:

$$(x_0 : x_1 : \ldots : x_n) \mapsto \left(\frac{x_0}{x_i}, \ldots, \frac{x_{i-1}}{x_i}, \frac{x_{i+1}}{x_i}, \ldots, \frac{x_n}{x_i} \right).$$

It is easy to see that this correspondence is correctly defined. It remains to consider the functions of the transition from an i-th chart to a j-th chart. Let $x_k^{(i)}$ be a k-th coordinate of the set $(x_0 : x_1 : \ldots : x_n)$ in an i-th chart, and $x_l^{(j)}$ an l-th coordinate in a j-th chart (let for simplicity $i < j$). Then

$$x_1^{(i)} = \frac{x_1^{(j)}}{x_{i+1}^{(j)}}, \ldots, x_i^{(i)} = \frac{x_i^{(j)}}{x_{i+1}^{(j)}}, x_{i+1}^{(i)} = \frac{x_{i+2}^{(j)}}{x_{i+1}^{(j)}}, \ldots, x_j^{(i)} = \frac{x_{j+1}^{(j)}}{x_{i+1}^{(j)}},$$

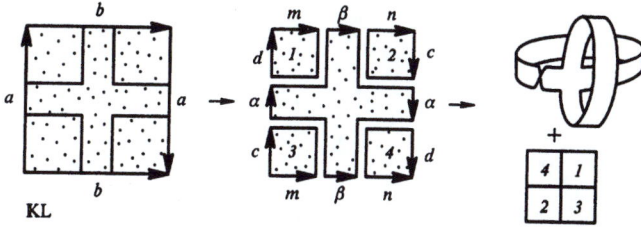

Figure 121

$$x^{(i)}_{j+1} = \frac{1}{x^{(j)}_{i+1}}, x^{(i)}_{j+2} = \frac{x^{(j)}_{j+2}}{x^{(j)}_{i+1}}, \ldots, x^{(i)}_n = \frac{x^{(j)}_n}{x^{(j)}_{i+1}}.$$

Thus, the transition functions are not only smooth, but also real-analytic.

7.4. See Problem 7.3.

7.5. (a) The atlas consists of one chart with the coordinate functions $(x_1, \ldots, x_n)$.

7.9. Using the local coordinates, calculate the rank of the Jacobian matrix of the mapping.

7.10. *Hint:* make use of the rule of differentiation of complicated functions.

7.12. The rank is equal to 1.

7.14. *Hint:* a composition of smooth mappings is a smooth mapping.

7.15. *Hint:* write formulas explicitly expressing the coordinates of the normal in the local coordinates of the torus.

7.16. Uniform coordinates of a line smoothly depend on linear coordinates on a sphere, and local coordinates on $\mathbb{R}P^2$, in turn, are expressed via uniform coordinates.

7.19. Represent elements of the group $SO(2)$ as rotations of the plane to some angle around the origin. The group $O(2)$ is homeomorphic to a union of two copies of S^1.

7.20. Represent elements of the group $SO(3)$ as rotations of the space around some axis to some angle.

7.27. Groups $GL(n, \mathbb{R})$ and $GL(n, \mathbb{C})$ are open sets in spaces of, respectively, all real and all complex matrices.

7.34. Any neighbourhood of the origin decomposes into no less than 4 connected components when the origin is deleted, which could not be on a mani- fold.

7.35. (a) Yes; (b) no.

7.37. *Hint:* use the rank properties of the product of two matrices.

7.38. $f: \mathbb{R}^1 \to \mathbb{R}^1$, $f(x) = x^3$.

7.41. Coordinate functions are a particular case of a smooth function on a manifold.

7.43. Make use of Problem 7.8.

7.51. See Fig. 88 from §21.

7.57. (a) See Fig. 119.

   (b) See Fig. 120.

   (d) See Fig. 121.

   (e) See Fig. 121.

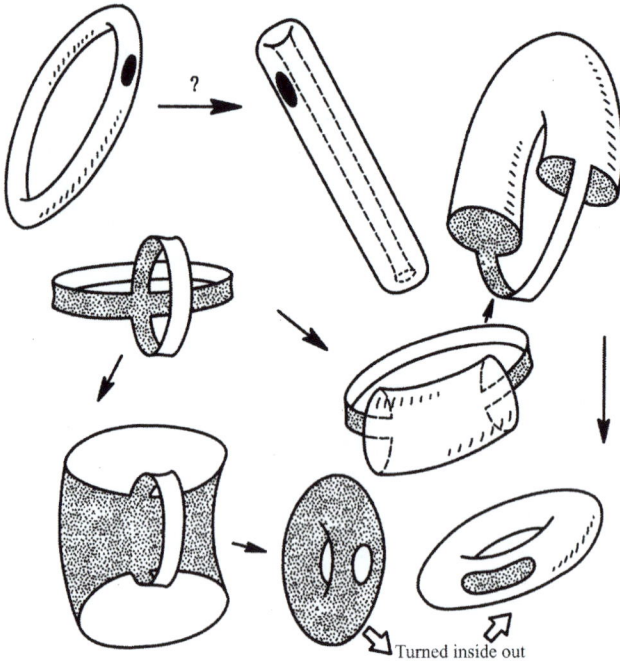

Figure 122

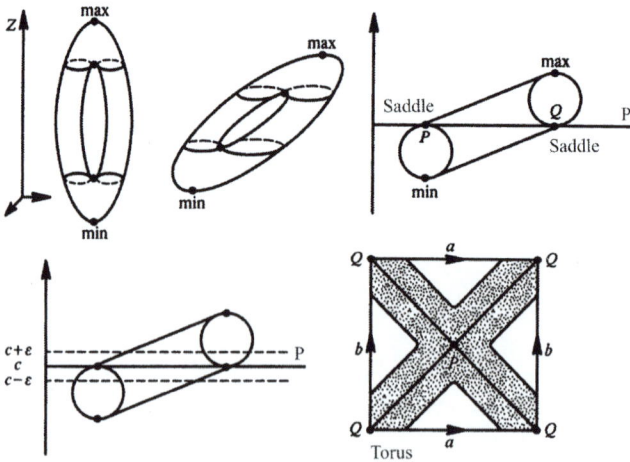

Figure 123

(f) *Hint:* the answer depends on the parity of numbers p and q. Only three surfaces from (a), (b) and (d) can be obtained.

7.58. See Fig. 122.

7.59. Consider cutting the torus with a plane passing through the axis of revolution. Let plane P be perpendicular to the cutting plane and pass as shown in the figure. The sought-for planes are at a distance of ε to both sides from P. See Fig. 123 and Fig. 124.

7.60. See Fig. 125.

7.64. Two variants of an answer are shown in Fig. 126 and in Fig. 127. Note that if a surface is immobile, and only a curve (by means of isotopy) can be deformed, one marked curve cannot be transformed into the other, as they represent different elements in the fundamental group of surface: a and $a \cdot b$.

7.65. Two variants of the sought-for deformation are shown in Fig. 128 and Fig. 129. The same way as in the previous problem, one pair of curves cannot be transformed into the other by their isotopy on an immobile surface, as they form different pairs of elements of the fundamental group of surface: a, b and $a, a \cdot b$.

7.66. (a) Yes, the deformation is shown in Fig. 130.

(b) no, see the explanation in Fig. 131.

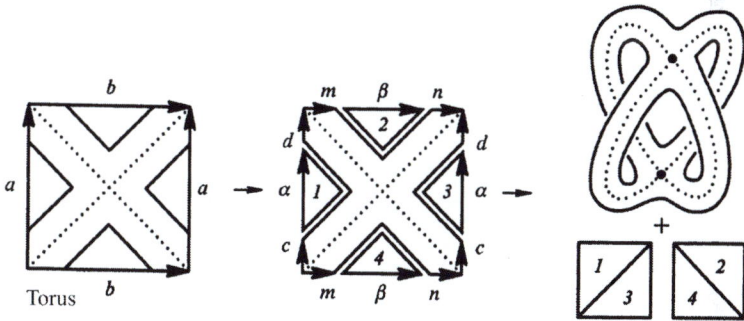

Figure 124

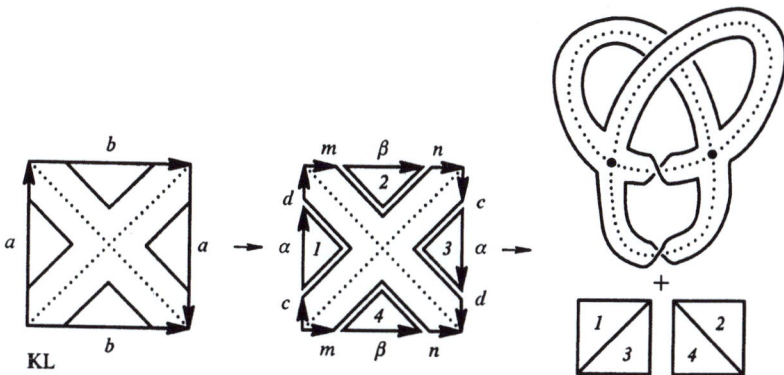

Figure 125

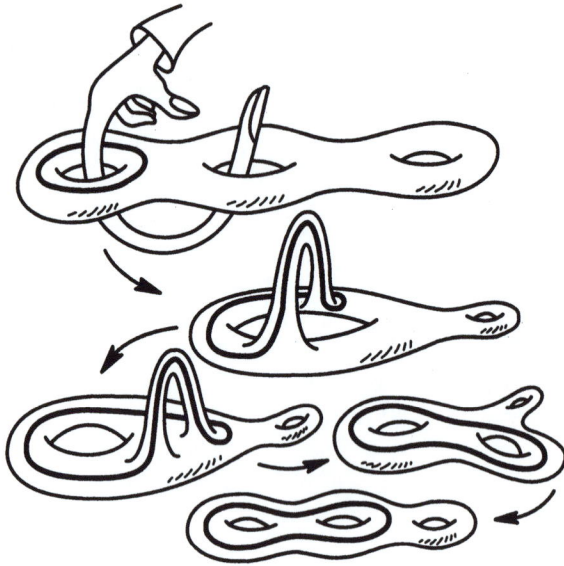

Figure 126

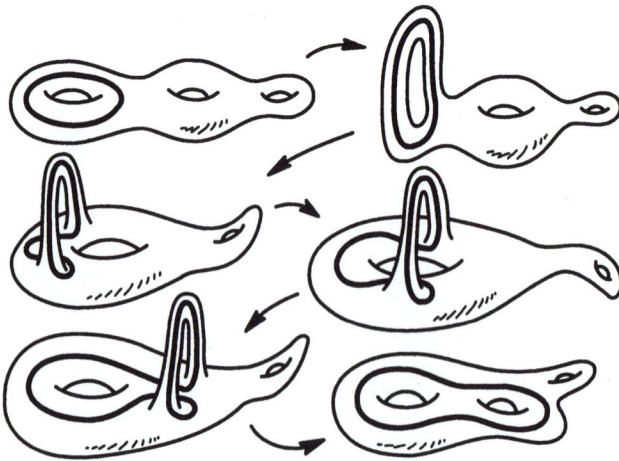

Figure 127

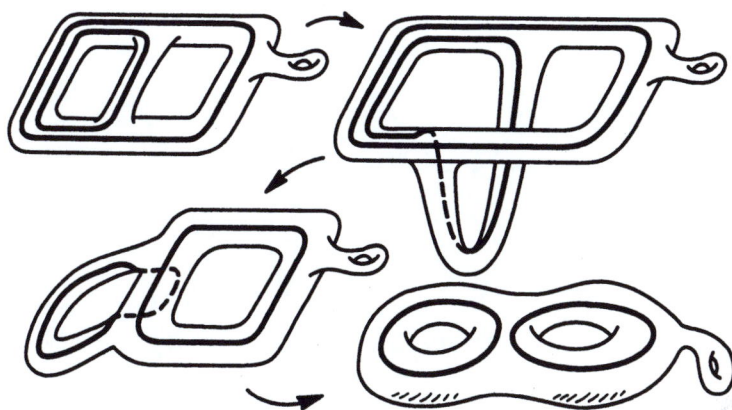

Figure 128

Figure 129

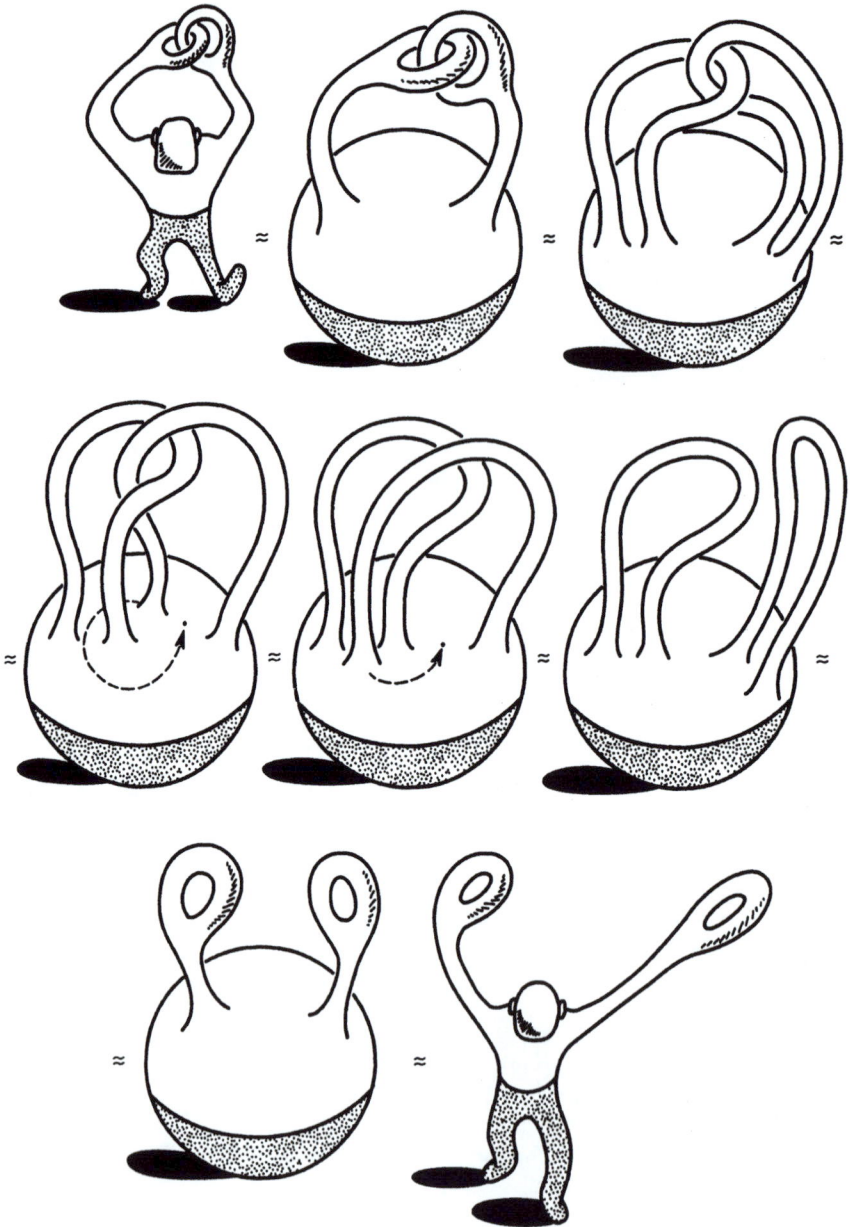

Figure 130

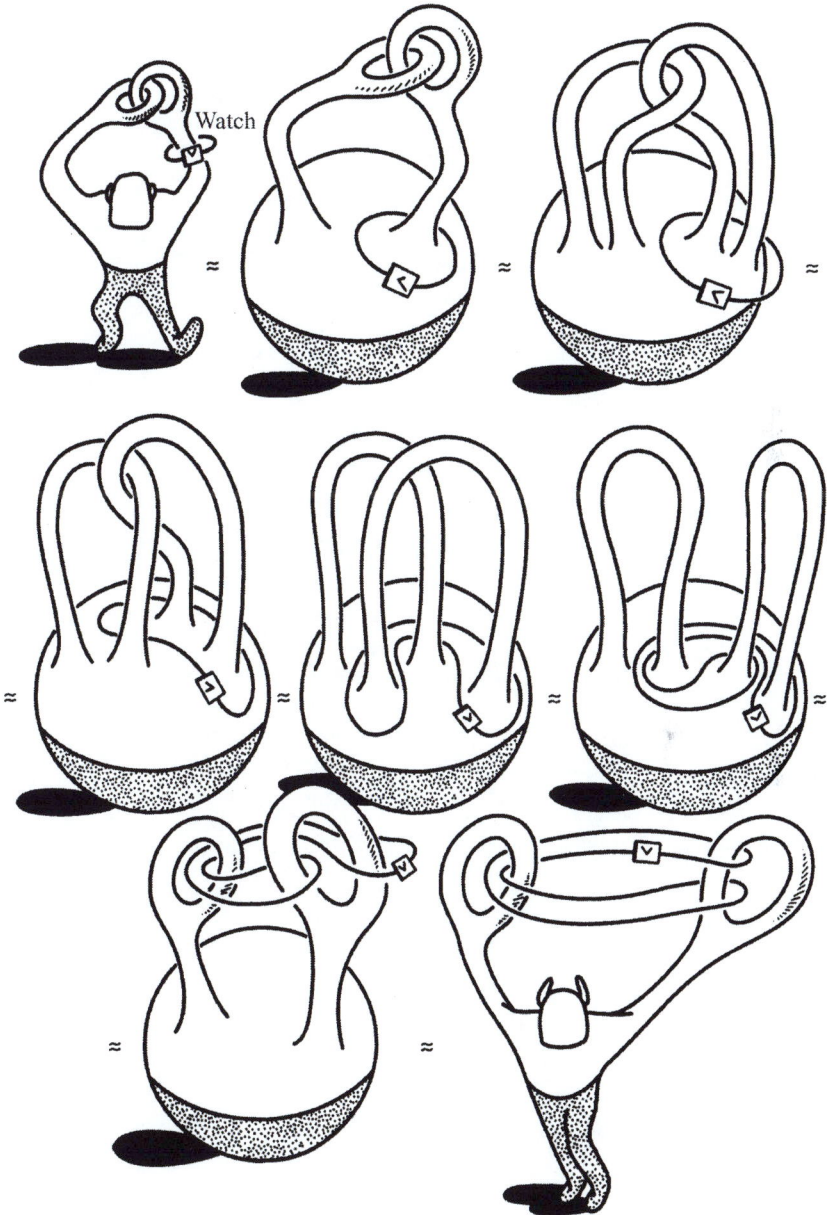

Figure 131

8.1. (a) $(0, 1)$; (b) $(0, 2)$; (c) $(1, 1)$; (d) $(0, 2)$.

8.8. If $\dim V = k$, then $\dim V_n^m = k^{(n+m)}$.

9.2. (a) $\dfrac{3}{\sqrt{15}}$; (b) $\dfrac{-7}{\sqrt{21}}$; (c) $\dfrac{e^3}{\sqrt{3}}$; (d) $-\dfrac{2}{5}$.

9.3. $1/r^2$.

9.4. 1.

9.11. Check the action of the commutator on the product of two smooth functions.

9.17. Consider the case when $\xi = \partial/\partial x^1$.

10.9. *Hint:* see Problem 10.8.

10.12. We shall consider coordinate r to be the first, and coordinate φ to be the second. Then $\Gamma_{11}^1 = \Gamma_{12}^1 = \Gamma_{21}^1 = 0$, $\Gamma_{22}^1 = -r$, $\Gamma_{11}^2 = \Gamma_{22}^2 = 0$, $\Gamma_{12}^2 = \Gamma_{21}^2 = \dfrac{1}{r}$.

10.13. We shall consider that coordinate u is the first, and coordinate v is the second, then $\Gamma_{11}^1 = \dfrac{\lambda_u}{2\lambda}$, $\Gamma_{12}^1 = \Gamma_{21}^1 = \dfrac{\lambda_v}{2\lambda}$, $\Gamma_{22}^1 = -\dfrac{\lambda_u}{2\lambda}$, $\Gamma_{11}^2 = -\dfrac{\lambda_v}{2\lambda}$, $\Gamma_{12}^2 = \Gamma_{21}^2 = \dfrac{\lambda_u}{2\lambda}$, $\Gamma_{22}^2 = \dfrac{\lambda_v}{2\lambda}$.

10.14. (a) $\Gamma_{11}^1 = \Gamma_{12}^1 = \Gamma_{21}^1 = 0$, $\Gamma_{22}^1 = -\sin\theta\cos\theta$, $\Gamma_{11}^2 = \Gamma_{22}^2 = 0$, $\Gamma_{21}^2 = \Gamma_{12}^2 = \cot\theta$. Here, coordinate θ is considered to be the first, and coordinate φ to be the second;

(b) $\Gamma_{11}^1 = -\dfrac{2x}{1 + x^2 + y^2}$, $\Gamma_{12}^1 = \Gamma_{21}^1 = -\dfrac{2y}{1 + x^2 + y^2}$, $\Gamma_{22}^1 = \dfrac{2x}{1 + x^2 + y^2}$, $\Gamma_{11}^2 = \dfrac{2y}{1 + x^2 + y^2}$, $\Gamma_{12}^2 = \Gamma_{21}^2 = -\dfrac{2x}{1 + x^2 + y^2}$, $\Gamma_{22}^2 = -\dfrac{2y}{1 + x^2 + y^2}$. Here, coordinate x is considered to be the first, and coordinate y to be the second;

(c) $\Gamma_{11}^1 = -\dfrac{2r}{1 + r^2}$, $\Gamma_{12}^1 = \Gamma_{21}^1 = 0$, $\Gamma_{22}^1 = \dfrac{r^3 - r}{1 + r^2}$, $\Gamma_{11}^2 = 0$, $\Gamma_{12}^2 = \Gamma_{21}^2 = \dfrac{1 - r^2}{r(1 + r^2)}$, $\Gamma_{22}^2 = 0$. Here, coordinate r is considered to be the first, and coordinate φ to be the second.

10.15. (a) $\Gamma_{12}^1 = \Gamma_{21}^1 = -\dfrac{1}{y}$, $\Gamma_{11}^2 = \dfrac{1}{y}$, $\Gamma_{22}^2 = -\dfrac{1}{y}$, $\Gamma_{11}^1 = \Gamma_{22}^1 = \Gamma_{12}^2 = \Gamma_{21}^2 = 0$. Here, coordinate x is considered to be the first, and coordinate y to be the second;

(b) $\Gamma_{11}^1 = \dfrac{2x}{1 - x^2 - y^2}$, $\Gamma_{12}^1 = \Gamma_{21}^1 = \dfrac{2y}{1 - x^2 - y^2}$, $\Gamma_{22}^1 = -\dfrac{2x}{1 - x^2 - y^2}$, $\Gamma_{11}^2 = -\dfrac{2y}{1 - x^2 - y^2}$, $\Gamma_{12}^2 = \Gamma_{21}^2 = \dfrac{2x}{1 - x^2 - y^2}$, $\Gamma_{22}^2 = \dfrac{2y}{1 - x^2 - y^2}$. Here, coordinate x is considered to be the first, and coordinate y to be the second;

(c) $\Gamma_{11}^1 = \dfrac{2r}{1 - r^2}$, $\Gamma_{12}^1 = \Gamma_{21}^1 = 0$, $\Gamma_{22}^1 = -\dfrac{r^3 + r}{1 - r^2}$, $\Gamma_{11}^2 = 0$, $\Gamma_{12}^2 = \Gamma_{21}^2 = \dfrac{1 + r^2}{r(1 - r^2)}$, $\Gamma_{22}^2 = 0$. Here, coordinate r is considered to be the first, and coordinate φ to be the second.

10.17. We shall consider coordinate u to be the first, and coordinate v to be the second. Then $\Gamma^1_{11} = \dfrac{f'f'' + g'g''}{(f')^2 + (g')^2}$, $\Gamma^1_{12} = \Gamma^1_{21} = 0$, $\Gamma^1_{22} = -\dfrac{f'f}{(f')^2 + (g')^2}$, $\Gamma^2_{11} = 0$, $\Gamma^2_{12} = \Gamma^2_{21} = \dfrac{f'}{f}$, $\Gamma^2_{22} = 0$.

10.18. We shall consider coordinate u to be the first, and coordinate v to be the second. Then $\Gamma^1_{11} = -\dfrac{1}{\cos u \sin u}$, $\Gamma^1_{12} = \Gamma^1_{21} = 0$, $\Gamma^1_{22} = -\dfrac{\sin^3 u}{\cos u}$, $\Gamma^2_{11} = 0$, $\Gamma^2_{12} = \Gamma^2_{21} = \cot u$, $\Gamma^2_{22} = 0$.

10.19. We shall consider coordinate u to be the first, and coordinate v to be the second. Then $\Gamma^1_{11} = \Gamma^1_{12} = \Gamma^2_{21} = \Gamma^2_{11} = \Gamma^2_{22} = 0$, $\Gamma^1_{22} = -\sinh u \cosh u$, $\Gamma^2_{12} = \Gamma^2_{21} = \coth u$.

10.20. We shall consider coordinate u to be the first, and coordinate v to be the second. Then $\Gamma^1_{11} = \dfrac{1}{a} \tanh \dfrac{u}{a}$, $\Gamma^2_{12} = \Gamma^2_{21} = \dfrac{1}{a} \tanh \dfrac{u}{a}$, $\Gamma^1_{12} = \Gamma^1_{21} = \Gamma^2_{11} = \Gamma^2_{22} = 0$, $\Gamma^1_{22} = -a \tanh \dfrac{u}{a}$.

10.21. We shall consider that coordinate u is the first, and coordinate v is the second. The metric has the form $ds^2 = du^2 + (u^2 + h^2)\, dv^2$. The Christoffel symbols are equal to $\Gamma^1_{11} = \Gamma^1_{12} = \Gamma^1_{21} = \Gamma^2_{11} = \Gamma^2_{22} = 0$, $\Gamma^1_{22} = -u$, $\Gamma^2_{12} = \Gamma^2_{21} = \dfrac{u}{u^2 + h^2}$.

10.24. (a) Parameterization of the line $\theta = \theta_0$, $\varphi = t$. The equations of the parallel translation:

$$\frac{d\xi^1}{dt} - \sin\theta_0 \cos\theta_0 \xi^2 = 0, \quad \frac{d\xi^2}{dt} + \cot\theta_0 \xi^1 = 0.$$

The general solution of the equations:

$$\xi^1 = -C_1 \sin\theta_0 \cos(t\cos\theta_0 + C_2), \quad \xi^2 = C_1 \sin(t\cos\theta_0 + C_2);$$

(b) parameterization of the line $\theta = t$, $\varphi = \varphi_0$. The equations of the parallel translation:

$$\frac{d\xi^1}{dt} = 0, \quad \frac{d\xi^2}{dt} = -\xi^2 \cot t.$$

The general solution of the equations: $\xi^1 = C_1 \dfrac{1}{\sin t}$, $\xi^2 = C_2$.

10.25. *Hint:* see Problem 10.23, see also Fig. 132, or else make use of the result of the previous problem.

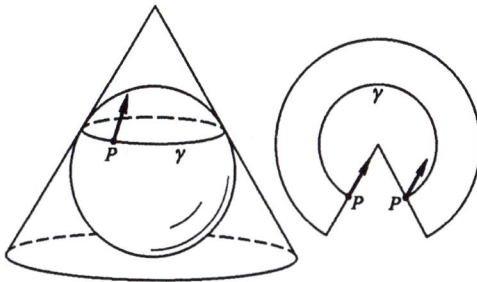

Figure 132

10.27. Let the surface be formed by revolution of the graph of function f. Then the angle α between the initial and final vectors is defined by the equality $\cos\alpha = \cos\dfrac{2\pi}{\sqrt{1+f'^2}}$.

10.28. (a) *Hint:* differentiate covariantly the vector field consisting of vectors tangent to the parallels.

10.29. The angle of revolution is equal to $\dfrac{2\pi}{\sqrt{1+h^2}}$.

10.30. (a) Parameterization of the curve: $x(t) = x_0$, $y(t) = t$, $t \in (0,+\infty)$. The equations of parallel translation: $\dfrac{d\xi^1}{dt} - \dfrac{1}{t}\xi^1 = 0$, $\dfrac{d\xi^2}{dt} - \dfrac{1}{t}\xi^2 = 0$. The general solution of the equations: $\xi^1 = C_1 t$, $\xi^2 = C_2 t$;

(b) parameterization of the curve: $x(t) = t$, $y(t) = y_0$, $t \in (-\infty,+\infty)$. The equations of parallel translation: $\dfrac{d\xi^1}{dt} - \dfrac{1}{y_0}\xi^2 = 0$, $\dfrac{d\xi^2}{dt} - \dfrac{1}{y_0}\xi^1 = 0$. The general solution of the equations:

$$\xi^1 = C_1 \sin\left(\frac{t}{y_0} + C_2\right), \quad \xi^2 = C_1 \cos\left(\frac{t}{y_0} + C_2\right).$$

10.31. (a) *Hint:* substitute a cone for the sphere and make use of the result of Problem 10.22.

(b) *Hint:* substitute a cylinder for the sphere.

(c) make use of the results of the two previous items.

10.36. $\pi + S = \alpha$.

10.37. $\pi - S = \alpha$.

11.7. By to the Meusnier theorem, the radius of curvature R of curve γ at some point is equal to the projection of the radius of the geodesic curvature $R_g = 1/k_g$ on the tangent plane of curve γ, i.e., $R = |R_g \cos\theta|$. Vector $\mathbf{e} = \mathbf{v} \times \mathbf{m}$ is a unit vector lying in the tangent plane to the surface and orthogonal to curve γ, vector $\mathbf{n}$ is a unit vector of the principal normal of curve γ. Hence, $|\cos\theta| = |\langle\mathbf{e},\mathbf{n}\rangle|$ and, therefore,

$$k_g = k\,|\cos\theta| = k\,|\langle\mathbf{e},\mathbf{n}\rangle| = |\langle\mathbf{e},\dot{\mathbf{v}}\rangle| = |\langle\mathbf{v}\times\mathbf{m},\dot{\mathbf{v}}\rangle| = |\langle\mathbf{m}\times\dot{\mathbf{r}},\ddot{\mathbf{r}}\rangle|.$$

11.16. Suppose that the rectilinear generators are parallel to the axis Oz. Then the equation of the surface can be taken in the form

$$\mathbf{r}(u,v) = f(u)\mathbf{e}_1 + \varphi(u)\mathbf{e}_2 + v\mathbf{e}_3,$$

where u is the natural parameter of the directrix. We shall seek for the equation of the geodesic in the form

$$v = v(u). \qquad (*)$$

Then

$$\mathbf{m} = [\mathbf{r}_u \times \mathbf{r}_v] = \varphi'\mathbf{e}_1 - f'\mathbf{e}_2,$$

$$dr = (f' \mathbf{e}_1 + \varphi' \mathbf{e}_2 + v' \mathbf{e}_3)\, du,$$

$$d^2 \mathbf{r} = (f'' \mathbf{e}_1 + \varphi'' \mathbf{e}_2 + v'' \mathbf{e}_3)\, du^2,$$

and the equation for defining geodesic curves has the form (see Problem 11.8):

$$\begin{vmatrix} \varphi' & -f' & 0 \\ f' & \varphi' & v' \\ f'' & \varphi'' & v'' \end{vmatrix} = 0,$$

or

$$\left(\varphi'^2 + f'^2\right) v'' - \left(\varphi'\varphi'' + f'f''\right) v' = 0.$$

But $\varphi'^2 + f'^2 = 1$, therefore,

$$\varphi'\varphi'' + f'f'' = \frac{1}{2}\left(\varphi'^2 + f'^2\right)' = 0.$$

Thus, $v'' = 0$, i.e., $v = c_1 u + c_2$. The vector equation of the family of geodesics has the form:

$$\rho(u) = f(u)\, \mathbf{e}_1 + \varphi(u)\, \mathbf{e}_2 + (c_1 u + c_2)\, \mathbf{e}_3.$$

Note that

$$\cos\theta = \cos\left(\widehat{\rho_u, \mathbf{e}_3}\right) = \frac{c_1}{\sqrt{1 + c_1^2}}.$$

Therefore, the found geodesics are generalized helical lines. Besides, the rectilinear generators are geodesics. They were not included into the found family of geodesics, as the equations of rectilinear generators cannot be represented in the form (*).

11.18. The equations of the geodesics have the form:

$$\mathbf{r}(v) = \left(\frac{C \cos v}{\sin\left((C_1 \pm v)/\sqrt{2}\right)}, \frac{C \sin v}{\sin\left((C_1 \pm v)/\sqrt{2}\right)}, \frac{C}{\sin\left((C_1 \pm v)/\sqrt{2}\right)} \right).$$

11.19. Consider the parametric equation of a cone in the form $\mathbf{r}(u, v) = u\rho(v)$ and take $|\rho| = 1$, $|\rho'| = 1$. Then the equations of the geodesics have the form:
$$\mathbf{r}(v) = \frac{C_1}{\sin(C - v)}\rho(v).$$

11.20. The proof is obtained from the consideration of the development of the cone, on which geodesics are rectilinear segments. Closed geodesics on the cone have the form shown in Fig. 65. At the point of self-intersection, a geodesic intersects itself at an angle distinct from 0 and π. This angle depends only on the angle at the vertex of the cone.

11.22. $v = C_1 + \displaystyle\int \frac{C\,du}{\sqrt{(u^2 + h^2)(u^2 + h^2 - C^2)}}.$

11.23. We shall consider v as the function u along the geodesic.

The Christoffel symbols have the form (see Problem 10.13):

$$\Gamma_{11}^1 = \frac{\varphi_u'}{2(\varphi + \psi)}, \quad \Gamma_{12}^1 = \Gamma_{21}^1 = \frac{\psi_v'}{2(\varphi + \psi)}, \quad \Gamma_{22}^1 = -\frac{\varphi_u'}{2(\varphi + \psi)},$$

$$\Gamma_{11}^2 = -\frac{\psi_v'}{2(\varphi + \psi)}, \quad \Gamma_{12}^2 = \Gamma_{21}^2 = \frac{\varphi_u'}{2(\varphi + \psi)}, \quad \Gamma_{22}^2 = \frac{\psi_v'}{2(\varphi + \psi)}.$$

Hence, we obtain the equations of the geodesics:

$$2\left(\psi+\varphi\right)\frac{d^2v}{du^2} = \frac{d\psi}{dv} - 2\frac{d\varphi}{du}\frac{dv}{du} - \frac{d\psi}{dv}\left(\frac{dv}{du}\right)^2,$$

$$0 = \frac{d\varphi}{du} + 2\frac{d\psi}{dv}\frac{dv}{du} - \frac{d\varphi}{du}\left(\frac{dv}{du}\right)^2.$$

We multiply the second equation by $\dfrac{dv}{du}$ and add to the first equation to obtain

$$2\left(\psi+\varphi\right)\frac{d^2v}{du^2} = \frac{d\psi}{dv} - \frac{d\varphi}{du}\frac{dv}{du} + \frac{d\psi}{dv}\left(\frac{dv}{du}\right)^2 - \frac{d\varphi}{du}\left(\frac{dv}{du}\right)^3.$$

It is easy to see that this condition is tantamount to the equality

$$\frac{d}{du}\left(\frac{\psi - \varphi\left(dv/du\right)^2}{1 + \left(dv/du\right)^2}\right) = 0.$$

Integrating this equality, we get the sought-for equations.

11.26. $\ddot{\varphi} + 2\dfrac{\rho'\left(\tau\right)}{\rho\left(\tau\right)}\dot{\varphi}\dot{\tau} = 0,$

$$\ddot{\tau} + \frac{\rho'\left(\tau\right)}{\rho\left(\tau\right)}\left(\dot{\tau}^2 - \dot{\varphi}^2\right) = 0.$$

11.29. *Hint:* make use of Clairaut's equation.

11.35. (a) $\dfrac{a}{a^2 + h^2}$; (b) 0.

11.36. $\dfrac{\sinh v}{\sqrt{2}\cosh^2 v}$.

11.37. The geodesic curvature is equal to 1.

11.40. The geodesic curvature of the meridians is equal to 0, and of the parallels, to $\dfrac{f'\left(u\right)}{f\left(u\right)\sqrt{1 + f'^2\left(u\right)}}$.

11.41. The geodesic curvature of curves $v = $ const is $\dfrac{1}{\sqrt{EG}}\dfrac{\partial\sqrt{E}}{\partial v}$, and of curves $u = $ const is equal to $\dfrac{1}{\sqrt{EG}}\dfrac{\partial\sqrt{G}}{\partial v}$.

12.5. $R = -\dfrac{1}{\lambda}\Delta\ln\lambda.$

12.8. (a) $\Gamma_{11}^1 = \Gamma_{12}^1 = \Gamma_{11}^2 = 0,\ \ \Gamma_{12}^2 = \dfrac{G_u}{2G},\ \Gamma_{22}^1 = -\dfrac{G_u}{2},\ \Gamma_{22}^2 = \dfrac{G_v}{2G}.$

(b) $K = -\dfrac{\left(\sqrt{G}\right)_{uu}}{\sqrt{G}}.$

12.9. $K = -\dfrac{\omega_{uv}}{\sin\omega}.$

13.11. (a) $2(z-1)\,dx \wedge dy \wedge dz$; (b) $yzdx \wedge dz + xzdy \wedge dz$; (c) $6y^2 dx \wedge dy \wedge dz$; (d) 0; (e) 0; (f) 0; (g) $df \wedge dg$; (h) 0.

13.12. Reduce the problem to the case of fixed factors.

13.22. (a) $(2r\cos\theta,\; -r\sin\theta,\; 0)$;

(b) $(6r\sin\theta + e^r \sin\varphi r \sin\theta)$;

(c) $\left(-\dfrac{2\cos\theta}{r^3},\; -\dfrac{\sin\theta}{r^3},\; 0\right)$.

13.39. *Hint:* consider the vector field $\mathbf{X}$ dual to the form ω. Show that there exists a system of local coordinates $(x^1, \ldots, x^n)$, in which $\mathbf{X} = \dfrac{\partial}{\partial x^1}$. Further, consider the forms Ω and ω in this system of coordinates.

14.30. As space X, we take space l_2 whose elements are sequences of real numbers $x = (x_1, x_2, \ldots, x_n, \ldots)$, satisfying the condition $||x||^2 = \sum_{n=1}^{\infty} |x_n|^2 < \infty$. As space $Y \subset X$, we take a sphere in X, i.e., a set of such x, for which $||x||^2 = 1$. We consider in Y a sequence of points x_i, which have 1 in the i-th position and 0 elsewhere. This infinite sequence has no accumulation point, as $||x_i - x_j|| = \sqrt{2}$ for any i,j. Therefore, Y is not a compact.

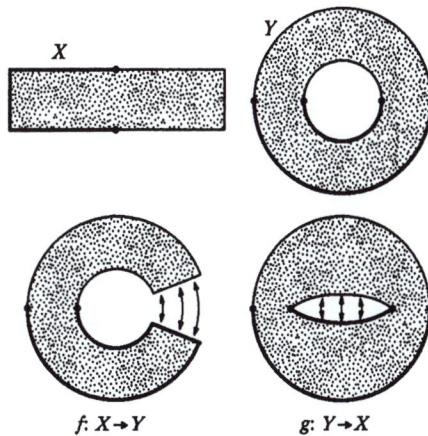

Figure 133

14.44. No. If a topological metric compact space is connected, it is not necessarily linearly connected. A known example is a tuple of points on the plane (x,y) given as:

$$\left\{y = \sin\frac{1}{x}\right\} \cup \{(x = 0;\; -1 \le y \le 1)\}.$$

14.57. See an example in Fig. 133.

15.18. (a) Curves of the level of conjugate function $\operatorname{Im} z^n = r^n \sin n\varphi$ are integral trajectories of vector field $\operatorname{grad}(\operatorname{Re} z^n)$. The unique singular point of the field

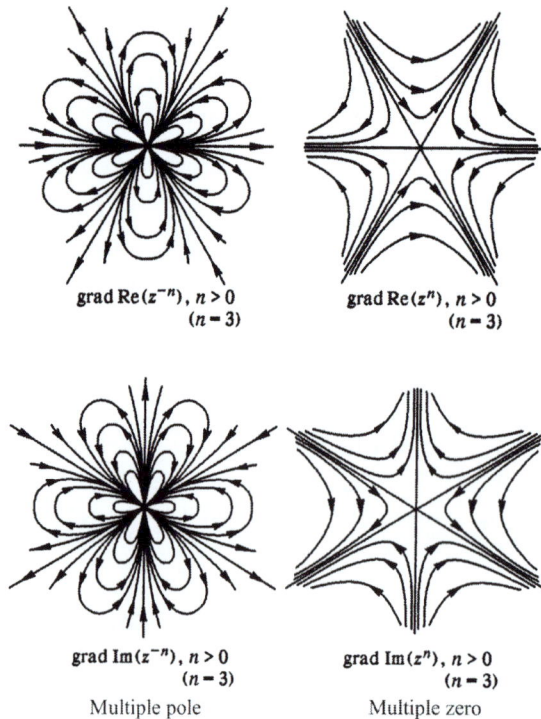

grad Re(z^{-n}), $n > 0$
($n = 3$)

grad Re(z^n), $n > 0$
($n = 3$)

grad Im(z^{-n}), $n > 0$
($n = 3$)

grad Im(z^n), $n > 0$
($n = 3$)

Multiple pole

Multiple zero

Figure 134

grad(Re z^n) is $z = 0$, as only at this point $f'(z) = 0$. The point $z = 0$ is a degenerate saddle. See Fig. 134.

Make a small perturbation of the function $z^n \rightarrow \prod\limits_{i=1}^{n} (z - \varepsilon_i)$. Then the singular point decomposes into $n - 1$ second-order nondegenerate saddles. Consider the behaviour of the integral trajectories near one of the singular points. Decompose the function into a Taylor series:

$$f(z) = f(a_i) + f'(a_i)(z - a_i) + \frac{f''(a_i)}{2}(z - a_i)^2 + \dots$$

The decomposition begins from a second-order term, as $f'(a_i) = 0$. Herewith, $f''(a_i) \neq 0$ (a nondegenerate critical point), as $f''(a_i) = 0$ then and only then, when a_i is a multiple root for $f(z)$. (See also Figs 137, 138.)

(b) For the function $f(z) = z + \dfrac{1}{z}$, going over to the polar coordinates $(z = \rho e^{i\varphi})$, we have:

$$\text{Re}(f(z)) = \left(\rho + \frac{1}{\rho}\right)\cos\varphi, \quad \text{Im}(f(z)) = \left(\rho - \frac{1}{\rho}\right)\sin\varphi.$$

The origin is a singular point, as the function $\dfrac{1}{z}$ is discontinuous. The derivative

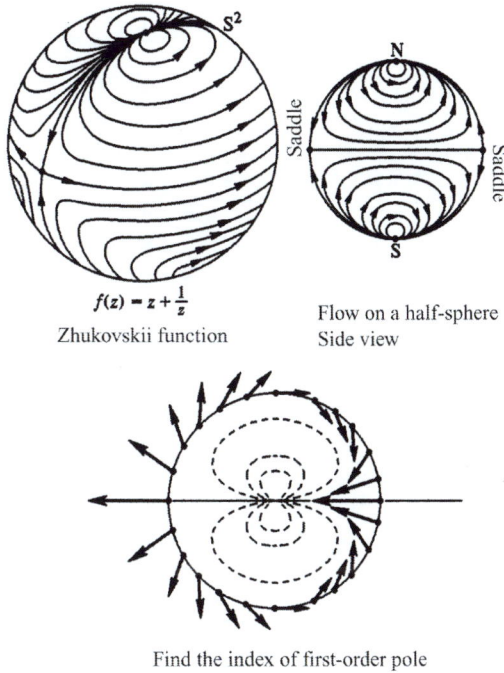

$f(z) = z + \frac{1}{z}$

Zhukovskii function

Flow on a half-sphere
Side view

Find the index of first-order pole

Figure 135

of the function $f(z)$ is equal to $1 - \dfrac{1}{z^2}$, i.e., the singular points are $z = 1$ and $z = -1$. Both points are nondegenerate. Consider the integral trajectories of the field $\operatorname{grad}(\operatorname{Re} f(z))$. They are given by the equations $\left(\rho - \dfrac{1}{\rho}\right) \sin \varphi = c$. For the integral trajectories entering the singular points and exiting them, i.e., for separatrices, $c = 0$. Hence, we get that the separatrices are given by the equations $\varphi = \pm \pi$ (a unit circle consisting of two separatrices) and $\rho = 1$ (a real axis consisting of four separatrices). In a similar manner, we can construct the separatrices of the field $\operatorname{grad}(\operatorname{Im} f(z))$, which are given by the equation $\left(\rho + \dfrac{1}{\rho}\right) \cos \varphi = 2$ and have the form of two tangent loops. See Figs 135, 136. See also Fig. 137.

(c) $f(z) = z + \dfrac{1}{z^2}$. Consider $\operatorname{grad} \operatorname{Re} f(z)$. The integral trajectories of this flow are curves of the level of the function $\operatorname{Im} f(z)$:

$$\operatorname{Im} f(z) = y - \frac{2xy}{(x^2 + y^2)^2} = r \sin \varphi - \frac{\sin 2\varphi}{r^2}.$$

Curves of the level of the function $\operatorname{Re} f(z) = r \cos \varphi - \dfrac{\cos 2\varphi}{r^2}$ are sought for in a similar manner.

(d) The singular points of the function $f(z) = z + \dfrac{1}{z - 2}$ are $z = 2$ (the pole)

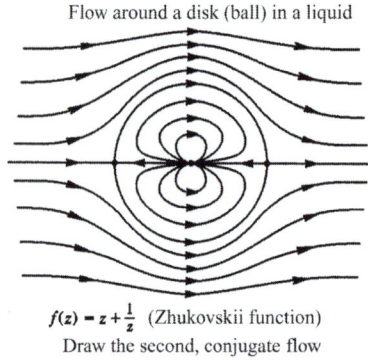

Flow around a disk (ball) in a liquid

$f(z) = z + \frac{1}{z}$ (Zhukovskii function)
Draw the second, conjugate flow

Figure 136

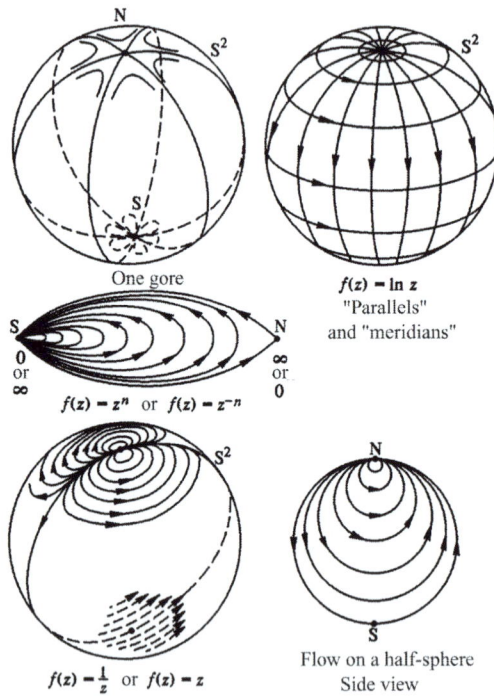

One gore

$f(z) = \ln z$
"Parallels"
and "meridians"

$f(z) = z^n$ or $f(z) = z^{-n}$

$f(z) = \frac{1}{z}$ or $f(z) = z$

Flow on a half-sphere
Side view

Figure 137

and $z = 1$, $z = 3$ (the zeros of the function $f'(z)$). The singular points $z = 1$ and $z = 3$ are nondegenerate saddles, as $f''(1) \neq 0$ and $f''(3) \neq 0$. In the neighbourhood of the point $z = 2$, the integral trajectories of the vector fields $\mathrm{grad}\,(\mathrm{Re}\,f(z))$ and $\mathrm{grad}\,(\mathrm{Im}\,f(z))$ are qualitatively organized in the same way as the respective trajectories for the function $f(z) = \dfrac{1}{z}$ in the neighbourhood of zero.

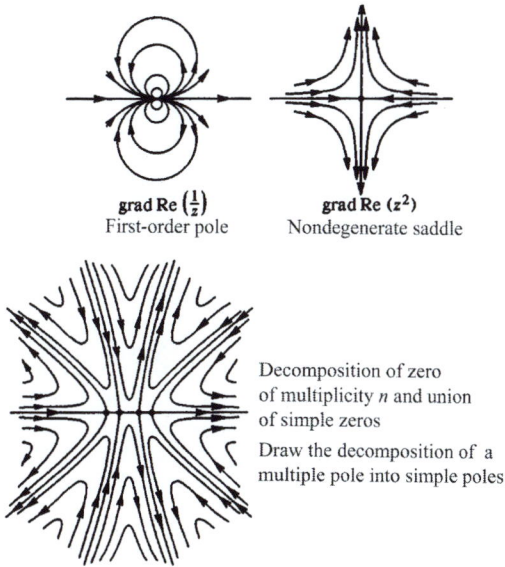

grad Re $\left(\frac{1}{z}\right)$
First-order pole

grad Re (z^2)
Nondegenerate saddle

Decomposition of zero
of multiplicity n and union
of simple zeros

Draw the decomposition of a
multiple pole into simple poles

Figure 138

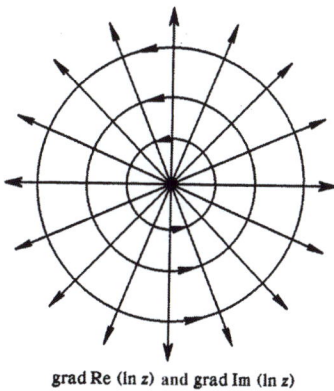

grad Re $(\ln z)$ and grad Im $(\ln z)$

Figure 139

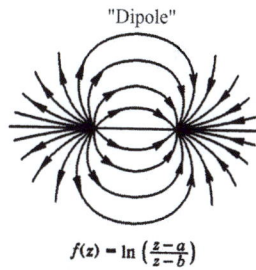

"Dipole"

$f(z) = \ln\left(\frac{z-a}{z-b}\right)$

Draw the second, conjugate flow

Figure 140

(e) See Fig. 139.

(f) See Fig. 140.

(g) The singular points of the function $f(z) = z^3 (z-1)^{100} (z-2)^{900}$, i.e., the zeros of the derivative $f'(z)$, are as follows: $z_1 = 0$, a second-order saddle; $z_2 = 1$, a 99-order saddle; $z_3 = 2$, a 899-order saddle; $z_4 \approx 0.005$ and $z_5 \approx 1.1$ (these are the roots of the quadratic equation $1003z^2 - 1109z + 6 = 0$) are nondegenerate saddles. Locally, in the neighbourhood of each singular point, the integral trajectories behave the same as in the neighbourhood of corresponding-order saddles.

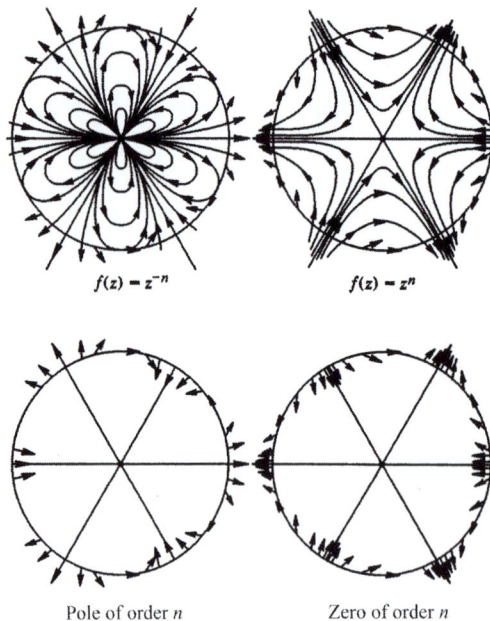

Figure 141

(h) The function $f(z) = 1 + z^4 \left(z^4 - 4\right)^{44} \left(z^{44} - 44\right)^{444}$ in the neighbourhood of the point $z = 0$ can be replaced by $\tilde{f}(z) = 1 + 4^{44} 44^{444} z^4$. The qualitative pattern of the behaviour of the curves in the neighbourhood of the point $z = 0$ would not change because of this. But the addition of a constant does not change the appearance of the trajectories, so we can consider the function $f_1(z) = cz^4$ in the neighbourhood of zero, where $c = 4^{44} 44^{444}$. The point $z = 0$ is a nondegenerate singular point for the function $f_1(z)$ (and, thus, for the function $f(z)$ too). At a suitably small perturbation, this singular point decomposes into three nondegenerate singular points.

(i) The function $f(z) = \dfrac{1}{100} \ln \left(\dfrac{z - 2i}{z - 4}\right)^3$ has logarithmic singularities at the points $z = 2i$ and $z = 4$. There are no singular points besides them.

(j) To simplify the representation of the function $f(z) = \dfrac{1}{z^2 + 2z - 1}$, we make the shift $w = z + 1$. Then $g(w) = f(w - 1) = \dfrac{1}{w^2 - 2}$. The points $w = \pm\sqrt{2}$ are singular points (poles) of the function $g(w)$. The singular points of the vector fields $\mathrm{grad}\,(\mathrm{Re}\,f(z))$ and $\mathrm{grad}\,(\mathrm{Im}\,f(z))$ coincide with zeros of the function $f'(z)$. Hence, $w = 0$ is a singular point; moreover, it is nondegenerate, as $g''(0) \neq 0$.

(k) The point $z = 0$ is a singular point for the function $f(z) = \dfrac{2}{z} + 21 \ln z^2$. Besides, $z = 1/21$ is a singular point of the vector fields $\mathrm{grad}\,(\mathrm{Re}\,f(z))$ and $\mathrm{grad}\,(\mathrm{Im}\,f(z))$, as $f'(1/21) = 0$.

l) The point $z = 0$ is a singular point for the function $f(z) = z^5 + 2\ln z$. The

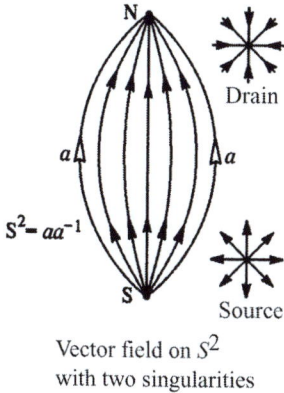

Vector field on S^2
with two singularities

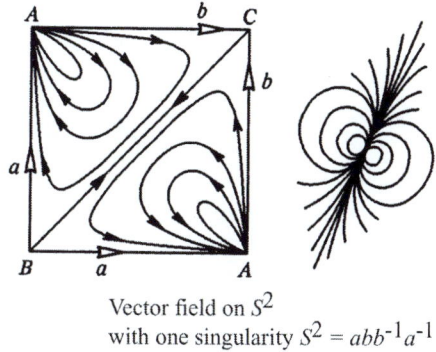

Vector field on S^2
with one singularity $S^2 = abb^{-1}a^{-1}$

Figure 142   **Figure 143**

zeros of the derivative $f'(z)$ are the roots of the equation $z^5 = -2/5$ (the vertices of a pentagon). At these points $z_1, \ldots, z_5$ we have: $f(z) \sim f(z_i) + k_i(z - z_i)^2 + \ldots$, where $k_i \neq 0, i = 1, \ldots, 5$.

m) The points $z = 1$ and $z = -10i$ are singular points of the functions

$$ f(z) = 2\ln(z-1)^2 - \frac{4}{3}\ln(z+10i)^3. $$

Herewith, $f'(z) \neq 0$ for all values of z.

n) The points $z = 0$ and $z = i$ are singular points of the function $f(z) = \dfrac{1}{z^3} - \dfrac{1}{(z-i)^3}$ (third-order poles). We differentiate $f'(z) = -\dfrac{3}{z^4} + \dfrac{1}{(z-i)^4} = 0$ to obtain four points, at which $f'(z) = 0$:

$$ \frac{i\sqrt[4]{3}}{\sqrt[4]{3}-1}, \; \frac{i\sqrt[4]{3}}{\sqrt[4]{3}+1}, \; \frac{i\sqrt{3}+\sqrt[4]{3}}{\sqrt{3}+1}, \; \frac{i\sqrt{3}-\sqrt[4]{3}}{\sqrt{3}+1}. $$

The integral trajectories of the vector fields

$$ \text{grad}\,(\text{Re}\,f(z)) \quad \text{and} \quad \text{grad}\,(\text{Im}\,f(z)) $$

on the infinity behave in the same way as the integral trajectories of respective vector fields for the function $1/z^4$.

15.19. (a), (b) See Fig. 141.

15.22. (a) See Fig. 142.

15.23. See: (a) Figs 143, 144; (b) Fig. 145; (c) Fig. 146; (d) Fig. 147; (e) Figs 148, 149; (f) Figs 150, 151; (g) Fig. 152.

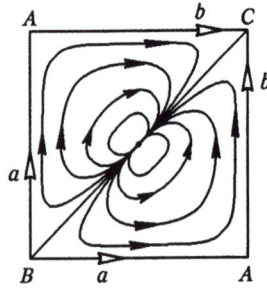

Figure 144 A vector field on a sphere with one singularity $S^2 = abb^{-1}a^{-1}$

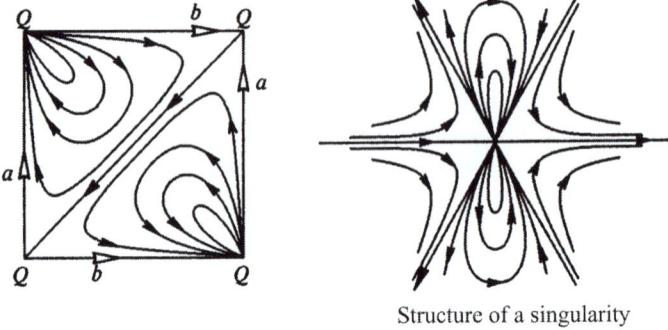

Structure of a singularity

Figure 145 A vector field on a torus with one singularity $T^2 = aba^{-1}b^{-1}$

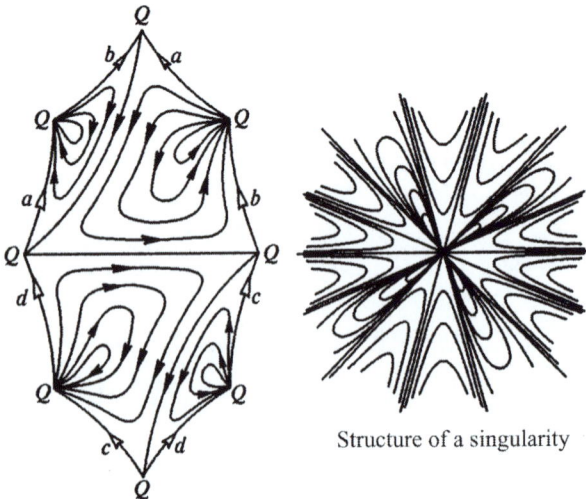

Structure of a singularity

Figure 146 A vector field on a pretzel $aba^{-1}b^{-1}c^{-1}d^{-1}cd$

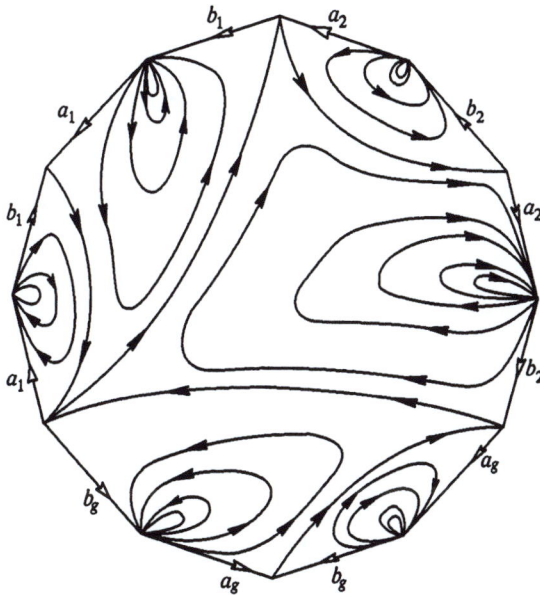

Figure 147 A vector field with one singular point on an orientable surface of genus g: $a_1 b_1 a_1^{-1} b_1^{-1} \cdots a_g b_g a_g^{-1} b_g^{-1}$

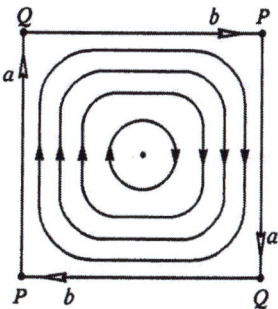

Figure 148

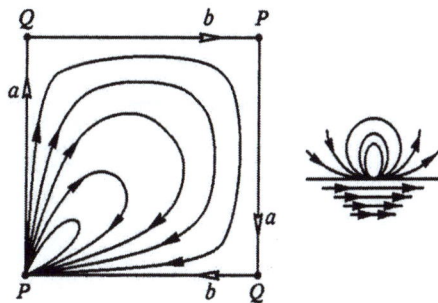

Figure 149 A field on $\mathbb{R}P^2$ with one singular point $\mathbb{R}P^2 = abab$

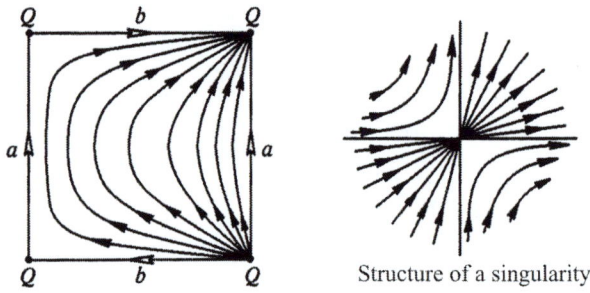

Structure of a singularity

Figure 150 A vector field on a Klein bottle with one singularity $\mathbb{K}L = aba^{-1}b$

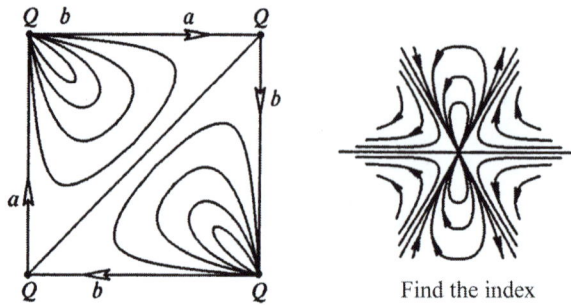

Find the index

Figure 151 A field with one singularity on $\mathbb{K}L = aabb$ (a Klein bottle)

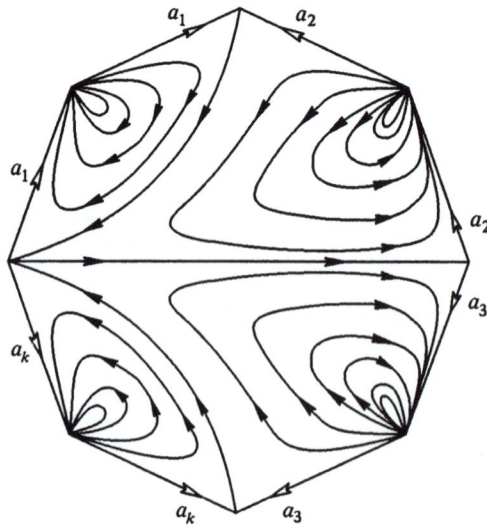

Figure 152 A field with one singular point on a nonorientable surface of a general form

Answers and Solutions

Part 2

16.1. (b) $\dfrac{-a^2}{\left(\cosh u_1 - \cos u_2\right)^2}$, $\dfrac{-\left(\cosh u_1 - \cos u_2\right)^2}{a^2}$; (c) yes.

16.2. (b) $\dfrac{c^3 \sin u_1}{\left(\cosh u_2 - \cos u_1\right)^3}$, $\dfrac{\left(\cosh u_2 - \cos u_1\right)^3}{c^3 \sin u_1}$; (c) yes.

16.3. (b) $c^3\left(u_2^2 - u_1^2\right)$, $\dfrac{1}{c^3\left(u_2^2 - u_1^2\right)}$; (c) yes.

16.4. (b) $c^3 u_1 u_2 \dfrac{u_1^2 - u_2^2}{\sqrt{\left(1 - u_2^2\right)\left(u_1^2 - 1\right)}}$, $\dfrac{\sqrt{\left(1 - u_2^2\right)\left(u_1^2 - 1\right)}}{c^3 u_1 u_2 \left(u_1^2 - u_2^2\right)}$; (c) yes.

16.5. $\dfrac{\partial^2 V}{\partial \varphi^2} + V$.

16.6. $\dfrac{\partial^2 V}{\partial u^2} + 2uv^2 \dfrac{\partial V}{\partial u} + 2v\left(1 - v^2\right) \dfrac{\partial V}{\partial v} + u^2 v^2 V = 0$, the domain $y \neq 0$, the range $v \neq 0$.

16.7. $\left(\dfrac{1}{4} \ln^2\left(u^2 + v^2\right) + \arctan^2 \dfrac{v}{u}\right)\left(u^2 + v^2\right)\left(\dfrac{\partial^2 V}{\partial u^2} + \dfrac{\partial^2 V}{\partial v^2}\right)$.

16.8. $\dfrac{1}{4} \ln^2 \dfrac{\left(u - u_1\right)^2 + \left(v - v_1\right)^2}{\left(u - u_2\right)^2 + \left(v - v_2\right)^2} +$

$\arctan^2 \dfrac{\left(u - u_2\right)\left(v - v_1\right) - \left(u - u_1\right)\left(v - v_2\right)}{\left(u - u_1\right)\left(u - u_2\right) + \left(v - v_1\right)\left(v - v_2\right)} \times$

$\dfrac{\left(\left(u - u_2\right)^2 + \left(v - v_2\right)^2\right)\left(\left(u - u_1\right)^2 + \left(v - v_1\right)^2\right)}{\left(u_2 - u_1\right)^2 + \left(v_2 - v_1\right)^2}\left(\dfrac{\partial^2 V}{\partial u^2} + \dfrac{\partial^2 V}{\partial v^2}\right)$.

16.9. $\dfrac{1}{9\left(u^2 + v^2\right)\left((u - 2)^2 + v^2\right)}\left(\dfrac{\partial^2 V}{\partial u^2} + \dfrac{\partial^2 V}{\partial v^2}\right)$.

17.1. (a) For instance: $x = t^2$, $y = t^3$. See Fig. 153.

17.3. (a) For instance: $x = \dfrac{t^n}{\ln\left(1/t^2\right)}$, $y = \dfrac{|t|^n}{\ln\left(1/t^2\right)}$.

17.4. (a) k is odd and $n \neq mk$, $m \in \mathbb{N}$; (b) k is odd and $n = km$, $m \in \mathbb{N}$; (c) k is even.

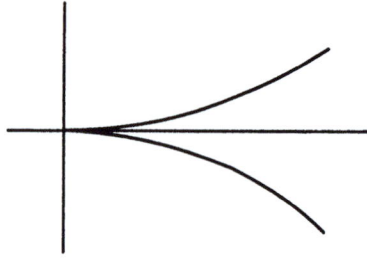

Figure 153 An irregular curve with analytical parameterization

17.5. If the curvature at the point of transition of the curve from one plane to the other is different from zero, then the smoothness is not greater than C^1; if, however, the curvature at this point can be equal to zero, then C^∞.

17.6. One of the possible answers for analytic parameterization:

$$x = a \cos \varphi \frac{\sqrt{1 + \cos^2 \varphi}}{\sqrt{2}}, \quad y = a \cos \varphi \frac{\sin \varphi}{\sqrt{2}}, \quad \varphi \in [0, 2\pi].$$

17.7. No. An obstacle is the rotation number of the regular closed curve.

17.8. (a) $y^2 = 2ax - 2C$ is a parabola with the axis OX and parameter $p = |a|$. The curve is turned by its convexity to the left-hand side in the case of $a < 0$ and to the right-hand side in the case of $a > 0$; (b) $y = Ce^{-x/a}$; (c) $(x - C)^2 + y^2 = a^2$, the circle of radius a with the centre on the axis OX.

17.9. The constant tangent length condition is written as $y\sqrt{1 + \left(\dfrac{dx}{dy}\right)^2} = a$. We shall consider the curve only in the upper half-plane and for this reason let $|y| = y > 0$. Consider an angle φ, $0 < \varphi < \pi$, defined from the condition $\tan \varphi = dy/dx$. By substituting dx/dy via $\cot \varphi$, we get $y/\sin \varphi = a$ or $y = a \sin \varphi$. Hence, $dy = a \cos \varphi d\varphi$. But from the condition defining the angle φ, it follows that $dx = \cot \varphi dy$. Substituting here the expression obtained for dy, we have $dx = a \dfrac{\cos^2 \varphi}{\sin \varphi} d\varphi$, or $dx = a \left(\dfrac{1}{\sin \varphi} - \sin \varphi\right) d\varphi$. Integrating termwise, we find $x = a \left(\ln \tan \dfrac{\varphi}{2} + \cos \varphi\right) + C$. The equation of this curve can be written in another way:

$$x = \frac{a}{2} \ln \frac{a + \sqrt{a^2 - x^2}}{a - \sqrt{a^2 - x^2}} - \sqrt{a^2 - x^2}.$$

17.10. $y = a^3 / (x^2 + a^2)$; $x = a \cot t$, $y = a \sin^2 t$.

17.11. *Hint:* apply Rolle's theorem to the function $\langle \mathbf{a}, \mathbf{r}(t) - \mathbf{r}(t_0) \rangle$.

17.12. Use the fact that $|\mathbf{r}_2(t) - \mathbf{r}_1(t)|^2 = $ const, where $\mathbf{r}_1, \mathbf{r}_2$ are the radius vectors of moving points, t is time.

17.13. Let $\dfrac{\mathbf{r}'}{\mathbf{r}} = \lambda$; $\lambda(t)$ be a function, continuous on a segment $[a, b]$, preserving a definite sign. We have $\mathbf{r}' - \lambda \mathbf{r} = 0$, from where $\mathbf{r} = \mathbf{a}e^{\int \lambda dt}$. As the derivative of the function $e^{\int \lambda dt}$ is equal to $\lambda e^{\int \lambda dt}$, it preserves its sign on the segment $[a, b]$, i.e., $e^{\int \lambda dt}$ is a monotone and continuous function of t.

17.14. Applying the method of solving the previous problem, we shall have $\mathbf{r}' = \mathbf{a}e^{\int \lambda dt}$, from where

$$\mathbf{r} = \mathbf{a} \int e^{\int \lambda dt} dt + \mathbf{b}.$$

The derivative of $\int e^{\int \lambda dt} dt$ is equal to $e^{\int \lambda dt} > 0$; therefore, $\int e^{\int \lambda dt} dt$ is a monotonously increasing function of $t \in [a, b]$.

Note. We shall denote by $[\mathbf{r}]$ the vector obtained on $\mathbf{r}$ by revolution to an angle $+\dfrac{\pi}{2}$.

17.15. The radius vector $\boldsymbol{\rho}$ of an arbitrary point of an immobile centroid can be defined by one of the relations: $\boldsymbol{\rho} = \mathbf{r}_1 + \lambda\,[\mathbf{r}_1'] = \mathbf{r}_2 + \mu\,[\mathbf{r}_2']$, $\mathbf{r}_1 - \mathbf{r}_2 + \lambda\,[\mathbf{r}_1'] = \mu\,[\mathbf{r}_2']$, $\langle \mathbf{r}_1 - \mathbf{r}_2, \mathbf{r}_2' \rangle + \langle \lambda\,[\mathbf{r}_1']\,\mathbf{r}_2' \rangle = 0$, $\lambda = \dfrac{\langle \mathbf{r}_2 - \mathbf{r}_1, \mathbf{r}_2' \rangle}{|\mathbf{r}_1' \times \mathbf{r}_2'|}$. Therefore, $\boldsymbol{\rho} = \mathbf{r}_1 + \dfrac{\langle \mathbf{r}_2 - \mathbf{r}_1, \mathbf{r}_2' \rangle}{|\mathbf{r}_1' \times \mathbf{r}_2'|}\,[\mathbf{r}_1']$.

In coordinates:

$$\xi = x_1 - \frac{(x_2 - x_1)\,x_2' + (y_2 - y_1)\,y_2'}{x_1'y_2' - x_2'y_1'}\,y_1',$$

$$\eta = y_1 - \frac{(x_2 - x_1)\,x_2' + (y_2 - y_1)\,y_2'}{x_1'y_2' - x_2'y_1'}\,x_1'.$$

17.16. Consider the vector $\lambda\,[\mathbf{r}_1']$, where $\lambda = \dfrac{\langle \mathbf{r}_2 - \mathbf{r}_1, \mathbf{r}_2' \rangle}{|\mathbf{r}_1' \times \mathbf{r}_2'|}$; if this vector is laid off from the endpoint M_1 of the rod, its endpoint will get into the instantaneous centre of revolution. The projections of the vector $\lambda\,[\mathbf{r}_1']$ to the vectors $\mathbf{r}_2 - \mathbf{r}_1$ and $[\mathbf{r}_2 - \mathbf{r}_1]$ are, respectively, equal to:

$$\frac{\langle \lambda\,[\mathbf{r}_1']\,,\mathbf{r}_2 - \mathbf{r}_1 \rangle}{|\mathbf{r}_2 - \mathbf{r}_1|} \quad \text{and} \quad \frac{\langle \lambda\,[\mathbf{r}_1']\,,[\mathbf{r}_2 - \mathbf{r}_1] \rangle}{|\mathbf{r}_2 - \mathbf{r}_1|}.$$

Therefore, the equations of a movable centroid are:

$$x = \frac{\langle \mathbf{r}_2 - \mathbf{r}_1, \mathbf{r}_1' \rangle}{|\mathbf{r}_1' \times \mathbf{r}_2'|}\,\frac{|\mathbf{r}_1' \times (\mathbf{r}_2 - \mathbf{r}_1)|}{|\mathbf{r}_2 - \mathbf{r}_1|}, \qquad y = \frac{\langle \mathbf{r}_2 - \mathbf{r}_1, \mathbf{r}_2' \rangle}{|\mathbf{r}_1' \times \mathbf{r}_2'|}\,\frac{\langle \mathbf{r}_1', \mathbf{r}_2 - \mathbf{r}_1 \rangle}{|\mathbf{r}_2 - \mathbf{r}_1|},$$

or

$$x = \frac{\left((x_2 - x_1)\,x_1' + (y_2 - y_1)\,y_1'\right) \begin{vmatrix} x_1' & y_1' \\ x_2 - x_1 & y_2 - y_1 \end{vmatrix}}{\begin{vmatrix} x_1' & y_1' \\ x_2' & y_2' \end{vmatrix} \sqrt{(x_2 - x_1)^2 + (y_2 - y_1)^2}},$$

$$y = \frac{\left((x_2 - x_1)\,x_2' + (y_2 - y_1)\,y_2'\right)\left((x_2 - x_1)\,x_1' + (y_2 - y_1)\,y_1'\right)}{\begin{vmatrix} x_1' & y_1' \\ x_2' & y_2' \end{vmatrix} \sqrt{(x_2 - x_1)^2 + (y_2 - y_1)^2}}.$$

17.17. $\mathbf{R} = \mathbf{r}_1 + \xi\mathbf{a} + \eta\,[\mathbf{a}]$, where $\mathbf{a} = \mathbf{r}_2 - \mathbf{r}_1$, $\xi = \text{const}$, $\eta = \text{const}$ (point M is rigidly connected with the rod); $\mathbf{R}' = \mathbf{r}_1' + \xi\mathbf{a}' + \eta\,[\mathbf{a}']$. As $|\mathbf{a}| = |\mathbf{r}_2 - \mathbf{r}_1| = \text{const}$, then $\mathbf{a}' \perp \mathbf{a}$; therefore, $\mathbf{a}' = s\,[\mathbf{a}]$, $\mathbf{r}_2' - \mathbf{r}_1' = s\,[\mathbf{r}_2 - \mathbf{r}_1]$, $\langle \mathbf{r}_2' - \mathbf{r}_1', [\mathbf{r}_1'] \rangle = s\,\langle [\mathbf{r}_2 - \mathbf{r}_1],$

$[\mathbf{r}'_1]$), $\langle [\mathbf{r}'_1], \mathbf{r}'_2 \rangle = \langle s (\mathbf{r}_2 - \mathbf{r}_1), \mathbf{r}'_1 \rangle$, $s = \dfrac{|\mathbf{r}'_1 \times \mathbf{r}'_2|}{\langle \mathbf{r}_2 - \mathbf{r}_1, \mathbf{r}'_1 \rangle} = \dfrac{1}{\lambda}$. Thus, $\mathbf{a}' = \dfrac{1}{\lambda} [\mathbf{a}]$, $[\mathbf{a}'] =$

$-\dfrac{1}{\lambda}\mathbf{a}$, $\mathbf{R}' = \mathbf{r}'_1 + \dfrac{\xi}{\lambda} [\mathbf{a}] - \eta \lambda \mathbf{a} = \dfrac{1}{\lambda} (\lambda \mathbf{r}'_1 + \xi [\mathbf{a}] - \eta \mathbf{a})$. On the other hand,

$$\mathbf{r} = \mathbf{R} - \rho = \mathbf{r}_1 + \xi \mathbf{a} + \eta [\mathbf{a}] - \mathbf{r}_1 - \lambda [\mathbf{r}'_1] = \xi \mathbf{a} + \eta [\mathbf{a}] - \lambda [\mathbf{r}'_1],$$

$$[\mathbf{r}] = \lambda \mathbf{r}'_1 - \xi [\mathbf{a}] - \eta \mathbf{a},$$

therefore, $\mathbf{R}' = \dfrac{1}{\lambda} [\mathbf{r}]$, $\omega = \dfrac{1}{\lambda}$.

17.21. $\dfrac{x^2}{\left(a/\sqrt{2}\right)^2} + \dfrac{y^2}{\left(b/\sqrt{2}\right)^2} = 1$, where a and b are the semiaxes of the given ellipse.

17.22. $xy = \pm s/2$, where s is the given area.

17.23. $y = ax^2 + \sqrt[3]{\dfrac{9as^2}{16}}$, where the parabola is given by the equation $y = ax^2$, and s is the area of the segment.

17.24. $\left(x - \dfrac{l}{\cos(\alpha/2)}\right)^2 + y^2 = \left(l \tan \dfrac{\alpha}{2}\right)^2$, where α is the given angle and l is the half-perimeter of the triangle.

17.25. $\dfrac{x^2}{a^2} + \dfrac{y^2}{2a^2} = 1$, where a is the radius of the given circle.

17.26. $\mathbf{r} = (l \cos^3 v, l \sin^3 v)$, where l is the given sum of semiaxes.

17.27. $x = \dfrac{a}{4} (3 \cos v - \cos 3v)$, $y = \dfrac{a}{4} (3 \sin v - \sin 3v)$ is the hypocycloid.

17.28. $xy = \pm \dfrac{1}{2} \sqrt{c}$, where c is the given area.

17.29. $(x - c)^2 + y^2 = 4a^2$, where a is the major semiaxis of the ellipse, $c = \sqrt{a^2 - b^2}$.

17.30. $\rho = \mathbf{r} \pm a \dfrac{[\mathbf{r}']}{|\mathbf{r}'|}$.

17.31. $\rho = \mathbf{r} + [\mathbf{r}'] \dfrac{|\mathbf{r}'|^2}{|\mathbf{r}' \times \mathbf{r}''|}$; in coordinates:

$$\xi = x - y' \dfrac{x'^2 + y'^2}{x'y'' - x''y'}, \qquad \eta = y + x' \dfrac{x'^2 + y'^2}{x'y'' - x''y'}.$$

17.32. Cardioid.

17.33. $(x + 1)/2 = (y - 13)/3 = z/6$; $2x + 3y + 6z - 37 = 0$.

17.34. For the given point A, we have: $t = -1$.

The tangent line: $(x - 3)/6 = (y + 7)/(-17) = (z - 2)/7$.

The normal plane: $6x - 17y + 7z - 151 = 0$.

17.35. For the given point A, we have: $t = 1$. As $\mathbf{r}'(1) = 0$, and $\mathbf{r}''(1) = (2, 2, 12) \neq 0$, then the direction of the tangent is defined by this latter vector or the collinear vector $(1, 1, 6)$.

The tangent line: $(x - 2)/1 = y/1 = (z + 2)/6$.

The normal plane: $x + y + 6z + 10 = 0$.

17.36. The equation of the tangent:

$$X = x + \lambda \begin{vmatrix} \dfrac{\partial F_1}{\partial y} & \dfrac{\partial F_1}{\partial z} \\ \dfrac{\partial F_2}{\partial y} & \dfrac{\partial F_2}{\partial z} \end{vmatrix}, \quad Y = y + \lambda \begin{vmatrix} \dfrac{\partial F_1}{\partial z} & \dfrac{\partial F_1}{\partial x} \\ \dfrac{\partial F_2}{\partial z} & \dfrac{\partial F_2}{\partial x} \end{vmatrix}, \quad Z = z + \lambda \begin{vmatrix} \dfrac{\partial F_1}{\partial x} & \dfrac{\partial F_1}{\partial y} \\ \dfrac{\partial F_2}{\partial x} & \dfrac{\partial F_2}{\partial y} \end{vmatrix}.$$

The normal plane:

$$\begin{vmatrix} X - x & Y - y & Z - z \\ \dfrac{\partial F_1}{\partial x} & \dfrac{\partial F_1}{\partial y} & \dfrac{\partial F_1}{\partial z} \\ \dfrac{\partial F_2}{\partial x} & \dfrac{\partial F_2}{\partial y} & \dfrac{\partial F_2}{\partial z} \end{vmatrix} = 0.$$

17.37. $s = 5at$.

17.38. $s = 8a\sqrt{2}$.

17.39. $s = 9a$.

17.40. $s = 10$. The curve has four cuspidal points with the change of sign ds/dt at the points $t = 0$, $\pi/2$, π, $3\pi/2$.

17.42. A necessary and sufficient condition: $\mathbf{e}' \neq 0$, $(\rho', \mathbf{e}, \mathbf{e}') = 0$. The equation of the envelope: $\mathbf{r} = \rho - \dfrac{\langle \rho', \mathbf{e}' \rangle}{|\mathbf{e}'|^2} \mathbf{e}$.

17.43. $\mathbf{r} = \rho + v\mathbf{e}$.

17.44. $\mathbf{r} = v\rho$.

17.45. $\mathbf{r} = \rho + v\rho'$.

17.46. $\mathbf{r}(s, \varphi) = \rho(s) + \mathbf{n}(s)\cos\varphi + \mathbf{b}(s)\sin\varphi$, where $\mathbf{n}(s)$ and $\mathbf{b}(s)$ are the principal normal and binormal, respectively.

17.47. $\mathbf{r}(u, v) = (\varphi(v)\cos u, \varphi(v)\sin u, \psi(v))$. In a particular case, $\mathbf{r} = (f(v)\cos u, f(v)\sin u, v)$.

17.48. If the equation of a helical line is given in the form

$$\rho = (a\cos u, a\sin u, bu),$$

then $\mathbf{n} = (-\cos u, -\sin u, 0)$ is the vector of the principal normal. Hence, the sought-for equation

$$\mathbf{r} = \rho - \lambda \mathbf{n} = ((a + \lambda)\cos u, (a + \lambda)\sin u, bu) = (v\cos u, v\sin u, bu)$$

defines a right helicoid.

17.49. $\mathbf{r} = \rho(s) + \lambda(\mathbf{n}(s)\cos\varphi(s) + \mathbf{b}(s)\sin\varphi(s))$, where $\varphi(s)$ is an arbitrary function of the variable s.

17.50. The vectors $\mathbf{n} = (\cos u, \sin u, 0)$ and $\mathbf{k} = (0, 0, 1)$ define the normal plane to the circle $\rho = (a\cos u, a\sin u, 0)$. The vector lying in the normal plane and inclined at an angle u to the vector $\mathbf{n}$ is the vector $\boldsymbol{\alpha} = \mathbf{n}\cos u + \mathbf{k}\sin u$. Therefore, the equation of the sought-for surface

$$\mathbf{r} = \rho + v\boldsymbol{\alpha} = \left(a\cos u + v\cos^2 u, a\sin u + v\sin u\cos u, v\sin u\right).$$

Excluding the parameters u and v, we find:

$$x = \cot u\,(a\sin u + z\cos u), \quad y = (a\sin u + z\cos u),$$

$$\frac{x}{y} = \cot u, \quad \frac{y^2}{\sin^2 u} = (a + z \cot u)^2, \quad y^2\left(1 + \frac{x^2}{y^2}\right) = \left(a + \frac{xz}{y}\right)^2.$$

Thus, $y^2(x^2 + y^2) = (ay + xz)^2$ is a fourth-order surface.

17.51. $\mathbf{R}(u, v) = \dfrac{1}{2}(\mathbf{r}(u) + \boldsymbol{\rho}(v))$.

17.52. The equation of the given line $\mathbf{r}_1 = (u, 0, h)$. The equation of the ellipse $\mathbf{r}_2 = (a \cos v, b \sin v, 0)$. Further, $\mathbf{r}_1 - \mathbf{r}_2 = (u - a \cos v, -b \sin v, h)$. Hence, at $u - a \cos v = 0$ we have $\mathbf{r}_1 - \mathbf{r}_2 = (0, -b \sin v, h)$. The sought-for equation of the conoid is:

$$\mathbf{r} = (a \cos v, b \sin v, 0) + \lambda(0, -b \sin v, h) = (a \cos v, b(1 - \lambda) \sin v, \lambda h).$$

Excluding the parameters λ and v, we get an equation of the conoid in an implicit form:

$$\left(1 - \frac{x^2}{a^2}\right)\left(\frac{z}{h} - 1\right)^2 - \frac{y^2}{b^2} = 0.$$

17.53. $\mathbf{r}_1 = (a, 0, u)$, $\mathbf{r}_2 = \left(0, v, \dfrac{v^2}{2p}\right)$, $\mathbf{r}_1 - \mathbf{r}_2 = \left(a, -v, u - \dfrac{v^2}{2p}\right)$. If $u - \dfrac{v^2}{2p} = 0$, then $\mathbf{r}_1 - \mathbf{r}_2 = (a, -v, 0)$. Hence,

$$\mathbf{r} = \left(0, v, \frac{v^2}{2p}\right) + \lambda(a, -v, 0) = \left(a\lambda, v(1 - \lambda), \frac{v^2}{2p}\right),$$

or $a^2 y^2 = 2pz(x - a)^2$.

17.54. The parametric equations of the given circles:

$$\mathbf{r}_1 = (a(1 + \cos u), 0, a \sin u), \quad \mathbf{r}_2 = (0, a(1 + \cos v), a \sin v).$$

We find: $\mathbf{r}_1 - \mathbf{r}_2 = (a(1 + \cos u), -a(1 + \cos v), a(\sin u - \sin v))$. We have $\sin u - \sin v = 0$, whence: 1) $v = u + 2k\pi$, 2) $v = \pi - u + 2k\pi$. In the former case, $\mathbf{r}_1 - \mathbf{r}_2 = (a(1 + \cos u), -a(1 + \cos u), 0)$ is parallel to $(1, -1, 0)$. Thus, we get an elliptic cylinder

$$\boldsymbol{\rho}(u, \lambda) = (a(1 + \cos u), 0, a \sin u) + \lambda(1, -1, 0) =$$
$$(a(1 + \cos u) + \lambda, -\lambda, a \sin u).$$

In the latter case, $\mathbf{r}_1 - \mathbf{r}_2 = (a(1 + \cos u), -a(1 - \cos u), 0)$ is parallel to $(1, -1, 0)$, and the second surface composing the given cylindroid is defined by the equation:

$$\mathbf{R}(u, \lambda) = (a(1 + \cos u), 0, a \sin u) + \lambda(a(1 + \cos u), -a(1 - \cos u), 0) =$$
$$(a(1 + \lambda)(1 + \cos u) + \lambda, -a\lambda(1 - \cos u), a \sin u).$$

Excluding the parameters λ and u, we get:

$$z^4 + z^2((x - y)^2 - 2a(x + y)) + 4a^2 xy = 0.$$

17.55. $\mathbf{r}_1 = \left(\dfrac{u^2}{2p}, u, 0\right)$, $\mathbf{r}_2 = \left(\dfrac{-v^2}{2p}, 0, v\right)$, $\mathbf{r}_1 - \mathbf{r}_2 = \left(\dfrac{u^2 + v^2}{2p}, u, -v\right)$. The collinearity condition of the vector $\mathbf{r}_1 - \mathbf{r}_2$ and of the plane $y - z = 0$ gives $u + v = 0$, $v = -u$, $\mathbf{r}_1 - \mathbf{r}_2 = \left(\dfrac{u^2}{p}, u, u\right)$. The sought-for equation:

$$\mathbf{r}(u, v) = \left(\frac{u^2}{2p}, u, 0\right) + v\left(\frac{u^2}{p}, u, u\right) = \left(\frac{u^2}{2p}(1 + 2v), u(1 + v), uv\right).$$

Excluding the parameters u and v, we have $y^2 - z^2 = 2px$, a hyperbolic paraboloid.

17.56. The equation of the axis Oz has the form $\mathbf{r}_1 = (0, 0, u)$; the equation of the given curve,

$$\mathbf{r}_2(v) = \left(b \cos v, b \sin v, \frac{a^3}{b^2 \cos v \sin v} \right).$$

Hence,

$$\mathbf{r}_2 - \mathbf{r}_1 = \left(b \cos v, b \sin v, \frac{a^3}{b^2 \cos v \sin v} - u \right), \quad u = \frac{a^3}{b^2 \cos v \sin v},$$

$$\mathbf{r}_2 - \mathbf{r}_1 = (b \cos v, b \sin v, 0),$$

then

$$\mathbf{r} = \left(0, 0, \frac{a^3}{b^2 \cos v \sin v} \right) + \lambda (b \cos v, b \sin v, 0) =$$

$$\left(\lambda b \cos v, \lambda b \sin v, \frac{a^3}{b^2 \cos v \sin v} \right).$$

Excluding the parameters λ and v, we get: $b^2 xyz = a^3 (x^2 + y^2)$.

17.57. From the condition $\langle \mathbf{a} + u\mathbf{b} - \boldsymbol{\rho}, \mathbf{n} \rangle = 0$, we find $u = \dfrac{\langle \mathbf{n}, \boldsymbol{\rho} - \mathbf{a} \rangle}{\langle \mathbf{n}, \mathbf{b} \rangle}$. Hence,

$$\mathbf{R} = \boldsymbol{\rho} + \lambda (\mathbf{a} + u\mathbf{b} - \boldsymbol{\rho}) = (1 - \lambda) \boldsymbol{\rho} + \lambda \left(\mathbf{a} + \frac{\langle \mathbf{n}, \boldsymbol{\rho} - \mathbf{a} \rangle}{\langle \mathbf{n}, \mathbf{b} \rangle} \mathbf{b} \right).$$

17.58. We take the equations of the given ellipses in the form

$$\mathbf{r}_1 = (a, b \cos u, c \sin u), \quad \mathbf{r}_2 = (-a, c \cos v, b \sin v).$$

The vector $\mathbf{r}_1 - \mathbf{r}_2 = (2a, b \cos u - c \cos v, c \sin u - b \sin v)$ is parallel to the plane xOy, therefore, $c \sin u - b \sin v = 0$. Hence,

$$\sin v = \frac{c}{b} \sin u, \quad \cos v = \pm \frac{1}{b} \sqrt{b^2 - c^2 \sin^2 u},$$

$$\mathbf{r}_1 - \mathbf{r}_2 = \left(2a, b \cos u \pm \frac{c}{b} \sqrt{b^2 - c^2 \sin^2 u}, 0 \right).$$

The sought-for equation:

$$\mathbf{R} = (a, b \cos u, c \sin u) + v \left(2a, b \cos u \pm \frac{c}{b} \sqrt{b^2 - c^2 \sin^2 u}, 0 \right)$$

or

$$\mathbf{R} = \left(a + 2av, b \cos u + v \left(b \cos u \pm \frac{c}{b} \sqrt{b^2 - c^2 \sin^2 u} \right), c \sin u \right).$$

17.59. The equation of the axis Oz: $\mathbf{p} = (0, 0, v)$. We find:

$$\boldsymbol{\rho} - \mathbf{p} = (u, u^2, u^3 - v), \quad u^3 - v = 0, \quad \mathbf{p} = (0, 0, u^3).$$

The sought-for equation:

$$\mathbf{r} = \mathbf{p} + v (\boldsymbol{\rho} - \mathbf{p}) = (0, 0, u^3) + v (u, u^2, 0) = (uv, u^2 v, u^3).$$

17.60. $\mathbf{r} = (bv, av \cos u, (b + a \cos u)(1 - v) + a \sin u)$.

17.61. The equations of the given lines: $\rho = (u, 1, 1)$ and $\mathbf{p} = (1, v, 0)$. The equation of the line passing through two arbitrary points of these lines has the form $\mathbf{r} = (1, v, 0) + \lambda (u, 1, 1)$. For the point of the intersection of this line with the plane xOz, we have: $v + \lambda = 0$, $\mathbf{r} = (1, v, 0) - v(u, 1, 1) = (1 - uv, 0, -v)$. This point should lie on the circle $x = \cos \varphi$, $y = 0$, $z = \sin \varphi$. Therefore, $1 - uv = \cos \varphi$, $v = -\sin \varphi$, hence, $u = \dfrac{1 - \cos \varphi}{-\sin \varphi} = -\tan \dfrac{\varphi}{2}$. It remains to compose the equation of the line passing through the points $\left(-\tan \dfrac{\varphi}{2}, 1, 1\right)$ and $(1, -\sin \varphi, 0)$. As the result, we get:

$$\mathbf{r} = (1 - \sin \varphi, 0) + \psi \left((1 - \sin \varphi, 0) - \left(-\tan \dfrac{\varphi}{2}, 1, 1\right)\right) =$$

$$\left(1 + \psi \left(1 + \tan \dfrac{\varphi}{2}\right), -\sin \varphi - \psi (1 + \sin \varphi), -\psi\right).$$

17.62. $\mathbf{r} = (a(\cos v - u \sin v), \ a(\sin v + u \cos v), b(u + v))$.

17.63. $c^2 (x^2 + y^2)^2 = a^2 (x^2 - y^2)(z + c)^2$.

17.64. We will consider that rectilinear Cartesian coordinates (ξ, η) are given in plane π. Then the equation of the curve $\rho = \rho(u)$ can be written down in the coordinate form: $\xi = \xi(u)$, $\eta = \eta(u)$. Besides, let us assume that the line AB is an axis z in space and that an axis η of the moving plane π glides on it. At a proper choice of the axes x, y and positive directions on the coordinate axes, we have:

$$\mathbf{R}(u, v) = (\xi(u) \cos v, \xi(u) \sin v, \eta(u) + av).$$

17.65. $\mathbf{R}(u, v) = \mathbf{r}(u) + a\mathbf{n}(u) \cos v + a\mathbf{b}(u) \sin v$, where $\mathbf{n}$ and $\mathbf{b}$ are the principal normal and binormal of the curve $\mathbf{r} = \mathbf{r}(u)$; the points (u, v) and $(u, v + 2\pi)$ are identified.

17.66. Take the point of intersection of the normals as the origin of the radius vectors. Then $\langle \mathbf{r}, \mathbf{r}_u \rangle = 0$, $\langle \mathbf{r}, \mathbf{r}_v \rangle = 0$, whence $|\mathbf{r}|^2 = \text{const}$. Therefore, this surface is a sphere or part of a sphere.

17.67. The volume of the tetrahedron is $9a^3/2$.

17.68. The tangent plane: $\dfrac{x}{u \sin v} + \dfrac{y}{u \cos v} + \dfrac{z}{\sqrt{a^2 - u^2}} = a^2$. The sought-for sum is a^6.

17.69. The equation of the line of intersection in curvilinear coordinates: $u = u_1 \cos(v + v_1) / \cos 2v_1$ (except the generatrix $v = v_1$, where u_1, v_1 are the coordinates of the tangent point; the parametric equations of the same line in Cartesian coordinates:

$$x = u_1 \dfrac{\cos(v + v_1)}{\cos 2v_1} \cos v, \quad y = u_1 \dfrac{\cos(v + v_1)}{\cos 2v_1} \sin v, \quad z = a \sin 2v.$$

The equation of its projection on the plane xy:

$$x^2 + y^2 = \dfrac{u_1}{\cos 2v_1} (x \cos v_1 - y \sin v_1).$$

As the projection is a circle, the (plane) line itself is an ellipse.

17.70. The equation of the tangent plane:

$$Z - xf = \left(f - \dfrac{y}{x} f'\right)(X - x) + (Y - y) f'$$

or $Z = \left(f - \dfrac{y}{x}f'\right)X + Yf'$, i.e., all tangent planes pass through one point, the origin of coordinates. It is clear, though, from the fact that this equation defines a cone with the vertex at the origin of coordinates (z is a homogeneous function of x and y).

17.71. The tangent plane: $kx \sin u - ky \cos u + vz - kuv = 0$; the normal: $\mathbf{r} = (v \cos u + \lambda k \sin u, v \sin u - \lambda k \cos u, ku + \lambda v)$.

17.72. $\dfrac{X}{x} + \dfrac{Y}{y} + \dfrac{Z}{z} = 3$.

17.73. Let the equation of the curve γ have the form $\boldsymbol{\rho} = \boldsymbol{\rho}(s)$. The equation of the surface: $\mathbf{r} = \boldsymbol{\rho} + \lambda \mathbf{v}$, where $\mathbf{v}$ is a unit vector tangent to the curve γ. We find:

$$\frac{\partial \mathbf{r}}{\partial s} = \mathbf{v} + \lambda k \mathbf{n}, \qquad \frac{\partial \mathbf{r}}{\partial \lambda} = \mathbf{v}, \qquad \frac{\partial \mathbf{r}}{\partial \lambda} \times \frac{\partial \mathbf{r}}{\partial s} = \lambda k \mathbf{b};$$

at $s = $ const (i.e., at points of the same tangent), this vector is unidirectional (because then $\mathbf{b} = $ const). Hence, it also follows that the osculating plane of this curve is a tangent plane to such a surface at all points of the curve γ.

17.74. The equation of the surface: $\mathbf{r} = \boldsymbol{\rho} + \lambda \mathbf{n}$;

$$\frac{\partial \mathbf{r}}{\partial s} = \mathbf{v} + \lambda(-k\mathbf{v} + \varkappa \mathbf{b}), \qquad \frac{\partial \mathbf{r}}{\partial \lambda} = \mathbf{n}, \qquad \frac{\partial \mathbf{r}}{\partial s} \times \frac{\partial \mathbf{r}}{\partial \lambda} = (1 - \lambda k)\mathbf{b} - \lambda \varkappa \mathbf{v}.$$

The equation of the tangent plane is as follows: $\langle \mathbf{R} - \boldsymbol{\rho} - \lambda \mathbf{n}, \mathbf{b} - \lambda k \mathbf{b} - \lambda \varkappa \mathbf{v} \rangle = 0$, or $\langle \mathbf{R}, \mathbf{b} - \lambda \varkappa \mathbf{v} \rangle - \langle \boldsymbol{\rho}, \mathbf{b} - \lambda \varkappa \mathbf{v} \rangle + \lambda^2 \varkappa = 0$. The equation of the normal: $\mathbf{R} = \boldsymbol{\rho} + \lambda \mathbf{n} + \xi(\mathbf{b} - \lambda \varkappa \mathbf{v})$.

17.75. The equation of the surface: $\mathbf{r} = \boldsymbol{\rho} + \lambda \mathbf{b}$;

$$\frac{\partial \mathbf{r}}{\partial s} = \mathbf{v} - \lambda \varkappa \mathbf{n}, \qquad \frac{\partial \mathbf{r}}{\partial \lambda} = \mathbf{b}, \qquad \frac{\partial \mathbf{r}}{\partial s} \times \frac{\partial \mathbf{r}}{\partial \lambda} = -\mathbf{n} - \lambda \varkappa \mathbf{v}.$$

The equation of the tangent plane is as follows: $\langle \mathbf{R} - \boldsymbol{\rho} - \lambda \mathbf{b}, \mathbf{n} + \lambda \varkappa \mathbf{v} \rangle = 0$, or $\langle \mathbf{R} - \boldsymbol{\rho}, \mathbf{n} + \lambda \varkappa \mathbf{v} \rangle = 0$. The equation of the normal: $\mathbf{R} = \boldsymbol{\rho} + \lambda \mathbf{b} + \xi(\mathbf{n} + \lambda \varkappa \mathbf{v})$.

17.77. If $\mathbf{a}$ is a directing vector of the given line and the origin of radius vectors is taken from this line, then the vectors $\mathbf{r}$, $\mathbf{a}$ and $\dfrac{\partial \mathbf{r}}{\partial u} \times \dfrac{\partial \mathbf{r}}{\partial v}$ lie in one plane, and

$$\left\langle \mathbf{r}, \mathbf{a} \times \left(\frac{\partial \mathbf{r}}{\partial u} \times \frac{\partial \mathbf{r}}{\partial v}\right) \right\rangle = 0.$$

Hence,

$$\left\langle \mathbf{r}, \frac{\partial \mathbf{r}}{\partial u} \right\rangle \left\langle \mathbf{a}, \frac{\partial \mathbf{r}}{\partial v} \right\rangle - \left\langle \mathbf{r}, \frac{\partial \mathbf{r}}{\partial v} \right\rangle \left\langle \mathbf{a}, \frac{\partial \mathbf{r}}{\partial u} \right\rangle = 0.$$

But this equality can be written as a zero equality of the functional determinant:

$$\frac{\partial |\mathbf{r}|^2}{\partial u} \frac{\partial \langle \mathbf{a}, \mathbf{r} \rangle}{\partial v} - \frac{\partial |\mathbf{r}|^2}{\partial v} \frac{\partial \langle \mathbf{a}, \mathbf{r} \rangle}{\partial u} = 0.$$

Hence, it follows that there is a functional dependence $|\mathbf{r}|^2 = f(\langle \mathbf{a}, \mathbf{r} \rangle)$ between the values $|\mathbf{r}|^2$ and $\langle \mathbf{a}, \mathbf{r} \rangle$. Choosing the axis Oz along the vector $\mathbf{a}$, we get the surface of revolution $x^2 + y^2 = f(z)$.

17.80. The envelope: $4z^2 \left(\dfrac{x^2}{a^2} + \dfrac{y^2}{b^2}\right) = 1$; the cuspidal edge is imaginary.

17.81. The envelope: $x^2 + \dfrac{a^2}{a^2 + b^2} y^2 + z^2 = a^2$.

17.82. The envelope: $\left(x^2 + y^2 + z^2 - x\right)^2 = x^2 + y^2$; the cuspidal edge degenerates into a point $(0, 0, 0)$.

17.83. Taking the equations of the parabolas in the form $y^2 = 2px$, $z = 0$ and $y^2 = 2qz$, $x = 0$, we get the equation of the envelope in the form $y^2 = 2px + 2qz$, a parabolic cylinder with the parameter $\sqrt{p^2 + q^2}$.

17.84. Differentiating the expression $|\mathbf{R} - \boldsymbol{\rho}|^2 = a^2$ with respect to s, we get $\langle \mathbf{R} - \boldsymbol{\rho}, \mathbf{v} \rangle = 0$. Hence, $\mathbf{R} - \boldsymbol{\rho} = \lambda \mathbf{b} + \mu \mathbf{n}$. As $|\mathbf{R} - \boldsymbol{\rho}|^2 = a^2$, then $\lambda^2 + \mu^2 = a^2$ and we can assume $\lambda = a \cos \varphi$, $\mu = a \sin \varphi$. Thus, the equation of the envelope is:

$$\mathbf{R} = \boldsymbol{\rho} + a \left(\mathbf{b} \cos \varphi + \mathbf{n} \sin \varphi\right).$$

17.85. The cuspidal edge is a curve whose points are obtained at the intersection of the curvature axes by the curve $\boldsymbol{\rho} = \boldsymbol{\rho}(s)$ with respective spheres of the given family of spheres.

17.86. The equation of the family: $(x - b \cos \varphi)^2 + y - b \sin \varphi + z^2 - a^2 = 0$. The envelope is a torus (a fourth-order surface):

$$\left(x^2 + y^2 + z^2 + b^2 - a^2\right)^2 - 4b^2 \left(x^2 + y^2\right) = 0.$$

This equation is obtained at the exclusion of φ from the equations $F = 0$ and $\dfrac{\partial F}{\partial s} \varphi = 0$. The cuspidal edge in the case $a > b$ degenerates into two points $\left(0, 0, \pm\sqrt{a^2 - b^2}\right)$, and in the case $a = b$ degenerates into one point $(0, 0, 0)$.

17.87. The equation of the family: $x^2 + y^2 + z^2 - 2u^3x - 2u^2y - 2uz = 0$. Excluding u from this equation and from the equation $3u^2x + 2uy + z = 0$, we find the envelope:

$$3x \left(9x \left(x^2 + y^2 + z^2\right) - 2zy\right)^2 + 2y \left(9x \left(x^2 + y^2 + z^2\right) - 2zy\right) -$$
$$\left(12xz - 4y^2\right) + z \left(12xz - 4y^2\right)^2 = 0.$$

The cuspidal edge is found by adding the equation $6ux + 2y = 0$ to two above equations. Hence, $u = -y/3x$, and the cuspidal edge equation:

$$27x^2 \left(x^2 + y^2 + z^2\right) - 4y^3 + 18xyz = 0, \quad y^2 - 3xz = 0.$$

The cuspidal edge can also be obtained in the parametric form:

$$\mathbf{r} = \left(\frac{2u^3}{9u^4 + 9u^2 + 1}, \frac{-6u^4}{9u^4 + 9u^2 + 1}, \frac{6u^5}{9u^4 + 9u^2 + 1}\right).$$

17.88. $x^{2/3} + y^{2/3} + z^{2/3} = l^{2/3}$.

17.89. $x^{2/3} + y^{2/3} + z^{2/3} = a^{2/3}$.

17.90. $xyz = 2a^3/9$.

17.91. The envelope: $y^2 = 4xz$; the cuspidal edge degenerates into a point coinciding with the origin.

17.92. The cuspidal edge is a helical line:

$$x = C \cos \alpha, \quad y = C \sin \alpha, \quad z = C\alpha.$$

17.93. Let $\rho = \rho(s)$ be the equation of this curve. The equation of the family of osculating planes: $\langle \mathbf{r} - \rho, \mathbf{b} \rangle = 0$. Differentiating with respect to s, we get $\langle \mathbf{r} - \rho, \mathbf{v} \rangle = 0$. A characteristic is the tangent: $\langle \mathbf{r} - \rho, \mathbf{b} \rangle = 0$, $\langle \mathbf{r} - \rho, \mathbf{n} \rangle = 0$. The envelope is $\mathbf{r} = \rho + \lambda \mathbf{v}$, a surface formed by tangents to this curve. Differentiating the expression $\langle \mathbf{r} - \rho, \mathbf{n} \rangle = 0$ once more, we get $\langle \mathbf{r} - \rho, \mathbf{b} \rangle = 0$. Hence, from the relations $\langle \mathbf{r} - \rho, \mathbf{b} \rangle = 0$, $\langle \mathbf{r} - \rho, \mathbf{n} \rangle = 0$, we also have $\mathbf{r} = \rho$, i.e., this curve is a cuspidal edge.

17.94. The characteristics are the curvature axes of this curve. The envelope is a surface formed by the curvature axes. The cuspidal edge is a curve described by the centres of osculating spheres of this curve.

17.95. $\langle \mathbf{r}, \mathbf{n} \rangle + D' = 0$, $\mathbf{r} = \alpha \mathbf{n} + \beta \mathbf{n}' + \lambda \mathbf{n} \times \mathbf{n}'$, $\alpha = \langle \mathbf{r}, \mathbf{n} \rangle = -D$, $\beta = \dfrac{\langle \mathbf{r}, \mathbf{n}' \rangle}{|\mathbf{n}'|^2} =$

$-\dfrac{D'}{|\mathbf{n}'|^2}$. The envelope equation: $\mathbf{r} = -D\mathbf{n} - \dfrac{D'\mathbf{n}'}{|\mathbf{n}'|^2} + \lambda \mathbf{n} \times \mathbf{n}'$ (the parameters u and λ). The characteristics are the lines $u = \text{const}$. The cuspidal point is defined by solving the equations $\langle \mathbf{r}, \mathbf{n} \rangle + D = 0, \langle \mathbf{r}, \mathbf{n}' \rangle + D' = 0$, $\langle \mathbf{r}, \mathbf{n}'' \rangle + D'' = 0$ relative to $\mathbf{r}$:

$$\mathbf{r} = \frac{\langle \mathbf{r}, \mathbf{n} \rangle \langle \mathbf{n}' \times \mathbf{n}'' \rangle + \langle \mathbf{r}, \mathbf{n}' \rangle \langle \mathbf{n}'' \times \mathbf{n} \rangle + \langle \mathbf{r}, \mathbf{n}'' \rangle \langle \mathbf{n} \times \mathbf{n}' \rangle}{\langle \mathbf{n}, \mathbf{n}', \mathbf{n}'' \rangle} =$$

$$- \frac{D\mathbf{n}' \times \mathbf{n}'' + D'\mathbf{n}'' \times \mathbf{n} + D''\mathbf{n} \times \mathbf{n}'}{\langle \mathbf{n}, \mathbf{n}', \mathbf{n}'' \rangle}.$$

17.96. The envelope of the family of planes tangent to both parabolas. The equation of the family:

$$X\alpha - 2aY - \left(\frac{\alpha^2}{2b} \frac{4a^3}{b\alpha} \right) Z + \frac{4a^3}{\alpha} = 0,$$

where α is the parameter of the family.

17.97. The normal vector $\mathbf{n} = (u + v)(\mathbf{i} \sin v - \mathbf{j} \cos v + \mathbf{k})$ is parallel to vector $\mathbf{i} \sin v - \mathbf{j} \cos v + \mathbf{k}$, which does not change if parameter v preserves its constant value. Hence, the lines $v = \text{const}$ are rectilinear generators of the surface, and line $u + v = 0$ is a cuspidal edge, because at any of its points the modulus of vector $\mathbf{n}$ turns zero.

17.99. The equation of the curve: $u = \text{const}$; the cuspidal edge:

$$x = 2(a - b) u \cos^2 v, \quad y = 2(a - b) u \sin^3 v,$$

$$z = 2u^2 \left((a - 2b) \cos^2 v + (b - 2a) \sin^2 v \right).$$

17.100. $x = 3t$, $y = -3t^2/b$, $z = -t^3/ab$.

17.101. The sought-for developable surface envelopes a family of planes

$$Xx + Y\sqrt{a^2 - x^2} + Z\sqrt{\frac{a^4}{b^2} - x^2} = a^2,$$

where x is the parameter of the family.

18.1. (a) $\dfrac{4am\,(1+m)}{1+2m}\sin\dfrac{t}{2}$; (b) $\dfrac{a\left(y^4 - 2by^3 + a^2b^2\right)^{3/2}}{y^3\left|2a^2b - y^3 - 3by^2\right|}$.

18.2. (a) $\left|\cos x\right|$; (b) $\dfrac{1}{6}$; (c) $\dfrac{c}{a\pi}$; (d) $\dfrac{3}{8a\left|\sin\left(t/2\right)\right|}$.

18.3. (a) $\dfrac{2+\varphi^2}{a\left(1+\varphi^2\right)^{3/2}}$; (b) $\dfrac{k\left(k+1\right)+\varphi^2}{a\varphi^{k-1}\left(k^2+\varphi^2\right)^{3/2}}$; (c) $\dfrac{1}{\sqrt{1+\left(\ln a\right)^2}}$.

18.4. (a) $\dfrac{\left|\sin x\right|}{\left(1+\cos^2 x\right)^{3/2}}$.

18.5. $k=\dfrac{1}{4}\sqrt{1+\sin^2\dfrac{t}{2}}$.

18.6. (a) $\mathbf{r}=\left(\dfrac{a}{2}\left(\sin 4t + 2\sin 2t\right),\,-\dfrac{a}{2}\left(\cos 4t + 2\cos 2t\right)\right)$;

(b) $\mathbf{r}=\left(a\left(2t+\sin 2t\right),\,a\left(2-\cos 2t\right)\right)$, cycloid;

(c) $\mathbf{r}=\left(a\cos t,\,a\ln\left(\tan\left|\dfrac{\pi}{4}+\dfrac{t}{2}\right|\right)-a\sin t\right)$, tractrix.

18.10. $9x - 27y - z + 7 = 0$.

18.11. For an osculating plane, we find the equation $cx - ay = bc - ad$ containing no parameter u. Substituting into this equation an expression for x, y via u, we obtain an identity, from where we conclude that the curve does indeed lie in its osculating plane.

18.12. Osculating plane: $6x - 8y - z + 3 = 0$. Principal normal: $x = 1 - 31\lambda$, $y = 1 - 26\lambda$, $z = 1 + 22\lambda$. Binormal: $x = 1 + 6\lambda$, $y = 1 - 8\lambda$, $z = 1 - \lambda$.

18.13. $y = 1$.

18.14. Tangent: $\mathbf{r}=\left(a\cos t - a\lambda\sin t, a\sin t + a\lambda\cos t, b(\lambda + t)\right)$. Normal plane: $ax\sin t - ay\cos t - bz + b^2 t = 0$. Binormal: $\mathbf{r}=\left(a\cos t + b\lambda\sin t, a\sin t - b\lambda\cos t, bt + a\lambda\right)$. Osculating plane: $bx\sin t - by\cos t + az - abt = 0$. Principal normal: $\mathbf{r}=\left((a+\lambda)\cos t, (a+\lambda)\sin t, bt\right)$.

18.15. Tangent: $x = 1 + 2\lambda$, $y = -\lambda$, $z = 1 + 3\lambda$. Normal plane: $2x - y + 3z - 5 = 0$. Binormal: $x = 1 - 3\lambda$, $y = -3\lambda$, $z = 1 + \lambda$. Osculating plane: $3x + 3y - z - 2 = 0$. Principal normal: $x = 1 - 8\lambda$, $y = 11\lambda$, $z = 1 + 9\lambda$.

18.17. (a) $d \sim \dfrac{1}{6}k_0\varkappa_0 s^3$, $s \to 0$; (b) $d \sim \dfrac{1}{2}k_0^2 s^2$, $s \to 0$.

18.18. $R = \dfrac{x}{2}\left(\dfrac{x^2}{a^2}+\dfrac{a^2}{2x^2}\right)^2$, $r = -\dfrac{x}{2}\left(\dfrac{x^2}{a^2}+\dfrac{a^2}{2x^2}\right)^2$.

18.19. $v = Ce^{\frac{u\cot\theta}{\sqrt{1+k^2}}}$.

18.20. (a) Let $\mathbf{a}$ be a unit vector of fixed direction. Then $\langle \mathbf{a}, \mathbf{v}\rangle = \cos v$ $(v = \text{const})$. As $\dfrac{d}{ds}\langle \mathbf{a}, \mathbf{v}\rangle = \langle \mathbf{a}, \dot{\mathbf{v}}\rangle = 0$, $k\langle \mathbf{a}, \mathbf{n}\rangle = 0$. Excluding the case when $k = 0$ (straight lines), we get $\langle \mathbf{a}, \mathbf{n}\rangle = 0$. Therefore, the normals are perpendicular to the fixed direction. Conversely, if $\mathbf{n}$ is perpendicular to the fixed direction, $\langle \mathbf{a}, \mathbf{v}\rangle = \text{const}$.

(b) Let $\varkappa \neq 0$. From $\langle \mathbf{a}, \mathbf{n}\rangle = 0$ and the third Frénet formula, it follows that $\langle \mathbf{a}, \dot{\mathbf{b}}\rangle = 0$, whence $\langle \mathbf{a}, \mathbf{b}\rangle = \text{const}$. Conversely, differentiating this equality, we get: $\langle \mathbf{a}, \mathbf{n}\rangle = 0$.

(c) Differentiating the equality $\langle \mathbf{a}, \mathbf{n}\rangle = 0$, we get $k\langle \mathbf{a}, \mathbf{v}\rangle = \varkappa\langle \mathbf{a}, \mathbf{b}\rangle$, whence

$\dfrac{k}{\varkappa} = \dfrac{\langle \mathbf{a}, \mathbf{b} \rangle}{\langle \mathbf{a}, \mathbf{v} \rangle} = \text{const}$. Conversely, from the first and third Frénet formulas, it follows

that $\dfrac{\dot{\mathbf{v}}}{k} + \dfrac{\dot{\mathbf{b}}}{\varkappa} = 0$, whence $\dfrac{\varkappa}{k}\dot{\mathbf{v}} + \dot{\mathbf{b}} = 0$, $\dfrac{\varkappa}{k}\mathbf{v} + \mathbf{b} = \text{const} = \mathbf{a}$. Multiplying scalarly
by $\mathbf{n}$, we get $\langle \mathbf{a}, \mathbf{n} \rangle = 0$.

18.21. Show that for this curve $\dfrac{k}{\varkappa} = \text{const}$, and make use of Problem 18.20.

18.22. Show that $\dfrac{k}{\varkappa} = 1$, and make use of Problem 18.20. The sought-for vector
$\mathbf{e} = \mathbf{v} + \mathbf{b} = (1, 0, 1)$.

18.24. According to Problem 18.20, $\left(\dfrac{d}{ds}\mathbf{b}, \dfrac{d^2}{ds^2}\mathbf{b}, \dfrac{d^3}{ds^3}\mathbf{b} \right) = 0$; for this reason, the
spherical indicatrix is a plane curve. And as it lies on a sphere, it is a circle.

18.33. (a) Let $\mathbf{r} = \mathbf{r}(s)$ be the equation of the curve γ in the natural parameter.
Then the curve γ^* can be given as $\mathbf{r}^* = \mathbf{r}(s) + a(s)\,\mathbf{n}(s)$, where $\mathbf{n}(s)$ is the principal
normal to the curve γ, and $a(s)$ is some scalar function. It is clear that, generally
speaking, s is not a natural parameter on the curve γ^*. We have $\dfrac{d}{ds}\mathbf{r}^* = \mathbf{v} +$
$\mathbf{n}\dfrac{d}{ds}a + a(-k\mathbf{v} + \varkappa\mathbf{b})$. Further, from the condition it follows that $\left\langle \dfrac{d}{ds}\mathbf{r}^*, \mathbf{n} \right\rangle =$
$\left\langle \dfrac{d}{ds}\mathbf{r}^*, \mathbf{n}^* \right\rangle$. As the vector $\dfrac{d}{ds}\mathbf{r}^*$ is parallel to the vector $\mathbf{v}^*$, $\left\langle \dfrac{d}{ds}\mathbf{r}^*, \mathbf{n}^* \right\rangle = 0$.
Thus, it has been proven that $\left\langle \dfrac{d}{ds}\mathbf{r}^*, \mathbf{n} \right\rangle = 0$. Substituting the expression for
$\dfrac{d}{ds}\mathbf{r}^*$, we get that $\dfrac{d}{ds}a = 0$, i.e., $a = \text{const}$.

(b) From the solution of the previous item it follows that

$$\frac{d}{ds}\mathbf{r}^* = \mathbf{v}(1 - ak) + a\varkappa\mathbf{b}.$$

Differentiate this equality with respect to s and make use of the Frénet formula. We
have:

$$\frac{d^2}{ds^2}\mathbf{r}^* = k\mathbf{n}(1 - ak) + \mathbf{v}\frac{d}{ds}(1 - ak) + a\left(\frac{d}{ds}\varkappa\right)\mathbf{b} - a\varkappa^2\mathbf{n}.$$

Note that in the osculating plane of the curve γ^* there are the vectors $\dfrac{d}{ds}\mathbf{r}$, $\dfrac{d^2}{ds^2}\mathbf{r}$,
$\mathbf{n}^*$. But by the condition $\mathbf{n} = \pm\mathbf{n}^*$, so the vectors $\left(\dfrac{d}{ds}\mathbf{r} \right) \times \mathbf{n} = (1 - ak)\mathbf{b} -$
$a\varkappa\mathbf{v}$, $\left(\dfrac{d^2}{ds^2}\mathbf{r} \right) \times \mathbf{n} = \left(\dfrac{d}{ds}(1 - ak) \right)\mathbf{b} - a\left(\dfrac{d}{ds}\varkappa \right)\mathbf{v}$ are collinear. Therefore, the
coordinates of these vectors are proportional, $\dfrac{(d/ds)(1 - ak)}{a - ak} = \dfrac{a(d/ds)\varkappa}{a\varkappa}$, which
is equivalent to the equality $\dfrac{d}{ds}\left(\dfrac{1 - ak}{\varkappa} \right) = 0$. Hence, $\dfrac{1 - ak}{\varkappa} = b = \text{const}$, i.e.,
$ak + b\varkappa = 1$, where a and b are some constants.

18.35. Let the initial curve be given as $\mathbf{r} = \mathbf{r}(s)$. We show that, as a curve making
a Bertrand couple with the initial curve, we can take the curve $\boldsymbol{\rho}(s) = \mathbf{r}(s) + a\mathbf{n}(s)$,
where $\mathbf{n}(s)$ is the normal vector to the initial curve.

It is easy to show that $\dfrac{d}{ds}\boldsymbol{\rho} = (1 - ak)\mathbf{v}(s) + a\varkappa\mathbf{b}(s)$ is perpendicular to $\mathbf{n}(s)$.

Further, it is easy to check that

$$\frac{d^2}{ds^2}\rho = (1 - ak)\, k\mathbf{v} + a\left(\frac{d}{ds}k\right)\mathbf{v} + a\frac{d}{ds}\varkappa\mathbf{b} - a\varkappa^2\mathbf{n}.$$

To prove the fact that $\rho(s) = \mathbf{r}(s) + a\mathbf{n}(s)$ makes a Bertrand couple with the initial curve, it is sufficient to show that the vectors $\dfrac{d}{ds}\rho$, $\dfrac{d^2}{ds^2}\rho$ and $\mathbf{n}$ are complanar, i.e., are parallel to one plane. For this, it is sufficient to verify that the vectors $\left(\dfrac{d}{ds}\rho\right)\times\mathbf{n}$ and $\left(\dfrac{d^2}{ds^2}\rho\right)\times\mathbf{n}$ are proportional. It is easy to show that

$$\left(\frac{d}{ds}\rho\right)\times\mathbf{n} = (1-ak)\,\mathbf{b} - a\varkappa\mathbf{v}, \quad \left(\frac{d^2}{ds^2}\rho\right)\times\mathbf{n} = -a\left(\frac{d}{ds}k\right)\mathbf{b} - \left(\frac{d}{ds}\varkappa\right)a\mathbf{v}.$$

As $\dfrac{1 - ak}{\varkappa} = b = \text{const}$, $\dfrac{d}{ds}\left(\dfrac{1 - ak}{\varkappa}\right) = 0$. This equality can be rewritten as $\dfrac{-a\,(d/ds)\,k}{1 - ak} = \dfrac{a\,(d/ds)\,\varkappa}{a\varkappa}$. Hence, it follows that the vectors $\left(\dfrac{d}{ds}\rho\right)\times\mathbf{n}$ and $\left(\dfrac{d^2}{ds^2}\rho\right)\times\mathbf{n}$ are proportional.

18.36. Make use of Problem 18.35. Let $b = 0$, $a = 1/k$.

18.39. Let the curve γ be given in natural parameterization $\mathbf{r} = \mathbf{r}(s)$. Then the curve γ^* is given as $\rho(s) = \mathbf{r}(s) + \lambda\mathbf{b}(s)$. As the binormal of the curve γ^* is parallel to $\mathbf{b}$, $\left\langle\mathbf{b}, \dfrac{d}{ds}\rho\right\rangle = 0$. Hence, $\dfrac{d}{ds}\lambda = 0$, i.e., λ is a constant value. Therefore, $\dfrac{d}{ds}\rho = \mathbf{v} - \lambda\varkappa\mathbf{n}$ and $\dfrac{d^2}{ds^2}\rho = k\mathbf{n} - \dot\lambda\varkappa\mathbf{n} - \lambda\dot\varkappa\mathbf{n} + \lambda\varkappa k\mathbf{v} - \lambda\varkappa^2\mathbf{b}$. Note now that $\left\langle\dfrac{d^2}{ds^2}\rho, \mathbf{b}\right\rangle = 0$, so $\lambda\varkappa^2 = 0$. As the curves γ and γ^* are different, $\varkappa = 0$.

18.40. Let $\mathbf{r} = \mathbf{r}(s)$ be a natural parameterization of the curve γ. We write down the equation of the curve γ^* as $\rho(s) = \mathbf{r}(s) + \lambda(s)\,\mathbf{n}(s)$, where $\mathbf{n}(s)$ is the vector of the principal normal of the curve γ, and $\lambda(s)$ is some scalar function. We have $\dfrac{d}{ds}\rho = \mathbf{v} + \lambda(-k\mathbf{v} + \varkappa\mathbf{b}) + \dot\lambda\mathbf{n}$. As $\left\langle\dfrac{d}{ds}\rho, \mathbf{b}^*\right\rangle = 0$, $\left\langle\dfrac{d}{ds}\rho, \mathbf{n}\right\rangle = 0$.

Hence, $\lambda = \text{const}$ and $\dfrac{d}{ds}\rho = \mathbf{v} + \lambda(-k\mathbf{v} + \varkappa\mathbf{b})$. Further, the vector $\dfrac{d^2}{ds^2}\rho = -\lambda\dot\varkappa\mathbf{v} + (k - \lambda k^2 - \lambda\varkappa^2)\,\mathbf{n} + \lambda\dot\varkappa\mathbf{b}$ is orthogonal to the vector $\mathbf{b}^* = \mathbf{n}$. Therefore, $k - \lambda k^2 - \lambda\varkappa^2 = 0$, i.e., $k = \lambda(k^2 + \varkappa^2)$.

18.54. Let k and $\varkappa$ be, respectively, the curvature and torsion of the initial curve parameterized by the natural parameter s, and k^* and $\varkappa^*$ be, respectively, the curvature and torsion of its evolute. Then

$$k^* = \frac{\sqrt{k^2 + \varkappa^2}}{|sk|}, \quad \varkappa^* = \frac{\varkappa^2}{sk\,(k^2 + \varkappa^2)}\frac{d}{ds}\left(\frac{k}{\varkappa}\right).$$

18.55. In slope curves, the ratio $\dfrac{k}{\varkappa}$ is constant. Hence, by Problem 18.54 we get the required.

18.56. We search for the equation of the pedal in the form $\rho(t) = \mathbf{r}(t) + \lambda(t)\,\mathbf{r}'(t)$. The scalar function λ shall be found from the condition $\langle \rho, \mathbf{r}' \rangle = 0$. We have $\lambda = -\dfrac{\langle \mathbf{r}, \mathbf{r}' \rangle}{|\mathbf{r}'|^2}$. Hence, $\rho(t) = \mathbf{r} - \mathbf{r}'\dfrac{\langle \mathbf{r}, \mathbf{r}' \rangle}{|\mathbf{r}'|^2}$.

18.57. $x = a\left(\cos t + \dfrac{h^2 t}{a^2 + h^2}\sin t\right)$, $y = a\left(\sin t - \dfrac{h^2 t}{a^2 + h^2}\cos t\right)$, $z = \dfrac{a^2 h t}{a^2 + h^2}$.

18.65. (a) The unit binormal vector $\mathbf{b} = (-1, 0, 0)$ is parallel to the axis Ox. The osculating plane of the curve in the considered point coincides with the plane xOy. The centre of the osculating circle is at the point $(0, 1, 0)$. The radius of this circle is equal to 1. The osculating circle can be given by the equations $(y - 1)^2 + z^2 = 1$, $x = 0$;

(b) $\left(x - \dfrac{1}{2}\right)^2 + (y + 3)^2 + (z - 4)^2 = \dfrac{81}{4}$, $\quad 2x - 2y - z - 3 = 0$.

18.66. Make use of Problem 18.64. The curvature radius of the curve is equal to

$$R = \frac{\left(1 + 4t^2 + 9t^4\right)^{3/2}}{\left(36t^4 + 36t^2 + 4\right)^{1/2}}.$$

Further,

$$\dot{R} = \frac{dR}{ds} = \frac{dR}{dt}\frac{dt}{ds}, \quad \frac{ds}{dt} = \left(1 + 4t^2 + 9t^4\right)^{1/2}.$$

Hence, we get that the radius of the osculating sphere is equal to $\dfrac{1}{2}$.

18.67. (a) $\dfrac{a^2 + h^2}{a}$; (b) $\left(e^t + e^{-t}\right)^2 \sqrt{\dfrac{1}{2} + (e^t - e^{-t})^2}$; (c) $3\sqrt{2}e^t$.

18.69. The helical line, which lies on the cylinder $x^2 + y^2 = \dfrac{h^4}{a^2}$ and the lead of which is 1.

18.72. Let the parameterization of the curve considered is $x = x(t)$, $y = y(t)$, $z = z(t)$. Consider the function

$$\varphi(t) = \begin{vmatrix} x(t) - x(t_0) & y(t) - y(t_0) & z(t) - z(t_0) \\ x'(t_0) & y'(t_0) & z'(t_0) \\ x''(t_0) & y''(t_0) & z''(t_0) \end{vmatrix}.$$

Decompose $x(t)$, $y(t)$ and $z(t)$ into Taylor series and in the expression for the function $\varphi(t)$ the coefficient at $(t - t_0)^3$ equate to zero. We get

$$\begin{vmatrix} x'''(t_0) & y'''(t_0) & z'''(t_0) \\ x''(t_0) & y''(t_0) & z''(t_0) \\ x'(t_0) & y'(t_0) & z'(t_0) \end{vmatrix} = 0.$$

Therefore, the torsion of the curve considered is 0.

18.74. $\mathbf{R} = \mathbf{r} + \dfrac{\langle \mathbf{r}, \mathbf{n} \rangle}{2k\,|\mathbf{r}|^2 + \langle \mathbf{r}, \mathbf{n} \rangle}\left(\langle \mathbf{r}, \mathbf{n} \rangle\,\mathbf{n} - \langle \mathbf{r}, \mathbf{v} \rangle\,\mathbf{v}\right)$.

18.75. $\mathbf{R} = \mathbf{r} + \dfrac{\mathbf{r}' \times \mathbf{r}}{2\,|\mathbf{r}|^2\,(\mathbf{r}' \times \mathbf{r}'') + (\mathbf{r}' \times \mathbf{r})\,|\mathbf{r}'|^2}\left(|\mathbf{r}' \times \mathbf{r}|\,[\mathbf{r}'] - \langle \mathbf{r}, \mathbf{r}' \rangle\,\mathbf{r}'\right)$.

18.76. $\mathbf{R} = \mathbf{r} + \dfrac{\langle \mathbf{e}, \mathbf{n} \rangle}{2k}\left(\mathbf{n}\,\langle \mathbf{e}, \mathbf{n} \rangle - \mathbf{v}\,\langle \mathbf{e}, \mathbf{v} \rangle\right)$. If the curve is given by the equation $\mathbf{r} = \mathbf{r}(t)$, then

$$\mathbf{R} = \mathbf{r} + \frac{\mathbf{r}' \times \mathbf{e}}{2\,|\mathbf{r}' \times \mathbf{r}''|}\left(|\mathbf{r}' \times \mathbf{e}|\,[\mathbf{r}'] - \langle \mathbf{r}', \mathbf{e} \rangle\,\mathbf{r}'\right).$$

If the curve is given by the equation $y = f(x)$, then

$$X = x - \frac{(m - lf'(x))^2}{2f''(x)} f'(x) - \frac{(m - lf'(x))(l + mf'(x))}{2f''(x)},$$

$$Y = f(x) + \frac{(m - lf'(x))^2}{2f''(x)} - \frac{(m - lf'(x))(l + mf'(x))}{2f''(x)} f'(x),$$

where $e = (l, m)$.

18.78. (a) Let the ellipse be given in a parametric form: $x(t) = a\cos t$, $y(t) = b\sin t$. Then the evolute is given as

$$x(t) = \frac{a^2 - b^2}{a} \cos^3 t, \quad y(t) = \frac{b^2 - a^2}{b} \sin^3 t.$$

The curvature radius is equal to $\dfrac{\left(a^2 \sin^2 t + b^2 \cos^2 t\right)^{3/2}}{ab}$. To find the equation for the ellipse evolvent, we write the expression for the length of the ellipse arc:

$$s(t) = \int \sqrt{a^2 \sin^2 t + b^2 \cos^2 t}\, dt.$$

Note that this integral is not taken in elementary functions. Hence, we get the evolvent equation:

$$x = a\cos t + (s(t) - C) \frac{a\sin t}{s'(t)}, \quad y = b\sin t + (s(t) - C) \frac{b\cos t}{s'(t)}.$$

Here C is an arbitrary constant.

(b) Let the hyperbola be parameterized as follows: $x(t) = a\cosh t$, $y(t) = b\sinh t$. Then the evolute equation has the form

$$x = \frac{a^2 + b^2}{a} \cosh^3 t, \quad y = -\frac{a^2 + b^2}{b} \sinh^3 t.$$

The hyperbola curvature radius is $\dfrac{\left(a^2 \sinh^2 t + b^2 \cosh^2 t\right)^{3/2}}{ab}$.

(c) Parameterize the parabola as follows: $x(t) = t$, $y(t) = \dfrac{t^2}{2p}$. The evolute is given as $x = -\dfrac{t^3}{p^2}$, $y = p + \dfrac{3t^2}{2p}$. The curvature radius of the parabola is equal to $p\left(1 + \dfrac{t^2}{p^2}\right)^{3/2}$.

(d) The evolute of this curve is given as $x = -a\pi + a(t' - \sin t')$, $y = -2a + a(1 - \cos t')$, $t' = t + \pi$. The curvature radius of this curve is equal to $R = 4a\left|\sin \dfrac{t}{2}\right|$.

(e) *Hint:* go over to Cartesian coordinates.

The evolute equation has the following form: $x = \dfrac{a}{3}\left(\cos\varphi - \cos^2 t + 2\right)$, $y = \dfrac{a}{3}(1 - \cos\varphi)\sin\varphi$. This curve is a cardioid. To verify this, it is sufficient to make the substitution $\varphi = \pi - t$ and to perform the substitution of the coordinates

$X = -\left(x - \dfrac{2}{3}a\right)$, $Y = y$. The curvature radius of the initial curve is equal to $\dfrac{4}{3}a\left|\cos\dfrac{\varphi}{2}\right|$.

(f) The evolute: $x = x - a\sinh\dfrac{x}{a}\cosh\dfrac{x}{a}$, $y = 2a\cosh\dfrac{x}{a}$. The curvature radius: $a\cosh^2\dfrac{x}{a}$. The evolvent of the catenary passing through its vertex is given as $\left(a\left(\ln\tan\dfrac{t}{2} + \cos t\right), a\sin t\right)$, i.e., is a tractrix.

(h) The evolute:

$$x = \frac{a\varphi\cos\varphi}{\varphi^2 + 2} - \frac{a\left(\varphi^2\right)}{\varphi^2 + 2}\sin\varphi, \quad y = \frac{a\varphi\sin\varphi}{\varphi^2 + 2} + \frac{a\left(\varphi^2 + 1\right)}{\varphi^2 + 2}\cos\varphi.$$

The curvature radius: $\dfrac{a\left(1 + \varphi^2\right)^{3/2}}{2 + \varphi^2}$.

The evolvent:

$$x = a\varphi\cos\varphi - (C + s)\frac{\cos\varphi - \varphi\sin\varphi}{\sqrt{1 + \varphi^2}},$$

$$y = a\varphi\sin\varphi - (C + s)\frac{\sin\varphi + \varphi\cos\varphi}{\sqrt{1 + \varphi^2}},$$

where $s = \dfrac{a}{2}\left(\varphi\sqrt{1 + \varphi^2} + \ln\left(\varphi + \sqrt{1 + \varphi^2}\right)\right)$, and C is an arbitrary constant.

18.79. The evolute of the logarithmic spiral is given as

$$x = -a^\varphi\ln a\sin\varphi, \quad y = a^\varphi\ln a\cos\varphi.$$

It is easy to verify that these formulas can be rewritten as follows:

$$x = \left(a^{-\pi/2}\ln a\right)a^{\pi/2+\varphi}\cos\left(\frac{\pi}{2} + \varphi\right),$$

$$y = \left(a^{-\pi/2}\ln a\right)a^{\pi/2+\varphi}\sin\left(\frac{\pi}{2} + \varphi\right).$$

It is now easy to check that these equations give a logarithmic spiral with the same parameter a turned by angle $\dfrac{\pi}{2}$ relative to the initial spiral.

18.82. $x = a\cos t + (at - C)\sin t$, $y = a\sin t - (at - C)\cos t$, where C is an arbitrary constant.

18.83. $8r\sin\dfrac{t}{2}$.

18.85. Set the astroid parametrically: $x = R\cos^3\dfrac{t}{4}$, $y = R\sin^3\dfrac{t}{4}$.

The evolute:

$$x = R\cos\frac{t}{4}\left(\cos^2\frac{t}{4} + 3\sin^2\frac{t}{2}\right), \quad y = R\sin\frac{t}{4}\left(\sin^2\frac{t}{4} + 3\cos^2\frac{t}{2}\right).$$

By the orthogonal transformation $X = \dfrac{1}{\sqrt{2}}(x - y)$, $Y = \dfrac{1}{\sqrt{2}}(x + y)$, i.e., by a turn to an angle $\dfrac{\pi}{4}$, the evolute equations are reduced to the form

$$x = 2R\cos^3\left(\frac{t}{4} + \frac{\pi}{4}\right), \quad y = 2R\sin^3\left(\frac{t}{4} + \frac{\pi}{4}\right).$$

18.88. Natural equations of such curves: $k = \dfrac{1}{-s+C}$, where C is an arbitrary constant.

18.89. $\rho = \varphi(s)\,\varphi'(s)$ and $\sigma = \varphi(s) + \text{const.}$

18.92. Let the curve γ be given in the natural parameter by the equation $\mathbf{r} = \mathbf{r}(s)$. In this case, the curve γ^* is given by the equation $\boldsymbol{\rho}(s) = \mathbf{r}(s) + \dfrac{1}{k}\mathbf{n}(s)$, where $\mathbf{n}(s)$ is the vector of the principal normal to the curve γ. Note that, generally speaking, s is not the natural parameter on the curve γ^*. We have:

$$\boldsymbol{\rho}' = \frac{\varkappa}{k}\mathbf{b}, \quad \boldsymbol{\rho}'' = \frac{d}{ds}\left(\frac{\varkappa}{k}\right) - \frac{\varkappa^2}{k}\mathbf{n}, \quad \boldsymbol{\rho}' \times \boldsymbol{\rho}'' = \frac{\varkappa^3}{k^2}\mathbf{v}.$$

Hence, $k^* = k$, $\varkappa^* = \dfrac{k^2}{\varkappa}$.

18.94. $\dfrac{1}{6}\dfrac{(\mathbf{r}',\mathbf{r}'',\mathbf{r}''')}{|\mathbf{r}' \times \mathbf{r}''|}$; in a particular case: $\dfrac{1}{6}k\,|\varkappa|$.

18.96. Show that the asymptotic direction can not be perpendicular to the axis of revolution.

18.110. We should take an arbitrary arc of the helical line with parallel tangents at the endpoints and connect these points by a plane curve. The ends of the helical-line arc should, certainly, be first slightly extended such that the torsion would continuously pass into zero.

18.133. Make use of the Gauss–Bonnet formula and the formula for the preimage of the sphere area form at a Gaussian mapping.

———————

19.1. Let a, b, c, x and β be the legs, the hypotenuse, the perpendicular on leg a and the angle subtending side b, respectively. Then, by the sine law,

$$\sinh x = \sinh \frac{c}{2}\sin\beta,$$

and, on the other hand,

$$\sinh b = \sinh c \sin\beta, \quad 2\sinh\frac{b}{2}\cosh\frac{b}{2} = 2\sinh\frac{c}{2}\cosh\frac{c}{2}\sin\beta.$$

Thus, $\dfrac{\sinh x}{\sinh(b/2)} = \dfrac{\cosh(c/2)}{\cosh(b/2)}$. It remains to note that the functions cosh and sinh ascend by $\mathbb{R}_+$.

19.4. (a) π; (b) $\pi/2$; (c) $\pi/2$; (d) $\pi/6$; (e) $\pi/2$; (f) $\pi/2$; (g) $\pi/4$.

19.5. $4\arcsin\dfrac{1}{\sqrt{5}} - \dfrac{\pi}{2}$.

19.7. $\pi/2$.

19.14. (b) $ds^2 = \dfrac{(1-y^2)\,dx^2 + 2xy\,dx\,dy + (1-x^2)\,dy^2}{(1-x^2-y^2)^2}$.

(c) $\Gamma_{11}^1 = 2\Gamma_{12}^2 = 2\Gamma_{21}^2 = \dfrac{2x}{1-x^2-y^2}$, $\Gamma_{22}^2 = 2\Gamma_{12}^1 = 2\Gamma_{21}^1 = \dfrac{2y}{1-x^2-y^2}$,

$\Gamma_{11}^2 = 2\Gamma_{22}^1 = 0$.

(e) The chords and diameters of the unit disk.

19.16. Consider $\mathbb{R}_1^4$ with the pseudo-Euclidean product given by the form $dx^2 + dy^2 + dz^2 - dw^2$. Let C be the cone $x^2 + y^2 + z^2 - w^2 = 0$. Then $S^2 = C \cap \{w = -1\}$. The circles on the sphere S^2 are its sections by the hyperplanes Π_i: $a_i x + b_i x + c_i z - d_i w = 0$, $i = 1, 2$. Let $\gamma_i = (a_i, b_i, c_i, d_i)$. It remains to verify that $\cos\varphi = \dfrac{\langle \gamma_1, \gamma_2 \rangle}{|\gamma_1||\gamma_2|}$.

19.17. (a) $\dfrac{1}{4}\left(v + \dfrac{1}{v}\right)^2 (a^2 \sin^2 u + b^2 \cos^2 u)\, du^2 +$

$\quad \dfrac{1}{2}(b^2 - a^2)\sin u \cos u \left(v - \dfrac{1}{v^3}\right) du\, dv +$

$\quad \dfrac{1}{4}\left(\left(1 - \dfrac{1}{v^2}\right)^2 (a^2 \cos^2 u + b^2 \sin^2 u) + c^2 \left(1 + \dfrac{1}{v^2}\right)^2\right) dv^2;$

(b) $\dfrac{1}{(u+v)^4}\left(a^2 (v^2 - 1)^2 + 4b^2 v^2 + c^2 (v^2 + 1)^2\right) du^2 +$

$\quad 2\left(a^2 (u^2 - 1)(v^2 - 1) - 4b^2 uv + c^2 (u^2 + 1)(v^2 + 1)\right) du\, dv +$

$\quad \dfrac{1}{(u+v)^4}\left(a^2 (u^2 - 1)^2 + 4b^2 u^2 + c^2 (u^2 + 1)^2\right) dv^2;$

(c) $\dfrac{1}{4}\left(v - \dfrac{1}{v}\right)^2 (a^2 \sin^2 u + b^2 \cos^2 u)\, du^2 +$

$\quad \dfrac{1}{2}(b^2 - a^2)\sin u \cos u \left(v - \dfrac{1}{v^3}\right) du\, dv +$

$\quad \dfrac{1}{4}\left(\left(1 + \dfrac{1}{v^2}\right)^2 (a^2 \cos^2 u + b^2 \sin^2 u) + c^2 \left(1 - \dfrac{1}{v^2}\right)^2\right) dv^2;$

(d) $(p\sin^2 u + q\cos^2 u)\, v^2 du^2 + 2(q - p)\sin u \cos u\, du\, dv + (p\cos^2 u + q\sin^2 u + v^2)\, dv^2;$

(e) $(p + q + 4v^2)\, du^2 + 2(p - q + 4uv)\, du\, dv + (p + q + 4u^2)\, dv^2;$

(f) $(a^2 \sin^2 u + b^2 \cos^2 u)\, du^2 + dv^2;$

(g) $\left(\dfrac{a^2}{4}\left(1 - \dfrac{1}{u^2}\right)^2 + \dfrac{b^2}{4}\left(1 + \dfrac{1}{u^2}\right)^2\right) du^2 + dv^2.$

19.21. The sphere: $ds^2 = du^2 + R^2 \cos^2 (u/R)\, dv^2.$

The torus: $ds^2 = du^2 + \left(a + b\cos\dfrac{u}{b}\right)^2 dv^2.$

The catenoid: $ds^2 = du^2 + (a^2 + u^2)\, dv^2.$

The pseudosphere: $ds^2 = du^2 + e^{-2u/a}\, dv^2.$

Hint: u is the natural parameter of the meridian.

19.22. $ds^2 = d\tilde{u}^2 + e^{-2\tilde{u}/a}\, d\tilde{v}^2.$ Assuming $u = \tilde{v}$, $v = ae^{\tilde{u}/a}$, we get $ds^2 = \dfrac{a^2}{v^2}(du^2 + dv^2).$

19.26. Let $r = F(\rho)$ be the sought-for dependence. We have:

$$ds^2 = (f'^2 + g'^2)\, dr^2 + f^2 d\varphi^2 = (f'^2 + g'^2) F'^2 d\rho^2 + f^2 d\varphi^2 =$$
$$\Lambda^2 (du^2 + dv^2) = \Lambda^2 (d\rho^2 + \rho^2 d\varphi^2),$$

whence we get the equation $\dfrac{\rho}{\rho'(r)} = \dfrac{f}{\sqrt{f'^2 + g'^2}}$, solving which we get the sought-for relation

$$\ln \rho = \int \frac{\sqrt{f'^2\,(r) + g'^2\,(r)}}{f\,(r)}\,dr.$$

Application of this formula to the catenoid yields $\rho = e^z$ and, setting $u = \rho \cos \varphi$, $v = \rho \sin \varphi$, we get

$$x = \frac{1}{2}\left(u + \frac{u}{u^2 + v^2}\right) = \frac{1}{2}\operatorname{Re}\left(w + \frac{1}{w}\right),$$

$$y = \frac{1}{2}\left(v + \frac{v}{u^2 + v^2}\right) = \frac{1}{2}\operatorname{Re}\left(-iw + \frac{i}{w}\right),$$

$$z = \frac{1}{2}\ln\left(u^2 + v^2\right) = \operatorname{Re}\ln w, \quad w = u + iv, \quad 0 < |w| < \infty.$$

19.28. Let the surface $M^2 \subset \mathbb{R}^3$ be given by the equations $x_i = x_i\,(p, q)$, $i = 1, 2, 3$, and the variables p and q be changing in some domain on the plane. Let the functions $x_i = x_i\,(p, q)$ be real-analytic. The couple (p, q) can be considered as the coordinates of a point on the surface M^2. The curve C on M^2 is given by the equations

$$p = p\,(t), \quad q = q\,(t), \quad a \le t \le b.$$

An element of arc length is expressed via the vector $\mathbf{x} = (x_1, x_2, x_3)$:

$$ds^2 = \langle d\mathbf{x}, d\mathbf{x} \rangle = \langle \mathbf{x}_p dp + \mathbf{x}_q dq, \mathbf{x}_p dp + \mathbf{x}_q dq \rangle,$$

or

$$ds^2 = \langle \mathbf{x}_p, \mathbf{x}_p \rangle\,dp^2 + 2\langle \mathbf{x}_p, \mathbf{x}_q \rangle\,dp\,dq + \langle \mathbf{x}_q, \mathbf{x}_q \rangle\,dq^2 = E\,dp^2 + 2F\,dp\,dq + G\,dq^2,$$

where $E = \langle \mathbf{x}_p, \mathbf{x}_p \rangle$, $F = \langle \mathbf{x}_p, \mathbf{x}_q \rangle$, $G = \langle \mathbf{x}_q, \mathbf{x}_q \rangle$.

Since the length element ds^2 is always positive, $W^2 = EG - F^2$ is also positive. We find the coordinate system (u, v) with the arc element $ds^2 = \lambda\,(u, v)\left(du^2 + dv^2\right)$. We have:

$$ds^2 = \left(\sqrt{E}\,dp + \frac{F + iW}{\sqrt{E}}\,dq\right)\left(\sqrt{E}\,dp + \frac{F - iW}{\sqrt{E}}\,dq\right).$$

Assume that we can find such an integrating factor $\sigma = \sigma_1 + i\sigma_2$ that

$$\sigma\left(\sqrt{E}\,dp + \frac{F + iW}{\sqrt{E}}\,dq\right) = du + i\,dv.$$

Then

$$\bar{\sigma}\left(\sqrt{E}\,dp + \frac{F - iW}{\sqrt{E}}\,dq\right) = du - i\,dv$$

and, finally, $|\sigma|^2\,ds^2 = du^2 + dv^2$. Assuming $\sigma^2 = 1/\lambda$, we obtained the isothermic coordinates (u, v) by finding an integrating factor, which transforms the expression $\sqrt{E}\,dp + \dfrac{F + iW}{\sqrt{E}}\,dq$ into the total differential. The differential $du + i\,dv$ can be written as

$$du + i\,dv = \left(\frac{\partial u}{\partial p} + i\frac{\partial v}{\partial p}\right)dp + \left(\frac{\partial u}{\partial q} + i\frac{\partial v}{\partial q}\right)dq.$$

Further,

$$\frac{\partial u}{\partial p} + i\frac{\partial v}{\partial p} = \sigma\sqrt{E}, \quad \frac{\partial u}{\partial q} + i\frac{\partial v}{\partial q} = \sigma\frac{F + iW}{\sqrt{E}}.$$

Excluding σ, we get

$$E\left(\frac{\partial u}{\partial q} + i\frac{\partial v}{\partial q}\right) = (F + iW)\left(\frac{\partial u}{\partial p} + i\frac{\partial v}{\partial p}\right),$$

or

$$E\frac{\partial u}{\partial q} = F\frac{\partial u}{\partial p} - W\frac{\partial v}{\partial p}, \quad E\frac{\partial v}{\partial q} = W\frac{\partial u}{\partial p} + F\frac{\partial v}{\partial p}.$$

Resolving this system relative to the unknowns $\partial v/\partial p$ and $\partial v/\partial q$, we get

$$\frac{\partial v}{\partial p} = \frac{F\left(\partial u/\partial p\right) - E\left(\partial u/\partial q\right)}{\sqrt{EG - F^2}}, \quad \frac{\partial v}{\partial q} = \frac{G\left(\partial u/\partial p\right) - F\left(\partial u/\partial q\right)}{\sqrt{EG - F^2}}. \quad (*)$$

Similarly,

$$\frac{\partial u}{\partial p} = \frac{E\left(\partial v/\partial q\right) - F\left(\partial v/\partial p\right)}{\sqrt{EG - F^2}}, \quad \frac{\partial u}{\partial q} = \frac{F\left(\partial v/\partial q\right) - G\left(\partial v/\partial p\right)}{\sqrt{EG - F^2}}.$$

Therefore, u satisfies the equation

$$\frac{\partial}{\partial q} = \left(\frac{F\left(\partial u/\partial p\right) - E\left(\partial u/\partial q\right)}{W}\right) + \frac{\partial}{\partial p}\left(\frac{F\left(\partial u/\partial q\right) - G\left(\partial u/\partial p\right)}{W}\right) = 0,$$

which is called the Beltrami–Laplace equation. If a second family of isothermic coordinates (x, y) in the neighbourhood of the point is known, then $ds^2 = \mu\left(dx^2 + dy^2\right)$. Using the coordinates (x, y) instead of the coordinates (p, q), we get $E = G = \mu$, $F = 0$ and $\dfrac{\partial v}{\partial x} = -\dfrac{\partial u}{\partial y}$, and $\dfrac{\partial v}{\partial y} = \dfrac{\partial u}{\partial x}$.

Thus, we obtained the Cauchy–Riemann equations, from which it is seen that the functions u and v are conjugate harmonic functions, and $f = u + iv$ is an analytical function of $z = x + iy$. The Beltrami equation takes the form of the known Laplace equation $\partial^2 u/\partial x^2 + \partial^2 u/\partial y^2 = 0$. It is said that the complex-valued function $f(p, q)$ defined on M^2 is called a complex potential on M^2 if its real and imaginary parts satisfy the equations (*). Thus, the real and imaginary parts of the complex potential on the manifold M^2 give isothermic coordinates in the neighbourhood of each point of M^2. Note that these coordinates are local: they are not defined, generally speaking, on the entire 2D manifold; in transition from one point to another, the complex potential will be changing.

19.30. (a) Consider some curve $\varphi = \varphi(\theta)$ on the surface of the sphere. In the movement along this curve, the compass arrow forms an angle ψ with the direction of the movement; the angle is defined by the relations $\tan\psi = \sin\theta\dfrac{d\varphi}{d\theta}$. Here the angle ψ is measured from the axis y clockwise. On the chart, we get $\dfrac{dy}{dx} = \tan\left(\psi + \dfrac{\pi}{2}\right) = -\dfrac{1}{\tan\psi}$. From these two relations, it follows that

$$\sin\theta\frac{d\varphi}{d\theta} = -\frac{dx/d\theta}{dy/d\theta} = -\frac{dx/d\theta + (dx/d\varphi)(d\varphi/d\theta)}{dy/d\theta + (dy/d\varphi)(d\varphi/d\theta)},$$

$$\left(\frac{dy}{d\theta} + \frac{dy}{d\varphi}\frac{d\varphi}{d\theta}\right)\frac{d\varphi}{d\theta}\sin\theta = -\frac{dx}{d\theta} - \frac{dx}{d\varphi}\frac{d\varphi}{d\theta}.$$

As this correlation should be fulfilled in the considered point at any value of $d\varphi/d\theta$, then, equating the coefficients at the same degrees of the derivative $d\varphi/d\theta$ in the right- and left-hand sides, we get

$$\frac{dy}{d\varphi} = 0, \quad y = y(\theta), \quad \frac{dx}{d\theta} = 0, \quad x = x(\varphi), \quad -\sin\theta\frac{dy}{d\theta} = \frac{dx}{d\varphi}.$$

From the two next-to-last correlations, it follows that the left-hand side of the last correlation depends only on θ, whereas the right-hand side depends only on φ, so both parts of this correlation should be constant. We let this constant be equal to unity. Thus, in the Mercator projection a mapping is given by the formulas

$$x = \varphi, \quad y = -\int \frac{d\theta}{\sin\theta} = \ln\cot\frac{\theta}{2}.$$

(b) $ds^2 = d\theta^2 + \sin^2\theta d\varphi^2 = \sin^2\theta\left(dx^2 + dy^2\right) = \dfrac{dx^2 + dy^2}{\cosh^2 y}$.

19.31. $ds^2 = \dfrac{dv^2}{(1 - v^2/c^2)^2} + \dfrac{v^2 d\varphi^2}{1 - v^2/c^2}$.

19.32. $ds^2 = d\chi^2 + \sinh^2\chi d\varphi^2$.

19.33. $ds^2 = \dfrac{d\rho^2 + \rho^2 d\varphi^2}{(1 - \rho^2)^2}$.

19.40. *Hint:* construct an isometric mapping of the three-dimensional space into the four-dimensional one in the form of a three-dimensional cylindrical surface.

19.49. *Hint:* it is better to prove an inverse course of bending: the surface of revolution S of the form

$$x = g(u)\cos\varphi, \quad y = g(u)\sin\varphi, \quad z = -\int\sqrt{1 - g'^2}\,du$$

with the metric $ds^2 = du^2 + g^2(u)\,d\varphi^2$ should be bent into helical surfaces S_h of the form $x = \sqrt{g^2 - h^2}\cos(\varphi + hF(u))$, $y = \sqrt{g^2 - h^2}\sin(\varphi + hF(u))$, $z = h\varphi + h^2 F(u) + G(u, h)$.

Solution. Items (a) and (b) are evident. To prove (c), we first compute the metric of the general helical surface. We obtain $ds^2 = (1 + f'^2(u))\,du^2 + 2hf'(u)\,du\,dv + (h^2 + u^2)\,dv^2$. Make the substitution of the variables $u = u(r)$, $v = H(r) + \varphi$. In new variables, the metric will be represented as

$$ds^2 = (1 + f'^2(u(r))\,u'^2(r) + 2hf'(r)\,u'(r)\,H'(r) + (h^2 + u^2(r))\,H'^2(r)) \times$$
$$dr^2 + 2(hf'(r)\,u'(r) + (h^2 + u^2(r))\,H'(r))\,dr\,d\varphi + (h^2 + u^2(r))\,d\varphi^2.$$

To arrive at the metric of revolution, it suffices to require that $hf'(r)u'(r) + (h^2 + u^2(r))H'(r) = 0$ and $(1 + f'^2(u(r))u'^2(r) + 2hf'(r)u'(r)H'(r) + (h^2 + u^2(r))H'^2(r)) = 1$. We get the required substitutions:

$$r(u) = \int\sqrt{1 + \frac{u^2 f'^2(u)}{h^2 + u^2}}\,du, \quad H(r) = \int\frac{hf'(u)\,du}{h^2 + u^2},$$

where $u = u(r)$ leading to the metric of revolution of the form $ds^2 = dr^2 + G^2(r)\,d\varphi^2$, where $G^2 = h^2 + u^2(r)$.

Conversely, having the metric of the form $ds^2 = dr^2 + G^2(r)\, d\varphi^2$, it is easy to choose the substitutions $r = r(u)$, $\varphi = v - H(r)$ such that to get the metric of the general helical surface with the corresponding function $f(u)$;

(d) the required bending is given by the formulas

$$x = \sqrt{g^2(u) - h^2}\, \cos(\varphi + hF(u)),$$

$$y = \sqrt{g^2(u) - h^2}\, \sin(\varphi + hF(u)),$$

$$z = h\varphi + h^2 F(u) + G(u, l),$$

where

$$F(u) = \int \frac{h\,R\,du}{g\,(g^2 - h^2)}, \qquad G(u) = -\int \frac{g\,R\,du}{g^2 - h^2}$$

and $R^2 = g^2 - h^2 - g^2 g'^2$; h is a numeric parameter.

(e) as $x(u, v; t) = u\cos(v + t)$, $y(u, v; t) = u\sin(v + t)$, $z(u, v; t) = h(v + t) + f(u)$, the bending of sliding is generated by helical motion of the entire space consisting of translational motion in parallel to the axis Oz with the constant velocity vector $\mathbf{v} = (0, 0, h)$ and rotational motion around the axis Oz at a constant angular velocity 1. Write out the law of transition from point (x, y, z) to its position at instant t, singling out the translational and rotational parts in it.

19.50. Let γ_0: $\mathbf{u} = \mathbf{u}(t; u_0, v_0)$, $\mathbf{v} = \mathbf{v}(t; u_0, v_0)$ be the trajectory of the point (u_0, v_0) at the sliding bending. Let the point (u_0, v_0) correspond to the value $t = 0$. At an isometry generated by the sliding bending at a given value of parameter t, each arc of the curve γ, beginning in (u_0, v_0), passes into the arc of the same curve beginning in its point corresponding to the value t. Through points of the curve γ_0, we draw geodesics Γ_t orthogonal to it, and mark on each of them the point $M_t(d)$ at an equal distance d from the corresponding point on γ_0. As the neighbourhood of the point (u_0, v_0) is isometric to the neighbourhood of the point

$$\mathbf{u} = \mathbf{u}(t; u_0, v_0), \qquad \mathbf{v} = \mathbf{v}(t; u_0, v_0)$$

at any small t, at this isometry the geodesic Γ_0 emanating from the point (u_0, v_0) orthogonally to γ_0 is in correspondence with the geodesic Γ_t also emanating orthogonally to γ_0 at a corresponding point, and, due to the preservation of distances, the point $M_0(d) \in \Gamma_0$ will prove to be in correspondence to the point $M_t(d)$. Therefore, at different d the points $M_t(d)$ will be the trajectories of the points $M_0(d)$, which we will designate as γ_d. Due to the correspondence of the arcs on γ_d by isometry, the geodesic curvature of each curve γ_d is constant. We introduce the semigeodesic coordinate system, taking the lines Γ_t for the family of parallel geodesics and the curves γ_d for lines orthogonal to them. For convenient writing, we leave the old designations (u, v) for the new coordinates. The length in semigeodesic coordinates has the form $ds^2 = du^2 + G(u, v)\, dv^2$; what is more, lines $u = \text{const}$ correspond to the curves γ_t. Computing the geodesic curvature k_g of these lines, we get that $k_g = \dfrac{1}{2G}\dfrac{\partial G}{\partial u}$. But, as it is constant along each line $u = \text{const}$, the expression $\dfrac{\partial \ln G}{\partial u}$ does not depend on v; therefore, in reality $G(u, v) = G(u)$, and this means that the metric ds^2 is a metric of revolution.

20.1. (a) We have:

$\mathbf{r}_u = (a \sinh u \cos v, a \sinh u \sin v, c \cosh u)$,

$\mathbf{r}_v = (-a \cosh u \sin v, a \cosh u \cos v, 0)$,

$\mathbf{r}_u \times \mathbf{r}_v = (-ac \cosh^2 u \cos v, -ac \cosh^2 u \sin v, a^2 \cosh u \sinh u)$,

$|\mathbf{r}_u \times \mathbf{r}_v| = a \cosh u \sqrt{c^2 \cosh^2 u + a^2 \sinh^2 u}$,

$\mathbf{n} = \dfrac{\mathbf{r}_u \times \mathbf{r}_v}{|\mathbf{r}_u \times \mathbf{r}_v|} =$

$\dfrac{1}{\sqrt{c^2 \cosh^2 u + a^2 \sinh^2 u}} (-c \cosh u \cos v, -c \cosh u \sin v, a \sinh u)$,

$\mathbf{r}_{uu} = (a \cosh u \cos v, a \cosh u \sin v, c \sinh u)$,

$\mathbf{r}_{uv} = (-a \sinh u \sin v, a \sinh u \cos v, 0)$,

$\mathbf{r}_{vv} = (-a \cosh u \cos v, -a \cosh u \sin v, 0)$,

$L = \langle \mathbf{r}_{uu}, \mathbf{n} \rangle =$

$\dfrac{1}{\sqrt{c^2 \cosh^2 u + a^2 \sinh^2 u}} ac \left(-\cosh^2 u \cos^2 v - \cosh^2 u \sin^2 v + \sinh^2 u \right) =$

$- \dfrac{ac}{\sqrt{c^2 \cosh^2 u + a^2 \sinh^2 u}}$,

$M = \langle \mathbf{r}_{uv}, \mathbf{n} \rangle = \dfrac{1}{\sqrt{c^2 \cosh^2 u + a^2 \sinh^2 u}} ((-c \cosh u \cos v)(-a \sinh u \sin v)$

$+ (-c \cosh u \sin v)(a \sinh u \cos v)) = 0$,

$N = \langle \mathbf{r}_{vv}, \mathbf{n} \rangle = \dfrac{1}{\sqrt{c^2 \cosh^2 u + a^2 \sinh^2 u}} ((-c \cosh u \cos v)(-a \cosh u \cos v) +$

$(-c \cosh u \sin v)(-a \cosh u \sin v)) = \dfrac{ac \cosh^2 u}{\sqrt{c^2 \cosh^2 u + a^2 \sinh^2 u}}$.

Answer: $\dfrac{ac}{\sqrt{c^2 \cosh^2 u + a^2 \sinh^2 u}} \left(-du^2 + \cosh^2 u \, dv^2 \right)$.

(b) We have:

$\mathbf{r}_u = (a \cosh u \cos v, a \cosh u \sin v, c \sinh u)$,

$\mathbf{r}_v = (-a \sinh u \sin v, a \sinh u \cos v, 0)$,

$\mathbf{r}_u \times \mathbf{r}_v = (-ac \sinh^2 u \cos v, -ac \sinh^2 u \sin v, a^2 \cosh u \sinh u)$,

$|\mathbf{r}_u \times \mathbf{r}_v| = a \sinh u \sqrt{a^2 \cosh^2 u + c^2 \sinh^2 u}$,

$\mathbf{n} = \dfrac{\mathbf{r}_u \times \mathbf{r}_v}{|\mathbf{r}_u \times \mathbf{r}_v|} =$

$\dfrac{1}{\sqrt{a^2 \cosh^2 u + c^2 \sinh^2 u}} (-c \sinh u \cos v, -c \sinh u \sin v, a \cosh u)$,

$\mathbf{r}_{uu} = (a \sinh u \cos v, a \sinh u \sin v, c \cosh u)$,

$\mathbf{r}_{uv} = (-a \cosh u \sin v, a \cosh u \cos v, 0)$,

$\mathbf{r}_{vv} = (-a \sinh u \cos v, -a \sinh u \sin v, 0)$,

$$L = \langle \mathbf{r}_{uu}, \mathbf{n} \rangle = \frac{1}{\sqrt{a^2 \cosh^2 u + c^2 \sinh^2 u}} ac \left(-\sinh^2 u \cos^2 v - \right.$$

$$\left. \sinh^2 u \sin^2 v + \cosh^2 u \right) = -\frac{ac}{\sqrt{a^2 \cosh^2 u + c^2 \sinh^2 u}},$$

$$M = \langle \mathbf{r}_{uv}, \mathbf{n} \rangle = \frac{1}{\sqrt{a^2 \cosh^2 u + c^2 \sinh^2 u}} \left((-c \sinh u \cos v)(-a \cosh u \sin v) + \right.$$

$$\left. (-c \sinh u \sin v)(a \cosh u \cos v) \right) = 0,$$

$$N = \langle \mathbf{r}_{vv}, \mathbf{n} \rangle = \frac{1}{\sqrt{a^2 \cosh^2 u + c^2 \sinh^2 u}} \left((-c \sinh u \cos v)(-a \sinh u \cos v) + \right.$$

$$\left. (-c \sinh u \sin v)(-a \sinh u \sin v) \right) = \frac{ac \sinh^2 u}{\sqrt{a^2 \cosh^2 u + c^2 \sinh^2 u}}.$$

Answer: $\dfrac{ac}{\sqrt{a^2 \cosh^2 u + c^2 \sinh^2 u}} \left(du^2 + \sinh^2 u \, dv^2 \right).$

20.2. Let for definiteness $u > 0$. We have:

$$\mathbf{r}_u = (\cos v, \sin v, 2u), \quad \mathbf{r}_v = (-u \sin v, u \cos v, 0),$$

$$\mathbf{r}_u \times \mathbf{r}_v = \left(-2u^2 \cos v, -2u^2 \sin v, u \right), \quad |\mathbf{r}_u \times \mathbf{r}_v| = u\sqrt{1 + 4u^2},$$

$$\mathbf{n} = \frac{\mathbf{r}_u \times \mathbf{r}_v}{|\mathbf{r}_u \times \mathbf{r}_v|} = \frac{1}{\sqrt{1 + 4u^2}} (-2u \cos v, -2u \sin v, 1),$$

$$\mathbf{r}_{uu} = (0, 0, 2), \quad \mathbf{r}_{uv} = (-\sin v, \cos v, 0), \quad \mathbf{r}_{vv} = (-u \cos v, -u \sin v, 0),$$

$$L = \langle \mathbf{r}_{uu}, \mathbf{n} \rangle = \frac{2}{\sqrt{1 + 4u^2}},$$

$$M = \langle \mathbf{r}_{uv}, \mathbf{n} \rangle = \frac{1}{\sqrt{1 + 4u^2}} \left((-2u \cos v)(-\sin v) + (-2u \sin v) \cos v \right) = 0,$$

$$N = \langle \mathbf{r}_{vv}, \mathbf{n} \rangle = \frac{1}{\sqrt{1 + 4u^2}} \left((-2u \cos v)(-u \cos v) + \right.$$

$$\left. (-2u \sin v)(-u \sin v) \right) = \frac{2u^2}{\sqrt{1 + 4u^2}},$$

Answer: $\dfrac{2}{\sqrt{1 + 4u^2}} \left(du^2 + u^2 dv^2 \right).$

20.3. We have:

$$\mathbf{r}_u = (0, 0, 1), \quad \mathbf{r}_v = (-R \sin v, R \cos v, 0),$$

$$\mathbf{r}_u \times \mathbf{r}_v = (-R \cos v, -R \sin v, 0), \quad |\mathbf{r}_u \times \mathbf{r}_v| = R,$$

$$\mathbf{n} = \frac{\mathbf{r}_u \times \mathbf{r}_v}{|\mathbf{r}_u \times \mathbf{r}_v|} = (-\cos v, -\sin v, 0),$$

$$\mathbf{r}_{uu} = (0, 0, 0), \quad \mathbf{r}_{uv} = (0, 0, 0), \quad \mathbf{r}_{vv} = (-R \cos v, -R \sin v, 0),$$

$$L = \langle \mathbf{r}_{uu}, \mathbf{n} \rangle = 0, \quad M = \langle \mathbf{r}_{uv}, \mathbf{n} \rangle = 0,$$

$$N = \langle \mathbf{r}_{vv}, \mathbf{n} \rangle = ((-\cos v)(-R\cos v) + (-\sin v)(-R\sin v)) = R,$$

Answer: $R\,dv^2$.

20.4. Let for definiteness $u > 0$. We have:

$$\mathbf{r}_u = (\cos v, \sin v, k), \quad \mathbf{r}_v = (-u\sin v, u\cos v, 0),$$

$$\mathbf{r}_u \times \mathbf{r}_v = (-ku\cos v, -ku\sin v, u), \quad |\mathbf{r}_u \times \mathbf{r}_v| = u\sqrt{1+k^2},$$

$$\mathbf{n} = \frac{\mathbf{r}_u \times \mathbf{r}_v}{|\mathbf{r}_u \times \mathbf{r}_v|} = \frac{1}{\sqrt{1+k^2}}(-k\cos v, -k\sin v, 1),$$

$$\mathbf{r}_{uu} = (0,0,0), \quad \mathbf{r}_{uv} = (-\sin v, \cos v, 0), \quad \mathbf{r}_{vv} = (-u\cos v, -u\sin v, 0),$$

$$L = \langle \mathbf{r}_{uu}, \mathbf{n} \rangle = 0,$$

$$M = \langle \mathbf{r}_{uv}, \mathbf{n} \rangle = \frac{1}{\sqrt{1+k^2}}((-k\cos v)(-\sin v) + (-k\sin v)\cos v) = 0,$$

$$N = \langle \mathbf{r}_{vv}, \mathbf{n} \rangle = \frac{1}{\sqrt{1+k^2}}((-k\cos v)(-u\cos v) +$$

$$(-k\sin v)(-u\sin v)) = \frac{ku}{\sqrt{1+k^2}}.$$

Answer: $\dfrac{ku\,dv^2}{\sqrt{1+k^2}}$.

20.5. We have:

$$\mathbf{r} = (u\cos v, u\sin v, f(v)), \quad \mathbf{r}_u = (\cos v, \sin v, 0),$$

$$\mathbf{r}_v = (-u\sin v, u\cos v, f'),$$

$$E = \langle \mathbf{r}_u, \mathbf{r}_u \rangle = 1, \quad F = \langle \mathbf{r}_u, \mathbf{r}_v \rangle = 0, \quad G = \langle \mathbf{r}_v, \mathbf{r}_v \rangle = u^2 + (f')^2,$$

$$\mathbf{r}_u \times \mathbf{r}_v = (f'\sin v, -f'\cos v, u), \quad |\mathbf{r}_u \times \mathbf{r}_v| = \sqrt{u^2 + (f')^2},$$

$$\mathbf{n} = \frac{\mathbf{r}_u \times \mathbf{r}_v}{|\mathbf{r}_u \times \mathbf{r}_v|} = \frac{1}{\sqrt{u^2 + (f')^2}}(f'\sin v, -f'\cos v, u), \quad \mathbf{r}_{uu} = (0,0,0),$$

$$\mathbf{r}_{uv} = (-\sin v, \cos v, 0), \quad \mathbf{r}_{vv} = (-u\cos v, -u\sin v, f''),$$

$$L = \langle \mathbf{r}_{uu}, \mathbf{n} \rangle = 0, \quad M = \langle \mathbf{r}_{uv}, \mathbf{n} \rangle = \frac{-f'}{\sqrt{u^2 + (f')^2}},$$

$$N = \langle \mathbf{r}_{vv}, \mathbf{n} \rangle = \frac{uf''}{\sqrt{u^2 + (f')^2}}, \quad K = \frac{LN - M^2}{EG - F^2} = -\frac{(f')^2}{(u^2 + (f')^2)^2}.$$

Thus, $K = \lambda_1\lambda_2 < 0$ and, therefore, the principal curvatures λ_1 and λ_2 have different signs.

20.6. Let $\mathbf{v}, \mathbf{n}, \mathbf{b}$ be the Frénet frame of this curve. Then the surface is given by the equation $\mathbf{r}(s, u) = \boldsymbol{\rho}(s) + u\mathbf{b}(s)$, where s is the natural parameter of the curve

ρ. Then $\mathbf{r}_s = \mathbf{v} - u\varkappa\mathbf{n}$, $\mathbf{r}_u = \mathbf{b}$, $E = \langle\mathbf{r}_s, \mathbf{r}_s\rangle = 1 + u^2\varkappa^2$, $F = \langle\mathbf{r}_s, \mathbf{r}_u\rangle = 0$, $G = \langle\mathbf{r}_u, \mathbf{r}_u\rangle = 1$, $\mathbf{r}_s \times \mathbf{r}_u = -\mathbf{n} - u\varkappa\mathbf{v}$, $|\mathbf{r}_s \times \mathbf{r}_u| = \sqrt{1 + u^2\varkappa^2}$,

$$\mathbf{m} = \frac{\mathbf{r}_s \times \mathbf{r}_u}{|\mathbf{r}_s \times \mathbf{r}_u|} = -\frac{1}{\sqrt{1 + u^2\varkappa^2}}\left(u\varkappa\mathbf{v} + \mathbf{n}\right),$$

$$\mathbf{r}_{ss} = k\mathbf{n} - u\dot\varkappa\mathbf{n} - u\varkappa\left(-k\mathbf{v} + \varkappa\mathbf{b}\right) = uk\varkappa\mathbf{v} + (k - u\dot\varkappa)\mathbf{n} - u\varkappa^2\mathbf{b},$$

$$\mathbf{r}_{su} = -k\mathbf{n}, \mathbf{r}_{uu} = 0, \quad L = \langle\mathbf{r}_{ss}, \mathbf{m}\rangle = -\frac{ku^2\varkappa^2 + k - u\dot\varkappa}{\sqrt{1 + u^2\varkappa^2}},$$

$$M = \langle\mathbf{r}_{su}, \mathbf{m}\rangle = -\frac{\varkappa}{\sqrt{1 + u^2\varkappa^2}}, \quad N = \langle\mathbf{r}_{uu}, \mathbf{m}\rangle = 0,$$

$$K = \frac{LN - M^2}{EG - F^2} = -\frac{\varkappa^2}{(1 + u^2\varkappa^2)^2},$$

$$H = \frac{EN + GL - 2MF}{EG - F^2} = -\frac{k + ku^2\varkappa^2 - u\dot\varkappa}{(1 + u^2\varkappa^2)^{3/2}}.$$

Answer: $K = -\dfrac{\varkappa^2}{(1 + u^2\varkappa^2)^2}$, $H = -\dfrac{k + ku^2\varkappa^2 - u\dot\varkappa}{(1 + u^2\varkappa^2)^{3/2}}$.

20.7. Let $\mathbf{v}$, $\mathbf{n}$, $\mathbf{b}$ be the Frénet frame of this curve. Then the surface is given by the equation $\mathbf{r}(s, u) = \boldsymbol{\rho}(s) + u\mathbf{n}(s)$, where s is the natural parameter of the curve $\boldsymbol{\rho}$. Then $\mathbf{r}_s = \mathbf{v} - ku\mathbf{v} + u\varkappa\mathbf{b} = (1 - ku)\mathbf{v} + u\varkappa\mathbf{b}$, $\mathbf{r}_u = \mathbf{n}$, $E = \langle\mathbf{r}_s, \mathbf{r}_s\rangle = (1 - ku)^2 + u^2\varkappa^2$, $F = \langle\mathbf{r}_s, \mathbf{r}_u\rangle = 0$, $G = \langle\mathbf{r}_u, \mathbf{r}_u\rangle = 1$,

$$\mathbf{r}_s \times \mathbf{r}_u = -u\varkappa\mathbf{v} + (1 - ku)\mathbf{b}, \quad |\mathbf{r}_s \times \mathbf{r}_u| = \sqrt{(1 - ku)^2 + u^2\varkappa^2},$$

$$\mathbf{m} = \frac{\mathbf{r}_s \times \mathbf{r}_u}{|\mathbf{r}_s \times \mathbf{r}_u|} = \frac{1}{\sqrt{(1 - ku)^2 + u^2\varkappa^2}}\left(-u\varkappa\mathbf{v} + (1 - ku)\mathbf{b}\right),$$

$$\mathbf{r}_{ss} = -u\dot{k}\mathbf{v} + k(1 - ku)\mathbf{n} + u\dot\varkappa\mathbf{b} - u\varkappa^2\mathbf{n} = -\dot{k}u\mathbf{v} + (k - k^2u - \varkappa^2u)\mathbf{n} + \dot\varkappa u\mathbf{b},$$

$$\mathbf{r}_{su} = -k\mathbf{v} + \varkappa\mathbf{b}, \quad \mathbf{r}_{uu} = 0, \quad L = \langle\mathbf{r}_{ss}, \mathbf{m}\rangle = \frac{\left(\dot{k}\varkappa - k\dot\varkappa\right)u^2 + \dot\varkappa u}{\sqrt{(1 - ku)^2 + u^2\varkappa^2}},$$

$$M = \langle\mathbf{r}_{su}, \mathbf{m}\rangle = \frac{\varkappa}{\sqrt{(1 - ku)^2 + u^2\varkappa^2}}, \quad N = \langle\mathbf{r}_{uu}, \mathbf{m}\rangle = 0,$$

$$K = \frac{LN - M^2}{EG - F^2} = -\frac{\varkappa^2}{\left((1 - ku)^2 + u^2\varkappa^2\right)^2},$$

$$H = \frac{EN + GL - 2MF}{EG - F^2} = \frac{\left(\dot{k}\varkappa - k\dot\varkappa\right)u^2 + \dot\varkappa u}{\left((1 - ku)^2 + u^2\varkappa^2\right)^{3/2}}.$$

Answer: $K = -\dfrac{\varkappa^2}{\left((1 - ku)^2 + u^2\varkappa^2\right)^2}$, $H = \dfrac{\left(\dot{k}\varkappa - k\dot\varkappa\right)u^2 + \dot\varkappa u}{\left((1 - ku)^2 + u^2\varkappa^2\right)^{3/2}}$.

20.9. We have: $\mathbf{r} = \left(\dfrac{a}{2}(u - v), \dfrac{b}{2}(u + v), \dfrac{uv}{2}\right)$, $\mathbf{r}_u = \left(\dfrac{a}{2}, \dfrac{b}{2}, \dfrac{v}{2}\right)$,

$$\mathbf{r}_v = \left(-\frac{a}{2}, \frac{b}{2}, \frac{u}{2}\right), \quad E = \langle \mathbf{r}_u, \mathbf{r}_u \rangle = \frac{1}{4}\left(a^2 + b^2 + v^2\right),$$

$$F = \langle \mathbf{r}_u, \mathbf{r}_v \rangle = \frac{1}{4}\left(b^2 - a^2 + uv\right), \quad G = \langle \mathbf{r}_v, \mathbf{r}_v \rangle = \frac{1}{4}\left(a^2 + b^2 + u^2\right),$$

$$\mathbf{r}_u \times \mathbf{r}_v = \frac{1}{4}\left(b\left(u - v\right), -a\left(u + v\right), 2ab\right),$$

$$\mathbf{n} = \frac{\mathbf{r}_u \times \mathbf{r}_v}{|\mathbf{r}_u \times \mathbf{r}_v|} = \frac{\left(b\left(u - v\right), -a\left(u + v\right), 2ab\right)}{\sqrt{a^2\left(u + v\right)^2 + b^2\left(u - v\right)^2 + 4a^2b^2}},$$

$$\mathbf{r}_{uu} = (0, 0, 0), \quad \mathbf{r}_{uv} = \left(0, 0, \frac{1}{2}\right), \quad \mathbf{r}_{vv} = (0, 0, 0), \quad L = \langle \mathbf{r}_{uu}, \mathbf{n} \rangle = 0,$$

$$M = \langle \mathbf{r}_{uv}, \mathbf{n} \rangle = \frac{ab}{\sqrt{a^2\left(u + v\right)^2 + b^2\left(u - v\right)^2 + 4a^2b^2}}, \quad N = \langle \mathbf{r}_{vv}, \mathbf{n} \rangle = 0.$$

We write the differential equation defining the curvature lines of the surface:

$$0 = \begin{vmatrix} dv^2 & -du\,dv & du^2 \\ E & F & G \\ L & M & N \end{vmatrix} = -M\left(Gdv^2 - Edu^2\right) =$$

$$\frac{ab\left(\left(a^2 + b^2 + v^2\right)du^2 - \left(a^2 + b^2 + u^2\right)dv^2\right)}{4\sqrt{a^2\left(u + v\right)^2 + b^2\left(u - v\right)^2 + 4a^2b^2}}.$$

Hence, we get:

$$\frac{du^2}{a^2 + b^2 + u^2} = \frac{dv^2}{a^2 + b^2 + v^2}, \quad \frac{du}{\sqrt{a^2 + b^2 + u^2}} = \pm\frac{dv}{\sqrt{a^2 + b^2 + v^2}},$$

$$\operatorname{arsinh}\frac{u}{\sqrt{a^2 + b^2}} = \pm\operatorname{arsinh}\frac{v}{\sqrt{a^2 + b^2}} + C,$$

$$\sinh\left(\operatorname{arsinh}\frac{u}{\sqrt{a^2 + b^2}} \mp \operatorname{arsinh}\frac{v}{\sqrt{a^2 + b^2}}\right) = \sinh C,$$

$$\frac{u}{\sqrt{a^2 + b^2}}\frac{\sqrt{a^2 + b^2 + v^2}}{\sqrt{a^2 + b^2}} \mp \frac{\sqrt{a^2 + b^2 + u^2}}{\sqrt{a^2 + b^2}}\frac{v}{\sqrt{a^2 + b^2}} = \sinh C,$$

as $\sinh\left(x \pm y = \sinh x \cosh y \pm \cosh x \sinh y\right)$ and $\cosh x = \sqrt{1 + \sinh^2 x}$. Thus, the curvature lines of this surface are given by the equation:

$$u\sqrt{a^2 + b^2 + v^2} \pm v\sqrt{a^2 + b^2 + v^2} = \text{const.}$$

20.10. As is known (see the respective problems for the helicoid in Part 1 of the book), we have $E = \langle \mathbf{r}_u, \mathbf{r}_u \rangle = 1$, $F = \langle \mathbf{r}_u, \mathbf{r}_v \rangle = 0$, $G = \langle \mathbf{r}_v, \mathbf{r}_v \rangle = a^2 + u^2$, $L = \langle \mathbf{r}_{uu}, \mathbf{n} \rangle = 0$, $M = \langle \mathbf{r}_{uv}, \mathbf{n} \rangle = -\frac{a}{\sqrt{a^2 + u^2}}$, $N = \langle \mathbf{r}_{vv}, \mathbf{n} \rangle = 0$. We write down the differential equation defining the curvature lines of the surface:

$$0 = \begin{vmatrix} dv^2 & -du\,dv & du^2 \\ E & F & G \\ L & M & N \end{vmatrix} = -M\left(G\,dv^2 - E\,du^2\right) = -\frac{a\left(du^2 - \left(a^2 + u^2\right)dv^2\right)}{\sqrt{a^2 + u^2}}.$$

Hence, we get: $dv^2 = \dfrac{du^2}{a^2 + u^2}$, $dv = \pm\dfrac{du}{\sqrt{a^2 + u^2}}$. Therefore,

$$v = \pm\ln\left(u + \sqrt{a^2 + u^2}\right) + C.$$

20.11. According to Problem 20.10, the curvature lines of the helicoid are given by the equations $v = \pm \ln\left(u + \sqrt{a^2 + u^2}\right) + C$. As is known, the second quadratic form of the catenoid has the form $-\dfrac{a\,du^2}{u^2 + a^2} + a\,dv^2$. Then the second quadratic form on the tangent vectors to the curves being the images of the curvature lines in the isometry, is

$$-\frac{a}{u^2 + a^2} + a\left(\frac{1}{\sqrt{a^2 + u^2}}\right)^2 = 0.$$

Thus, the curvature lines at the application pass to the asymptotic lines of the catenoid.

20.12. From the condition, it follows that $\rho(s)$ is a plane curve. Let $\mathbf{n}, \mathbf{b}$ be the vectors of the normal and binormal of the curve $\rho(s)$. Then $\mathbf{a} = \pm \mathbf{b}$. We will consider that $\mathbf{a} = \mathbf{b}$. Hence, it follows that $\mathbf{v} \times \mathbf{n} = \mathbf{a}$, $\mathbf{v} \times \mathbf{a} = -\mathbf{n}$, $\mathbf{n} \times \mathbf{a} = \mathbf{v}$.
$\mathbf{r}_s = \mathbf{v} + kg\mathbf{v} = (1 + kg)\,\mathbf{v}$, $\mathbf{r}_u = f'\mathbf{a} - g'\mathbf{n}$,

$$E = \langle \mathbf{r}_s, \mathbf{r}_s \rangle = (1 + kg)^2, \quad F = \langle \mathbf{r}_s, \mathbf{r}_u \rangle = 0,$$

$$G = \langle \mathbf{r}_u, \mathbf{r}_u \rangle = (f')^2 + (g')^2, \quad \mathbf{r}_s \times \mathbf{r}_u = -(1 + kg)\,g'\mathbf{a} - (1 + kg)\,f'\mathbf{n},$$

$$\mathbf{m} = \frac{\mathbf{r}_s \times \mathbf{r}_u}{|\mathbf{r}_s \times \mathbf{r}_u|} = -\frac{1}{\sqrt{(f')^2 + (g')^2}}\,(f'\mathbf{n} + g'\mathbf{a}),$$

$$\mathbf{r}_{ss} = k'g\mathbf{v} + k(1 + kg)\,\mathbf{n}, \quad \mathbf{r}_{su} = kg'\mathbf{v}, \quad \mathbf{r}_{uu} = f''\mathbf{a} - g''\mathbf{n},$$

$$L = \langle \mathbf{r}_{ss}, \mathbf{m} \rangle = -\frac{kf'(1 + kg)}{\sqrt{(f')^2 + (g')^2}}, \quad M = \langle \mathbf{r}_{su}, \mathbf{m} \rangle = 0,$$

$$N = \langle \mathbf{r}_{uu}, \mathbf{m} \rangle = \frac{f'g'' - f''g'}{\sqrt{(f')^2 + (g')^2}}.$$

The equations for the curvature lines have the form

$$0 = \begin{vmatrix} du^2 & -ds\,du & ds^2 \\ E & F & G \\ L & M & N \end{vmatrix} = (EN - GL)\,du\,ds.$$

From here, it is seen that the curvature lines are lines defined by the equations $s = $ const, $u = $ const.

20.13. We have: (a), (b) $\mathbf{r}_s = \mathbf{v} - \mathbf{v}a\cos\varphi k = \mathbf{v}(1 - ak\cos\varphi)$,

$$\mathbf{r}_\varphi = -\mathbf{n}a\sin\varphi + \mathbf{b}a\cos\varphi,$$

$$E = \langle \mathbf{r}_s, \mathbf{r}_s \rangle = (1 - ak\cos\varphi)^2, \quad F = \langle \mathbf{r}_s, \mathbf{r}_\varphi \rangle = 0, \quad G = \langle \mathbf{r}_\varphi, \mathbf{r}_\varphi \rangle = a^2,$$

$$\mathbf{r}_s \times \mathbf{r}_\varphi = -\mathbf{b}a(1 - ak\cos\varphi)\sin\varphi - \mathbf{n}a(1 - ak\cos\varphi)\cos\varphi,$$

$$|\mathbf{r}_s \times \mathbf{r}_\varphi| = a(1 - ak\cos\varphi), \quad \mathbf{m} = \frac{\mathbf{r}_s \times \mathbf{r}_\varphi}{|\mathbf{r}_s \times \mathbf{r}_\varphi|} = -\mathbf{b}\sin\varphi - \mathbf{n}\cos\varphi,$$

$$\mathbf{r}_{ss} = \mathbf{n}k(1 - ak\cos\varphi) - \mathbf{v}a\dot{k}\cos\varphi, \quad \mathbf{r}_{s\varphi} = \mathbf{v}ak\sin\varphi,$$

$$\mathbf{r}_{\varphi\varphi} = -\mathbf{n}a\cos\varphi - \mathbf{b}a\sin\varphi, \quad L = \langle \mathbf{r}_{ss}, \mathbf{m} \rangle = -k(1 - ak\cos\varphi)\cos\varphi,$$

$$M = \langle \mathbf{r}_{s\varphi}, \mathbf{m} \rangle = 0, \quad N = \langle \mathbf{r}_{\varphi\varphi}, \mathbf{m} \rangle = a,$$

$$K = \frac{LN - M^2}{EG - F^2} = -\frac{k\cos\varphi}{a(1 - ak\cos\varphi)},$$

$$H = \frac{EN + GL - 2MF}{EG - F^2} =$$

$$\frac{(1 - ak\cos\varphi)^2\, a + a^2\,(-k\,(1 - ak\cos\varphi)\cos\varphi)}{a^2\,(1 - ak\cos\varphi)^2} = \frac{1 - 2ak\cos\varphi}{a\,(1 - ak\cos\varphi)};$$

(c) we write down the equation for the curvature lines:

$$0 = \begin{vmatrix} d\varphi^2 & -ds\,d\varphi & ds^2 \\ E & F & G \\ L & M & N \end{vmatrix} = (EN - GL)\,ds\,d\varphi = a\,(1 - ak\cos\varphi)\,ds\,d\varphi.$$

As $1 - ak\cos\varphi > 0$, then $ds\,d\varphi = 0$ and the curvature lines have the form $s = \text{const}$, $\varphi = \text{const}$.

20.14. We have:

$$\mathbf{r}_s = \mathbf{v} + a\,(-k\mathbf{v} + \varkappa\mathbf{b})\cos\varphi - a\varkappa\mathbf{n}\sin\varphi =$$
$$\mathbf{v}\,(1 - ak\cos\varphi) - \mathbf{n}a\varkappa\sin\varphi + \mathbf{b}a\varkappa\cos\varphi,$$

$$\mathbf{r}_\varphi = -\mathbf{n}a\sin\varphi + \mathbf{b}a\cos\varphi, \quad E = \langle\mathbf{r}_s, \mathbf{r}_s\rangle = (1 - ak\cos\varphi)^2 + a^2\varkappa^2,$$
$$F = \langle\mathbf{r}_s, \mathbf{r}_\varphi\rangle = a^2\varkappa, \quad G = \langle\mathbf{r}_\varphi, \mathbf{r}_\varphi\rangle = a^2,$$

$$\mathbf{r}_s \times \mathbf{r}_\varphi = \mathbf{v}\,(-a^2\varkappa\sin\varphi\cos\varphi + a^2\varkappa\cos\varphi\sin\varphi) +$$
$$\mathbf{n}\,(ak\cos\varphi - 1)\,a\cos\varphi + \mathbf{b}\,(ak\cos\varphi - 1)\,a\sin\varphi =$$
$$- a\,(1 - ak\cos\varphi)\,(\mathbf{n}\cos\varphi + \mathbf{b}\sin\varphi),$$

$$\mathbf{m} = \frac{\mathbf{r}_s \times \mathbf{r}_\varphi}{|\mathbf{r}_s \times \mathbf{r}_\varphi|} = -\mathbf{n}\cos\varphi - \mathbf{b}\sin\varphi,$$

$$\mathbf{r}_{ss} = \mathbf{v}\,(-a\dot{k}\cos\varphi + ak\varkappa\sin\varphi) +$$
$$\mathbf{n}\,(-a\dot{\varkappa}\sin\varphi + k\,(1 - ak\cos\varphi) - a\varkappa^2\cos\varphi) + \mathbf{b}\,(a\dot{\varkappa}\cos\varphi - a\varkappa^2\sin\varphi),$$

$$\mathbf{r}_{s\varphi} = \mathbf{v}ak\sin\varphi - \mathbf{n}a\varkappa\cos\varphi - \mathbf{b}a\varkappa\sin\varphi,$$
$$\mathbf{r}_{\varphi\varphi} = -\mathbf{n}a\cos\varphi - \mathbf{b}a\sin\varphi, \quad L = \langle\mathbf{r}_{ss}, \mathbf{m}\rangle = a\varkappa^2 - k\cos\varphi + ak^2\cos^2\varphi,$$
$$M = \langle\mathbf{r}_{s\varphi}, \mathbf{m}\rangle = a\varkappa, \quad N = \langle\mathbf{r}_{\varphi\varphi}, \mathbf{m}\rangle = a,$$

We write down the equation for the curvature lines:

$$0 = \begin{vmatrix} d\varphi^2 & -ds\,d\varphi & ds^2 \\ E & F & G \\ L & M & N \end{vmatrix} =$$

$$\begin{vmatrix} d\varphi^2 & -ds\,d\varphi & ds^2 \\ (1 - ak\cos\varphi)^2 - a^2\varkappa^2 & a^2\varkappa & a^2 \\ a\varkappa^2 - k\cos\varphi + ak^2\cos^2\varphi & a\varkappa & a \end{vmatrix} =$$

$$\begin{vmatrix} d\varphi^2 & -ds\,d\varphi & ds^2 \\ (1 - ak\cos\varphi)^2 + a^2\varkappa^2 & a^2\varkappa & a^2 \\ k\cos\varphi - \dfrac{1}{a} & 0 & 0 \end{vmatrix} = a\,(1 - ak\cos\varphi)\,(d\varphi + \varkappa ds)\,ds.$$

As $1 - ak \cos\varphi > 0$, we have $ds = 0$ or $\dfrac{d\varphi}{ds} = -\varkappa$. Therefore, the curvature lines have the form $s = \text{const}$, $\varphi = -\int \varkappa ds + \text{const}$.

20.15. We have:

$$\mathbf{r}_v = \left(3v^2 - 3u^2 - \frac{1}{3}, 6uv, 2v\right), \quad \mathbf{r}_u = \left(6uv, 3u^2 - 3v^2 - \frac{1}{3}, 2u\right),$$

$$E = \langle \mathbf{r}_u, \mathbf{r}_u \rangle = \left(3u^2 - 3v^2 - \frac{1}{3}\right)^2 + 36u^2v^2 + 4u^2 = \left(3u^2 + 3v^2 + \frac{1}{3}\right)^2,$$

$$F = \langle \mathbf{r}_u, \mathbf{r}_v \rangle = \left(3v^2 - 3u^2 - \frac{1}{3}\right)6uv + 6uv\left(3u^2 - 3v^2 - \frac{1}{3}\right) + 2v2u = 0,$$

$$G = \langle \mathbf{r}_v, \mathbf{r}_v \rangle = 36u^2v^2 + \left(3u^2 - 3v^2 - \frac{1}{3}\right)^2 + 4u^2 = \left(3u^2 + 3v^2 + \frac{1}{3}\right)^2,$$

$$\mathbf{r}_u \times \mathbf{r}_v = \left(6u^2v + 6v^3 + \frac{2}{3}v, 6u^3 + 6uv^2 + \frac{2}{3}u,\right.$$
$$\left. \left(\frac{1}{3} + 3u^2 + 3v^2\right)\left(\frac{1}{3} - 3u^2 - 3v^2\right)\right),$$

$$|\mathbf{r}_u \times \mathbf{r}_v| = \left(3u^2 + 3v^2 + \frac{1}{3}\right)^2,$$

$$\mathbf{n} = \frac{\mathbf{r}_u \times \mathbf{r}_v}{|\mathbf{r}_u \times \mathbf{r}_v|} = \frac{(2v, 2u, 1/3 - 3u^2 - 3v^2)}{3u^2 + 3v^2 + 1/3},$$

$$\mathbf{r}_{uu} = (-6u, 6v, 0), \quad \mathbf{r}_{uv} = (6v, 6u, 2), \quad \mathbf{r}_{vv} = (6u, -6v, 0),$$

$$L = \langle \mathbf{r}_{uu}, \mathbf{n} \rangle = \frac{-6u2v + 6v2u}{3u^2 + 3v^2 + 1/3} = 0,$$

$$M = \langle \mathbf{r}_{uv}, \mathbf{n} \rangle = \frac{6v2v + 6u2u + 2\left(1/3 - 3u^2 - 3v^2\right)}{3u^2 + 3v^2 + 1/3} = 2,$$

$$N = \langle \mathbf{r}_{vv}, \mathbf{n} \rangle = \frac{6u2v - 6v2u}{3u^2 + 3v^2 + 1/3} = 0.$$

We write down the equation for the curvature lines:

$$0 = \begin{vmatrix} dv^2 & -du\,dv & du^2 \\ E & F & G \\ L & M & N \end{vmatrix} = 2\left(3u^2 + 3v^2 + \frac{1}{3}\right)^2 (d^2u - d^2v).$$

From here, we have

$$d^2u - d^2v = (du + dv)(du - dv) = 0 \Rightarrow u \pm v = \text{const}.$$

Answer: $u \pm v = \text{const}$.

20.16. As $E = \langle \mathbf{r}_u, \mathbf{r}_u \rangle$, then $\mathbf{r}_u = \sqrt{E}\mathbf{l}$, $|\mathbf{l}| = 1$. Similarly, $\mathbf{r}_v = \sqrt{G}\mathbf{m}$, $|\mathbf{m}| = 1$. The Gaussian curvature of the surface is expressed through the coefficients of the first and second quadratic forms as follows: $K = \dfrac{LN - M^2}{EG}$, where $L = \langle \mathbf{r}_{uu}, \mathbf{n} \rangle$,

$M = \langle \mathbf{r}_{uv}, \mathbf{n} \rangle$, $N = \langle \mathbf{r}_{vv}, \mathbf{n} \rangle$, $\mathbf{r}_{uu} = \dfrac{1}{2} \dfrac{E_u}{\sqrt{E}} \mathbf{l} + \sqrt{E} \mathbf{l}_u$. As $\langle \mathbf{r}_u, \mathbf{n} \rangle = 0$, then $L = \sqrt{E} \langle \mathbf{l}_u, \mathbf{n} \rangle$. As $\langle \mathbf{l}, \mathbf{l} \rangle = 1$, then $\langle \mathbf{l}_u, \mathbf{l} \rangle = 0$ and hence $\mathbf{l}_u = p_1 \mathbf{m} + q_1 \mathbf{n}$.

Similarly, $N = \sqrt{G} \langle \mathbf{m}_v, \mathbf{n} \rangle$, $\mathbf{m}_v = p_2 \mathbf{l} + q_2 \mathbf{n}$. Then $LN = \sqrt{EG} \langle \mathbf{l}_u, \mathbf{n} \rangle \cdot \langle \mathbf{m}_v, \mathbf{n} \rangle = \sqrt{EG} q_1 q_2 = \sqrt{EG} \langle \mathbf{l}_u, \mathbf{m}_v \rangle$.

In the same way, it is shown that

$$M^2 = \sqrt{EG} \langle \mathbf{l}_v, \mathbf{m}_u \rangle \quad \text{and} \quad K = \frac{\langle \mathbf{l}_u, \mathbf{m}_v \rangle - \langle \mathbf{l}_v, \mathbf{m}_u \rangle}{\sqrt{EG}},$$

$$\langle \mathbf{l}_u, \mathbf{m}_v \rangle = \langle \mathbf{l}_u, \mathbf{m} \rangle_v - \langle \mathbf{l}_{uv}, \mathbf{m} \rangle, \ \langle \mathbf{l}_v, \mathbf{m}_u \rangle = \langle \mathbf{l}_v, \mathbf{m} \rangle_u - \langle \mathbf{l}_{uv}, \mathbf{m} \rangle.$$

Then $K = \dfrac{\langle \mathbf{l}_u, \mathbf{m} \rangle_v - \langle \mathbf{l}_v, \mathbf{m} \rangle_u}{\sqrt{EG}}$,

$$\frac{\partial}{\partial v} \langle \mathbf{l}_u, \mathbf{m} \rangle = \frac{\partial}{\partial v} \left(\frac{\langle \mathbf{r}_{uu}, \mathbf{r}_v \rangle}{\sqrt{EG}} \right) = -\frac{\partial}{\partial v} \left(\frac{\langle \mathbf{r}_u, \mathbf{r}_{uv} \rangle}{\sqrt{EG}} \right) =$$

$$-\frac{1}{2} \frac{\partial}{\partial v} \left(\frac{(\partial/\partial v) \langle \mathbf{r}_u, \mathbf{r}_u \rangle}{\sqrt{EG}} \right) = -\frac{1}{2} \frac{\partial}{\partial v} \left(\frac{\partial E/\partial v}{\sqrt{EG}} \right).$$

In the same computations, two times we made use of the fact that $\langle \mathbf{r}_u, \mathbf{r}_v \rangle = 0$. Similarly,

$$\frac{\partial}{\partial u} \langle \mathbf{l}_v, \mathbf{m} \rangle = \frac{1}{2} \frac{\partial}{\partial u} \left(\frac{\partial G/\partial u}{\sqrt{EG}} \right).$$

As the result, we get

$$K = -\frac{1}{2\sqrt{EG}} \left\{ \frac{\partial}{\partial v} \left(\frac{\partial E/\partial v}{\sqrt{EG}} \right) + \frac{\partial}{\partial u} \left(\frac{\partial G/\partial u}{\sqrt{EG}} \right) \right\}.$$

20.17. Make use of the formula for the Gaussian curvature of the surface with the quadratic form $ds^2 = E \, du^2 + G \, dv^2$ (see Problem 20.16). The Gaussian curvature of such a surface is

$$K = -\frac{1}{2\sqrt{EG}} \left\{ \frac{\partial}{\partial v} \left(\frac{\partial E/\partial v}{\sqrt{EG}} \right) + \frac{\partial}{\partial u} \left(\frac{\partial G/\partial u}{\sqrt{EG}} \right) \right\}.$$

Letting in the expression for the curvature $E \equiv 1$, we get

$$K = -\frac{1}{2\sqrt{G}} \frac{\partial}{\partial u} \left(\frac{\partial G/\partial u}{\sqrt{G}} \right) = -\frac{1}{2\sqrt{G}} \frac{\partial}{\partial u} \left(2 \frac{\partial \sqrt{G}}{\partial u} \right) = -\frac{(\partial^2/\partial u^2) \sqrt{G}}{\sqrt{G}}.$$

Answer: $K = -\dfrac{1}{\sqrt{G}} \dfrac{\partial^2 \sqrt{G}}{\partial u^2}$.

20.18. $K = -\dfrac{1}{2B} \Delta \ln B$, where Δ is the Laplace operator.

20.19. To compute the curvature, we can make use of the formula deduced in Problem 20.17. We have $G(u, v) = e^{2u}$, whence

$$K = -\frac{1}{e^u} \frac{\partial^2 e^u}{\partial u^2} = -1.$$

20.20. Show that the inner products $\langle \mathbf{r}_{ij}, \mathbf{r}_k \rangle$ and differences $\langle \mathbf{r}_{ij}, \mathbf{r}_{kl} \rangle - \langle \mathbf{r}_{ik}, \mathbf{r}_{jl} \rangle$ are expressed through the metric.

20.24. Consider the surface of revolution M with the equation

$$x = r(u)\cos\varphi, \quad y = r(u)\sin\varphi, \quad z = h(u),$$

where $a \le u \le b$, $0 \le \varphi \le 2\pi$. The element of area of this surface is equal to $d\sigma = r(u)\sqrt{r'^2 + z'^2}\,du\,d\varphi$. We shall seek for the deformation of the surface of revolution in the form $x = R(u,\varepsilon)\cos\varphi$, $y = R(u,\varepsilon)\sin\varphi$, $z = H(u,\varepsilon)$ with the condition of equality of the elements of area

$$r\sqrt{r'^2 + h'^2}\,du\,d\varphi = R\sqrt{R'^2 + H'^2}\,du\,d\varphi.$$

Assume $R(u,\varepsilon) = \varepsilon r(u)$, $\varepsilon \to 1 - 0$. Then this equation admits the solution

$$H(u,\varepsilon) = h(0) + \frac{1}{\varepsilon}\int_0^u \sqrt{(1-\varepsilon^4)\,r'^2(t) + h'^2(t)}\,dt,$$

which does determine the sought-for family of surfaces of revolution satisfying the conditions of the problem.

20.27. (a) $F_1'^2(w) + F_2'^2(w) + F_3'^2(w) = 0$; (b) $ds^2 = \Lambda^2(w)\left(du^2 + dv^2\right)$, where

$$\Lambda^2 = \frac{\left|F_1'^2(w)\right| + \left|F_2'^2(w)\right| + \left|F_3'^2(w)\right|}{2}.$$

20.29. *Hint:* use the relation between three quadratic forms of the surface.

20.38. We compute the curvature K by the formula $K = -\dfrac{1}{2\Lambda}\Delta\ln\Lambda$, where $\Lambda = e^{-2u^2}$, and find that $K = 2e^{2u^2} > 0$. The incompleteness of the metric follows from the existence of a final-length path running to infinity, namely, $v = 0$. The isometric embedding of the metric into $\mathbb{R}^3$ in the form of a convex surface is impossible, because the integral curvature

$$\iint K\,dS = 2e^{2u^2}\,du\,dv = \infty,$$

and the integral curvature of the convex surface equal to the area of its spherical image does not exceed 4π.

22.3. No, as otherwise a continuous field of normals would have existed on the Möbius band. Indeed, as the Gaussian curvature is positive, the surface lies locally on one side of the osculating plane. Therefore, a smooth field of "external normals" can be defined, which contradicts the nonorientability of the Möbius band.

22.5. (a) Under the given conditions, a smooth field of "external normals" (see Problem 22.3) can be defined, which contradicts the nonorientability of the surface.

22.6. (a) *Hint:* from the infinity, move the plane up to its osculation with the surface. Consider the Gaussian curvature of the surface at the point of contact.

(b) *Hint:* choose a sphere of a sufficiently large radius such that it contains the surface inside. Then continuously reduce the radius of the sphere up to its osculation with the surface. Consider the Gaussian curvature of the surface at the point of contact.

22.7. *Hint:* see Problem 22.6.

22.8. *Hint:* apply the Gauss–Bonnet theorem.

22.10. (a) Consider part of the sphere $x^2 + y^2 + z^2 = R^2$, $|z| \leq a$, $a < R$. Identify the opposite points.

(b) *Hint:* find the surface of revolution of constant negative curvature invariant at the central symmetry relative to the origin of coordinates (see Problem 20.23). Identify with respect to an appropriate action of the group $\mathbb{Z}/2$.

Another variant of the solution. Consider on the rectangle Π: $-a \leq u \leq a$, $-b \leq v \leq b$ the metric $ds^2 = \cosh^2 v\, du^2 + dv^2$. Identify now the sides $u = -a$ and $u = a$ by the equality of the lengths of the glued segments such that to obtain a Möbius band. It is easy to show that on the Möbius band we get an analytic metric of curvature -1. At sufficiently small b, this metric is realized on a Minding "coil" having the equation $x = \cosh v \cos u$, $y = \cosh v \sin u$, $z = \int\limits_0^v \sqrt{1 - \sinh^2 t}\, dt$.

22.14. $\int\limits_{v_1}^{v_2} dv \int\limits_{u_1}^{u_2} \left(\sqrt{G}\right)_{uu} du$.

22.15. 2π.

22.17. A doubly covered sphere.

22.20. The surface of revolution of the curve $z = \dfrac{1}{8}x^2$, $0 \leq x \leq 2$, $z = x - \ln x$, $x \geqslant 2$.

22.21. Let the surface of revolution be parameterized as follows: $(u \cos v, u \sin v, f(u))$. Its parallels are given by the equations $u = $ const, and the meridians by the equations $v = $ const. The first quadratic form of this surface has the form $ds^2 = \left(1 + (f'(u))^2\right) du^2 + u^2 dv^2$. Replace the variable u by the variable U, where

$$U = \int \frac{\sqrt{1 + (f'(u))^2}}{u}\, du.$$

Designate the inverse dependence of u on U as $u = \lambda(U)$. The first quadratic form in this substitution of the variables takes the form $ds^2 = \lambda^2(U)(dU^2 + dv^2)$. Note that the parallels of the surface are given by the equations $U = $ const, and the meridians by the earlier equations $v = $ const.

Consider a Euclidean plane with the coordinates (x, y) and the metric $dx^2 + dy^2$. It is easy to verify that the mapping given in the coordinates by the formula $(U, v) \mapsto (x, y)$ possesses the required properties.

22.23. (a) 4π; (b) 2π; (c) 4π.

22.24. Let the rectangle P be given in the spherical coordinates φ, ψ by the inequalities $-a \leq \varphi \leq \pi$, $-\dfrac{\pi}{2} < b \leq \psi \leq c < \dfrac{\pi}{2}$. Then the isometric embedding of P in $\mathbb{R}^3$ is given as follows:

$$x = \frac{a}{\pi} \cos \frac{\pi}{a} \varphi \cos \psi, \quad y = \frac{a}{\pi} \sin \frac{\pi}{a} \varphi \cos \psi,$$

$$z = f(\psi) = \int \sqrt{1 - \frac{a^2}{\pi^2} \sin^2 \psi}\, d\psi,$$

from where it is easy to establish the analyticity of both the embedding and metric.

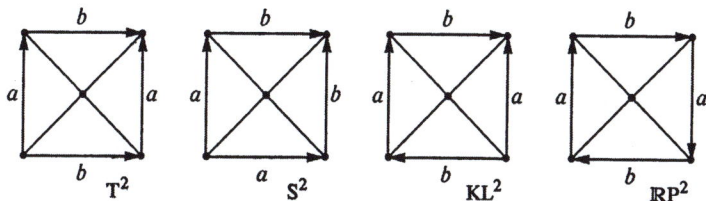

Figure 154 Triangulations of a torus, a sphere, a Klein bottle and a projective plane

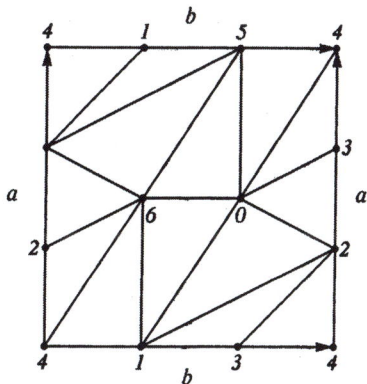

Figure 155 Triangulation of a torus

22.25. The metric of a unit sphere in spherical coordinates has the form $ds^2 = \cos^2\varphi\,d\varphi^2 + d\psi^2$. Identification is done by the mapping $(-a, \psi) \rightarrow (a, b + c - \psi)$. What remains is evident.

23.1. The partitionings are shown in Fig. 154.

23.3. *Hint:* the required estimate follows from the system of inequalities

$$V\,(V-1) \geq 2E,$$
$$V - E + F = \chi\,(M),$$
$$3F = 2E.$$

23.4. See (a) Figs 155, 156; (b) Fig. 157.

23.5. Minimal triangulation of the torus can be constructed by way of considering a plane lattice consisting of regular triangles (see Fig. 158). We should take the second lattice generated by vectors $\mathbf{e}_1$, $\mathbf{e}_2$ shown in the figure. The parallelogram with sides $\mathbf{e}_1$, $\mathbf{e}_2$ is the fundamental domain of this second lattice. By identifying its opposite sides, we get a torus on which the triangles of the initial regular lattice give the sought-for triangulation.

23.9. See Fig. 159.

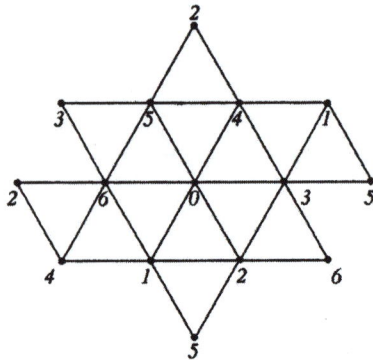

Figure 156 Triangulation of a torus (another variant)

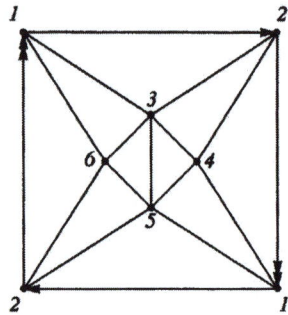

Figure 157 Triangulation of a projective plane

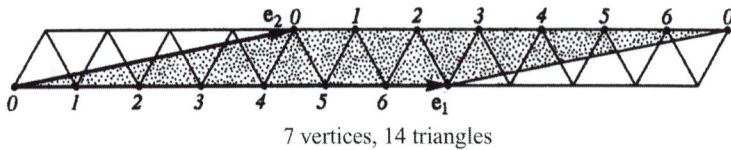

7 vertices, 14 triangles

Figure 158 Lattices generating the triangulation of a torus

23.11. *Hint:* construct the homeomorphisms shown in Fig. 160.
23.12. See Fig. 161.
23.13. See Fig. 162.
23.14. See Fig. 163.

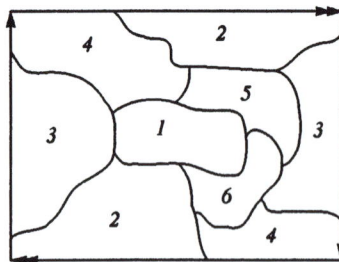

Figure 159 A chart on a projective plane

23.21. See Fig. 164. In a more symmetric form, the sought-for embedding is shown in Fig. 165.
23.25. See Fig. 166.
23.26. See Fig. 167.

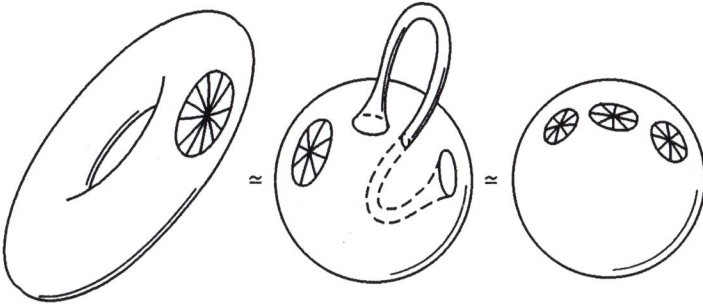

Figure 160

24.1. Differentiate the equation giving this family of lines to get $c = \dfrac{u}{v}\dfrac{du}{dv}$. Substitute this expression for c into the equation of the family. The differential equation of the lines of the family is obtained: $\dfrac{dv}{du} = \dfrac{1+v^2}{uv}$. Substituting into the differential equation defining the conjugate family

$$L\,du\,\delta u + M\,(du\,\delta v + dv\,\delta u) + N\,du\,\delta u = 0,$$

we get $\dfrac{\delta v}{\delta u} = -1$. Here L, M and N are the coefficients of the second quadratic form of the surface; (du, dv) is the direction of the velocity vector of a line of the initial family; and $(\delta u, \delta v)$ is the direction of the velocity vector of a line of the conjugate family. Therefore, the conjugate family consists of the lines $u + v = C_1$, where C_1 is an arbitrary constant.

24.3. The differential equation of lines of the conjugate family has the form $\dfrac{du}{u} = -\dfrac{1}{4}\sin 2v\,dv$. Integrating, we get the equations of lines of the conjugate family $u = Ce^{(1/8)\cos 2v}$, where C is an arbitrary constant.

24.4. The condition of conjugacy of families of lines indicated in the condition of the problem is reduced to the differential equation $f_{xy} = 0$. Hence, $f(x,y) = g(x)+h(y)$, where g and h are arbitrary smooth functions of one variable.

24.5. Introduce on the surfaces the coordinates such that respective points on the surfaces are given by the same pair of numbers. From the condition of the problem, we get

$$L\,du\,\delta u + M\,(du\,\delta v + dv\,\delta u) + N\,dv\,\delta v =$$
$$\lambda\,(E_1 du\,\delta u + F_1\,(du\,\delta v + dv\,\delta u) + G_1 dv\,\delta v)\,,$$

$$L_1 du\,\delta u + M_1\,(du\,\delta v + dv\,\delta u) + N_1 dv\,\delta v =$$
$$\mu\,(E\,du\,\delta u + F\,(du\,\delta v + dv\,\delta u) + G\,dv\,\delta v)\,.$$

Here, λ and μ are some proportionality factors. Hence, we obtain that

$$\frac{L}{E_1} = \frac{M}{F_1} = \frac{N}{G_1}, \quad \frac{L_1}{E} = \frac{M_1}{F} = \frac{N_1}{G}.$$

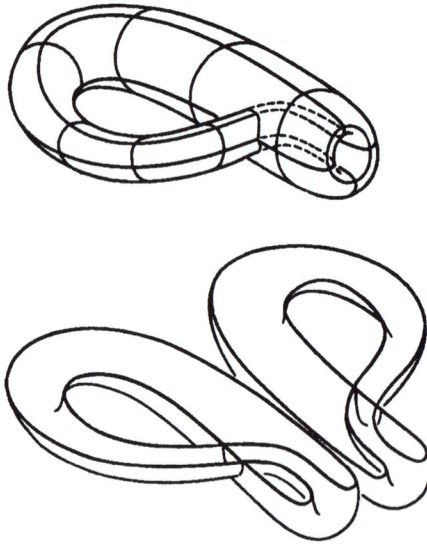

Figure 161 Glueing of a Klein bottle from two Möbius bands

24.6. The differential equation of the asymptotic lines has the form

$$y^2 - 2xy\frac{dy}{dx} + x^2\left(\frac{dy}{dx}\right)^2 = b^2 + a^2\left(\frac{dy}{dx}\right)^2.$$

Hence, $y = x\dfrac{dy}{dx} \pm \sqrt{b^2 + a^2\left(\dfrac{dy}{dx}\right)^2}$. It is easy to see that the obtained differential equation is a Clairaut equation. Solving it, we get

$$y = C_1x + \sqrt{b^2 + a^2C_1^2}, \quad y = C_2x - \sqrt{b^2 + a^2C_2^2},$$

where C_1 and C_2 are arbitrary constants. Note that the rectilinear generators of a one-sheeted hyperboloid are its asymptotic lines.

24.7. The differential equation of the asymptotic lines has the form $f''(x)\cdot dx^2 - f''(y)\, dy^2 = 0$. The condition of orthogonality of the asymptotic net is written as the equality $\dfrac{f''(x)}{1 + (f'(x))^2} = \dfrac{f''(y)}{1 + (f'(y))^2}$. From here, we immediately get that

$$\frac{f''(x)}{1 + (f'(x))^2} = a, \quad \frac{f''(y)}{1 + (f'(y))^2} = a,$$

where a is some constant. Solving these equations, we get

$$f'(x) = \tan(ax + b), \quad f(x) = -\frac{1}{a}\ln|\cos(ax + b)| + c,$$

$$f'(y) = \tan(ay + b), \quad f(y) = -\frac{1}{a}\ln|\cos(ay + b)| + c.$$

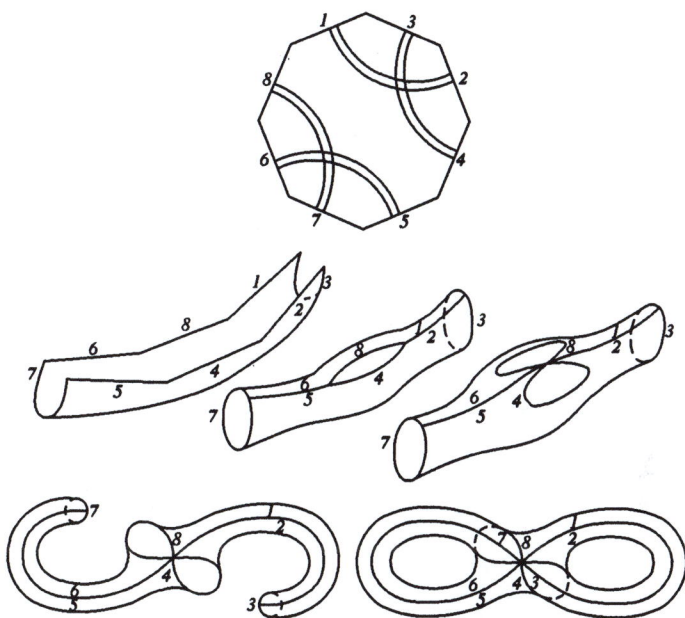

Figure 162

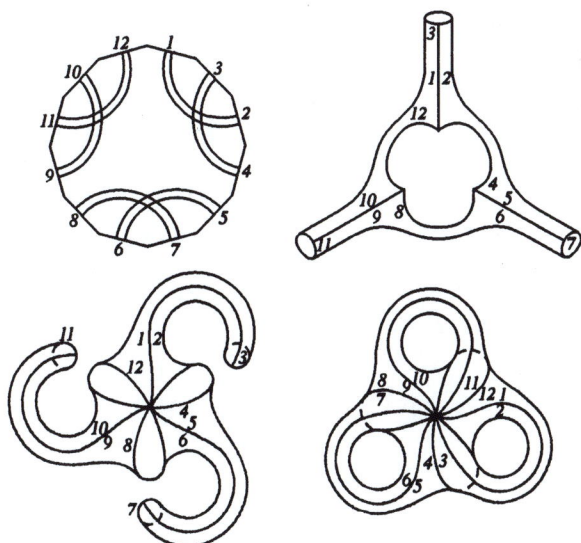

Figure 163

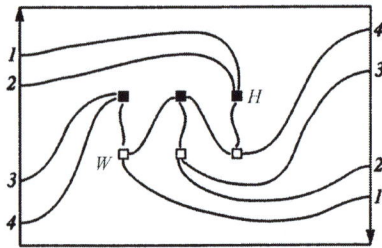

Figure 164 Graph "3 houses and 3 wells" on a Möbius band

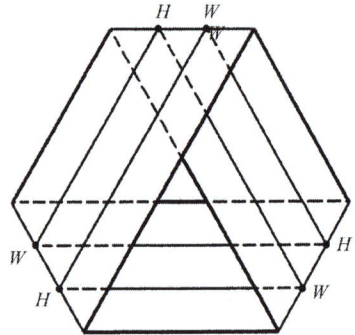

Figure 165 Improved graph "3 houses and 3 wells" on a Möbius band

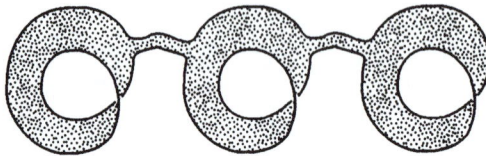

Figure 166

Hence, we obtain the equation of the sought-for surface:

$$z = \frac{1}{a} \ln \left| \frac{\cos{(ax+b)}}{\cos{(ay+b)}} \right|.$$

24.8. It is easy to show that the differential equation of asymptotic lines has the form

$$xy\,dx^2 + \left(x^2 - y^2\right)dx\,dy - xy\,dy^2 = 0.$$

This equation decomposes: $(x\,dx - y\,dy)(x\,dy + y\,dx) = 0$. Hence, we immediately get that the asymptotic lines are $x^2 - y^2 = C_1$, $xy = C_2$, where C_1 and C_2 are arbitrary constants. The asymptotic lines passing through the point M_0 have the equations $x^2 - y^2 = -3$, $xy = 2$.

24.9. $u + v = C_1$, $u - v = C_2$, where C_1 and C_2 are arbitrary constants.

24.10. Parameterize the helicoid as follows:

$$(u\cos v, u\sin v, av).$$

Show that the asymptotic lines are given by the equations $u = C_1$ and $v = C_2$, where C_1 and C_2 are arbitrary constants.

24.15. Parameterize the surface as follows:

$$x = \frac{a}{2}\left(\cos u + \cos v\right), \quad y = \frac{a}{2}\left(\sin u + \sin v\right), \quad z = \frac{b}{2}\left(u + v\right).$$

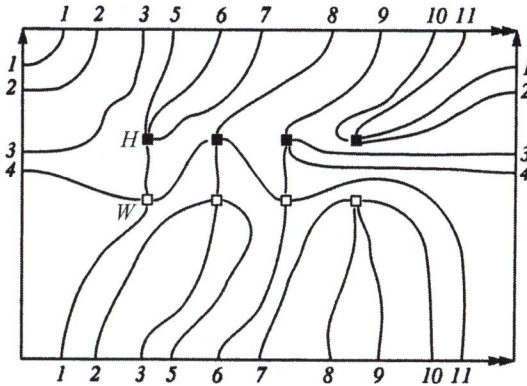

Figure 167 Graph "4 houses and 4 wells" on a torus

Then the differential equation of the asymptotic lines has the form: $du^2 - dv^2 = 0$. Hence, we get the equations of the asymptotic lines: $u + v = C_1$, $u - v = C_2$, where C_1 and C_2 are arbitrary constants.

24.17. Note that the osculating plane to the surface at points of the curve coincides with the osculating plane of this curve. Hence, the required assertion follows immediately.

24.18. Take a net consisting of asymptotic lines for the coordinate net on the surface. Then the cosine of the angle between the asymptotic lines is $\cos \varphi = \dfrac{F}{\sqrt{EG}}$. Besides,

$$K = \frac{-M^2}{EG - F^2}, \qquad H = \frac{-FM}{EG - F^2}.$$

Hence, we get that $\cos^2 \varphi = \dfrac{H^2}{H^2 - K}$. Therefore, $K = -H^2 \tan^2 \varphi$.

24.19. $v = C_1$, $u^2 = C_2 \sin \lambda v$.

24.20. $u + v = C_1$, $u - v = C_2$.

24.21. $x = C_1$, $y^2 \sin x = C_2$.

24.22. $x = \dfrac{a}{2}(u + v)$, $y = \dfrac{b}{2}(u - v)$, $z = \dfrac{1}{2}uv$.

24.24. If the coordinate net is asymptotic, then $L = N = 0$. In this case, $K = \dfrac{-M^2}{EG - F^2}$, and the Peterson–Codazzi equations take the form

$$2\left(EG - F^2\right)M_u - \left(2F\left(E_v - F_u\right) - \left(EG_u - GE_u\right)\right)M = 0,$$

$$2\left(EG - F^2\right)M_v - \left(2F\left(F_v - G_u\right) - \left(EG_v - GE_v\right)\right)M = 0.$$

Hence, it is easy to get the required formulas.

24.25. Note that the net consisting of coordinate lines is Chebyshev then and only then when $E_u = G_v = 0$. Take for the coordinate lines of the surface its asymptotic lines and make use of the result of Problem 24.24.

24.26. It is easy to show that the osculating plane to the surface at some point is osculating for each of the asymptotic lines passing through this point. There-fore, binormals to asymptotic lines at the point of their intersection are normals to the surface, i.e., $\mathbf{m} = \pm\mathbf{b}$, where $\mathbf{b}$ is the vector of the binormal to the asymp-totic lines, and $\mathbf{m}$ is the vector of the normal to the surface. We differentiate this equality along the arc of the asymptotic line and, using the Frénet formulas, we get $\dfrac{d}{ds}\mathbf{m} = \pm\dfrac{d}{ds}\mathbf{b} = \mp\varkappa\mathbf{n}$. Here, $\mathbf{n}$ is the vector of the principal normal of the asymptotic curve. Hence, we get that

$$\varkappa^2 = \left\langle \frac{d}{ds}\mathbf{m}, \frac{d}{ds}\mathbf{m} \right\rangle = \frac{\mathbf{III}}{\mathbf{I}}.$$

As $\mathbf{I}$ and $\mathbf{III}$, we denote here the values of the first and, respectively, the third quadratic form on the velocity vector of the asymptotic curve. From the relation between the quadratic forms of the surfaces, $\mathbf{III} = H\mathbf{II} - K\mathbf{I}$, and the fact that in the asymptotic direction $\mathbf{II} = 0$, we get $\dfrac{\mathbf{III}}{\mathbf{I}} = -K$. Hence, $\varkappa^2 = -K = |K|$.

24.28. The differential equation of the curvature lines, as it is easy to verify, has the form $du^2 - (u^2 + a^2)\, dv^2 = 0$. Integrating, we find

$$v = C_1 - \ln\left(u + \sqrt{u^2 + a^2}\right), \qquad v = C_2 + \ln\left(u + \sqrt{u^2 + a^2}\right),$$

where C_1 and C_2 are arbitrary constants.

24.29. The curvature lines of a cylindrical surface different from a plane are its rec-tilinear generators and lines by which the surface is intersected by planes orthogonal to the generatrices of the surface.

24.30. The curvature lines of a conical surface different from a plane are its recti-linear generators and lines by which the surface is intersected by all possible spheres with the centre at the vertex of the conical surface.

24.31. Make use of the fact that only in a sphere and in a plane the respective coefficients of the first and second quadratic forms are proportional.

24.44. *Hint:* show that all geodesics are curvature lines. Make use of the fact that if on a surface a curvature line passes in each direction, then all points are umbilical.

24.47. Let $\mathbf{r}(s) + t\mathbf{l}(s)$ be the radius vector of a ruled surface in some asymptotic parameterization. We introduce new coordinates s, τ, where $\tau = t + t(s)$, $t(s) = \int\limits_0^s \langle \mathbf{l}, \dot{\mathbf{r}}' \rangle\, ds$. Then we get the required parameterization $\mathbf{R}(s) + \tau\mathbf{l}(s)$, in which the new generator has the form $\mathbf{R}(s) = \mathbf{r}(s) - t(s)\mathbf{l}(s)$.

24.48. $ds^2 = du^2 + 2F\,du\,dv + dv^2$.

24.49. A cylinder of revolution or a domain on it.

24.52. Deduce the formula for Gaussian curvature in asymptotic coordinates.

24.56. (a) Evidently, any surface with constant extrinsic geometry has constant Gaussian and mean curvatures. Consider two cases.

1) $K = 0$. A developable surface. If on it $H = 0$, then it is a plane. Let $H \neq 0$. We write an equation of the surface in orthogonal asymptotic parameterization $\mathbf{r}(s) + t\mathbf{l}(s)$ in which the mean curvature is computed by the formula (in Rashevsky's notation)

$$H = C_0 = \frac{A(s)\, t^2 + B(s)\, t + C(s)}{1 + 2t\, \langle \mathbf{r}', \mathbf{l}' \rangle + t^2\, \langle \mathbf{l}', \mathbf{l}' \rangle}.$$

As at a fixed s this equality can be regarded as an algebraic equality relative to t, we have (at $(t \to \infty)$) that $\langle l', l' \rangle = 0$, i.e., the vector l is constant. And as $\langle l, r' \rangle = 0$ due to the orthogonality of parameterization, the generatrix is plane. Further, we get that $A(s) = (r'', r', l) = \text{const}$, therefore, the curvature of the plane curve $r(s)$ is constant, i.e., it is a circle.

2) $K = \text{const} = C_1 \neq 0$. We introduce on the surface the coordinates in curvature lines. Then $E = M = 0$ and $2H = \dfrac{N}{G} + \dfrac{L}{E} = 2C_2$, $K = \dfrac{LN}{EG} = C_1$. Having solved this system of algebraic equations, we get

$$L = E\left(C_2 \mp \sqrt{H^2 - K}\right), \quad N = G\left(C_2 \pm \sqrt{H^2 - K}\right).$$

Now, from one of the Peterson–Codazzi equations, e.g., from $L_v = C_2 E_v$, we get: $E_v\left(C_2 \mp \sqrt{H^2 - K}\right) = C_2 E_v$, i.e., $H^2 - K = 0$. If $E_v = 0$, then another of the Peterson–Codazzi equations is applicable. Therefore, all points of the surface are umbilical, i.e., the surface is a sphere;

(b) rotation of the space $\mathbb{R}^4$ with the matrix

$$\begin{pmatrix} \cos\alpha & \sin\alpha & 0 & 0 \\ -\sin\alpha & \cos\alpha & 0 & 0 \\ 0 & 0 & \cos\beta & \sin\beta \\ 0 & 0 & -\sin\beta & \cos\beta \end{pmatrix},$$

where $\alpha = u_1 - u_2$, $\beta = v_1 - v_2$, takes the neighbourhood of the point

$$(R_1 \cos u_1, R_1 \sin u_1, R_2 \cos v_1, R_2 \sin v_1)$$

to the neighbourhood of the point $(R_1 \cos u_2, R_1 \sin u_2, R_2 \cos v_2, R_2 \sin v_2)$.

24.64. Make use of the Beltrami–Enneper theorem on the torsion of an asymptotic line on the surface of negative curvature.

24.65. See the previous problem.

24.67. Apply the formula for the Gaussian curvature K.

25.17. See Fig. 168 with three variants of the answer.

25.19. See Fig. 169.

25.24. A set of self-intersection points is homeomorphic to the union of three circles: $S^1 \vee S^1 \vee S^1$. The vertex of this union is a threefold point of self-intersection, and any point of the union different from the vertex is twofold.

25.25. The boundary M^2 of the constructed normal tubular neighbourhood of radius ε is, evidently, projected on $\mathbb{R}P^2$ (two endpoints of a normal segment pass into its centre lying on $\mathbb{R}P^2$). Thus, M^2 is a smooth two-dimensional compact closed manifold covering the projective plane in a two-sheeted manner. If we prove that this manifold is connected, we thereby prove that it is a two-dimensional sphere, as S^2 is the only two-sheeted connected cover over $\mathbb{R}P^2$.

To establish the connectivity, it is sufficient to consider two points on M^2 being the endpoints of the same normal segment, and show on M^2 the path connecting these two points. To construct such a path, it is sufficient to consider a point T on $\mathbb{R}P^2$, which is the centre of the segment considered, and take on $\mathbb{R}P^2$ a closed

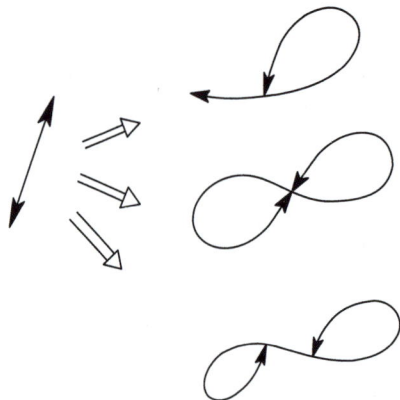

Figure 168

path, which begins and ends at the point T and in transition along which the two-place reference, sliding along the path and remaining all the time tangent to $\mathbb{R}P^2$, changes its orientation. Then, having supplemented this reference by a third vector orthogonal to $\mathbb{R}P^2$, and having considered the trace of this vector cut in continuous transition of the reference along the closed path, we do get on M^2 a continuous path connecting two points we chose.

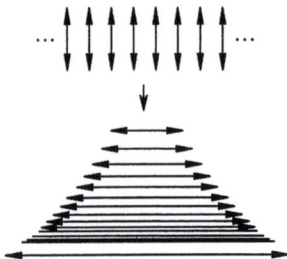

Figure 169

Additional note. The constructed immersion of a two-dimensional sphere in a three-dimensional Euclidean space makes it possible to prove a remarkable topological fact, the possibility of "turning inside out" a two-dimensional sphere in $\mathbb{R}^3$. This problem is beyond the limits of our course, so we would only restrict ourselves to a brief explanation. The given immersion of S^2 is such that it enables, remaining in the class of regular immersions, an interchange of the exterior and interior of a two-dimensional sphere. Indeed, it is sufficient to consider the smooth deformation of a two-dimensional sphere along a normal vector field defined by the above described normal segments. Herewith, the internal and external surfaces of the sphere would change places.

25.50. The structures of fibre bundles are shown in Figs 170 and 171. The bold line shows the fibre bundle base and thin lines indicate layers.

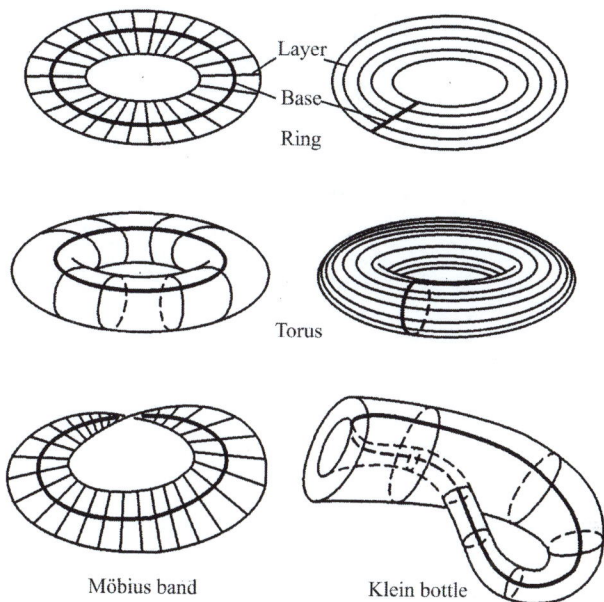

Layer
Base
Ring

Torus

Möbius band

Klein bottle

Figure 170 A torus, a ring, a Klein bottle

25.58. Consider the mapping of $\mathbb{R}^2$ in $\mathbb{R}^2$, which is given by the formulas $(x, y) \mapsto (x^2, y^3)$. At this mapping, the line $x = y$ passes into the semicubical parabola $y^3 = x^2$.

29.1. $3\sqrt{2}/5$.

29.2. $1/4$.

29.3. (a) 0; (b) $\dfrac{2\sqrt{3}}{3}\left(\sqrt{2} + 3\right)$; (c) 0; (d) -2; (e) $\dfrac{\pi a^2}{\sqrt{a + R^2}}$.

29.5. $\langle \operatorname{grad} f, \operatorname{grad} g \rangle$.

29.10. (a) $(0, x, y - x)$; (b) $\left(0, 0, y^2 - 2xz\right)$; (c) $(0, e^x - xe^y, 0)$;
(d) $\left(0, 3x^2, 2y^3 - 6xz\right)$; (e) $\left(0, -x\left(x + y^2\right), x^3 + y^3\right)$;
(f) $\left(0, xz^2 + yze^{x^2}, -2xyz\right)$; (g) $(\sin xz/x, 0, -\sin xz/y)$;
(h) $\left(xz/\left(x^2 + y^2\right), yz/\left(x^2 + z^2\right), -1\right)$.

29.11. Use the theorem on the existence and uniqueness of the solution of a system of ordinary differential equations.

29.15. Present the sphere S^3 as a group of quaternions of the unit module.

29.20. Let $z_0 = x_0 + iy_0$ be a singular point of the vector field $\operatorname{grad}(\operatorname{Re} f)$, where $f(z) = u(z) + iv(z)$, i.e.,

$$\left.\frac{\partial u}{\partial x}\right|_{(x_0, y_0)} = \left.\frac{\partial u}{\partial y}\right|_{(x_0, y_0)} = 0.$$

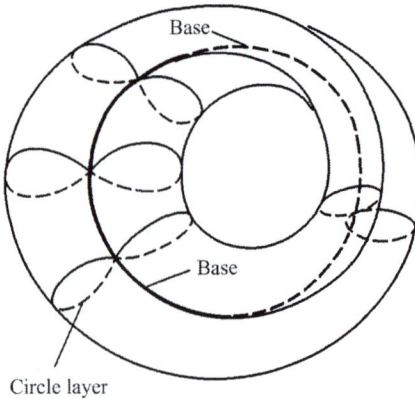

Figure 171 A Klein bottle

According to the Cauchy–Riemann condition

$$\frac{\partial u}{\partial x} \equiv \frac{\partial v}{\partial y}, \quad \frac{\partial u}{\partial y} \equiv \frac{\partial v}{\partial x},$$

therefore, $f'_{\bar{z}}(z_0) = 0$. Conversely, let $f'_{\bar{z}}(z_0) = 0$. As $f'_{\bar{z}}(z_0) \equiv 0$, then $\dfrac{\partial u}{\partial x}(z_0) = \dfrac{\partial u}{\partial y}(z_0) = 0$, i.e., $\operatorname{grad}(\operatorname{Re} f)(z_0) = 0$. For the field $\operatorname{grad}(\operatorname{Im} f)$, the proof is similar.

29.21. Seek for the integral trajectories only in the half-plane lying above the line AB. Level lines for the function $f(x)$ are arcs of the circles for which the segment AB is a chord. The vector $\operatorname{grad} f(x)$ is orthogonal to the level line. Therefore, the vector orthogonal to it is tangent to the level line, i.e., to the circle, and all circumscribed arcs of circles are integral trajectories of the flow $v_1(x)$.

29.22. Let the flow $\mathbf{v} = (P, Q)$ be irrotational, i.e.,

$$\operatorname{rot} \mathbf{v} = \frac{\partial P}{\partial y} - \frac{\partial Q}{\partial x} \equiv 0 \quad \text{or} \quad \frac{\partial P}{\partial y} \equiv \frac{\partial Q}{\partial x}.$$

Find such a function f that $P = \partial f/\partial x$, $Q = \partial f/\partial y$. For this, integrate the first correlation over x from 0 up to x:

$$f(x, y) = \int_0^x P\,dx + g(y).$$

To find $g(y)$, differentiate the latter correlation with respect to y:

$$Q(x, y) = \frac{\partial f}{\partial s} y(x, y) = \int_0^x \frac{\partial P}{\partial s} y\,dx + g'(y) =$$

$$\int_0^x \frac{\partial Q}{\partial s} x\,dx + g'(y) = Q(x, y) - Q(0, y) + g'(y).$$

Thus, $Q(x, y) = Q(x, y) - Q(0, y) + g'(y)$. Hence, $g'(y) = Q(0, y)$, i.e.,

$$g(y) = \int_0^y Q(0, y) \, dy + C.$$

Therefore,

$$f(x, y) = \int_0^x P(x, y) \, dy + \int_0^y Q(0, y) \, dy + C.$$

Let γ_1 and γ_2 be two paths from $(0, 0)$ to (x, y) in the plane (x, y). Then, if rot $\mathbf{v} = 0$, then

$$\int_{\gamma_1} (P \, dx + Q \, dy) = \int_{\gamma_2} (P \, dx + Q \, dy).$$

Therefore,

$$f(x, y) = \int_\gamma (P \, dx + Q \, dy) + C,$$

where γ is an arbitrary path from $(0, 0)$ to (x, y).

Let the flow be also incompressible, i.e., $\dfrac{\partial P}{\partial x} + \dfrac{\partial Q}{\partial y} = 0$. Consider the flow $\mathbf{v}' = (-Q, P)$. As is evident, rot $\mathbf{v}' = 0$ and, therefore, the field $\mathbf{v}'$ is potential. Thus, there exist such functions $a(x, y)$ and $b(x, y)$ that $\mathbf{v} = \operatorname{grad} a(x, y)$, $\mathbf{v}' = \operatorname{grad} b(x, y)$. In view of the fact that div $\mathbf{v} = \operatorname{div} \mathbf{v}' = 0$, we get that $a(x, y)$ and $b(x, y)$ are harmonic functions, i.e., $\Delta a \equiv \Delta b \equiv 0$. Consider the function $f = a + ib$. It is complex-analytic, as the Cauchy–Riemann equations

$$\frac{\partial a}{\partial x} = \frac{\partial b}{\partial y} = P(x, y), \qquad \frac{\partial a}{\partial y} = -\frac{\partial b}{\partial x} = Q(x, y)$$

take place. Such a function f is called a complex potential of the flow.

29.25. *Hint:* $d\varphi_2$ is homotopic to $d\varphi_1$.

29.32. Consider in $\mathbb{R}^4$ the linear differential equation $\dot{x} = Ax$, where

$$A = \begin{pmatrix} 0 & -1 & & \\ 1 & 0 & & \\ & & 0 & -1 \\ & & 1 & 0 \end{pmatrix},$$

and $x = (x_0, x_1, x_2, x_3)$. The integral trajectories of this equation, pertaining to the sphere $S^3 = \{x : |x| = 1\}$, are the sought-for set. It is clear that $x(t) = e^{At} x(0)$. If we consider $\mathbb{R}^4$ as $\mathbb{C}^2 (z_1, z_2)$, where $z_1 = x_1 + ix_0$, $z_2 = x_3 + ix_2$, then the integral trajectory passing through the point $(z_1, z_2) \in S^3$ has the form $(e^{it} z_1, e^{it} z_2)$, because e^{At} has the complex representation $\begin{pmatrix} e^{it} & 0 \\ 0 & e^{it} \end{pmatrix}$. The point $(z_1 : z_2)$ pertaining to the space $\mathbb{C}P^1$ (which is homeomorphic to S^2) corresponds to this trajectory. The correspondence is correct, as any other pair of $(z_1' : z_2')$ lying on the same trajectory is different from $(z_1 : z_2)$ only by the factor e^{it} and, therefore, sets the same point in $\mathbb{C}P^1$. It remains to note that the constructed mapping is one-to-one and continuous.

30.21. Fix the orthonormal reference $\{e_1, e_2, e_3\}$ in $\mathbb{R}^3$. An arbitrary state of the described system is uniquely given by the point $x \in S^2$ and the velocity vector $\mathbf{v}(x) \in T_x(S^2)$, where $|\mathbf{v}(x)| = c = \text{const} \neq 0$. The mapping $\mathbf{x} \rightarrow \mathbf{x}$, $\mathbf{v}(x) \rightarrow \mathbf{v}(x)/c$ is, evidently, a homeomorphism; $\mathbf{x}$ is the unit vector in $\mathbb{R}^3$ emanating from the point 0, $\mathbf{v}(x)$ is the unit vector in $\mathbb{R}^3$ (transfer the origin $\mathbf{v}(x)$ to the point 0). This transformation is identical on the vectors $\mathbf{x}$ and $\mathbf{v}(x)$; $\mathbf{x}$ and $\mathbf{v}$ are orthogonal. Let $y \in \mathbb{R}^3$ be such a vector that $|\mathbf{y}| = 1$, it is orthogonal to the vectors $\mathbf{x}$ and $\mathbf{v}$, and the system $\{e_1, e_2, e_3\}$ is in like manner oriented with the system $\{\mathbf{x}, \mathbf{v}, \mathbf{y}\}$. The mapping $(\mathbf{x}, \mathbf{v}) \rightarrow \{\mathbf{x}, \mathbf{v}, \mathbf{y}\}$ is, evidently, a homeomorphism. Each system $\{\mathbf{x}, \mathbf{v}, \mathbf{y}\}$ is given one-to-one and continuously by the matrix corresponding to the linear transformation in $\mathbb{R}^3$ transferring the orthonormal reference $\{e_1, e_2, e_3\}$ into the orthonormal reference $\{\mathbf{x}, \mathbf{v}, \mathbf{y}\}$. These matrices form the group $SO(3) = \{A : AA^\mathsf{T} = E, \det A = 1\}$. Thus, the space of states of our system is homeomorphic to the manifold $SO(3)$. Any orthogonal transformation $\mathbb{R}^3$ preserving the orientation is the revolution around some axis by an angle φ, where $-\pi < \varphi < \pi$.

30.22. *Hint:* roll the ball such that its point of osculation with the lower plane circumscribe the boundary of the rectangle with the vertices in the given order: $(0,0), (\pi/2, 0), (\pi/2, \varphi), (0, \varphi), (0,0)$. Show that herewith the ball turns to an angle φ around the axis passing through its centre parallel to the axis Ox. Similarly, realize a turn to an angle ψ around the axis Oy. Designate these turns, respectively, $A_x(\varphi)$, $A_y(\psi)$. Show that by compositions of these turns we can get any element of the group $SO(3)$. See Problem 30.10 about Euler angles.

30.39. Let G be the final group acting effectively on $\mathbb{R}^n$, i.e., if $gx = x$ for some $x \in \mathbb{R}^n$, then $g = e$. The group $\mathbb{Z}_k$ generated by some element $g \neq e$ also effectively acts on $\mathbb{R}^n$. Consider the space $X = G/\mathbb{Z}_k$, where $x \sim y$, if $y = g^l x$. As $\mathbb{R}^n \rightarrow X$ is a covering mapping, then $\pi_1(X) = \mathbb{Z}_k$, $\pi_i(X) = \pi_i(\mathbb{R}^n) = 0$ at $i > 1$. This means that the space X is homotopically equivalent to the space $K(\mathbb{Z}_k, 1)$, i.e., to the lens space. But homologies of $K(\mathbb{Z}_k, 1)$ are nontrivial in an infinite number of dimensions, whereas X has no cells of dimensionality greater than n. Thus, to complete the proof, it is required only to substantiate the following assertions: (a) if a discrete group G acts on $\mathbb{R}^n$ without immobile points and $X_G = \mathbb{R}^n/G$ is a space of orbits, then the natural mapping $p: \mathbb{R}^n \rightarrow X_G$ is a cover; (b) $\pi_1(X_G) = G$. The proof of these assertions is left to the reader.

32.3. Recall the definition of the topology of CW complexes. If K is a CW complex, the set $F \subset K$ is closed then and only then, when for all cells e_i^q the complete preimage $(f_i^q)^{-1}(F) \subset B^q$ is closed in B^q. We shall designate this topology by the symbol (W).

Assume that in space X there are two topologies $\{U_\alpha\}$ and $\{V_\beta\}$. We say that $\{V_\beta\} \geq \{U_\alpha\}$ (stronger) if for each point $x \in X$ and for each $V_{\beta_0} \ni x$ there is such $U_{\alpha_0} \ni x$ that $U_{\alpha_0} \subset V_{\beta_0}$.

Assume that in the CW complex, besides the (W) topology, there is also some topology $\{U_\alpha\}$. Take an arbitrary point $x \in K$, i.e., a point pertaining to the CW complex and to $U_{\alpha_0} \ni x$. The neighbourhood is a union of pairwise disjoint open intersections $(e_i^q \cap U_{\alpha_0})$. Consider the complete image $(f_i^q)^{-1}(U_{\alpha_0})$. It is open in B^q (this follows from the continuity of mappings f_i^q). This means that for the complement $(K \backslash U_{\alpha_0})$ the complete preimage $(f_i^q)^{-1}(K \backslash U_{\alpha_0})$ is closed in B^q for all e_i^q. By the axiom of (W), it follows that $(K \backslash U_{\alpha_0})$ is closed in K, therefore, U_{α_0} is open in the topology of (W), i.e., U_{α_0} pertains to the system of open sets (W).

32.25. A Klein bottle.

32.32. Let $\alpha_1, \alpha_2 \in H(X', Y)$, $\alpha_1 \sim \alpha_2$. This means that there exists the homotopy $F: \ X' \times I \to Y$ such that $F(x, 0) = \alpha_1(x)$, $F(x, 1) = \alpha_2(x)$. Let $F' = F \circ \varphi$. Then $F': X \times I \to Y$, $F'(x, 0) = F(h(x), 0) = \alpha_1(h(x))$, $F'(x, 1) = F(h(x), 1) = \alpha_2(h(x))$. Therefore,

$$\alpha_1 \circ h \sim \alpha_2 \circ h.$$

32.34. Let $S^\infty = \lim_{n \to \infty} S^n$, where S^{n+1} is a suspension over S^n. The sphere S^∞ thus defined is a CW complex. Consider $\alpha \in \pi_i(S^\infty)$ and $f \in \alpha: f: S^i \to S^\infty$, and f transfers the distinguished point in S^i to the distinguished point of S^∞. Let $f: K \to L$ be a continuous mapping of the complex K in the complex L, on the subcomplex $K_1 \subset K$ the mapping being cellular. Then there exists such a mapping $g: K \to L$ that (a) f is homotopic to g; (b) g is cellular on K; (c) $f|_{k_1} \equiv g|_{k_1}$; (d) the homotopy connecting f and g is identical on K_1. This fact implies that for f there exists a mapping, homotopic to it, which is transferred by S^i into an i-dimensional skeleton S^∞, i.e., into S^i, but $S^i \subset S^{i+1} \subset S^\infty$. This means that $g: S^i \to S^{i+1}$. As $\pi_i(S^n) = 0$ at $i < n$, then any mapping $i \in \alpha \in \pi_i(S^\infty)$ is homotopic to a mapping translating all S^i into a distinguished point S^n (constant mapping). This does mean that the mapping $f: S^i \to S^\infty$ is homotopic to the constant mapping. The mapping f was chosen arbitrarily, therefore, $\pi_i(S^\infty) = 0$.

If X and Y are cell complexes, and the mapping $f: \ X \to Y$ induces the isomorphism of all homotopic groups, then f is a homotopy equivalency. As f, we take the mapping $S^\infty \to *$. The isomorphism of homotopic groups is induced, as all of them are equal to zero. This means that the sphere S^∞ is homotopically equivalent to a point. Therefore, S^∞ collapses to a point.

32.36. Let $p^{-1}(x_0) = F_0$, $p^{-1}(x_1) = F_1$. Let $\varphi_0: F_0 \to X$ be an embedding. Then $p \circ \varphi_0: F_0 \to x_0 \in Y$. Join x_0 and x_1 by the path, i.e., arrange a homotopy between the mapping F_0 in x_0 and x_1, namely: $\psi_t: \ F_0 \to Y$, $\psi_t(F_0) = \gamma(t)$, where γ is our path. Then from the axiom about covering homotopy it follows that there exists a covering homotopy (a family of mappings $\varphi_t: F_0 \to X$) such that $(p \circ \varphi_t)(F_0) = \gamma(t)$, i.e. $p \circ \varphi_1(F_0) = \gamma(1) = x_1$. Hence, it follows that $\varphi_1(F_0) \subset F_1$. Thus, along the path γ we constructed the mapping $_\gamma\varphi_1: (F_0) \to F_1$. We prove that $_\gamma\varphi_1$ depends only on the homotopic class of the path γ, i.e., if γ_1 is homotopic to γ_2, then $_{\gamma_1}\varphi_1$ is homotopic to $_{\gamma_2}\varphi_1$. Note that the constructed mapping $F_0 \to F_1$ does not depend on the choice of the covering homotopy in the sense that any two such mappings are homotopic. Indeed, let φ_t and ξ_t cover ψ_t. Then the mapping $\varphi_1: F_0 \to F_1$ is homotopic to $\varphi_0: F_0 \to F_0$, $\varphi_0 = \xi_0$, and the latter, in turn, is homotopic to $\xi_1: F_0 \to F_1$. Let now a family of γ_t paths be given. Show that $_{\gamma_0}\varphi_1$ is homotopic to $_{\gamma_1}\varphi_1$. We have a mapping

$$_{\gamma_0}\varphi : F_0 \times I \to X; \quad (p \circ {}_{\gamma_0}\varphi)(F_0 \times I) = \gamma_0.$$

In Y, there is a homotopy γ_0 in γ_1, which can be covered by such a mapping $\Phi: (F_0 \times I) \times I \to X$ that $\Phi|_{(F_0 \times I) \times 0} = {}_{\gamma_0}\varphi$. Let $\Phi|_{(F_0 \times t) \times 1} = f_t$; clearly,

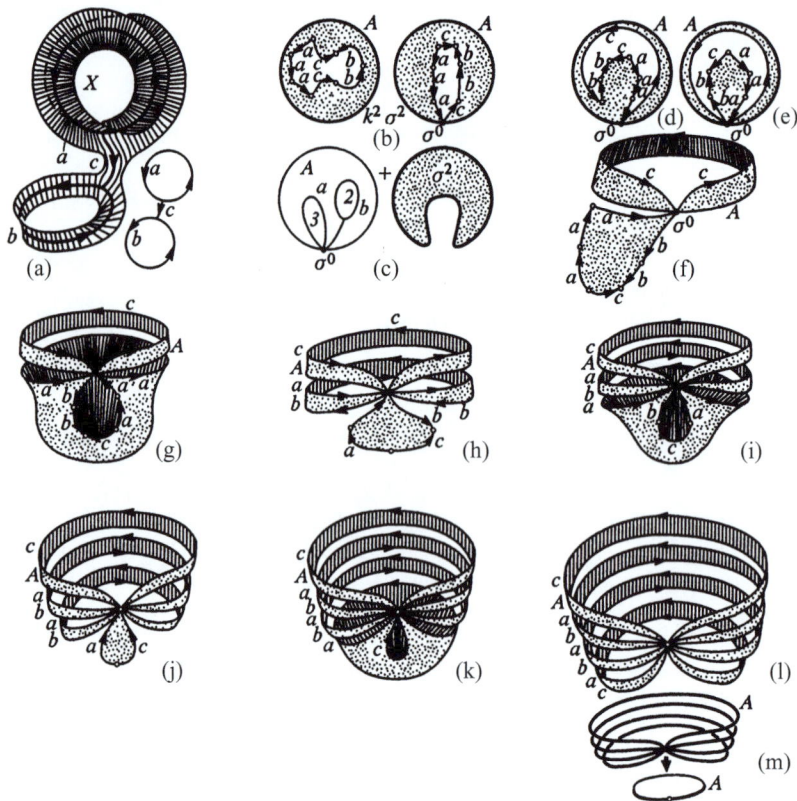

Figure 172

$(p \circ f)(F_0 \times I) = \gamma_1$. Therefore, the mapping f_t can be taken as a covering mapping for γ_1, $f_1 = {}_{\gamma_1}\varphi_1$. Then, $\Phi|_{(F_0 \times 1) \times I}$ is a homotopy between ${}_{\gamma_1}\varphi_1$ and ${}_{\gamma_0}\varphi_1$. We note further that in a similar way we can construct along the path $(-\gamma)$ a mapping $_{(-\gamma)}\chi_1 \colon F_1 \to F_0$. It remains to prove that the mapping $\big(_{(-\gamma)}\chi_1 \cdot {}_{\gamma}\varphi_1 \big) \colon F_0 \to F_0$ is homotopic to the identical one. But this mapping can be considered as a mapping induced by the path $\gamma + (-\gamma)$, which, evidently, is homotopic to a mapping into a point.

32.43. The solution is clear from Fig. 172.

32.44. Let $S^k \times S^{n-k}$ be a cell complex, which has only four cells: e^0, e^k, e^{n-k}, e^n. Consider the mapping $f \colon S^k \times S^{n-k} \to S^n$, $f\left(e^0\right) = *$, some point in S^n; take it for a zero-dimensional cell in S^n.

By the theorem on cell approximation, there exists a mapping $g \colon S^k \times S^{n-k} \to S^n$, which is already cellular and which is homotopic to f; moreover, on e^0 $f\left(e^0\right) = g\left(e^0\right)$ and all homotopy connecting f and g coincides on e^0 with f. As S^n consists of only two cells – zero-dimensional (*) and n-dimensional – in mapping g cells e^k and e^{n-k} pass into a point on S^n. We get that the mapping g can be not equal to a constant mapping only on an n-dimensional cell. This means that all mappings $S^k \times S^{n-k} \to S^n$ are different only by some mapping of an n-dimensional cell in $S^k \times S^{n-k}$, and then in S^n, which translates all the boundary into a point on S^n (due to the linear connectivity of S^n, the choice of point makes no difference). But these mappings are mappings $S^n \to S^n$, which establishes a one-to-one correspondence between $\pi\left(S^k \times S^{n-k}, S^n\right)$ and $\pi\left(S^n, S^n\right)$.

32.56. Let $\left(x^1, \ldots, x^n\right) \to \left(x^1, \ldots, x^n, 0\right)$ be the standard embedding $\mathbb{R}^n \to \mathbb{R}^{n+1}$ in the form of a hyperplane. Consider two points $A = (0, \ldots, 0, 1)$, $B = (0, \ldots, 0, -1)$ in $\mathbb{R}^{n+1}$ and construct the cones $C_A M$ and $C_B M$ with the vertices at points A and B, respectively, and with the common base $H \subset \mathbb{R}^n$. After this, any deformation of the subset $\mathbb{R}^n \backslash H$ in $\mathbb{R}^n$ can be continued up to the deformation of the subset $\sum\left(\mathbb{R}^n \backslash H\right)$ in $\mathbb{R}^{n+1}$.

32.57. Assume the inverse: let $\text{Cat}\left(M^n\right) < l\left(M^n; G\right)$, i.e., let there exist a covering of M^n with closed sets $X_1, \ldots, X_k$, $k < l\left(M^n; G\right)$, each of which collapses via M^n to a point. Due to Poincaré duality $H_k\left(M^n; G\right) \cong H^{n-k}\left(M^n; G\right)$, the cocycles $y_1, \ldots, y_l$ correspond to the cycles $x_1, \ldots, x_l$; herewith, the cycle $\alpha = y_1 \cap \ldots \cap y_l$ being the intersection of all cycles $y_1, \ldots, y_l$ corresponds to the product $h = x_1 \wedge \ldots \wedge x_l$ of the cocycles $x_1, \ldots, x_l$. As the operator D of Poincaré duality is a homomorphism, the intersection $y_1 \cap \ldots \cap y_l = \alpha$ is different from zero (i.e., the cycle α is not homological to zero). As any subset X_i $(1 \le i \le k)$ collapses via M^n into a point, $H^*\left(M^n; X_i\right) = H^*\left(M^n\right)$ (where $* > 0$). For this reason, it can be considered that the cycle $y_i \in H_*\left(M^n\right)$ is homological to the cycle $y_i \in H_*\left(M^n; X_i\right)$, i.e., the cycle carrier $\bar{y}_i$ lies in $M^n \backslash X_i$, $1 \le i \le k$. Hence, it follows that the intersection $\bar{y}_1 \cap \ldots \cap \bar{y}_k$ (homological to the intersection $y_1 \cap \ldots \cap y_k$) lies in the complement to (the union of) $X_1 \cup \ldots \cup X_k$, the more so that $\bar{y}_1 \cap \ldots \cap \bar{y}_k \cap \ldots \cap \bar{y}_l \subset M^n \backslash \left(X_1 \cup \ldots \cup X_k\right) = \varnothing$ as $X_1, \ldots, X_k$ forms the covering M^n. As the intersection of cycle carriers $\bar{y}_1 \cap \ldots \cap \bar{y}_l = \varnothing$, then the corresponding product of cocycles $x_1 \wedge \ldots \wedge x_l = 0$, which contradicts the condition: $x_1 \wedge \ldots \wedge x_l \ne 0$. The theorem is proved.

32.58. Consider the fibration (E, p, X), where E is the space of all paths of space X, which begin at the point x_0, and p is a mapping, which assigns each path its endpoint. Space E is considered herewith with respect to compact open topology. A layer of this fibration is space $\Omega X = \Omega_{x_0}$ of all loops of space X at the point x_0. It is easy to see that space E collapses on itself into a point (each path collapses on itself into a point x_0). Therefore, $\pi_n\left(E\right) = 0$ and, therefore, the homotopic sequence of this fibration

$$\ldots \to \pi_{n+1}\left(E\right) \to \pi_{n+1}\left(X\right) \to \pi_n\left(\Omega_{x_0}\right) \to \pi_n\left(E\right) \to \ldots$$

generates the isomorphism $\pi_n\left(\Omega_{x_0}\right) \approx \pi_{n+1}\left(X\right)$, in particular, $\pi_1\left(\Omega_{X_0}\right) \approx \pi_2\left(X\right)$. The group $\pi_n\left(X\right)$ is Abelian at $n \ge 2$.

32.59. We recall two definitions.

Space X is called *contractible* if the identity mapping $X \to X$ is homotopic to the mapping $X \to X$ bringing all X to a point.

Connected space X is called *simply connected*, if $\pi_1\left(X\right) = 0$.

As X is contractible, there exists $\varphi_t \colon X \to X$, φ_0 – identity mapping $X \to X$, φ_1 – mapping $X \to x_0 \in X$. As the definition of the fundamental group does not depend on the distinguished point (to an accuracy of the isomorphism), then let $\gamma \colon I \to X$ be an arbitrary path on X, $\gamma\left(0\right) = \gamma\left(1\right) = x_0$; $\delta\left(\tau\right) \equiv x_0$, $\delta \colon I \to X$. The same homotopy $\varphi_t\colon X \to X$ establishes the homotopy of loops γ and δ. Thus, any two paths on X are homotopic, i.e., $\pi_1(X) = 0$.

32.60. The assertion of the problem appears from the following two assertions:

(a) any element from $\pi_1(B_A^1)$ (B_A^1 is a bouquet of circles) is representable as a finite product of elements η_α^{-1} and η_α, where $\eta_\alpha \in \pi_1(B_A^1)$ is the class of a mapping i_α, which is the standard embedding;

(b) such a representation is unique to an accuracy of the cancellation of the consecutive factors η_α and η_α^{-1}.

Prove assertion (a). Consider the mapping $f\colon S^1 \to B_A^1$. Represent each circle S^1 in $S_\alpha^1 \in B_A^1$ as a sum of three one-dimensional simplexes P, Q, R and $P_\alpha, Q_\alpha, R_\alpha$. By the theorem of the simplicial approximation, the mapping f is homotopic to the simplicial mapping F of some subdivision of the complex S^1 in B_A^1. The mapping F should be multiplied on the right by the homotopy φ_t, where φ_0 is an identity mapping, φ_1 takes P_α, R_α to the distinguished point and stretches Q_α to all S_α^1. We get the mapping F_1 homotopic to the initial mapping. The mapping F_1 either takes each of the equal parts, into which S^1 is divided, into a point or winds on one of S_α^1, $\alpha \in A$. The class of such a mapping in $\pi_1(B_A^1)$ is the product of elements of the form η_α, η_α^{-1}, e are units of the fundamental group, i.e., of the class of constant mapping.

We pass to assertion (b). The product $\eta_{\alpha_1}^{\varepsilon_1} \ldots \eta_{\alpha_k}^{\varepsilon_k}$ ($\varepsilon_s = \pm 1$), $k \geq 1$, in which η_α and η_α^{-1} do not occur without interruption, is not equal to unity in $\pi_1(B_A^1)$, i.e., there exist no correlations in $\pi_1(B_A^1)$. At the covering $p\colon T \to X$, the preimage of each point $p^{-1}* = D$ is in one to one correspondence with residue classes of the group $\pi_1(X)$ with respect to subgroup $p_*(\pi_1(T))$. In particular, if $x_1, x_2 \in T$, $x \in X$ and $p(x_1) = p(x_2) = x$, S is any path from x_1 to x_2, then the loop $p(S)$ with the vertex at point x is not homotopic to zero, as otherwise $x_1 = x_2$. Let $\eta = \eta_{\alpha_1}^{\varepsilon_1} \ldots \eta_{\alpha_k}^{\varepsilon_k}$, $\eta_{\alpha_i}^{\varepsilon_i}$ be a loop passing in the direction of the bouquet circle depending on the sign of ε_i. Take $k+1$ copies of the bouquet and arrange them one over another. Take η_{α_i} in the first and second bouquets, cut a segment each from both copies, and join the ends crosswise, extending on them the projection π. Similarly, join the second bouquet with the third, using $\eta_{\alpha_2}^{\varepsilon_2}$ etc. If in the word η there are the same two letters in succession, two segments should be cut from the same circle. Herewith, the second operation precedes the first, if $\varepsilon_i = 1$, and follows it if not. We get a $(k+1)$-sheeted covering over B_A^1. Herewith, the path η is covered with the path beginning in the lower point and ending in the upper point. This loop is not homotopic to zero.

32.61. Let $f\colon Y_1 \to Y_2$ and $g\colon Y_2 \to Y_1$ be homotopic equivalencies, i.e., $g \circ f \sim \mathrm{Id}_{Y_1}$; $f \circ g \sim \mathrm{Id}_{Y_2}$. Define the mapping $f_*\colon \pi_1(Y_1) \to \pi_1(Y_2)$ and $g_*\colon \pi_1(Y_2) \to \pi_1(Y_1)$. (If $\alpha\colon S^1 \to Y_1$, $\alpha \in \bar\alpha \in \pi_1(Y_1)$, then f^α is the class of the loop $f_* \circ \alpha\colon S^1 \to Y_2$.) As $f_*g_* = (f \circ g)_*$, then $f_*g_*\colon pi_1(Y_2) \to \pi_1(Y_2)$ and $g_*f_*\colon \pi_1(Y_1) \to \pi_1(Y_1)$ are isomorphisms. Hence, $\pi_1(Y_1) = \pi_1(Y_2)$.

32.62. Let $\pi_1(X) * \pi_1(Y)$ be the free product of $\pi_1(X)$ and $\pi_1(Y)$. Let $\widehat{X}$ and $\widehat{Y}$ be the universal coverings over X and Y, respectively. Let x_0 be a distinguished point of X, Y and of the bouquet $X \vee Y$. Construct the following space Z: take $\widehat{X}$, consider $p^{-1}(x_0)$, where $p\colon \widehat{X} \to X$ is a covering, and in each point $x_0^i \in p^{-1}(x_0)$ glue $\widehat{Y}$. Identify x_0^i with $x_0^{i'}$, where $x_0^{i'}$ is some point from $p_1^{-1}(x_0)$, $p_1\colon \widehat{Y} \to Y$ is a covering. In each remaining point from $p_1^{-1}(x_0)$ in each copy of the "glued" $\widehat{Y}$, glue in this manner $\widehat{X}$ etc. The projection $p''\colon Z \to X \vee Y$ should be defined in a natural way: each copy of $\widehat{Y}$ by means of p' is mapped in Y and each copy of $\widehat{X}$ by means of p is mapped in X. It is evident that the space obtained is a covering over $X \vee Y$. Consider the fundamental group $X \vee Y$, the points t_1, t_2 from Z are such that $t_1, t_2 \in (p'')^{-1}(x_0)$ and the path connecting t_1 and t_2. At the projection p'' this path will pass into some loop α, which represents the class $\tilde{\alpha}$ in $\pi_1(X \vee Y)$. Note that from the construction of the covering and from the simple connectivity of $\widehat{X}$ and $\widehat{Y}$ it follows that the path from t_1 to t_2 is unique to an accuracy of homotopy.

Let $\tilde{\alpha} \in \pi_1(X \vee Y)$ be decomposed along the generators $\tilde{c}_i \in \pi_1(X)$ and $\tilde{b}_j \in \pi_1(Y)$, i.e., $\tilde{\alpha} = \tilde{c}_{i_1}^{\varepsilon_1} \tilde{b}_{j_1}^{\sigma_1} \tilde{c}_{i_2}^{\varepsilon_1} \ldots \tilde{b}_{j_n}^{\sigma_n}$. Then this representation is one to one to an accuracy of up to the correlations in $\pi_1(X)$ and $\pi_1(Y)$. Indeed, let $\tilde{\beta} = \tilde{c}_{i_1}^{\varepsilon_1} \tilde{b}_{j_2}^{\varepsilon_2} \ldots \tilde{c}_{i_n}^{\varepsilon_n} \tilde{b}_{j_n}^{\sigma_n} \sim 1$, where 1 is a constant loop at the point x_0 and not all ε_k and σ_s are equal to zero (we take a reduced word). Then $\tilde{\beta}$ can be covered by the path in Z, which, as it evidently follows from the form of coating, will not be closed and, therefore, $\tilde{\beta} = 1$. Thus, we obtained that $\pi_1(X \vee Y) = \pi_1(X) * \pi_1(Y)$. The same result follows from the van Kampen theorem on the expression of the fundamental group of the complex through the fundamental groups of its subcomplexes and intersections.

32.63. D e f i n i t i o n . If K is a knot, then the fundamental group $\pi_1(\mathbb{R}^3 \backslash K)$ is called a *knot group*.

Find the corepresentation of this group. Consider the upper (lower) corepresentation of a trifolium. Let PK be its projection. The points K_i ($i = 1, \ldots, 6$) divide the knot into two classes of closed connected arcs, the class of transitions and the class of passages alternating one with the other. Let A_1, A_2, A_3 be transitions; B_1, B_2, B_3, passages; F_3, a free group with generators x, y, z. We call the path v in $\mathbb{R}^2$ simple, if it is the union of a finite number of closed rectilinear segments, its initial and final points do not belong to PK, it intersects PK in a finite number of points not being the vertices of PK or v. Each path v is assigned $v^{\#} \in F_3$: $v^{\#} = x_{i_1}^{\varepsilon_1} \ldots x_{i_l}^{\varepsilon_l}$, x_{i_k} the generators of the free group, $\varepsilon_k = 1$ or -1 depending on how v passes under A_{i_k}. The upper corepresentation of the group $\pi_1(\mathbb{R}^3 \backslash K)$ has the form $(x, y, z; r_1, r_2, r_3)$, where $r_i = v_i^{\#}$ are correlations. It is known that the upper corepresentation given by this formula is the corepresentation of $\pi_1(\mathbb{R}^3 \backslash K)$. The loops v_1, v_2, v_3 around the transitions (x, y, z are the generators) satisfy the equalities:

$$v_1^{\#} = x^{-1} y z y^{-1}, \quad v_2^{\#} = y^{-1} z x z^{-1}, \quad v_3^{\#} = z^{-1} x y x^{-1}.$$

We get the corepresentation $\left(xyz; x = yzy^{-1}, y = zxz^{-1}, z = xyx^{-1}\right)$. Substitute $z = xyx^{-1}$ to get

$$\pi_1(\mathbb{R}^3 \backslash K) = \left(x, y; x = yxyx^{-1}y^{-1}, y = xyx\right).$$

Thus, $\pi_1(\mathbb{R}^3 \backslash K) = (x, y; xyx = yxy)$. The trifolium cannot be untied, as its type is different from the type of a trivial knot. If the knots K' and K'' are of the same type, their additional spaces possess coinciding fundamental groups. The group $G = (x, y; xyx = yxy)$ is not an infinite cyclic group $\mathbb{Z}$. Indeed, we can construct a homeomorphism $\theta\colon G \to S_3$, where S_3 is generated by cycles (12), (23).

Let K' and K'' be connected subcomplexes of the connected n-dimensional simplicial complex K, each complex from K pertaining to at least one of these subcomplexes. Let $D = K' \cap K''$ be an intersection, which is not empty and not connected. Let F, F', F'', F_D be fundamental groups of complexes K, K', K'', D.

As the initial point of the closed paths, take $0 \in D$. Then each closed path of the subcomplex D is simultaneously a path of the complexes K' and K''. We refer here to the known van Kampen theorem. The group F is obtained from the free product $F' \times F''$ if we identify each two elements F' and F'', corresponding to the same element F_D, i.e., assuming these elements to be equal, thereby adding correlations between the forming groups F' and F''.

32.64. Find the fundamental group of a helical knot defined as follows: on a side surface of a circular cylinder, draw generatrices at a distance of $2\pi/m$ one from another, and then turn the lower and upper bases by $2\pi n/m$ relative each other. After that, identify the bases. Attach to $\mathbb{R}^3$ one improper point (∞), transforming $\mathbb{R}^3$ into S^3. Delete from S^3 all points pertaining to the tubular neighbourhood of the knot. We get a polyhedron K, which is a knot complement. Split S^3 into two parts by a torus, on which the helical knot lies. The complex K breaks into two solid tori, in each of which the tubular neighbourhood of the knot on the surface is discarded. Take one solid torus as K', the other (with the improper point) as K''. The fundamental group $F'(F'')$ of the polyhedron $K'(K'')$ is a free group with one generator $A(B)$. The generator A can be represented as the median of the solid torus of polyhedron K (treat B similarly). The intersection of D of both solid tori is a twisted ring. The fundamental group D is also free with one generator, for which take the median of the circular ring. The group $F' \odot F''$ is a free group from the generators A and B. At a proper orientation of the paths A and B the path C considered as an element of the group F' is equal to A^m, and as an element of the group F'' it is equal to B^n. We get the correlation $A^m = B^n$. Thus, the corepresentation of the group $\pi_1\left(S^3 \backslash \gamma\right) = \{A, \ B; \ A^2 = B^3\}$, where γ is a trifolium.

Two obtained corepresentations of the fundamental group of a trifolium are equivalent. The verification of this is left to the reader.

32.65. Choose as the initial point of the closed paths a point 0 pertaining to W. Then each closed path of the complex W is simultaneously a path of the complexes Z, Y, i.e., each element of the group $\pi_1(W)$ is assigned an element of the group $\pi_1(X)$ and an element of the group $\pi_1(Y)$. Represent Z, Y, W as simplicial complexes. Join each vertex X by a path with 0. If the vertex lies in W, then the path can be completely drawn in W due to the connectivity. The simplex of an arbitrary dimension of the complex X pertains either to Z (but not to Y) or $Y \backslash Z$ or else $Y \cap Z$. The set of all simplexes breaks into three nonintersecting subsets $\overline{Z}$, $\overline{Y}$, $\overline{W}$. The generators a_i of the group $\pi_1(X)$ can be assigned to the edges of the complex X. Depending on which simplicial complex this edge pertains to ($\overline{Z}$, $\overline{Y}$ or $\overline{W}$), rename a_i into z_i, y_i or w_i. Thus, $\pi_1(X)$ has the fundamental groups $\pi_1(Y)$ and $\pi_1(Z)$ as its generators (the generators $\pi_1(W)$ are included in the generators $\pi_1(Z)$ and $\pi_1(Y)$). The correlations in the group $\pi_1(X)$ are in one to one correspondence with the edges and triangles of the complex X. As a consequence of breaking the complex X into three subsets, these correlations also break into three classes. We write down the correlations:

$$\varphi_j\left(w_i, z_i\right) = 1 \ \left(\text{in } \overline{Z}\right), \quad \varphi_j\left(w_i, y_i\right) = 1 \ \left(\text{in } \overline{Y}\right), \quad \psi_j\left(w_i\right) = 1 \ \left(\text{in } \overline{W}\right).$$

The correlations of the third type are defining for the group $\pi_1(W)$; the second- and third-type correlations define the groups $\pi_1(Y)$ and $\pi_1(W)$; the first- and third-type correlations define the groups $\pi_1(Z)$ and $\pi_1(W)$; and, finally, the correlations of all three types define the group $\pi_1(X)$. These correlations can be rewritten as follows:

$$\varphi_j\left(w_i', z_i\right) = 1, \quad \psi_j\left(w_i'\right) = 1, \quad \varphi_j\left(w_i'', y_i\right) = 1, \quad \psi_j\left(w_i''\right) = 1, \quad w_i' = w_i''.$$

The correlations of the first four types define the free product of the groups $\pi_1(Z)$ and $\pi_1(Y)$. The latter correlation means that elements of the groups $\pi_1(Z)$ and $\pi_1(Y)$, corresponding to the same element w_i of the group $\pi_1(W)$, should be identified. The proof makes use of the fact that W is connected, as otherwise the obtained assertion about the group $\pi_1(X)$ is not true.

Example. $Z = Y = I$ is a segment; $W = S^0$, $X = S^1$, $\pi_1(X) = Z$, $\pi_1(Z) = \pi_1(Y) = e$.

32.85. As is known, for any subgroup $G \subset \pi_1(X)$ there exists a covering $p\colon \overline{X}_G \to X$ such that $\operatorname{Im}\pi_*\left(\pi_1\left(\overline{X}_G\right)\right) = G$. Introduce on $\overline{X}_G$ a multiplication. Let $\hat{e} \in p^{-1}(e)$, where e is a unit in X, $\hat{x},\hat{y} \in X_G$. Connect $\hat{e}$ with $\hat{x}$ and $\hat{y}$ by paths $\hat{x}_t$ and $\hat{y}_t$: $\hat{x}_0 = e$, $\hat{x}_1 = \hat{x}$, $\hat{y}_0 = e$, $\hat{y}_1 = \hat{y}$. Let $p(\hat{x}) = x$, $p(\hat{y}) = y$. Then x and y are connected with e by paths $p(\hat{x}_t) = x_t$ and $p(\hat{y}_t) = y_t$, respectively. These two paths we can multiply in X, i.e., can consider the path $z_t = \hat{x}_t \times \hat{y}_t$, which connects e with the point $z_1 = z = xy$. Assume that z_t can be raised in X_G to the path $\hat{z}_t$. Let $\hat{x} \times \hat{y} = \hat{z}_1$. It remains to verify the correctness of the definition. The following assertion can be proved. Let X be a groupoid with unit, $\alpha, \beta \in \pi_1(X,e)$. Then $\alpha\beta = \alpha \times \beta$, where on the left we have a multiplication in $\pi_1(X,e)$ and on the right a multiplication in X.

We omit the proof, leaving it to the reader. The correctness of the definition follows immediately from this assertion.

32.86. At $p > 0$ and $q > 0$ for any $n < p+q-1$, the isomorphism $\pi_n(S^p \vee S^q) \approx \pi_n(S^p) + \pi_n(S^q)$ takes place. As the pair $(S^p \times S^q, \ S^p \vee S^q)$ is a relative $(p+q)$-dimensional cell, it follows thence that $\pi_m(S^p \times S^q, \ S^p \vee S^q) = 0$ at $m < p+q$. Therefore, $\pi_m(S^p \vee S^q) = \pi_m(S^p) + \pi_m(S^q)$.

If for the triple (X, A, x_0) the couple (X, A) is a relative n-dimensional cell, then $\pi_m(X, A, x_0) = 0$ at $0 < m < n$. The proof is left for the reader.

32.88. From the Freudenthal theorem, an assertion follows: the excision homomorphism of $\pi_m(U, S^n) \to \pi_m(S^{n+1}, V)$ is an isomorphism at $m < 2n$ and an epiisomorphism at $m = 2n$, where U and V are the northern and southern half-spheres of S^{n+1}. Find $\pi_3(D^2, \partial D^2)$:

$$\ldots \to \pi_n(\partial D^2) \to \pi_n(D^2) \to \pi_n(D^2, \partial D^2) \to \pi_{n-1}(\partial D^2) \to \ldots$$

At $n = 3$, we have $\pi_3(\partial D^2) = \pi_3(S^1) = 0$, $\pi_3(D^2) = 0$, $\pi_2(\partial D^2) = 0$. Hence, $\pi_3(D^2) \approx \pi_3(D^2, \partial D^2) = 0$. From the exact sequence, $\pi_3(S^2) = \pi_3(S^2, D^2) = \mathbb{Z}$.

32.89. The proof follows from the exact homotopic sequence of Serre fibration.

32.90. The proof follows from the cell representation of projective space and from the consideration of the standard covering.

32.92. Prove that $\pi_1(\mathbb{C}P^n) = 0$. Here $\mathbb{C}P^n$ is a cell complex, which has one cell each in each even sequence, i.e., has no one-dimensional cells. By the theorem of the fundamental group of the cell complex with one zero-dimensional cell, we get that $\pi_1(\mathbb{C}P^n) = 0$. Further, the sphere S^{2n+1} is fibered over $\mathbb{C}P^n$ with the layer S^1. Indeed, let $S^{2n+1} \subset \mathbb{C}^{n+1}$ (the standard embedding). The point $(z_1, \ldots, z_{n+1})$ pertains to S^{2n+1} then and only then, when $\sum |z_i|^2 = 1$. Further,

$$\mathbb{C}P^n = \{(z_1, \ldots, z_{n+1}) \text{ accurate to multiplication by } \lambda\},$$

i.e., $\lambda(z_1, \ldots, z_{n+1})$ and $(z_1, \ldots, z_{n+1})$ define the same point in $\mathbb{C}P^n$. Perform the mapping $p \colon S^{2n+1} \to \mathbb{C}P^n$, $p(z_1, \ldots) = (z_1, \ldots)$. It is continuous and its image is the entire space $\mathbb{C}P^n$. Over the point of $\mathbb{C}P^n$, there "hangs" the following set of points of S^{2n+1}: let $(z_1, \ldots, z_{n+1} \in \mathbb{C}P^n)$, then $f^{-1}(z_1, \ldots, z_{n+1}) = \{e^{i\varphi}(z_1, \ldots, z_{n+1})\} \subset S^{2n+1}$, where f^{-1} is the complete preimage, $0 \le \varphi \le 2\pi$. Indeed, $e^{i\varphi_1}(z_1, \ldots, z_{n+1})$ and $e^{i\varphi_2}(z_1, \ldots, z_{n+1})$ are the same point in $\mathbb{C}P^n$, but if $\varphi_1 \ne \varphi_2$, then in S^{2n+1} they are a fibration. It remains to use the exact homotopic sequence of fibration.

32.93. The assertions follow from the theorem of cell approximation.

32.94. Let $p_X \colon X \times Y \to Y$, $p_Y \colon X \times Y \to Y$ be projections. Perform the homomorphism $\varphi \colon \pi_i(X \times Y) \to \pi_i(X) \oplus \pi_i(Y)$, namely: $\varphi(\alpha) = (p_{X*}\alpha, p_{Y*}\alpha)$. Prove that φ is a homomorphism:

$$\varphi(\alpha + \beta) = (p_{X*}(\alpha + \beta), p_{X*}(\alpha + \beta)) =$$
$$(p_{X*}\alpha, p_{Y*}\alpha) \oplus (p_{X*}\beta, p_{Y*}\beta) = \varphi(\alpha) \oplus \varphi(\beta).$$

Prove that φ is a monomorphism. Let $\varphi(\alpha) = 0$, i.e., $p_{X*}\alpha = 0$, $p_{Y*}\alpha = 0$. Therefore, $\psi_X = p_X \circ \alpha \colon S^n \to X$ is homotopic to the constant one, i.e., there exists $\psi_{Xt} \colon S^n \to X$, such that $\psi_{X^0} = \psi_X$, $\psi_{X^1} = *$. Then perform the homotopy Φ_t as follows: $\Phi_t(\alpha) = (\psi_{Xt}(\alpha), p_{Y*}\alpha)$. $\Phi_1(\alpha)$ is a mapping of S^n in $(*) \times Y \subset Y$, $(*, p_{Y*}(\alpha)) \in \pi_n((*) \times Y) = \pi_n(Y)$. But $p_{Y*}(\alpha)$ is a contractible spheroid, i.e., α is contractible.

Prove that φ is an epimorphism. Let $\beta \in \pi_n(X)$, $\gamma \in \pi_n(Y)$. Then $\alpha = (\beta, \gamma)$ at φ passes into $\beta \oplus \gamma$.

Let there be universal coverings: $E_1 \xrightarrow{p_1} X$, $E_2 \xrightarrow{p_2} Y$. Consider the mapping $p_1 \times p_2 \colon E_1 \times E_2 \to X \times Y$, $(p_1 \times p_2)(e_1 \times e_2) = (p_1 e_1 \times p_2 e_2)$. It is asserted to be a covering. We omit the proof, leaving it to the reader.

Let $\gamma_1 \in \pi_1(X, x_0)$, $\gamma_2 \in \pi_1(Y, y_0)$, $\alpha_1 \in \pi_n(X, x_0)$, $\alpha_2 \in \pi_n(Y, y_0)$, and let some homotopy F_t along the path γ_1 of the spheroid α_1 be given such that $F_0(\alpha_1) = \alpha_1$, $F_1(\alpha_1) = \gamma_1[\alpha_1]$, $F_t(\alpha_1) \in \pi_n(X, \gamma_1(t))$. Similarly, for $\Phi_t(\alpha) \colon \Phi_0(\alpha_2) = \alpha_2$, $\Phi_1(\alpha_2) = \gamma_2[\alpha_2]$, $\Phi_t(\alpha_2) \in \pi_n(Y, \gamma_2(t))$. Define the homotopy along the loop $\gamma = (\gamma_1 \oplus \gamma_2)(t)$ in $X \times Y$ of the spheroid $\alpha = (\alpha_1 \oplus \alpha_2)$ as $F_t(\alpha_1) \times \Phi_t(\alpha_2)$. Then $F_1(\alpha_1) \times \Phi_1(\alpha_2) = \gamma[\alpha_1] \oplus \gamma_2[\alpha_2]$. Thus, $[\gamma_1 \oplus \gamma_2][\alpha_1 \oplus \alpha_2] = \gamma_1[\alpha_1] \oplus \gamma_2[\alpha_2]$, but as any loop γ and any spheroid from $X \times Y$ have the form $\gamma_1 \oplus \gamma_2$ and $\alpha_1 \oplus \alpha_2$ for some $\gamma_1 \in \pi_1(X)$, $\gamma_2 \in \pi_n(Y)$, $\alpha_1 \in \pi_n(X)$, $\alpha_2 \in \pi_n(Y)$, the action of $\pi_1(X \times Y)$ on $\pi_n(X \times Y)$ is totally defined.

32.96. Make use of the Hopf fibration $S^3 \to S^2$. It is performed as follows:

$$S^3 = \{(z_1, z_2) : |z_1|^2 + |z_2|^2 = 1\} \in \mathbb{C}^2, \quad S^2 = \mathbb{C}P^1,$$

i.e., $S^2 = \{(z_1, z_2) : (\lambda z_1, \lambda z_2) \sim (z_1, z_2)\}$. We get the fibration $S^3 \to S^2$. For this fibration, write down the exact sequence:

$$\ldots \to \pi_i(S^1) \to \pi_i(S^3) \to \pi_{i-1}(S^1) \to \ldots$$

From the property of consistency, $\pi_i(S^3) = \pi_i(S^2)$ at $i \ge 3$. By the Freudenthal theorem, the homomorphism $\pi_{i-1}(S^{n-1}) \to \pi_i(S^n)$ is an epimorphism at $i \le 2n-2$ and an isomorphism at $i < 2n - 2$, i.e., the homomorphisms $\pi_1(S^1) \to \pi_2(S^2) \to \pi_3(S^3)$ are isomorphisms. Therefore, $\pi_3(S^3) = \mathbb{Z}$. As $\pi_i(S^3) = \pi_i(S^2)$, $i \ge 3$, then $\pi_3(S^2) = \mathbb{Z}$.

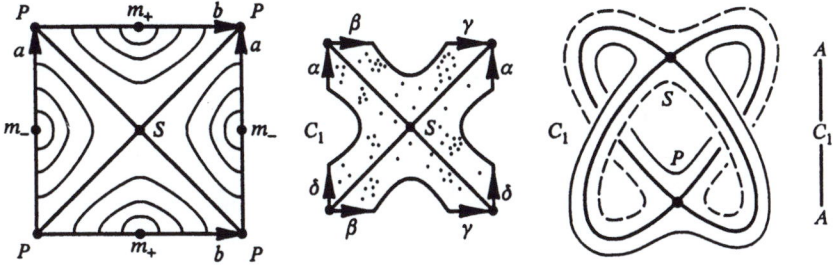

Figure 173

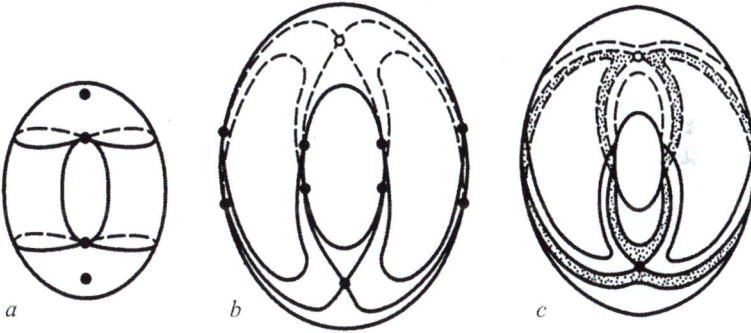

Figure 174

34.21. (a) See Fig. 173. (b) See Fig. 174.

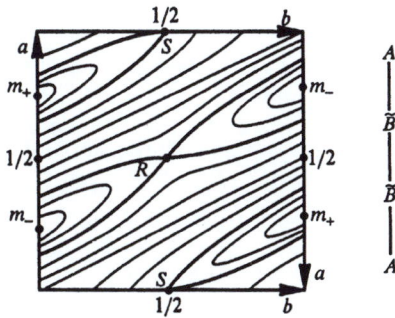

Figure 175

Consider the embedding of a torus in $\mathbb{R}^3$, at which its axis of revolution is horizontal, i.e., the torus "stands vertically". Figure 174a shows the critical levels of the function of height. Begin to tumble the torus to one side. Herewith, the critical

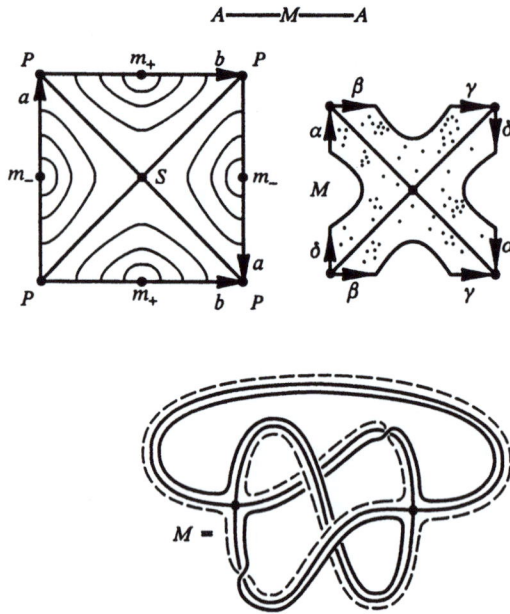

Figure 176

values corresponding to the saddle critical points approach. Figure 174b shows a moment when the saddle critical points proved on one level. A corresponding critical level of the function of height is depicted. Figure 174c presents an atom, i.e., the neighbourhood of the saddle critical level.

34.22. See Fig. 175.

34.23. See Fig. 176.

34.24. See Fig. 177.

35.1. It is sufficient to consider the Euler equations of the action functional and write these equations in local coordinates. Herewith, the explicit formulas for Christoffel symbols should be used.

35.2. We should write out the Euler equations for both functionals, then consider what happens with the functionals in the substitution of extremal solutions for the parameter (time). The sought-for assertion follows from the fact that the length functional is invariant at the substitution of the parameter, and the action functional is not invariant.

35.3. The proof is reduced to direct computation. We should write down the Euler equation in Cartesian coordinates and make use of the explicit formulas for mean curvature computed for the graph of the smooth function.

35.4. The proof is analogous to the proof of the previous problem. The analogy is based on that in both problems the codimensionality of the graph for the function

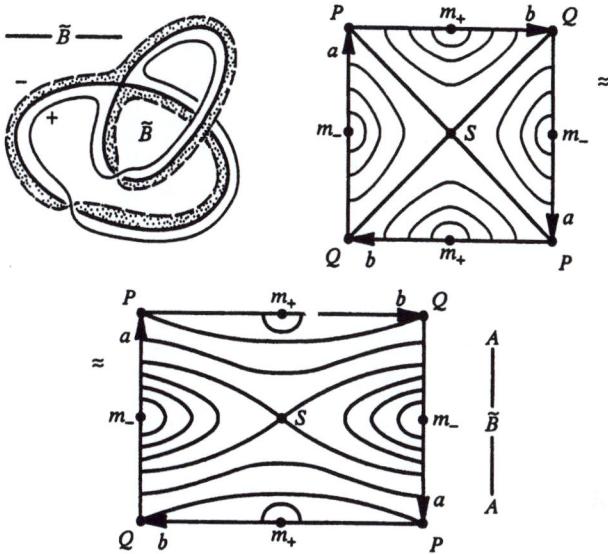

Figure 177

is unity, so the mean curvature tensor is given by one function, namely, by mean curvature.

35.5. Use the classical inequality

$$\left(\int\limits_a^b fg\,dt\right)^2 \le \left(\int\limits_a^b f^2 dt\right)\left(\int\limits_a^b g^2 dt\right).$$

35.6. Square the expressions under integral sign and compare them.

35.7. Use the local theory of the Beltrami–Laplace equation written in a curvilinear coordinate system. From this theory, it follows that on a two-dimensional surface given by analytical functions we can locally always introduce conformal coordinates. Addition of the condition that the mean curvature is equal to 0 transforms these coordinates into harmonic ones.

35.8. For instance, the function $\mathbf{r}\,(u,v) = \left(u, v, u^2 - v^2\right)$.

35.9. (a) *Hint:* let $r = 1$, ω_1, ω_2 be an orthonormal pair of vectors. We should prove that $\omega\,(\omega_1, \omega_2) \le 1$, where $\omega\,(\omega_1, \omega_2) = A\,(\omega_1, \omega_2)$. Consider

$$H = (\omega_1, \omega_2) = (S + iA)\,(\omega_1, \omega_2) = S\,(\omega_1, \omega_2) + iA\,(\omega_1, \omega_2) = iA\,(\omega_1, \omega_2).$$

Hence, $|H\,(\omega_1, \omega_2)| = |A\,(\omega_1, \omega_2)| \le |\omega_1| \cdot |\omega_2| = 1$. Let now $A\,(\omega_1, \omega_2) = 1$. Then $H\,(\omega_1, \omega_2) = i$, i.e., $S\,(i\omega_1, \omega_2) = 1$. As $|\omega_2| = |i\omega_1| = 1$, hence it follows that $\omega_2 = i\omega_1$, i.e., a two-dimensional plane spanned on ω_1, ω_2 is a complex line. For $r > 1$, we should use the correlation: $\Omega^{(2r)}\,(\omega_1, \omega_2) = \sqrt{\det g_{ij}}$, where g_{ij} is the skew symmetric scalar product defined by the 2-form $\omega^{(2)}$.

(b) *Hint:* the assertion follows from Wirtinger's inequality (see above) and Stokes' formula. Indeed, consider the external 2-form

$$\omega = \sum_{k=1}^{n} dz^k \wedge d\bar{z}^k \quad \text{and} \quad \Omega^{(2r)} = \frac{1}{r!}\omega \wedge \ldots \wedge \omega \ (r \text{ times})\,.$$

As $d\omega = 0$, $d\,\Omega^{(2r)} = 0$. From Stokes' formula, it follows that

$$\int_W \Omega^{(2r)} = \int_V \Omega^{(2r)}\,.$$

When integrating the formula $\Omega^{(2r)}$ over the $2r$-dimensional submanifold, consider (in local coordinates $x^1, \ldots, x^{2r}$) the expression of the form $\Omega^{(2r)}(\omega_1, \ldots, \omega_{2r})dx^1 \wedge \ldots \wedge dx^{2r}$, where $\omega_1, \ldots, \omega_{2r}$ is the orthonormal basis of an osculating surface to the submanifold (in the Riemannian metric induced by the ambient Euclidean metric in $\mathbb{C}^n = \mathbb{R}^{2n}$). If the submanifold W is complex, then

$$\Omega^{(2r)}(\omega_1, \ldots, \omega_{2r}) = 1, \quad \text{and} \quad \mathrm{vol}\,(W) = \int_W \Omega^{(2r)}\,.$$

If the submanifold V is of the general form (is real), then $\Omega^{(2r)}(\omega_1, \ldots, \omega_{2r}) \le 1$, i.e. $\int_V \Omega^{(2r)} \le \mathrm{vol}(V)$, which the assertion does prove.

(c) Computations performed in (b) are of local character, which makes it possible to conduct them in the neighbourhood of each point on a Kähler manifold. Stokes' formula, in contrast, takes place on any smooth manifold. Recall that the Kähler external 2-form is closed.

35.10. By the explicit function theorem, one of the coordinates, i.e., x^n, can be expressed (on a surface of the level of $F_0 = \mathrm{const}$) as a smooth function of other coordinates. After this, the function should be substituted in Euler's equation for the extremal of the functional J.

––––––––––

36.9. Note that the open ball B^n and the sphere S^n with the deleted point are homeomorphic. This assertion is proved by induction over the dimension of the complex. If the dimension $n = 0$, the assertion is evident. Let the assertion to be true for all numbers less than n. Then by the assumption of induction, the $(n-1)$-dimensional skeleton K^{n-1} of our complex is embedded in the Euclidean space $\mathbb{R}^N$. This means that continuous real functions $f_1(x), \ldots, f_N(x)$ are given on K^{n-1} such that $(f_1(x), \ldots, f_N(x)) \neq (f_1(y), \ldots, f_r(y))$ at $x \neq y$. Let e_j^n, $j = 1, \ldots, k$ be all n-dimensional cells of our complex. Then the functions $f_i(x)$ are defined at the boundary of each cell e_j^n (we denote it e_j^n). Let e_j^n be homeomorphic to the interior B^n of the closed ball D^n. Then we can consider the functions $f_i(x)$ to be given on $D^n \backslash B^n$. Their continuity in this case is preserved, the mutual one-to-oneness can be lost. Extend these functions from $D^n \backslash B^n$ to B^n (i.e., from ∂e_j^n to e_j^n) as follows. Let $z \in B^n$ and $z \neq 0$. We assume $f_i(z) = |z| f_i(z/|z|)$. If $z = 0$, we assume $f_i(z) = 0$. Thus, we extended the functions f_i to the continuous functions on all the complex K. Now, we define the functions $g_1^j(x),\ldots,g_{n+1}^j(x)$. External to e_j^n, let $g_s^j(x) \equiv 0$, $s = 1,\ldots,n+1$; on e_j^n, let

$$\left(g_1^j(x),\ldots,g_n^j(x),g_{n+1}^j(x)\right) = \left(\frac{x_1}{|x|}\sin \pi |x|,\ldots,\frac{x_n}{|x|}\sin \pi |x|,\cos \pi |x|+1\right).$$

Define $F\colon K \to \mathbb{R}^{N+k(n+1)}$ by the equality

$$F(x) = \left(f_1(x),\ldots,f_N(x);\; g_1^1(x),\ldots,g_{n+1}^1(x),g_1^k(x),\ldots,g_{n+1}^k(x)\right).$$

The mapping F is one-to-one. The assertion is proved.

36.11. Let $e_1,\ldots,e_s$ be the vertices of complex K. Take in $\mathbb{R}^{2n+1}$ points $e'_1,\ldots,e'_s$ in general position, i.e., at $j \leq 2n+2$ any j of points will be linearly independent. Assign each skeleton $T = |e_{i_0},\ldots,e_{i_r}| \in K$ to simplex $T' = |e'_{i_0},\ldots,e'_{i_r}| \in \mathbb{R}^{2n+1}$. This simplex exists, as due to the general position of the points $e'_1,\ldots,e'_s$ in $\mathbb{R}^{2n+1}$ and the inequality $r \leq n$ the points $e_{i_0},\ldots,e_{i_r}$ are linearly independent. The simplexes form a complex isomorphic to the complex T', as each vertex is assigned with one and only one vertex of K.

The complex K is a triangulation. To prove this, it is sufficient to show that no two simplexes $T'_i, T'_j \in K'$ intersect. Let $e'_{i_0},\ldots,e'_{i_p}$ be vertices T'_i; $e'_{j_0},\ldots,e'_{j_q}$, vertices T'_j (some vertices can be common). Let $e'_{k_0},\ldots,e'_{k_r}$ be all points being the vertices of at least one of the simplexes T'_i and T'_j. The number $r+1$ of these points satisfies the inequality

$$r+1 \leq (p+1) + (q+1) \leq (n+1) + (n+1) = 2n+2.$$

Due to the general position of points $e'_1,\ldots,e'_s$ in $\mathbb{R}^{2n+1}$, points $e'_{k_0},\ldots,e'_{k_r}$ are vertices of some nondegenerate simplex T_0 of dimension not higher than $2n+1$. The simplexes T'_i and T'_j are faces of T_0 and, thus, do not intersect if they are different.

36.13. Verify that for any $F \in \exp X$ a set of the form $O\langle U_1,\ldots,U_n\rangle$ can be found such that $F \in O\langle U_1,\ldots,U_n\rangle$. This is evident: $F \in O\langle X\rangle$. Verify now that for any $O_1 = O\langle U_1,\ldots,U_n\rangle$, $O_2 = O\langle V_1,\ldots,V_k\rangle$ such that $O_1 \cap O_2 \neq \varnothing$, and for any $F \in \exp X$ such that $F \in O_1 \cap O_2$, we can find $O_3 = O\langle W_1,\ldots,W_m\rangle$ such that $O_3 \subset O_1 \cap O_2$, where $F \in O_3$.

Let $F \in O_1 \cap O_2$. For any $i = 1,\ldots,n$ consider a set $F_i = U_i \cap F$. As $F \in O_1$, $F_i \neq \varnothing$. As $F \in O_2$, we can find $V_{j_1(i)},\ldots,V_{j_l(i)}$ such that $F_i \subset \bigcup_{p=1}^{l} V_{j_p(i)}$. We can consider that $\{V_{j_p(i)}\}$ are all those from the sets $V_1,\ldots,V_k$, which intersect with F_i. Let $W_1,\ldots,W_m$ be the numeration of the sets $\{V_{j_p(i)}\}$ for all i. By the construction of the sets W_i, $F \in O\langle W_1,\ldots,W_m\rangle$. Besides, it is easy to see that $O\langle W_1,\ldots,W_m\rangle \subset O_1 \cap O_2$. Thus, it is verified that sets of the form $O\langle U_1,\ldots,U_n\rangle$ form the basis of some topology, which is what was required to be proved.

36.14. As X is a T_1 space, all its one-point submanifolds are closed, therefore, the mapping f is defined correctly. Clearly, f is a bijection to its image $f(X) = \exp_1 X$. Further, let $O\langle U_1,\ldots,U_n\rangle$ be the neighbourhood of the point $\{x\}$ in $\exp X$. Let $U = \bigcap_{i=1}^{n} U_1$. The set U is non-empty, as $x \in U_i$ for all i. The set $O\langle U\rangle$ is the neighbourhood of the point $\{x\}$ in space $\exp X$; moreover, $O\langle U\rangle \subset O\langle U_1,\ldots,U_n\rangle$.

Evidently, $f^{-1}\left(O\left\langle U\right\rangle\right) = U$, therefore, f is continuous, as $U = f^{-1}\left(O\left\langle U\right\rangle\right) \subset f^{-1}\left(O\left\langle U_1,\ldots,U_n\right\rangle\right)$. The openness of f is evident, as $f\left(U\right) = O\left\langle U\right\rangle \cap \exp_1 X$ for any open U in X.

36.15. (a) Let X be regular. Let $F_1 \in \exp X$, $F_2 \in \exp X$ and $F_1 \neq F_2$. It can be considered that $F_2 \backslash F_1 \neq \varnothing$. Due to the regularity of space X, such neighbourhoods U of point x and V of set F_1 will be found that $U \cap V = \varnothing$. If $F_2 = \{x\}$, then $O\left\langle U\right\rangle$ and $O\left\langle V\right\rangle$ are disjunctive neighbourhoods of points F_1 and F_2 in space $\exp X$. If $F_2 \neq \{x\}$, then $O\left\langle U, X \backslash \{x\}\right\rangle$ and $O\left\langle V\right\rangle$ are the required neighbourhoods. Thus, $\exp X$ is Hausdorff.

(b) Let $\exp X$ be Hausdorff. Let $x \in X$, F be a closed subset X such that $x \notin F$. Points F and $F \cup \{x\}$ are various points in space $\exp X$. Therefore, non-intersecting neighbourhoods $O\left\langle U_1,\ldots,U_n\right\rangle$ and $O\left\langle V_1,\ldots,V_k\right\rangle$ of points F and $F \cup \{x\}$, respectively, will be found. Let $V = \cap\{V_i | x \in V_i\}$. Show that $V \cap \left(\bigcup_{i=1}^{n} U_i\right) = \varnothing$.

Assume that point $y \in V \cap \left(\bigcup_{i=1}^{n} U_i\right)$ will be found. Then $y \in \bigcup_{i=1}^{n} U_i$. Therefore, $\{y\} \cup F \in O\left\langle U_1,\ldots,U_n\right\rangle \cap O\left\langle V_1,\ldots,V_k\right\rangle$, which contradicts the choice of sets $O\left\langle U_1,\ldots,U_n\right\rangle$ and $O\left\langle V_1,\ldots,V_k\right\rangle$. Thus, $V \cap \left(\bigcup_{i=1}^{n} U_i\right) = \varnothing$, besides, $x \in V$ and $F \subset \bigcup_{i=1}^{n} U_i$. Therefore, V and $U = \bigcup_{i=1}^{n} U_i$ are the disjunctive neighbourhoods of point x and set F, respectively, hence, X is regular.

36.16. Prove the equivalency of conditions (a) and (b). Let X be normal, F be a closed subset of X, $O\left\langle U_1,\ldots,U_n\right\rangle$ be its neighbourhood in space $\exp X$. Note that the family $\{U_1,\ldots,U_n,X \backslash F\}$ is an open covering of X. As X is normal, according to the lemma of finite open covering compression there exists a closed finite covering $\{F_1,\ldots,F_n,F_{n+1}\}$ of space X; moreover, $F_i \subset U_i$, $i = 1,\ldots,n$, $F_{n+1} \subset X \backslash F$. Note that F_i can be supplemented with one point such that the condition $F_i \cap F \neq \varnothing$ be satisfied (the point can be chosen from the set $U_i \cap F \neq \varnothing$), so it can be considered that condition $F_i \cap F \neq \varnothing$ is satisfied.

As X is normal, by Urysohn's lemma functions $f_i \colon X \to [0,1]$, $i = 1,\ldots,n+1$ continuous on X will be found, moreover, $f_i|_{F_i} \equiv 0$, $f_i|_{X \backslash U_i} \equiv 1$. For an arbitrary $t > 0$, $i = 1,\ldots,n$ let

$$V_{i,t} = \{x | f_i\left(x\right) < t\} \quad W_t = O\left\langle V_{1,t},\ldots,V_{n,t}\right\rangle.$$

For any q and r of the segment $[0,1]$ such that $q < r$ the condition $\overline{V}_{i,q} \subset V_{i,r}$ is true (here and further, by $\overline{A}$ we denote the closure of set A).

For further reasoning, we will need two auxiliary lemmas. For subset X_0 of space X we denote through $\exp\left(X_0, X\right)$ the set of all $F \subseteq X_0$ such that $F \neq \varnothing$ and F is closed in X.

L e m m a   1. *Let* $X_0 \subset X$, *where* X *is a* T_1 *space. Then*

$$\overline{\exp\left(X_0, X\right)}^{\exp X} = \exp\left(\overline{X_0}^X, X\right).$$

P r o o f. Evidently, $\exp\left(X_0, X\right) \subseteq \exp\left(\overline{X}_0, X\right)$. Note that $\exp\left(\overline{X}_0, X\right)$ is closed in $\exp X$, therefore, $\overline{\exp\left(X_0, X\right)} \subset \exp\left(\overline{X}_0, X\right)$. Verify the reverse inclusion. Let F_0 be an arbitrary point of $\exp\left(\overline{X}_0, X\right)$, $O\left\langle U_1,\ldots,U_n\right\rangle$ be an arbitrary neighbourhood of this point. We have $F_0 \subset \overline{X}_0$ and for all $i = 1, \ldots, n$ $U_i \cap F_0 \neq \varnothing$. Therefore, $U_i \cap \overline{X}_0 \neq \varnothing$, so $U_i \cap X_0 \neq \varnothing$ for all i.

Choose $X_i \in U_i \cap X_0$. Let $F = \{X_1, \ldots, X_n\}$. Then $F \in \exp(X_0, X) \cap O\langle U_1, \ldots, U_n \rangle$, i.e., $\exp(X_0, X) \cap O\langle U_1, \ldots, U_n \rangle \neq \varnothing$. Thus, the inclusion $\exp(\overline{X}_0, X) \subset \overline{\exp(X_0, X)}$ is verified, which proves the lemma.

L e m m a 2 . *Let $\{U_1, \ldots, U_n\}$ and $\{V_1, \ldots, V_n\}$ be two families of sets open in T_1 space X, and for any $i = 1, \ldots, n$ it is true that $\overline{V}_i \subset U_i$, then*

$$\overline{O\langle V_1, \ldots, V_n \rangle}^{\exp X} \subset O\langle U_1, \ldots, U_n \rangle.$$

P r o o f .

$$\overline{O\langle V_1, \ldots, V_n \rangle} = \overline{\exp\left(\bigcup_{i=1}^{n} V_i, X\right) \cap \left(\bigcap_{i=1}^{n} (\exp X \setminus \exp(X \setminus V_i, X))\right)} \subset$$

as the closure of intersection is contained in the intersection of closures

$$\subset \exp\left(\overline{\bigcup_{i=1}^{n} V_i}^{X}, X\right) \cap \left(\bigcap_{i=1}^{n} \overline{(\exp X \setminus \exp(X \setminus V_i, X))}\right) \subset$$

by lemma 1

$$\subset \exp\left(\bigcup_{i=1}^{n} \overline{V}_i, X\right) \cap \left(\bigcap_{i=1}^{n} (\exp X \setminus \exp(X \setminus V_i, X))\right) \subset$$

by the condition of the lemma

$$\subset \exp\left(\bigcup_{i=1}^{n} U_i, X\right) \cap \left(\bigcap_{i=1}^{n} (\exp X \setminus \exp(X \setminus U_i, X))\right) = O\langle U_1, \ldots, U_n \rangle.$$

Lemma 2 is proved.

By lemma 2, the inclusion $[W_q] \subset W_r$ is performed. This means that space $\exp X$ admits a tight splitting between an arbitrary point F and an arbitrary closed set

$$F_1 = \exp X \setminus O\langle U_1, \ldots, U_n \rangle \, ;$$

therefore, $\exp X$ is quite regular.

It is evident that condition (c) follows from condition (b).

Prove that (a) follows from (c). Let F_0 and F_1 be nonempty closed nonintersecting subsets X. Consider

$$W = \exp(X \setminus F_1, X) \, ;$$

W is open in $\exp X$, moreover, $F_0 \in W$. Due to the regularity of space $\exp X$ a neighbourhood $O\langle U_1, \ldots, U_n \rangle$ of point F_0 will be found such that $\overline{O\langle U_1, \ldots, U_n \rangle}^{\exp X} \subset W$. Therefore, $F_0 \subset \bigcup_{i=1}^{n} U_i$. Show that $\overline{\bigcup_{i=1}^{n} U_i} \cap F_1 = \varnothing$. Assume the converse. Then

$$\bigcup_{i=1}^{n} (\overline{U}_i \cap F_1) = \bigcup_{i=1}^{n} (\overline{U}_i) \cap F_1 = \overline{\bigcup_{i=1}^{n} U_i} \cap F_1 \neq \varnothing \text{ and at least one index } j = 1, \ldots, n$$

will be found such that $\overline{U}_j \cap F_1 \neq \varnothing$. For each $i = 1, \ldots, n$ we choose a point $x_i \in \overline{U}_i$ if $i \neq j$, and $x_i \in \overline{U}_i \cap F_1$ if $i = j$. Let $F = \{x_i | i = 1, \ldots, n\}$. It is easy to see that any neighbourhood of point F in space $\exp X$ contains points from the set $O\langle U_1, \ldots, U_n \rangle$, so $F \in \overline{O\langle U_1, \ldots, U_n \rangle}^{\exp X} \subset W = \exp(X \setminus F_1, X)$. But $F \cap F_1 \neq \varnothing$. A contradiction.

Thus, $\bigcup\limits_{i=1}^{n} U_i \cap F_1 = \varnothing$. Due to the arbitrariliness of F_0 and F_1 this means that space X is normal.

36.17. Show that $\exp X \setminus \exp_n X$ is open in $\exp X$. Let $F \in \exp X \setminus \exp_n X$. Then $|F| > n$. Let $x_1, \ldots, x_{n+1}$ be various points pertaining to F. As X is Hausdorff, for any i a neighbourhood U_i of point x_i will be found such that $U_i \cap U_j = \varnothing$ at $i \neq j$. Let $W = O\langle X, U_1, \ldots, U_{n+1} \rangle$. Then $F \in W$. Besides, $W \cap \exp_n X = \varnothing$ because if $G \in W$, then $|G| > n$ by the construction of sets $U_1, \ldots, U_{n+1}$. Thus, $\exp X \setminus \exp_n X$ is open, therefore, $\exp_n X$ is closed.

36.18. Let x and y be two various points in space X. Let $F = \{x, y\}$. Then $F \in \exp_2 X \setminus \exp_1 X$. As $\exp_1 X$ is closed in $\exp_2 X$, $\exp_2 X \setminus \exp_1 X$ is open. Therefore, a neighbourhood $W = O\langle U_1, \ldots, U_n \rangle$ will be found for F in space $\exp_2 X$ such that $W \cap \exp_1 X = \varnothing$.

Let $O_x = \cap\{U_i | x \in U_i\}$, $O_y = \cap\{U_i | y \in U_i\}$. It is easy to see that $x \in O_x$, $y \in O_y$ and $O_x \cap O_y = \varnothing$. Thus, X is Hausdorff, which is what was required to be proved.

36.19. (a) Let $V = O\langle V_1, \ldots, V_k \rangle$, where V_i are open in X. Prove that $\pi_n^{-1}V$ is open in X^n. Let $x = (x_1, \ldots, x_n) \in \pi_n^{-1}V$. As $\pi_n x \in V$, without limiting the generality of reasoning, we can consider that $x_1, \ldots, x_{j_1} \in V_1, \ldots, x_{j_{k-1}+1}, \ldots, x_{j_k} \in V_k$. Let

$$O = \underbrace{V_1 \times \ldots \times V_1}_{j_1} \times \ldots \times \underbrace{V_k \times \ldots \times V_k}_{j_k - j_{k-1}}.$$

Then O is the neighbourhood of point x in X^n such that $O \subseteq \pi_n^{-1}V$. Therefore, $\pi_n^{-1}V$ is open.

(b) Let $\pi_n^{-1}V$ be open. Prove that V is open. Let $x = (x_1, \ldots, x_m) \in V$, $m \leq n$, where $x_1, \ldots, x_m$ are various points in X. Let $y \in \pi_n^{-1}x$. Then $y = (y_i^j)$, where $j = 1, \ldots, m$, $y_i^j = x_j$. As $\pi_n^{-1}V$ is open, then for any $y \in \pi_n^{-1}x$ a neighbourhood $O_y = \prod\limits_{i=1}^{n} O_i^j(y)$ will be found such that $O_y \subseteq \pi_n^{-1}V$. Let

$$O_j = \bigcap_{y \in \pi_n^{-1}x} \left(\bigcap_i O_i^j(y) \right).$$

Then $W = O\langle O_1, \ldots, O_m \rangle$ is an open neighbourhood of point x; herewith, $\pi_n^{-1}W \subseteq \pi_n^{-1}V$, therefore, $W \subset V$, so V is open in $\exp_n X$.

36.20. Let $[x] \in SP_{G'}^n X = X^n/G'$. Consider a point $(x_1, \ldots, x_n)$, any representative of the equivalence class $[x]$. To satisfy the condition $\pi_{G'G}^n \circ \pi_{G'}^n = \pi_G^n$, it is necessary that $\pi_{G'G}^n([x]) = [x]_G$ be fulfilled, where $[x]_G = \{(x_{\sigma(1)}, \ldots, x_{\sigma(n)}) | \sigma \in G\}$. Thus, the uniqueness of the mapping $\pi_{G'G}^n$ is verified. It remains to prove the independence of the choice of the representative $(x_1, \ldots, x_n)$. Let $(x_{g(1)}, \ldots, x_{g(n)})$ be another representative of the class $[x]$, where $g \in G'$. As $G' \subset G$, then $g \in G$, whence $\left(x_{g(1)}, \ldots, x_{g(n)}\right) \in \left\{\left(x_{\sigma(1)}, \ldots, x_{\sigma(n)}\right) \mid \sigma \in G\right\}$, which is what was required to be proved.

36.21. Let $O \subseteq X^n$ be open, $O = \prod_{i=1}^{n} O_i$. Prove that $\pi_G^n O$ is open. For this, according to the definition of topology in factor space X^n/G, it is sufficient to verify that $(\pi_G^n)^{-1} (\pi_G^n O)$ is open in X^n. Indeed,

$$(\pi_G^n)^{-1} (\pi_G^n O) = \bigcup_{\sigma \in G} O_{\sigma(1)} \times \ldots \times O_{\sigma(n)} = \bigcup_{\sigma \in G} \prod_{i=1}^{n} O_{\sigma(i)}.$$

Thus, the openness of the mapping π_G^n is verified.

Let $F \subseteq X^n$ be closed. It is necessary to verify that $\pi_G^n F$ is closed. Let $[x] \in SP_G^n X \setminus \pi_G^n F$. Then $[x] = \left[\left(x_{\sigma(1)}, \ldots, x_{\sigma(n)}\right)\right]$, where $(x_1, \ldots, x_n)$ is a representative of class $[x]$; herewith, $\left(x_{\sigma(1)}, \ldots, x_{\sigma(n)}\right) \notin F$ for any $\sigma \in G$. As F is closed, for any $g \in G$ a set Ox_g, a neighbourhood of point $x_g = \left(x_{g(1)}, \ldots, x_{g(n)}\right)$, will be found such that $Ox_g \cap F = \varnothing$. We can consider that $Ox_g = \prod_{i=1}^{n} O_i^g$; herewith, $x_{g(i)} \in O_i^g$ for all $i = 1, \ldots, n$. Let $V_i = \bigcap_{g \in G} O_i^g$, $V = \prod_{i=1}^{n} V_i$. Then for any $g \in G$ we have $\prod_{i=1}^{n} V_{g(i)} \cap F = \varnothing$. Therefore, $\pi_G^n V \cap \pi_G^n F = \varnothing$. Besides, according to the already proven, $\pi_G^n V$ is open in $SP_G^n (X)$ and, by the construction, $[x] \in \pi_G^n V$. Thus, $SP_G^n (X) \setminus \pi_G^n F$ is open, therefore, $\pi_G^n F$ is closed.

36.22. Prove the following general fact:

L e m m a . *Let X, Y, Z be topological spaces; $f: X \to Y$, $g: Y \to Z$, $h: X \to Z$, continuous mappings, where $h = g \circ f$ and the mapping f is surjective. Then, if h is open (closed), g is open (closed) too.*

P r o o f . Let h be open. Prove that g is open. Let V be an open subset in Y. Then $f^{-1}V$ is open due to the continuity of f. Besides, $g \left(f \left(f^{-1}V\right)\right) = g (V)$, as f is surjective. On the other hand, $g \left(f \left(f^{-1}V\right)\right) = h \left(f^{-1}V\right)$ is open, as h is open. In the case when h is closed, the proof is similar. The lemma is proved.

In our case, $f = \pi_{G'}^n$, $g = \pi_{G'G}^n$, $h = \pi_G^n$, where $\pi_{G'G}^n$ is a mapping existing in consequence of Problem 36.20. The openness and closedness of mappings $\pi_{G'}^n$ and π_G^n follow from Problem 36.21. Using the lemma, we get that $\pi_{G'G}^n$ is open and closed.

36.23. It should be verified that $V \subset \exp_n X$ is open then and only then when $q_n^{-1}V$ is open in $SP^n X$. Let V be open in $\exp_n X$. Verify that $q_n^{-1}V$ is open in $SP^n X$. For this, according to the definition of factor topology in $SP^n X$, we should prove that $(\pi^n)^{-1} \left(q_n^{-1}V\right)$ is open in X^n. But $(\pi^n)^{-1} \left(q_n^{-1}V\right) = \pi_n^{-1}V$, as $\pi_n = q_n \circ \pi^n$, and the set $\pi_n^{-1}V$ is open due to the factorization of the mapping π_n. As $q_n^{-1}V$ is open in $SP^n X$, then $(\pi^n)^{-1} \left(q_n^{-1}V\right)$ is open in X^n. But $(\pi^n)^{-1} \left(q_n^{-1}V\right) = \pi_n^{-1}V$, so because of the factorization of the mapping π_n, V is open in $\exp_n X$.

36.24. *Hint:* use the Alexander lemma.

S o l u t i o n . Let X be a compact. Then X is regular; therefore, $\exp X$ is Hausdorff (see Problem 36.15). We will need the following auxiliary lemma (the Alexander lemma): let in space X there exist such subbase $\mathcal{B}$ that from each covering of space X any final subcovering can be chosen as elements of this subbase, then X is bicompact.

Note that sets of the form $O\langle U\rangle$ and $O\langle X,U\rangle$ form the subbase of Vietoris topology. Let $\mathcal{D}$ be a covering consisting of elements of this subbase. Show that $\mathcal{D}$ contains the finite subcovering. Let $W = X\setminus\cup\,\Omega$, where $\Omega = \{V|O\langle X,U\rangle \in \mathcal{D}\}$. If $W = \varnothing$, then Ω is an open covering X. As X is bicompact, Ω contains the finite sibcovering $\{V_1,\ldots,V_n\}$. Then the family $\{O\langle X,V_1\rangle,\ldots,O\langle X,V_n\rangle\}$ is a covering of space $\exp X$, since for each point $F \in \exp X$ the respective closed set F of space X intersects with any V_i, therefore, $F \in O\langle X,V_i\rangle$.

If, however, W is not empty, then W does not intersect with any element of the family Ω. Therefore, an open set U will be found such that $O\langle U\rangle \in \mathcal{D}$, where $W \subset U$.

Note that Ω is a covering of the compact $X\setminus U$. Choose from it a finite subcovering $\{V_1,\ldots,V_n\}$. Then the family $\{O\langle U\rangle,\,O\langle X,V_1\rangle,\ldots,O\langle X,V_n\rangle\}$ is a covering of space $\exp X$, because for each point $F \in \exp X$ or $F \in O\langle X,V_i\rangle$ for some i, or $F\cap\left(\bigcup\limits_{i=1}^{n}V_i\right) = \varnothing$ and then $F \subset U$ and, therefore, $F \in O\langle U\rangle$.

Thus, from $\mathcal{D}$ we can choose a finite subcovering, therefore, according to the Alexander lemma, $\exp X$ is a compact.

36.25. According to Problem 36.24, $\exp X$ is a compact. According to Problem 36.17, $\exp_n X$ is closed in $\exp X$. Therefore, $\exp_n X$ is a compact as a closed subset of a compact.

36.26. As X is compact, then by the Tikhonov theorem X^n is a compact. Note that $SP_G^n X$ is an image of X^n at a continuous mapping of π_G^n. Thus, $SP_G^n X$ is a compact as a continuous image of the compact X^n.

36.27. Let $\mathcal{B}$ be a base of the topology of space X, and F be a closed subset of X. Let $F \in O\langle V_1,\ldots,V_n\rangle$, where the sets V_i, $i = 1,\ldots,n$, are open in X. For each $i = 1,\ldots,n$, we can find $U_i^{\alpha_i} \in \mathcal{B}$, $\alpha_i \in \mathcal{A}_i$ (where $\mathcal{A}_i$ are a set of indices) such that $V_i = \bigcup\limits_{\alpha_i\in\mathcal{A}_i} U_i^{\alpha_i}$ (by the definition of the base of topology). Then $\{U_i^{\alpha_i}|i = 1,\ldots,n,\alpha_i \in \mathcal{A}_i\}$ is an open covering of compact F. Select from it a finite subcovering $W_1,\ldots,W_m$. Without limiting the generality of reasoning, we can consider that $W_i\cap F \neq \varnothing$ for any $i = 1,\ldots,m$.

It is evident that $F \in O\langle W_1,\ldots,W_m\rangle \subset O\langle V_1,\ldots,V_n\rangle$. Thus, for any point of space $\exp X$ and for any its neighbourhood there exists another inlying neighbourhood of the form $O\langle U_1,\ldots,U_n\rangle$, where $U_i \in \mathcal{B}$, and $\mathcal{B}$ is the base of space X, which is what was required to be proved.

36.28. As the compact $\exp X$ contains $\exp_n X$ and, in particular, the subspace $\exp_1 X$ homeomorphic to X, the inequalities $w\exp X \geq w\exp_n X \geq wX$ are satisfied. Let $wX = \tau$. According to Problem 36.27, on $\exp X$ there exists a base of topology of power τ. Therefore, $w\exp X \leq wX$, from where we conclude that $w\exp X = w\exp_n X = wX$.

Finally, $SP_G^n X$ is a continuous image of X^n and contains X, therefore,

$$wX = wX^n \geq wSP_G^n X \geq wX,$$

which is what was required to be proved.

36.29. 1) The inequality $\rho_H(F_1,F_2) \geq 0$ is evident. Let $\rho_H(F_1,F_2) = 0$. Assume that $F_1 \neq F_2$. It can be considered that $F_2\setminus F_1 \neq \varnothing$. Let $x \in F_2\setminus F_1$. Set $d = \inf\limits_{y\in F_1}\rho(x,y)$. As F_1 is closed and $x \notin F_1$, then $d > 0$. But $F_2 \subset O_\varepsilon F_1$ for any $\varepsilon > 0$, in particular, $F_2 \subset O_{d/2} F_1$. This means that a point $x' \in F_1$ will be found such that $\rho(x, x') < \dfrac{d}{2}$. A contradiction with the choice of d, therefore, $F_1 = F_2$.

2) It is evident that $\rho_H(F_1, F_2) = \rho_H(F_2, F_1)$.

3) Verify the inequality of the triangle $\rho_H(F_1, F_3) \leq \rho_H(F_1, F_2) + \rho_H(F_2, F_3)$. The set F_2 is contained in the neighbourhood $(\rho_H(F_1, F_2) + \varepsilon)$ of the set F_1. The set F_3 is contained in the neighbourhood $(\rho_H(F_1, F_2) + \varepsilon)$ of the set F_2. From the triangle axiom for ρ it follows that any point of F_3 is removed from F_1 by no more than $\rho_H(F_1, F_2) + \varepsilon + \rho_H(F_2, F_3) + \varepsilon$. As ε is arbitrary, this implies the inequality of the triangle for ρ_H.

4) Finally, it is evident that $\rho(x, y) = \rho_H(\{x\}, \{y\})$.

36.30. Let $S = \{s_1, \ldots, s_m\} \in \exp X$, $m \leq n$.

(a) Show that each neighbourhood of point S in the sense of Vietoris topology contains the neighbourhood of the same point in the sense of Hausdorff metric. Let $O\langle U_1, \ldots, U_k \rangle$ be the neighbourhood of the point S. Choose $\varepsilon > 0$ such that for all $i = 1, \ldots, m$ and $j = 1, \ldots, k$ from $s_i \in U_j$ follows the inclusion $O_\varepsilon(s_i) \subset U_j$. Let $T \in \exp_n X$ be such a point that $\rho_H(S, T) < \varepsilon$. Then for all $t \in T$ there exists $s_i \in S$ such that $\rho(s_i, t) < \varepsilon$, i.e., $t \in O_\varepsilon(s_i) \subset U_j$ for some j. Therefore, $T \subset \bigcup\limits_{i=1}^{k} U_j$. Besides, for each U_j we can choose a point $s_i \in U_j$ such that $t \in T$ will be found for which $\rho(s_i, t) < \varepsilon$. Then $t \in O_\varepsilon(s_i) \subset U_j$. For this reason, $T \cap U_j \neq \varnothing$ for each $j = 1, \ldots, k$. Therefore, $T \in O\langle U_1, \ldots, U_k \rangle$. Thus, the neighbourhood ε (in the sense of the metric ρ_H) of point S is contained in $O\langle U_1, \ldots, U_k \rangle$.

(b) Show that each neighbourhood of point S in the sense of Hausdorff metric contains the neighbourhood of the same point in the sense of Vietoris topology. Let $\varepsilon > 0$. Set $U_i = O_\varepsilon(s_i)$, $i = 1, \ldots, m$. Then from the condition $T \in O\langle U_1, \ldots, U_m \rangle$ it follows that $\rho_H(S, T) < \varepsilon$, i.e., $O\langle U_1, \ldots, U_m \rangle$ is contained in the neighbourhood ε of point S.

36.31. Let (X, ρ) be a metric space. For points $[x]$ and $[y]$ from $SP_G^n X$, set

$$\rho_G([x], [y]) = \min_{\sigma \in G} \max_i \rho(x_i, y_{\sigma(i)}),$$

where $(x_1, \ldots, x_n)$, $(y_1, \ldots, y_n)$ are representatives of classes $[x]$ and $[y]$, respectively. It is evident that ρ_G does not depend on the choice of representatives of classes $[x]$ and $[y]$.

(a) Verify that ρ_G is a metric.

1) It is evident that $\rho_G \geq 0$. Prove that the condition $\rho_G([x], [y]) = 0$ is tantamount to the coincidence of the classes $[x]$ and $[y]$. Indeed, if $[x] = [y]$, then for $\sigma = e$ we have $\max\limits_i \rho(x_i, y_{\sigma(i)}) = \max\limits_i \rho(x_i, y_i) = \max\limits_i \rho(x_i, x_i) = 0$ (it can be considered that for the coincident classes $[x]$ and $[y]$ the same representatives are chosen, i.e., $x_i = y_i$). Conversely, let $\rho_G([x], [y]) = 0$. Then such a permutation $\sigma \in G$ will be found that $x_i = y_{\sigma(i)}$ for all $i = 1, \ldots, n$, whence follows the coincidence of the classes $[x]$ and $[y]$.

2) Prove that $\rho_G([x], [y]) = \rho_G([y], [x])$. Indeed,

$$\rho_G\left([x],[y]\right) = \min_{\sigma \in G} \max_i \rho\left(x_i, y_{\sigma(i)}\right) = \min_{\sigma \in G} \max_i \rho\left(y_{\sigma(i)}, x_i\right) =$$

$$\min_{\sigma \in G} \max_j \rho\left(y_j, x_{\sigma^{-1}(j)}\right) = \min_{g \in G} \max_j \rho\left(y_j, x_{g(j)}\right) = \rho_G\left([y],[x]\right).$$

3) Verify the inequality of the triangle $\rho_G([x],[y]) + \rho_G([y],[x]) \geq \rho_G([x],[z])$.
Let $\rho_G([x],[y]) = \rho_G(x_s, y_m)$, $\rho_G([y],[z]) = \rho_G(y_p, z_l)$, $\rho_G([x],[z]) = \rho_G(x_k, z_j)$.
Let further $\rho(x_t, z_l) = \max_i \rho(x_i, z_l)$. Then, according to the definition of ρ_G, the following inequality is performed:

$$\rho\left(x_k, z_j\right) \leq \rho\left(x_t, z_l\right).$$

Similarly, we have $\rho\left(y_p, z_l\right) \geq \rho\left(y_m, z_l\right)$, $\rho\left(x_s, y_m\right) \geq \rho\left(x_t, y_m\right)$. Proceeding from these three inequalities, we have a chain of inequalities:

$$\rho_G\left([x],[y]\right) + \rho_G\left([y],[z]\right) = \rho\left(x_s, y_m\right) + \rho\left(y_p, z_l\right) \geq \rho\left(x_t, y_m\right) + \rho\left(y_m, z_l\right) \geq$$
$$\rho\left(x_t, z_l\right) \geq \rho\left(x_k, z_j\right) = \rho_G\left([x],[z]\right),$$

which is what was required to be proved.

Verify that ρ_G generates the factor topology in $SP_G^n X$. Let $[x] \in SP_G^n X$, $O_\varepsilon[x]$ be an open neighbourhood ε of the point $[x]$ in the metric ρ_G. Show that $O_\varepsilon[x]$ is open in $SP_G^n X$. For this, it is required to show that $(\pi_G^n)^{-1} O_\varepsilon[x]$ is open in X^n. Let $y = (y_1, \ldots, y_n) \in (\pi_G^n)^{-1} O_\varepsilon[x]$. Then $\min_\sigma \max_i \rho(x_i, y_{\sigma(i)}) = \varepsilon_1 < \varepsilon$. Set $\varepsilon_2 = \varepsilon - \varepsilon_1 > 0$. Let $V = \{z \in X^n \mid \rho(y_i, z_i) < \varepsilon_2\}$. Note that V is an open neighbourhood of the point y in space X^n. It is easy to verify that $V \subset (\pi_G^n)^{-1} O_\varepsilon[x]$.

Let now U be an open subset in $SP_G^n X$ and the point $[x]$ lie in U. As $(\pi_G^n)^{-1} U$ is open in X^n, for any substitution $\sigma \in G$ we will find $\varepsilon_\sigma > 0$ such that

$$\left\{y \in X^n \mid \rho\left(x_{\sigma(i)}, y_i\right) < \varepsilon_\sigma\right\} \subset (\pi_G^n)^{-1} U.$$

Set $\varepsilon = \min_{\sigma \in G} \varepsilon_\sigma$. Then $O_\varepsilon[x] \subset U$. Thus, ρ_G is a metric generating the topology on $SP_G^n X$ defined by the factorization X^n/G, which is what was required to be proved.

36.32. Verify that the identity mapping $i\colon \exp X \to (\exp X, \rho_H)$ is continuous. The statement of the problem will follow therefrom, as the continuous bijection of the compacts is a homeomorphism (recall that $\exp X$ is a compact according to Problem 36.24).

Let the point F_0 lie in $(\exp X, \rho_H)$. Consider an arbitrary $\varepsilon > 0$. For each point $x \in F_0$, fix $O_{\varepsilon/2}(x)$ – an open $\frac{\varepsilon}{2}$ neighbourhood of the point x in the metric ρ. The family $\{O_{\varepsilon/2}(x) \mid x \in F_0\}$ is an open covering of the compact F_0. Choose from it a finite subcovering $(U_1, \ldots, U_n)$. Note that $F_0 \in O \langle U_1, \ldots, U_n \rangle$. Let F be an arbitrary point on $O \langle U_1, \ldots, U_n \rangle$. If $y \in F$, then $y \in U_j$ for some j. Note now that $F_0 \cap U_j \neq \varnothing$ at all j and diam $U_j = \frac{\varepsilon}{2}$. Therefore, $\rho(y, F_0) < \frac{\varepsilon}{2}$. As the point y was chosen arbitrarily, and also due to the compactness of the set F, it follows that $F \subseteq O_{\varepsilon/2}(F_0)$. Similarly, $F_0 \subset O_\varepsilon(F)$. Therefore, $\rho_H(F_0, F) < \varepsilon$. Thus, the set $i(O \langle U_1, \ldots, U_n \rangle)$ is contained in the neighbourhood ε of the point F_0, therefore, the mapping i is continuous in the point F_0. Due to the arbitrariliness of the point F_0, it follows therefrom that the mapping i is continuous, which is what was required to be proved.

36.33. This assertion follows from Problems 36.24–36.26 and 36.29–36.32.

36.34. Let $I = [0,1]$. Note that the mapping $\pi_2\colon I^2 \to \exp_2 I$ identifies points of the square I^2, symmetric relative to the diagonal $\Delta = \{(x,x)\,|x \in I\}$. As the result of this identification, we get a triangle, which is, evidently, homeomorphic to I^2.

36.35. The mapping $\pi_2\colon H^1 \times H^1 \to \exp_2 H_1$ identifies the points of the first quarter of the plane $\mathbb{R}^2$, symmetric relative to the ray $x = y$. As the result of this identification, we get a space consisting of such points $(x,y) \in \mathbb{R}^2$ that $x \geq 0$ and $x \leq y$, which, evidently, is homeomorphic to H^2.

36.36. The mapping $\pi_2\colon \mathbb{R} \times \mathbb{R}^2 = \mathbb{R}^2 \to \exp_2 \mathbb{R}$ identifies points of the plane $\mathbb{R}^2$ symmetrical relative to the straight line $x = y$. As the result, we get a half-plane consisting of such points $(x,y) \in \mathbb{R}^2$ that $x \leq y$. This half-plane is homeomorphic to H^2.

36.37. Realize a circle as a segment $I = [0,1]$ with identified ends. It is easy to see that $S^1 \times S^1$ is obtained from the square $I \times I$ by identification of points of the form $(0,x)$ with points of the form $(1,x)$ and points of the form $(x,0)$ with points of the form $(x,1)$. Denote this identification as p. The projection $\pi_2\colon S^1 \times S^1 \to \exp S_1$ is realized as a composition of identification p and projection $\pi_2\colon I \times I \to \exp_2 I$.

Let $T\{(x,y) \in I^2 | y \geq x\}$. It is easy to see that π_2 is realized as an identification of points of set T which have the form $(x,1)$ with points of the same set of the form $(0,x)$. Denote this identification as $\sim$. Thus, $\exp_2 S^1$ is homeomorphic to $T/\sim$. It is easy to verify that space $T/\sim$ is homeomorphic to a Möbius band.

36.38. Let T be a set of points of the cube I^3, whose coordinates satisfy the condition $x_1 \geq x_2$ and $x_1 \geq x_3$. It is easy to see that $I(3)$ is obtained from the set T by identification of points lying on the faces of the set T, satisfying the conditions $x_1 = x_2$ and $x_1 = x_3$ and passing into one another at the turn around the line $x_1 = x_2 = x_3$ by an angle $2\pi/3$. Note that the result of this identification is homeomorphic to a cone over the square I^2, and this cone is in turn homeomorphic to I^3, which is what was required to be proved.

36.39. See the solution of Problem 36.55.

36.40. Let T be a set of points of the cube I^3, the coordinates of which satisfy the condition $x_1 \geq x_2 \geq x_3$. It is evident that T is a three-dimensional simplex. Note that $\exp_3 I$ is obtained from T by the identification of points lying on the faces $x_1 = x_2$ and $x_2 = x_3$ of this simplex, at which the point (t,t,z) passes into the point (z,z,t). As the result of this identification, we get a cone, which is homeomorphic to I^3.

36.41. Note that H_1 is homeomorphic to the half-interval $[0,1)$. Set

$$T = \left\{(x_1, x_2, x_3) \in [0,1)^3 \,|x_1 \geq x_2,\; x_1 \geq x_3\right\}.$$

Note that $H^1(3)$ is obtained from T by the identification of points of the faces $x_1 = x_2$ and $x_1 = x_3$ of the set T, passing one into another at the turn to an angle $2\pi/3$ around the straight line $x_1 = x_2 = x_3$. As the result of this identification, we obtain a set homeomorphic to a cone without base points. It is easy to see that such a cone is homeomorphic to H^3.

36.42. Note that H_1 is homeomorphic to the half-interval $[0,1)$. Set

$$T = \left\{(x_1, x_2, x_3) \in [0,1)^3 \,|x_1 \geq x_2 \geq x_3\right\}.$$

It is easy to see that T is homeomorphic to $SP^3\left(H^1\right)$. On the other hand, T is a three-dimensional simplex, where points of one of the faces are deleted, therefore, T is homeomorphic to H^3.

36.43. Note that H_1 is homeomorphic to the half-interval $[0,1)$. Set

$$T = \left\{(x_1, x_2, x_3) \in [0,1)^3 \,|\, x_1 \geq x_2 \geq x_3 \right\}.$$

Then $\exp_3\left(H^1\right)$ is obtained from T by the identification of points of the faces $x_1 = x_2$ and $x_2 = x_3$ of the set T, at which the half-interval $\{x_2 = x_3 = 0\}$ passes into the half-interval $\{x_1 = x_2, \, x_3 = 0\}$, the interval $\{x_2 = x_3, \, x_1 = 0t\}$ into the interval $\{x_1 = x_2 = 1\}$. As the result of this identification, we obtain a set homeomorphic to a cone without the base points, and, in turn, it is homeomorphic to H^3, which is what was required to be proved.

36.44. Note that $\mathbb{R}$ is homeomorphic to $(0,1)$. Further, similar to the solution of Problem 36.41, we get that $\mathbb{R}(3)$ is homeomorphic to a cone without points of the boundary, i.e., to an open ball, which, evidently, is homeomorphic to $\mathbb{R}^3$.

36.45. Note that $\mathbb{R}$ is homeomorphic to $(0,1)$. Further, repeating verbatim the reasoning of Problem 36.42, we see that $SP^3\mathbb{R}$ is homeomorphic to a three-dimensional simplex, where points of two faces are deleted, but such a simplex is homeomorphic to H^3.

36.46. Note that $\mathbb{R}$ is homeomorphic to $(0,1)$. Further, similar to the solution of Problem 36.43, we get that $\exp_3(\mathbb{R})$ is homeomorphic to a cone, where points of the base and side surface are deleted. This set is homeomorphic to an open ball, which, in turn, is homeomorphic to $\mathbb{R}^3$.

36.47. Let X be a cell space. By X^n, designate the n-dimensional skeleton of X. We will need the following auxiliary assertion.

T h e o r e m   o f   t h e   f u n d a m e n t a l   g r o u p   o f   c e l l   s p a c e. *Let X be a cell space with unique zero-dimensional cell e^0, one-dimensional cells $\left\{e_i^1\right\}_{i=1}^N$ and two-dimensional cells $\left\{e_i^2\right\}_{i=1}^M$. Let $\beta_j \in \pi_1\left(X^1\right)$ be an element of the group $\pi_1\left(X^1\right)$, generated by the "glueing" mapping $g_i : S^1 \to X^1$ $(j = 1, \ldots, M)$. Then $\pi_1\left(X, e^0\right)$ is a group with a system of generators $\left\{e_i^1\right\}$ and a system of defining correlations $\beta_j = 1$ $(j = 1, \ldots, M)$.*

The proof of this theorem can be found in standard reference books on homotopic topology.

Realize a circle as a segment $J = [0,1]$ with identified ends. Then $\left(S^1\right)^3$ is obtained from I^3 by the identification of points of the opposite faces of the cube I^3.

Let $T = \left\{(x_1, x_2, x_3) \in I^3 \,|\, x_1 \geq x_2 \geq x_3 \right\}$. It is easy to see that T is a three-dimensional simplex. Set $A = (0,0,0)$, $B = (1,0,0)$, $C = (1,1,0)$, $D = (1,1,1)$. Then A, B, C and D are the vertices of the simplex T.

It is easy to see that $\exp_3\left(S^1\right)$ is obtained from T by the identification of points of the faces ABC and BCD, point B being identified with point A, point D being identified with C, point C being identified with B, as well as by the identification of points of the faces ADC and ABD, point B being identified with point C. Note that in such identification, points A, B, C, D pass into one point. Designate it as e^0. Besides, the edges AB, BC, CD, AC and BD pass into a circle, which we designate as α, and the edge BD passes into a circle, which we designate as β. Now it is easy to see that $\exp_3\left(S^1\right)$ is a cell space with the zero-dimensional cell e^0 and one-dimensional cells α and β. According to the theorem of the fundamental group of cell space, $\pi_1\left(\exp_3\left(S^1\right),e^0\right)$ is a group with a system of generators $\{\alpha,\beta\}$ and a system with defining correlations $\alpha^2\circ\alpha^{-1}=1$, $\alpha^2\circ\beta^{-1}=1$. Hence, we conclude that $\alpha=\beta=1$, therefore, the group $\pi_1\left(\exp_3\left(S^1\right),e^0\right)$ is trivial. It remains only to note that due to the connectivity of $\exp_3\left(S^1\right)$ the fundamental group does not depend on the choice of the distinguished point e^0. Thus, $\exp_3\left(S^1\right)$ is simply connected, which is what was required to be proved.

36.48. As in the solution of Problem 36.47, realize S^1 as a segment $I=[0,1]$ with identified ends and assume $T=\left\{(x_1,x_2,x_3)\in I^3|x_1\geq x_2\geq x_3\right\}$ to be a simplex with the vertices $A=(0,0,0)$, $B=(1,0,0)$, $C=(1,1,0)$, $D=(1,1,1)$. It is easy to see that $SP^3\left(S^1\right)$ is obtained from T by the identification of points of the faces ABC and BCD, edge AB being identified with edge BC, edge BC being identified with edge DC, and edge AC being identified with edge BD. Herewith, as it is easy to see, all points A, B, C, D pass into one point, which we designate as e^0. Further, the circle into which edges AB, BC and CD pass, will be designated as α; the circle into which edges AC and BD pass, will be designated as β and, finally, the circle into which edge AD passes, will be designated as γ.

It is easy to see that $SP^3\left(S^1\right)$ is a cell space with the vertex e^0 and one-dimensional cells α, β, γ. According to the theorem of the fundamental group of cell space (see the solution of Problem 36.47), $\pi_1\left(SP^3\left(S^1\right),e^0\right)$ is a group with generators α, β, γ and with defining correlations $\alpha^2\beta^{-1}=1$, $\beta\alpha\gamma^{-1}=1$. It is easy to see that the system of defining correlations is tantamount to the system $\alpha^2=\beta$, $\alpha^3=\gamma$. Thus, any element of the group $\pi_1\left(SP^3\left(S^1\right),e^0\right)$ will be represented (uniquely) in the form α^n, $n\in\mathbb{Z}$, whence we conclude that $\pi_1\left(SP^3\left(S^1\right),e^0\right)\approx\mathbb{Z}$. Therefore, $SP^3\left(S^1\right)$ is not simply connected.

36.49. Assume

$$T=\left\{(x_1,x_2,x_3)\in I^3|x_1\geq x_2,\,x_1\geq x_3\right\}.$$

It is easy to see that T is a polygon with the vertices $A=(0,0,0)$, $B=(1,0,0)$, $C=(1,1,0)$, $D=(1,1,1)$, $E=(1,0,1)$. Note that $S^1(3)$ is obtained from T by the identification of faces AED and ACD, in which points obtained one from another by the turn to an angle of $2\pi/3$ are "glued together", in particular, point E is identified with point C, as well as by the identification of points of the triangles ABE, BED, ABC and BCD. Designate this identification as $\sim$. Then $AB\sim BC\sim CD\sim BE\sim ED$, $AC\sim BD$.

Designate the circles into which edges AB, AC and AD pass in this identification as, respectively, α, β and γ; as e^0, designate the point into which the vertices T pass. It is easy to see that $S^1(3)$ is a cell space with the vertex e^0 and one-dimensional cells α, β and γ. By the theorem of the functional group of cell space (see the solution of Problem 36.47), the group $\pi_1\left(S^1(3),e^0\right)$ is a group with the generators α, β, γ and defining correlations $\alpha^2\beta^{-1}=1$, $\beta\alpha\gamma^{-1}=1$. This system of defining correlations is tantamount to the system $\alpha^2=\beta$ and $\alpha^3=\gamma$, therefore, any element of the group $\pi_1\left(S^1(3),e^0\right)$ is uniquely representable as α^n $n\in\mathbb{Z}$. Thus, $\pi_1\left(S^1(3),e^0\right)\approx\mathbb{Z}$, in particular, $S^1(3)$ is not simply connected.

36.53. Assume the opposite: let there exist an embedding of space $\exp_4 I$ into $\mathbb{R}^4$. Let $U_1=\left(1,\dfrac{1}{3}\right)$, $U_2=\left(\dfrac{1}{3},\dfrac{2}{3}\right)$, $U_3=\left(\dfrac{2}{3},1\right)$ be open disjunctive subsets of the segment I. Let

$$V=\left\{x\in\exp_4 I\,|\,x=\{x_1,x_2,x_3\}\quad\text{and}\quad x_i\in U_i\quad\text{for}\quad i=1,2,3\right\}.$$

It is easy to see that V is homeomorphic to $\mathbb{R}^3$. Let $\mathcal{O} = \langle U_1, U_2, U_3 \rangle$. Assume $W_i = \{x \in \exp_4 I \,|\, x \in \mathcal{O}, \ |x| = 4 \text{ and } |x \cap U_i| = 2\}$. It is easy to notice that W_i is homeomorphic to $\mathbb{R}^4$ for each i and $W_i \cap W_j = \varnothing \ \forall i \neq j$. Besides, it is evident that the set V is a common boundary for sets W_i, $i = 1, 2, 3$. Therefore, in $\mathbb{R}^4$ a set was found homeomorphic to $\mathbb{R}^3$ and being a common boundary for three nonintersecting sets, each of which is homeomorphic to $\mathbb{R}^4$, which contradicts the theorem of Jordan. Thus, $\exp_4 I \not\subset \mathbb{R}^4$, which is what was required to be proved.

36.54. Assume the opposite: let $\exp_4 I$ be a 4-manifold. Consider an arbitrary point $x = \{x_1, x_2, x_3\} \in \exp_4 I$, where $0 < x_1 < x_2 < x_3 < 1$. According to our assumption, a neighbourhood homeomorphic to $\mathbb{R}^4$ will be found in x. We can consider that this neighbourhood has the form $\mathcal{O} = O\langle U_1, U_2, U_3 \rangle$, where U_1, U_2 and U_3 are disjunctive neighbourhoods of points x_1, x_2 and x_3, respectively, herewith, $0 \notin U_1$ and $1 \notin U_3$. Let $V = \{x \in \mathcal{O} |\, |x| = 3\}$. Let $W = \mathcal{O} \setminus V$. It is easy to notice that $W = \bigcup\limits_{i=1}^{3} W_i$; herewith, W_i are open-closed in W. It is easy to see that $W_i \cap W_j = \varnothing \ \forall i \neq j$. Thus, W has at least three components of connectivity, from where it follows that a set homeomorphic to $\mathbb{R}^3$ splits the space $\mathbb{R}^4$ into more than two components of connectivity, which contradicts the theorem of Jordan. The obtained contradiction shows that $\exp_4 I$ is not a manifold, which is what was required to be proved.

36.55. Construct a mapping $h \colon SP^n I \to I^n$ as follows. Let $[x] \in SP^n I$. Then assume $h([x]) = (x_1, \ldots, x_n)$, where $(x_1, \ldots, x_n)$ is such a representative of the class $[x]$ that $x_1 \leq x_2 \leq \ldots \leq x_n$. Let $T^n = \{x \in X^n |\, x_1 \leq x_2 \leq \ldots \leq x_n\}$. Notice that T^n is an n-dimensional closed simplex. Thus, h maps $SP^n I$ on T^n.

Evidently, T^n is homeomorphic to an n-dimensional cube I^n. Therefore, it is sufficient for us to verify that h is a homeomorphism. The injectivity and surjectivity of the mapping h are evident. Note that h^{-1} is continuous, as $h^{-1} \equiv \pi^n|_{T^n}$. Besides, according to Problem 36.26, $SP^n I$ is a compact. Therefore, h^{-1} is a homeomorphism (as a continuous bijection of compacts), so h is a compact, too.

36.56. Note that H_1 is homeomorphic to the half-interval $[0, 1)$. Let $[x] \in SP^n H^1$, $(x_1, \ldots, x_n) \in [0, 1)^n$ be such a representative of class $[x]$ that $x_1 \leq x_2 \leq \ldots \leq x_n$. Set $h([x]) = (x_1, \ldots, x_n)$. Then h is a bijection of space $SP^n H^1$ on the space

$$Y = \{x \in [0, 1)^n |\, x_1 \leq x_2 \leq \ldots \leq x_n\}.$$

Show that h is a homeomorphism. Indeed, h^{-1} is continuous, as $h^{-1} \equiv \pi^n|_Y$. Verify that h is continuous. Let U be an open subset in Y. Then $h^{-1}(U) = \pi^n(U)$ is open, as the mapping π^n is closed. Thus, h is a homeomorphism.

Further, it is easy to see that Y is an n-dimensional simplex, in which points of the face $x_n = 1$ are deleted, and such a simplex is homeomorphic to H^n, which is what was required to be proved.

36.57. Note that $\mathbb{R}$ is homeomorphic to $(0, 1)$. As in the solution of Problem 36.56, construct a homeomorphism $h \colon SP^n \mathbb{R} \to Y$, where $Y = \{x \in (0, 1)^n |\, x_1 \leq x_2 \leq \ldots \leq x_n\}$. It is easy to see that Y is an n-dimensional simplex, where points of the faces $x_n = 1$ and $x_1 = 0$ are deleted; therefore, Y is homeomorphic to H^n, which is what was required to be proved.

36.59. Let M be an arbitrary n-manifold, $n \geq 1$. Assume the opposite, namely, that $M(4)$ is a 4n-manifold. Let a be an arbitrary point of the manifold M, $U(a)$ be the neighbourhood of this point, homeomorphic to $\mathbb{R}^4$. Denote as M' a set of points $[x] \in M(4)$, in which a representative (x_1, x_2, x_3, x_4) of the class $[x]$ will be found such that $x_i \in U(a)$ for all i. Note that by the definition of topology in $M(4)$ the subspace M' is open in $M(4)$ and, in particular, is also a manifold. On the other hand, by the choice of the neighbourhood $U(a)$, we have $M' = \mathbb{R}^n(4)$. Thus, it is sufficient for us to reduce to a contradiction the assumption that $\mathbb{R}^n(4)$ is a manifold.

36.60. Assume the contrary. Consider any different points $x, y \in M^2$. Then $\{x, y\} \in \exp_3 M^2$. By our assumption, a neighbourhood of these points in space $\exp_3 M^2$ homeomorphic to $\mathbb{R}^6$ will be found. We can consider that this neighbourhood has the form $O \langle U, V \rangle$, where U and V are disjunctive neighbourhoods of points x and y, respectively. Moreover, as M^2 is a two-dimensional manifold, we can consider that U and V are homeomorphic to $\mathbb{R}^2$. As $U \cap V = \varnothing$, then $\exp_1 M^2 \cap O \langle U, V \rangle = \varnothing$. Set

$$A = \exp_2 M^2 \cap O \langle U, V \rangle, \quad B = \{z \exp_3 M^2 \,||z| = 3\} \cap O \langle U, V \rangle.$$

Then it is easy to see that A is homeomorphic to $\mathbb{R}^4$. Besides, $O \langle U, V \rangle \setminus A$ is disjoint. Indeed, $O \langle U, V \rangle \setminus A = B_1 \cup B_2$, where

$$B^1 = \{z = \{x_1, x_2, x_3\} \in B \mid x_1 \in U, \, x_2 \in U, \, x_3 \in V\}, \quad B_2 = B \setminus B_1$$

and the point of B_1 cannot be linked with the point of B_2 by the path lying in B.

We will say that space X is split by its subspace Y if $X \setminus Y$ is disjoint. Then from the above said it follows that $O \langle U, V \rangle$ is split by set A. Recall that $O \langle U, V \rangle \approx \mathbb{R}^6$, A is homeomorphic to $\mathbb{R}^4$. Therefore, space $\mathbb{R}^6$ is split by the set homeomorphic to $\mathbb{R}^4$, which contradicts the following known theorem:

T h e o r e m . *Space $\mathbb{R}^n$ is not split by any set of dimension not exceeding $n - 2$ lying in it.*

From the contradiction obtained we conclude that $\exp_3 M^2$ is not a manifold, which is what was required to be proved.

36.63. Let $(x, y) \in \mathbb{R}^2$. Assuming $z = x + iy$, identify $\mathbb{R}^2$ with a complex plane $\mathbb{C}$. Thus, it should be shown that $SP^n \mathbb{C} = \mathbb{C}^n$. Let $[(z_1, \ldots, z_n)] \in SP^n \mathbb{C}$. Assign a set of polynomial coefficients $P(z) = (z - z_1) \ldots (z - z_n)$ to the class $[(z_1, \ldots, z_n)]$:

$$[(z_1, \ldots, z_n)] \mapsto (a_{n-1}, \ldots, a_0) \in \mathbb{C}^n.$$

It is easy to see that the obtained mapping $f \colon SP^n \mathbb{C} \to \mathbb{C}^n$ is surjective and injective (indeed, any polynomial over a field of complex numbers has exactly n roots, which are uniquely defined by its coefficients). Besides, the coefficients of a polynomial continuously depend on its roots and, vice versa, the roots of a polynomial are continuous functions of its coefficients, from where we conclude that the constructed mapping f is a homeomorphism, which is what was required to be proved.

36.64. Denote as $\overline{\mathbb{C}}$ an extended complex plane. Note that the sphere S^2 is homeomorphic to $\overline{\mathbb{C}}$, therefore, $SP^n S^2$ is homeomorphic to $SP^n \overline{\mathbb{C}}$. Construct the mapping $f \colon \overline{\mathbb{C}}^n \to \mathbb{C}P^n$ which assigns the vector $(z_1, \ldots, z_n) \in \overline{\mathbb{C}}^n$ to the class of equivalency

$$[a_0, \ldots, a_n] = \{(ca_0, \ldots, ca_n) \mid c \in \mathbb{C} \setminus \{0\}\} \in \mathbb{C}P^n$$

such that $z_1, \ldots, z_n$ are the roots of the polynomial

$$P(z) = a_0 + a_1 z + \cdots + a_n z^n,$$

and the vector of the form $(z_1, z_2, \ldots, z_k, \underbrace{\infty, \ldots, \infty}_{n-k})$, to the class of equivalency

$$[a_0, \ldots, a_k, 0, \ldots, 0];$$

herewith, $z_1, \ldots, z_k$ are the roots of the polynomial $P(z) = a_0 + a_1 z + \cdots + a_k z^k$.

Prove that the mapping h is continuous, closed and surjective. The surjectivity of f is evident, as any polynomial of power n has exactly n roots over the field $\mathbb{C}$. The continuity of f in points $(z_1, \ldots, z_n)$, where $z_i \neq \infty$ for all i follows from the fact that the vector $(a_0, \ldots, a_{n-1}, 1)$ composed of the coefficients of the polynomial $P(z) = (z - z_1) \ldots (z - z_n)$ and considered as a function of the set $z_1, \ldots, z_n$ is continuous with respect to all the variables $\{z_1, \ldots, z_n\}$.

We show by the example of $\overline{\mathbb{C}}^2$ that f is continuous in point (z_1, ∞). Let $z_2 \neq \infty$, $z_2 \neq 0$. Then

$$f(z_1, z_2) = [z_1 z_2, -(z_1 + z_2), 1] = \left[z_1, -\left(\frac{z_1}{z_2} + 1\right), \frac{1}{z_2} \right],$$

$$\lim_{z_2 \to \infty} f(z_1, z_2) = [z_1, -1, 0] = f(z_1, \infty),$$

i.e., f is continuous in point $(z_1, \infty) \in \overline{\mathbb{C}}^2$. In the case of arbitrary n the continuity is proved similarly.

The mapping f generates on $\overline{\mathbb{C}}^n \cong (S^2)^n$ a natural relation of equivalency $\sim_f$: $x \sim y \leftrightarrow f(x) = f(y)$. It is easy to see that this relation coincides with the relation of equivalency generated by the mapping π^n: $\overline{\mathbb{C}}^n \cong (S^2)^n \to SP^n \mathbb{C}$. Recall that $\mathbb{C}P^n$ is the image of f. Thus,

$$SP^n S^2 = SP^n \overline{\mathbb{C}} \cong \overline{\mathbb{C}}^n / \sim_{\pi^n} \cong \overline{\mathbb{C}}^n \sim_f \cong \mathbb{C}P^n,$$

which is what was required to be proved.

36.65. As in the solution of Problem 36.63, we identify $\mathbb{R}^2 \backslash \{0\}$ with $\mathbb{C} \backslash \{0\}$ and construct the mapping f: $SP^n (\mathbb{R}^2 \backslash \{0\}) \to \mathbb{C}^n$. The image of this mapping are the vectors $(a_{n-1}, \ldots, a_0)$ satisfying the condition $a_0 \neq 0$ and only they. The set of these vectors is naturally homeomorphic to the product $\mathbb{C}^{n-1} \times (\mathbb{C} \backslash \{0\})$, which, in turn, is homeomorphic to $(\mathbb{R}^2 \backslash \{0\}) \times \mathbb{R}^{2n-2}$, which is what was required to be proved.

Bibliography

[1] Mishchenko, A.S., Solovyev, Yu.P. and Fomenko, A.T. (1981) *A Collection of Problems in Differential Geometry and Topology*, Moscow: Moscow University Press (in Russian).

[2] Kovantsov, N.I., et al. (1989) *Differential Geometry, Topology, Tensor Analysis*, Kiev (in Russian).

[3] Vasilyev, A.M. and Solovyev, Yu.P. (1988) *Differential Geometry (Teaching Guidelines)*, Moscow: Moscow University Press (in Russian).

[4] Trofimov, V.V. (1990) *Problems in the Theory of Lie Groups and Algebras*, Moscow: Moscow University Press (in Russian).

[5] Rokhlin, V.A. and Fuks, D.B. (1977) *An Introductory Course of Topology (geometrical chapters)*, Moscow: Nauka (in Russian).

[6] Rozendorn, E.R. (1971) *Problems in Differential Geometry*, Moscow: Nauka (in Russian).

[7] Dubrovin, B.A., Novikov, S.P. and Fomenko, A.T. (1986) *Modern Geometry*, Moscow: Nauka (in Russian).

[8] Novikov, S.P. and Fomenko, A.T. (1987) *Elements of Differential Geometry and Topology*, Moscow: Nauka (in Russian).

[9] Mishchenko, A.S. and Fomenko, A.T. (2000) *A Course of Differential Geometry and Topology*, Moscow: Factorial Press (in Russian).

[10] Schwarz, J. (1970) *Differential Geometry and Topology*, Moscow: Mir (in Russian).

[11] Blaschke, W. (1957) *Introduction to Differential Geometry*, Moscow: Mir (in Russian).

[12] Rashevskii, P.K. (1967) *Riemannian Geometry and Tensor Analysis*, Moscow: Nauka (in Russian).

[13] Rashevskii, P.K. (1956) *A Course of Differential Geometry*, Moscow: Gostekhizdat (in Russian).

[14] Thorpe, J. (1979) *Elementary Topics in Differential Geometry*, New York: Springer.

[15] Manturov, O.V. (1991) *Elements of Tensor Analysis*, Moscow: Prosveshchenie (in Russian).

[16] Sternberg, S. (1964) *Lectures on Differential Geometry*, Englewood Cliffs, New Jersey: Prentice-Hall.

[17] Busemann, H. (1955) *The Geometry of Geodesics*, New York: Academic Press.

[18] Pogorelov, A.V. (1967) *Differential Geometry, 2nd edition*, Groningen: Nurdhoff.

[19] Pogorelov, A.V. (1973) *Extrinsic Geometry of Convex Surfaces*, Providence: American Mathematical Society.

[20] Alexandrov, A.D. (2006) *Intrinsic Geometry of Convex Surfaces*, Boca Raton: Chapman & Hall/CRC.

[21] Finikov, S.P. (1952) *A Course of Differential Geometry*, Moscow: Gostekhizdat (in Russian).

[22] Gromoll, D., Klingenberg, W. and Myer, W. (1968) *Riemannsche Geometrie im Grossen*, Berlin: Springer.

[23] Gilbert, D., Cohn-Vossen, S. (1990) *Geometry and the Imagination*, New York: Chelsea Publishing Company.

[24] Fischer, G. (1986) *Mathematical Models, in two volumes*, Braunschweig/Wiesbaden: Vieweg & Son.